IV International Conference of Computational Methods in Engineering Science (CMES'19)

IOP Conference Series: Materials Science and Engineering
Volume 710

Kazimierz Dolny, Poland
21 - 23 November 2019

ISBN: 978-1-7138-1039-1
ISSN: 1757-8981

Printed from e-media with permission by:

Curran Associates, Inc.
57 Morehouse Lane
Red Hook, NY 12571

Some format issues inherent in the e-media version may also appear in this print version.

Institute of Physics
Dirac House, Temple Back
Bristol BS1 6BE UK

Phone: 44 1 17 929 7481
Fax: 44 1 17 920 0979

techtracking@iop.org

Additional copies of this publication are available from:

Curran Associates, Inc.
57 Morehouse Lane
Red Hook, NY 12571 USA
Phone: 845-758-0400
Fax: 845-758-2633
Email: curran@proceedings.com
Web: www.proceedings.com

TABLE OF CONTENTS

IOP Conf. Series: Materials Science and Engineering **710** (2019) 011001 doi:10.1088/1757-899X/710/1/011001

CMES
CMES.POLLUB.PL

IV International Conference of Computational Methods in Engineering Science

21st-23rd November 2019 CMES'19 Kazimierz Dolny, Poland

ABOUT THE CONFERENCE

4th International Conference of Computational Methods in Engineering Science (CMES 2019), 21-23 November 2019, Kazimierz Dolny, Poland.

The main objective of the 4th CMES conference is development of engineering sciences, numerical and experimental techniques, broaden experience and good practices by representatives of the scientific community and industry. Conference allows exchange of knowledge between different research domestic and foreign centres representing various fields of engineering science and technology. The organizers of the conference are:

- Department of Thermodynamics, Fluid Mechanics and Aircraft Propulsion (Mechanical Engineering Faculty of Lublin University of Technology, Poland),
- Department of Machine Engineering Fundamentals and Mechatronics (Mechanical Engineering Faculty of Lublin University of Technology, Poland),
- Department of Materials Engineering (Mechanical Engineering Faculty of Lublin University of Technology, Poland),
- Department of Computer Science (Faculty of Electrical Engineering and Computer Science of Lublin University of Technology, Poland),
- Department of Organisation of Enterprise (Faculty of Management of Lublin University of Technology, Poland),
- Department of Renewable Energy Engineering (Faculty of Environmental Engineering of Lublin University of Technology, Poland),
- Department of Biomass and Waste Conversion in Biofuels (Faculty of Environmental Engineering of Lublin University of Technology, Poland),
- Self-Government of Doctoral Students of Lublin University of Technology, Poland,
- Department of Airframe and Engine (Aeronautics Faculty of Polish Air Force University, Poland).

TOPICS

- Analysis of engineering processes,
- Application of computer programs in technology,
- Artificial and computational intelligence,
- Computational fluid dynamics (CFD),
- Computer simulations of processes and phenomena,
- Finite Element Method (FEM),
- Material properties and structure research methods,
- Production engineering and quality control.

IOP Conf. Series: Materials Science and Engineering **710** (2019) 011001 doi:10.1088/1757-899X/710/1/011001

HONORARY PATRONAGE

Prof. Piotr Kacejko, Ph.D., D. Sc. (Eng.)

Rector of the Lublin University of Technology

Polish Information Processing Society

Stowarzyszenie Inżynierów i Techników Mechaników Polskich

POLISH ACADEMY of SCIENCES

IOP Conf. Series: Materials Science and Engineering **710** (2019) 011001 doi:10.1088/1757-899X/710/1/011001

SCIENTIFIC COMMITTEE

ORGANIZING COMMITTEE

IOP Conf. Series: Materials Science and Engineering **710** (2019) 011003 doi:10.1088/1757-899X/710/1/011003

Peer review statement

All papers published in this volume of *IOP Conference Series: Materials Science and Engineering* have been peer reviewed through processes administered by the proceedings Editors. Reviews were conducted by expert referees to the professional and scientific standards expected of a proceedings journal published by IOP Publishing.

IOP Conf. Series: Materials Science and Engineering **710** (2019) 012001 doi:10.1088/1757-899X/710/1/012001

The use of data by smart systems for price forecasting in the context of building customer relationships on the Lublin real estate market

A Bojanowska[1] and J Lipski[2]

[1] Lublin University of Technology, Faculty of Management, Department of Marketing, Nadbystrzycka St. 38, 20-618 Lublin, Poland
[2] Lublin University of Technology, Faculty of Management, Department of Organisation of Enterprise, Nadbystrzycka St. 38, 20-618 Lublin, Poland

a.bojanowska@pollub.pl

Abstract. This article aims to show how to design a functional application as an intelligent tool for forecasting real estate prices, among others, based on historical data. Real estate dealers or clients in this market are rarely aware of the possibility of using such a tool in the real estate valuation. This paper presents how the artificial neural network can be used for this purpose. The selected area of the Lublin real estate market has been covered by research carried out for this study. As input data for the smart system, historical data of transactions and variables characterising real estate were used, such as location of the property, its condition or size. Using the developed model, in the neural network, the sensitivity of the real estate price to selected features was carried out. It was also indicated how the use of this tool can facilitate purchasing decisions and customer relationship management on such a market.

1. Introduction

The real estate market in Poland and the world is a market with great value and high development dynamics. For example, data from the Central Statistical Office show that in the period January-November 2018, 163,800 apartments were commissioned, which is 2.5% more than last year [1]. The demand for housing and supply remain at a very high level. Customer relationship management on this market is extremely important as it is competitive and impacts not only real estate agencies but also on banks, notary offices, developers and many other important industries.

An important aspect of choosing a property by the customer is undoubtedly its price [2]. Various methods for estimating property prices can be found in the literature. Wang [3] proposed real estate price forecasting by PSO (particle swarm optimisation) and SVM (support vector machine), where PSO is chosen to determine the parameters of SVM. The real estate price forecasting cases are used to testify the forecasting performance of the proposed PSO–SVM model. Aiken [4] already points out in 2006 that trends in the real estate market can be predicted using neural networks. Other authors highlight the possibility of employing the method for spatial-temporal forecasting with an application to real estate prices [5]. Some authors have also tried to apply the theory of buyer behaviour to discover their motivation in real estate activities [6, 7]. This is a direct reference to the philosophy of customer relationship management, which is also mentioned in this article.

In this article, the goal was to design a functional application as an intelligent tool for forecasting the price of real estate, including based on historical data. The research carried out by the authors is

IOP Conf. Series: Materials Science and Engineering **710** (2019) 012001 doi:10.1088/1757-899X/710/1/012001

both literature and empirical. As part of the preparation of data for teaching the model for the purposes of this article, a survey of real estate prices was carried out together with features important to the customer, such as the area of real estate, plots, construction year, etc. For the cognitive purposes, 50 houses from the secondary and primary market located in Lublin, put up for sale on the otodom.pl portal, were randomly selected. The study was conducted in April 2019.

The computational or artificial intelligence [8], which was used in this article to achieve the goal, has already found many applications in numerous areas of engineering, data analysis, forecasting, biomedicine and others. The essence of systems based on computational intelligence is the survival and interpretation of data of a very different nature. It can be numeric, symbolic, binary, logical data or, for example, undecoded images read straight on the camera screen [9].

2. Building relationships with customers on the real estate market

Shaping customer relations and managing them has now become one of the crucial issues for entrepreneurs in almost all markets. It is not important only when the market is monopolistic or closed by special concessions and legal norms. Customers are more and more demanding [10] and are looking for newer solutions to meet their needs.

In order to effectively manage customer relations in the real estate market, CRM (Customer Relationship Management) solutions dedicated to this market can be used. Such IT solutions are built and "enclosed" with key external websites for the sale of real estate – such as, for example, Gratka, oferty.net or Vimeo [11]. to the offered solutions enable intelligent search, advanced tagging of objects and dynamic lists of offer parameters. This combination creates a system that perfectly combines the two most important elements of the real estate sale process – effective service and building relationships with customers and managing the offer process. It should be remembered that CRM – or customer relationship management – is not just software. CRM is a comprehensive strategy, which, thanks to appropriate tools, strives to best meet customer needs. CRM includes, among others marketing, sales and after-sales service. These activities lead to the acquisition, maintenance and – which in the era of easy exchange of recommendations on the Internet becomes crucial – maintaining good relations with the customer also after the transaction. CRM IT systems implemented on the real estate market allow that the search for offers can be managed both by the company's representatives during business talks and – through the service – by the customers themselves. As a result, additional information about the needs and preferences of customers is gained. Users of CRM systems on the real estate market pay special attention to such features of these systems as:

- Lead/contact management system – a strong contact management platform.
- List management – as a key source of revenue, this feature is of paramount importance.
- E-mail platform –social media are used for marketing whereas the email is regarded a more professional way of communicating.
- Drip e-mail – this enables sending automatic replies such as when a user adds a newsletter subscription.
- Compatibility with other platforms – compatibility with such products as MS Office and Google Docs is a great help.
- Mobile capability – agents are frequently on the road so cloud-based CRMs compatible with all mobile devices, including both Apple and Android are in demand. It extends the control and protects sensitive data such as financials [12].

An extremely important factor when searching for and choosing a property is its price. It is this aspect of shaping the relationship that is the primary focus of this paper.

Price is one of the many factors determining consumer behaviour in various markets. It belongs to the group of marketing factors [13]. Other factors that undoubtedly influence consumer decisions are: economic (e.g. savings owned), demographic (e.g., consumer sex), psychological (e.g. personality) and socio-cultural (e.g. opinion leaders) factors. This article focuses on price because it is very important for consumers on the real estate market (it is taken into account by 67% of buyers in this market [14].The research carried out on different markets indicated that a significant and positive relationship exists between customers' price perceptions and their purchase intentions and that the formation of price perceptions is significantly influenced by satisfaction with pricing and services [15]. In the case of real estate price modelling, it typically employs models equipped with a single neuron at the ANN (Artificial Neural Network) exit. The forecasting model built in this way generates a single answer on its output. Such a solution enforces the use of MLP (Multilayer Perception) containing, among others, one hidden layer, because a model with only one learning neuron at the output would not generate the expected results [16]. The solution proposed in this article is an alternative based on this solution.

3. Data usage by intelligent systems

In 1955, John McCarthy formulated the definition of artificial intelligence [17]. Recently, the notion of explainable artificial intelligence has seen a resurgence, after having slowed since the burst of work on explanation in expert systems over three decades ago; for example, see Chandrasekaran et al. [18], Swartout et al. [19] and Buchanan and Shortliffe [20]. He included these activities carried out by computers, whose implementation by a man would require the involvement of his intelligence. This definition is valid to this day and in light of it, intelligent systems should be understood as those offered by modern computer science, which can completely or partially replace the processes implemented by the human brain. This group of methods and IT tools includes, among others: data mining analysis systems allowing automatic detection of regularities in large data sets [21]. It is these systems, according to the assumptions of the authors of this article, that can be used to achieve the set goal, which is to design a functional application as an intelligent tool for forecasting the price of real estate, among others based on historical data. Specifically, a static neural network will be used [22]. Artificial neural networks find application in recognising and classifying patterns (assigning to category patterns), predicting time series, analysing statistical data, demining and compressing image and sound as well as in control and automation issues [23].

In order to execute the task set in the article, the price of real estate in the function of its essential features is predicted, and then serves to develop an application for forecasting property prices, finally a model using an artificial neural network was built. As input to the model the following features were selected:

- usable floor area of the property in square meters;
- plot area on which the property was built in square meters;
- number of rooms;
- year of construction;
- number of floors;
- primary (1) or secondary (2) market;
- nature of the building: detached (1), semi-detached (2), terraced (3);
- condition: living (1), renovation (2), finishing (3);
- the district in which the property is located:

City Centre (1), Szerokie (2), Ponikwoda (3), Czechow dolny (4), Nowy Slawin (5), Jakubowice Konskie (6), Kalinowka (7), Konstantynow (8), Zemborzyce (9), Botanik (10), Abramowice (11), Rudnik (12), Slawinek (13), Weglin (14), Konopnica (15), Prawiedniki (16), Slawin(17), Wrotkow (18), Marysin (19).

The above parameters of the real estate offers were taken into account as they are considered significant by the creators of many sales sites and real estate offices. These factors are also taken into account in the valuation of real estate by banks and independent experts, for instance for the purposes of preparing an appraisal report.

The research covered 50 randomly selected houses, put up for sale on a popular portal with sales and rental ads, limited to the area of Lublin. The collected data were used to build the model. The validation made use of15% input data, whereas15% was used for the purpose of testing.

The developed model is a black box (Figure 1) because a set of input parameters and their variability range are known, along with the corresponding set of output values representing the transaction price for a given subset of input values. The nature of relationships between individual input features and the output of the model is unknown, as well as the nature of the impact of the interaction of individual features. Some of the features are of quantitative nature, e.g. living space, the number of rooms, etc., while the qualitative part included, for example: for living (1), for renovation (2), for finishing (3), etc. For qualitative features, numerical values have been assigned to unify the format of input data.

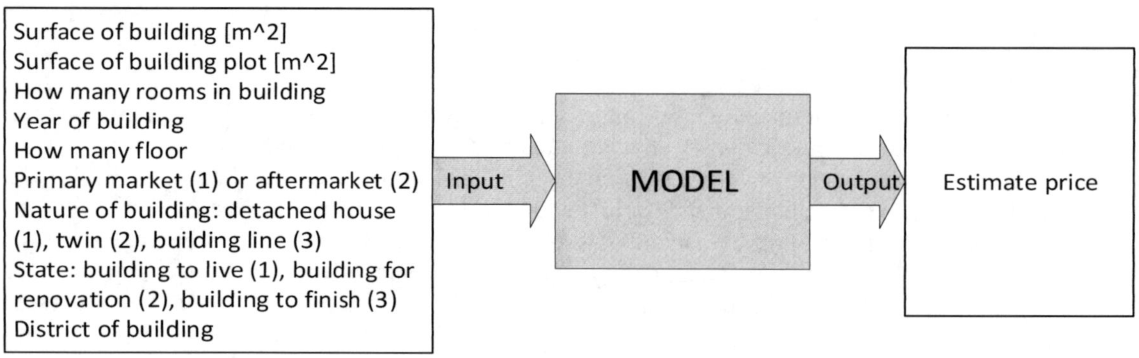

Figure 1. The developed model as a black box.

The process of building a model using a neural network is carried out in three phases:

- defining the structure of the neural network;
- network learning;
- validation and testing.

The modelling process was carried out in the MatLab computing environment using the Toolbox Neural Networks library. The neural network learning process in the MatLab environment starts with calling the function that builds the network structure:

net = feedforwardnet (hiddenSizes, trainFcn) (1)

where:
net – the structure of the neural network;
hiddenSizes – Row vector of one or more hidden layer sizes (default = 10);
trainFcn– training function (default = 'trainlm');
the function:

net = train(net,x,t) (2)

launches the network learning phase on the input data contained in matrix x. This matrix has the number of rows equal to the number of input data and the number of columns equal to the number of the training stream. The t parameter is the property price values corresponding to the next input data combinations. The structure of the network constructed in this way is shown in Figure 2. Designed a feed-forward network includes an input layer, a hidden layer of 10 neurons and one output.

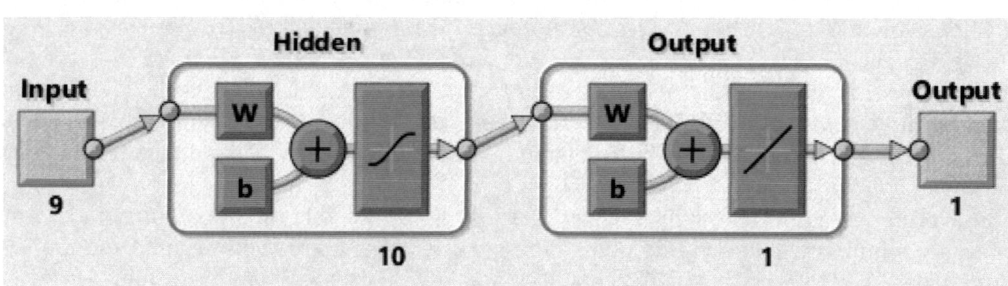

Figure 2. Neural network structure.

As a result of learning, the following model parameters shown in the regression graphs of Figure 3 were obtained.

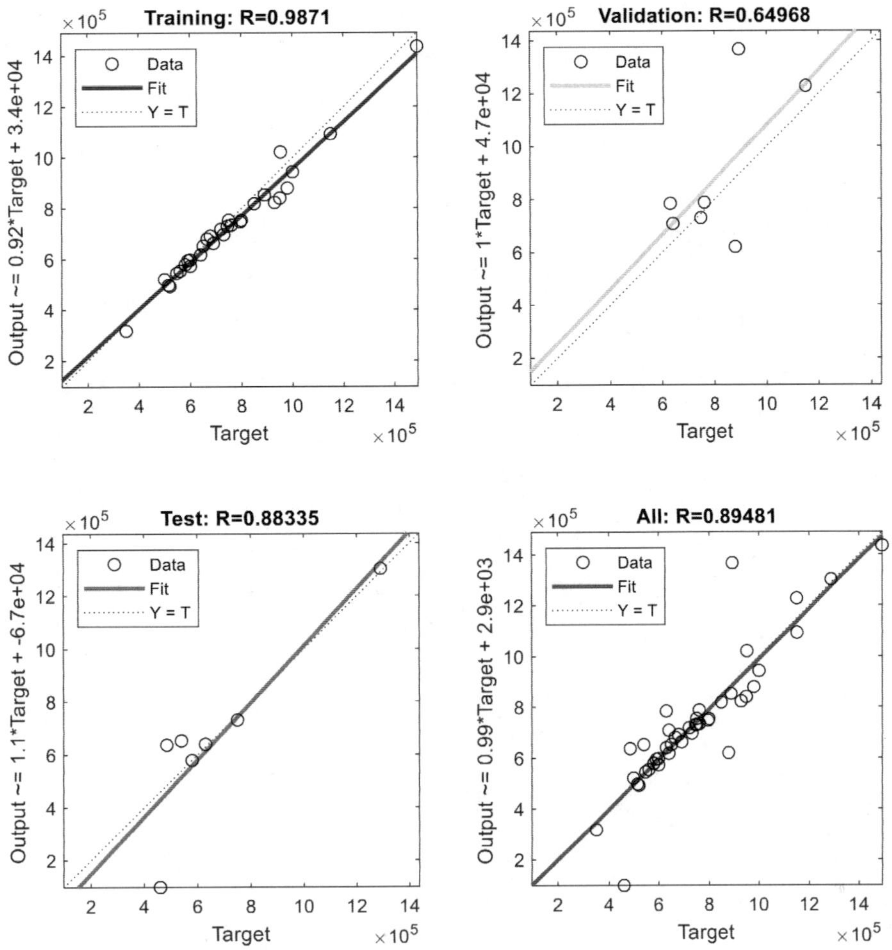

Figure 3. Network learning results on the training data set. Regression R-Values measure the correlation between outputs and targets. R-value of 1 means a close relationship, 0 a random relationship.

On the basis of the results of the regression analysis, it can be stated that the model largely maps the relationship between the price of real estate and the characteristics selected for research. The value of the regression coefficient $R = 0.88$ in the testing process allows to accept that the model maps the relationship between price and property features with acceptable accuracy. The model obtained in this way was used in an application developed in the MatLab computing environment. This application allowed for forecasting real estate prices on the basis of defined parameters consistent with the character with the model being taught. A number of studies on the sensitivity of the price to changes in selected parameters of the facility were also conducted. The results are shown in Figures 4 and 5.

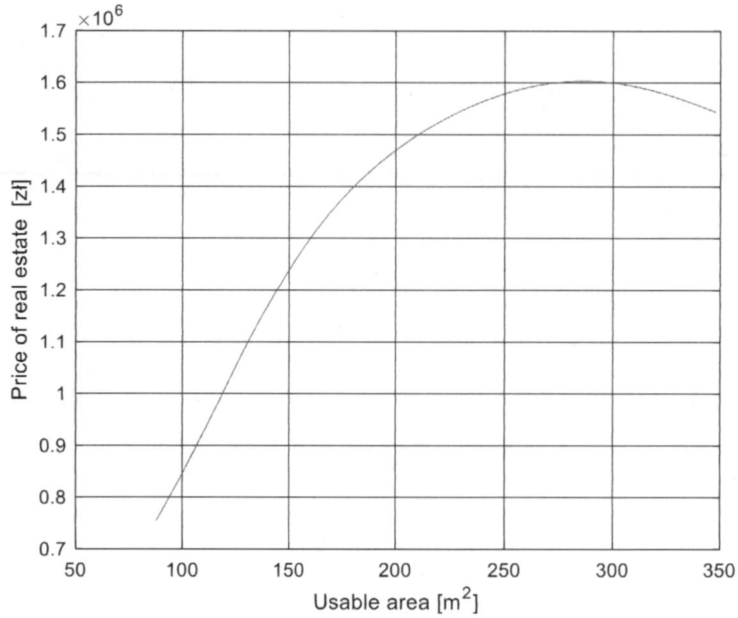

Figure 4. The nature of changes in the price of real estate from the usable area for the set other parameters.

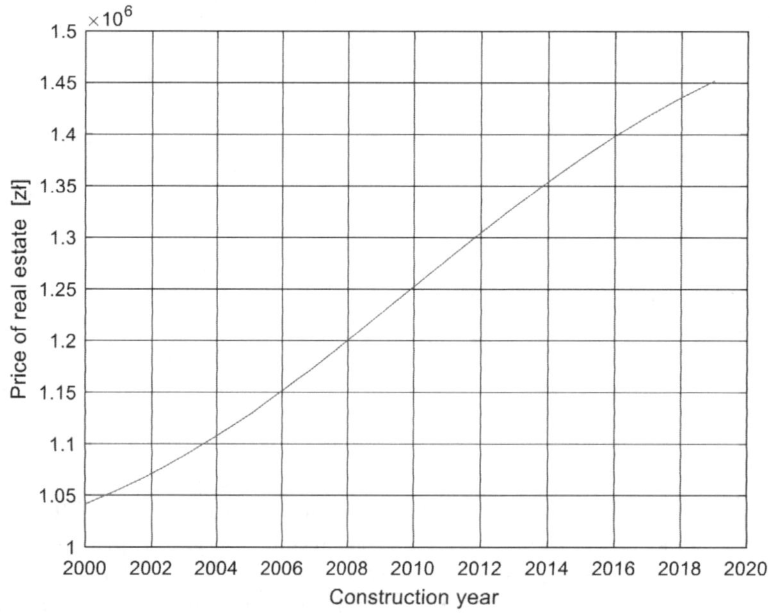

Figure 5. The dependence of the real estate price on the construction year, assuming other features at a constant level.

Using the forecasting model in the neural network, the calculator of the predicted property price for the introduced parameters was developed. The user interface of the application is shown in Figure 6. After selecting the property features and entering the desired value or option number into the text field, the simulation process starts and displays the expected value of the property.

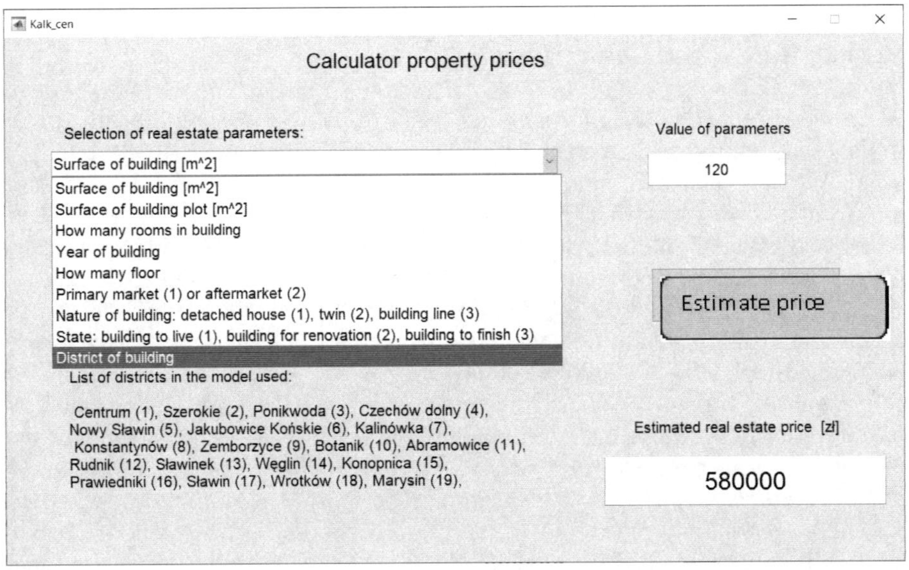

Figure 6. The interface of the real estate price calculator determined as a result of simulation in the learned neural network.

4. Conclusions

The conducted research on the use of the neural network for modelling causal relationships between property features and their price confirmed the effectiveness of the method. The process of collecting data for network learning was limited to a single are and selected districts. Therefore, the use of the learned network applies to the city of Lublin. The calculator developed using the model immediately estimates the price of the property with an accuracy rate of around 80 to 85%.The training data did not take into account the time of the transaction, because the interest rate on loans was stable and the inflation was around 2%.Modelling was carried out, based on data from 2019, in the MATLAB R2018b computing environment using Toolbox Neural Net Fitting – Fit data by training and two-layered-forward network. The MATLAB GUIDE (Graphic User Interface Design Environment) programming environment was used for the software of the "Real estate price calculator" application. The developed calculator allows you to analyse the impact of the value of individual property features on the expected price. This improves the experience of a potential customer, in the conditions of limiting the investment budget, to make a decision regarding the allowed changes in the area of parameters of the purchased object. Many properties of real estate, as shown by simulation analyses, are related to the price of non-linear relationships. This means that the selection of seemingly similar available offers may differ significantly in price with relatively small changes in selected parameters such as, for example, the size of the plot, the number of rooms or usable space. The calculator facilitates optimisation of the property selection. It can be useful for customers but can also support the process of customer relationship management for many entities operating within the real estate market. Offering such a tool to customers will certainly encourage them to cooperate and open up new areas for building relationships with him. It can also be used to collect data on customer preferences and encourage them to take self-service activities. The solution proposed in the article may operate independently or constitute an extension to the CRM software module.

The neural network used to build the model is an intelligent system applied in the developed application. Modelling and research undertaken in this article are set for further development, for example using more input for the model. They can also be the basis for building more effective CRM IT tools for the real estate market.

IOP Conf. Series: Materials Science and Engineering **710** (2019) 012001 doi:10.1088/1757-899X/710/1/012001

5. References

[1] Real estate sales in 2018. Statistics Poland

[2] Ariyawansa R G 2009 *Housing Market: A Review of Purchase Decision of Potential Buyers, Department of Estate Management and Valuation* (Sri Lanka: Faculty of Management Studies and Commerce University of Sri Jayewardenpura) doi: 10.4038/besl.v8i1.1906

[3] Wang X, Wen J, Yhang Z and Wang Z 2014 Real estate price forecasting based on SVM optimized by PSO *Optik* **125(**3) doi: 10.1016/j.ijleo.2013.09.017

[4] Aiken M and Bsat M 2006 Forecasting Market Trends with Neural Networks *Information Systems Management* **16**(4) doi: 10.1201/1078/43189.16.4.19990901/31202.6

[5] Pace R K, Barry R, Gilley O W and Sirmans C F 2000 A method for spatial–temporal forecasting with an application to real estate prices *International Journal of Forecasting* **16(**2) doi: 10.1016/S0169-2070(99)00047-3

[6] Chia J, Harun A, Kassim A W M, Martin D and Kepal N 2016 Understanding factors that influence house purchase intention among consumers in Kota Kinabalu: An application of buyer behavior model theory *Journal of Technology Management and Business* **3**(2)

[7] Zeng R 2013 *Attributes influencing home buyers' purchase decisions: a quantitative study of the Wuhan residential housing market* (DBA thesis, Lismore NSW: Southern Cross University)

[8] Miller T 2019 Explanation in artificial intelligence: Insights from the social sciences *Artificial Intelligence* **267** doi: 10.1016/j.artint.2018.07.007

[9] Rutkowski L 2019 *Metody sztucznej inteligencji* (Warszawa: Wydawnictwo Naukowe PWN)

[10] Hyken S 2012 Today's Customers Demand Customer Service On Their Terms https://www.forbes.com/sites/shephyken/2016/11/12/todays-customers-demand-customer-service-on-their-terms/ (12.09.2019)

[11] Real Estate CRM Platform https://makolab.com/pl-pl/rozwiazania/dedykowane-rozwiazania-biznesowe/platforma-crm-dedykowana-bran%C5%BCy-nieruchomo%C5%9Bci (13.05.2019 in:Polish)

[12] Tareen J 2019 Top features for real estate CRM https://www.discovercrm.com/real-estate-crm-features.html (20.08.2019)

[13] Wawrzyniak A 2015 Modelowanie wieloagentowe jako narzędzie do badania zachowania konsumentów *Zeszyty Naukowe SGGW w Warszawie Polityki Europejskie Finanse i Marketing* **14(**63) 134-46 doi:10.7172.1644-9584.38.7

[14] Cena, koszty utrzymania i bezpieczeństwo najważniejsze przy zakupie mieszkania https://www.wirtualnemedia.pl/artykul/najwazniejsze-warunki-przy-zakupie-mieszkania (23.05.2019 in:Polish)

[15] Munnukka J 2008 Customers' purchase intentions as a reflection of price perception *Journal of Product & Brand Management* **17**(3) 188-96 doi: 10.1108/10610420810875106

[16] Jasiński T and Bochenek A 2016 Prognozowanie cen nieruchomości lokalowych za pomocą sztucznych sieci neuronowych *Studia i Prace WNEiZ US* **45(**1) 317-27 doi: 10.18276/sip.2016.45/1-25

[17] McCarthy J 2007 What is Artificial Intelligence? *Computer Science Department* http://www-formal.stanford.edu/jmc/whatisai/ (12.09.2019)

[18] Chandrasekaran B, Tanner M C and Josephson J R 1989 Explaining control strategies in problem solving *IEEE Expert* **4**(1) 9-15 doi: 10.1109/64.21896

[19] Swartout W R and Moore J D 1993 Explanation in second generation expert systems *Second Generation Expert Systems* eds J M David, J P Krivine and R Simmons (Berlin, Heidelberg: Springer) pp 543-85 doi: 10.1007/978-3-642-77927-5_24

[20] Buchanan B and Shortliffe E 1984 *Rule-based Expert Systems: the MYCIN Experiments the Stanford Heuristic Programming Project* (Addison-Wesley*)* doi:10.1016/0004-3702(85)90067-0

[21] Hołyński M 1979 *Sztuczna inteligencja* (Warszawa: Wiedza Powszechna)

[22] Rojas R 2013 *Neural Networks: A Systematic Introduction* (Berlin, Heidelberg: Springer)

[23] Tadeusiewicz R 1993 *Sieci neuronowe* (Warszawa: Akademicka Oficyna Wydawnicza)

Water losses analysis based on FEFLOW FEM simulation and EPANET hydraulic modelling

P Suchorab and M Iwanek

Lublin University of Technology, Faculty of Environmental Engineering, Nadbystrzycka 38D, 20-618 Lublin, Poland

p.suchorab@pollub.pl

Abstract. Water losses occur in every water distribution system (WDS) and are one of the most onerous problems for water distribution network's exploiters. Water losses can cause not only financial and operating costs but also negative ecological and social issues. Therefore, there is a great need to individually assess the water losses level in every WDS. The aim of this paper is to analyse water losses in a selected water supply system group, serving approx. 5 000 inhabitants. The water losses analysis was performed on the basis of Finite Element Method simulation in FEFLOW v.5.3 software and numerical hydraulic modelling in EPANET software. The water outflow from a distribution pipeline was simulated for variable conditions: different pipe diameters and dynamic time-varying pressure values. The obtained results enabled to present differences in approaches to including water losses level into WDS models.

1. Introduction

The main goal of every water supply distribution system (WDS) is to deliver water to all customers in an unlimited amount and at any time. In a theoretically ideal situation, a total volume of water pumped into the water supply network (System Input Volume SIV) equals the total volume of water demanded by customers. However, as water losses occur in every WDS and often are one of the most onerous problems for network's operators, such theoretically ideal situation does not exist in complex or semi-complex water systems. Real water losses through leaking pipes are phenomena of a random and unpredictable character, which cause not only financial and operating costs but also can have negative ecological and social effects. Therefore, there is a great need to individually assess the water losses level in each WDS.

The water losses analysis can be performed on the basis of total volumes of water produced, imputed and sold, most often in accordance to IWA (International Water Association) methodology [1,2]. This widely used analysis method allows to define the approximate level of real water losses in a water supply system. However, as the water supply companies introduce the digital models of their WDSs more, there is a need to include water losses in hydraulic modelling. Most often, the process of WDS model creation includes two steps in defining the amount of water demanded by customers. First of them, is the amount of water actually demanded in accordance to billing meter system. The second is the calculation of water losses level in separate DMAs (District Meter Areas) and further multiplying base demands by water losses coefficients estimated on the basis of water flow analysis [3,4]. Such approach can lead to inexact assessment of hydraulic conditions due to the fact that customers' demands are calculated as Demand Driven Analysis (DDA) (pressure independently), while the amount of water leaking from a pipe depends on the pressure and therefore should be calculated as Pressure Driven Demand (PDD) [5].

The aim of this paper is to analyse water losses in a selected water supply system group in accordance to IWA methodology, to compare them with FEM calculations results in FEFLOW software and finally, to include real water losses in hydraulic modelling in EPANET, in order to present differences in approaches to including water losses level into WDS models.

2. Materials and methods

2.1. Water system characteristics

The water losses analysis was performed for an exemplary water supply system group located in Lublin Voivodship, Poland. The water supply system consists of three different zones with three individual underground intakes. The total length of water pipes is equal to 94.23 km (including 1 651 household connections the total length is 171.77 km). The material structure of WDS is homogenous, nearly all pipes are plastic (93.66% PVC, 5.65% PE, 0.69% steel), with diameters DN/OD: 90, 110, 125, 160 and two steel pipes DN 150 and DN 250. In 2015, the WDS served approx. 5500 inhabitants in 3 different settlement units (100% water supply coverage) with average operating pressure 0.38 MPa.

In recent years, one of the main issues hindering the proper operation of the analysed WDS was the significant level of water losses. Due to that fact, the water losses reduction strategy was implemented in 2010 and resulted in noticeable improvement. In 2015, the calculated Water Network Intensity Indicator WNII for the whole network was 7.50 $m^3/(day \cdot km)$ [6], while the average water production in 2010÷2015 years was 245 266 m^3/year, with average water sale 198 563 m^3/year. The average water consumption for own company purposes was 3 125.8 m^3/year. In accordance to [7], in 2015 the national average water demand for an individual consumer in Poland was 94 dm^3/day, while in the analysed system that value was significantly bigger (115.26 dm^3/day), which may suggest a noticeable influence of water losses. Detailed information and geometrical structure of WDS is presented in figure 1.

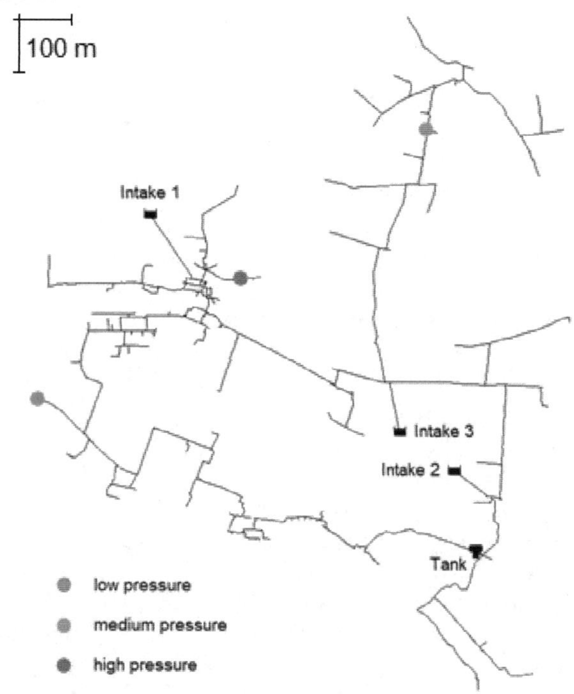

PARAMETER	UNIT	WDS
Number of individual consumers	ppl.	**5 502**
Total length of water supply network	km	**94.23**
Number of households connections	pcs.	**1 651**
Total length of households connections	km	**77.54**
Total length of water network with household connections	km	**171.77**
WNII	$m^3/(day \cdot km)$	**7.50**
Average operating pressure	MPa	**0.38**
Average water production in 2010÷2015	m^3/year	**245 266**
Average annual water consumption for own purposes	m^3/year	**3 125.8**
Average water sale in 2010÷2015	m^3/year	**198 563**
Water sale in 2015	m^3/year	**231 467**

Figure 1. Geometrical structure and detailed information about the analysed WDS [5].

IOP Conf. Series: Materials Science and Engineering **710** (2019) 012002 doi:10.1088/1757-899X/710/1/012002

The analysed WDS can be divided into three different pressure zones of low (0.27÷0.33 MPa), medium (0.29÷0.39 MPa) and high (0.39÷0.43 MPa) pressure. The representative nodes of each pressure zone are presented in figure 1 (blue node – represents low pressure zone, orange – medium pressure zone, red – high pressure zone). In each zone pipes of all possible diameters (DN/OD 90, 110, 125, 160) do exist.

2.2. *IWA water balance*
As a reference basis for numerical modelling in FEFLOW software, the water balance calculations over the period 2010÷2015 were used. The water balance was calculated in accordance to the IWA methodology. The details about IWA water balance for analysed water supply system are presented in paper [5]. In table 1 only selected components of water balance are presented.

Table 1. Selected components of the water balance over the period 2010÷2015 [5].

Parameter	Unit	2010	2011	2012	2013	2014	2015
System Input Volume (*SIV*)	m^3/year	251 156	245 486	259 532	232 733	237 061	258 000
Water Losses (*WL*)	m^3/year	67 296.0	50 745.0	53 807.0	33 153.0	27 804.0	24 533.0
Real Losses (*RL*)	m^3/year	57 249.8	40 925.6	47 318.7	27 334.7	21 877.5	18 083.0
Non-Revenue Water (*NRW*)	m^3/year	77 046.0	60 495.0	63 557.0	35 153.0	29 804.0	26 533.0

2.3. *FEFLOW modelling methodology*
The aim of the numerical simulations of water losses in the analysed WDS was to determine the possible amount of water lost through a single pipe break and later to compare it with the water losses calculated by the IWA water balance. Numerical simulations were performed in FEFLOW v.5.3 software, using the finite element method analysis (FEM). The application can solve the groundwater flow equation of both unsaturated and saturated media and can be successfully used for modelling of natural water flow in porous media of different kind and outflows from damaged pipes [8-11]. Numerical simulations of water outflow from damaged pipes were performed for 9 different variants: three diameters (DN/OD: 90, 125, 160) in three zones (low, medium and high pressure). In all cases, the conditions of the simulation were analogical – water was leaking through the one tenth of the circumference of the pipe. The 2-dimentional soil profile used in simulation consisted of two parts: the excavation (2,0 m × 0,7 m) and natural ground surrounding the excavation (20 m × 5 m). The pipes were located in the bottom of excavations on a sand base layer. Soil profiles were further automatically meshed with triangle polygon mesh. The soil model was implemented with two kinds of boundary conditions: the first type (Neumann boundary condition), reflecting the impermeable and evaporation layers, and the second type (Dirichlet boundary condition), reflecting the leaking part of the pipe. For each pressure zone variant, the time-varying pressure values over the 24-hours simulation were given as a boundary pressure condition. The flow in unsaturated media was calculated in accordance to standard Richards equation, with the automatic time step control via aggressive target-based strategies (backward Euler scheme) [12]. The graph of pressure varying in representative nodes of each zone is presented in figure 2. The scheme of cross section of soil profile with water pipe is presented in figure 3. The Mualem model parameters of used soils are given in table 2.

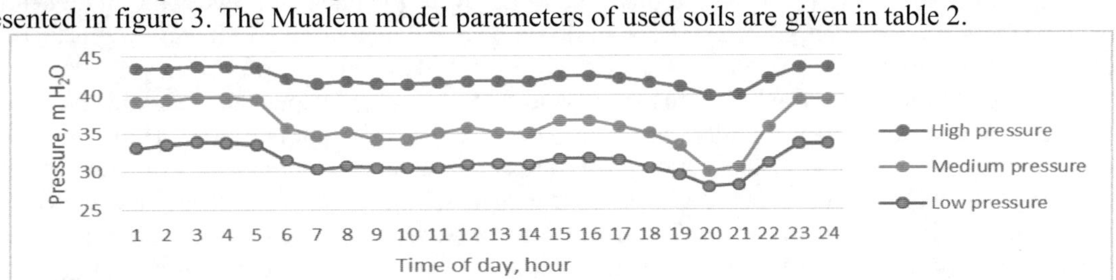

Figure 2. Pressure values in representative nodes of high, medium and low pressure zone.

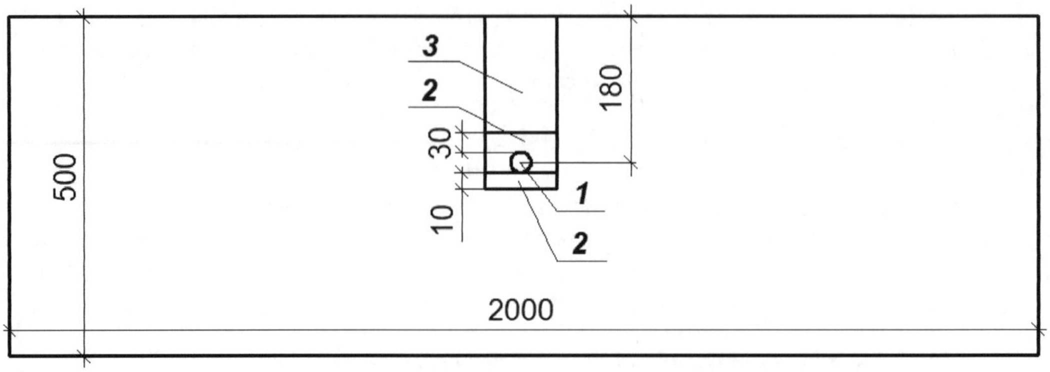

Figure 3. Scheme of the cross section of soil profile with the water pipe (1 – inner pipe diameter, 2 – sand, 3 – natural ground in and around the excavation; dimensions in cm).

Table 2. Hydraulic parameters of soils used in the simulation in accordance to Mualem model (θ – volumetric water content (m^3/m^3); K_S – saturated hydraulic conductivity coefficient (m/s); P – porosity; α, n – fitting parameters).

Soil	θ % (vol.)	K_S $\cdot 10^{-4}$ m/s	P % (vol.)	α 1/m	n -
in the excavation	0.177	3.34	36.12	1.662	1.334
around the excavation	0.146	1.40	26.56	2.170	2.133

2.4. EPANET modelling methodology

EPANET is a popular water system modelling software application developed by the United States Environmental Protection Agency (EPA U.S.) for performing hydraulic and water-quality extended-period simulations (EPS) of pressurized water supply networks [13]. Hydraulic modelling, using EPANET or any other modelling software, can be a comprehensive tool supporting both the designing and operating of WDS [14-16]. In this paper, on the basis of numerical simulations in FEFLOW software, the modelling in EPANET software (v. 2.00.12) was used to present different approaches to including real water losses in hydraulic modelling.

The numerical model of analysed WDS includes 1387 junctions and 1394 pipes, three reservoirs representing three water intakes, four pumps and one tank. The model has a general character – includes only distribution pipes without household connections. The Darcy-Weisbach head loss formula was used in calculations. The total duration of simulation was 24 hours, with one hour hydraulic calculation time step. Three different stepwise demand patterns were included in a model: general (common for all customers) and two specific ones for a school and a breeding farm. The minimum demand hour for the whole WDS: 03:00, maximum demand hour: 19:00.

The real water losses occurring in the analysed WDS were included in a hydraulic model in two ways. First, base demands in all nodes were multiplied by calculated factor. In this way, all nodes generated the increased demand. The second approach included determination of the emitter coefficient for selected nodes, which caused the additional pressure dependent demand equalled to calculated leakages from pipes. Emitters are model devices that can simulate the outflow through an orifice or nozzle discharging to the atmosphere. Most often, emitters are used to model sprinkler systems and irrigation networks, but they can also be successfully used to simulate the pipe leakages or fire flows at hydrants [17]. In EPANET software, emitters are not separate network components but are treated as a property of a junction. The emitter flow rate varies as a function of pressure available at the associated node, according to the formula (1) [18-19]. In this paper, the pressure exponent was assumed as 0.5 and leakage volume was determined by fitting the discharge coefficient C.

$$q = C \cdot p^{\gamma} \tag{1}$$

where:
q – flow rate (dm^3/s), C – discharge coefficient ($dm^3/s/m^{\gamma}$), p – pressure (m H_2O), γ – pressure exponent (-).

3. Results

The results of possible amount of water lost through a single pipe break (calculated in FEFLOW software) in different pressure zones in the analysed WDS are presented in table 3. According to the IWA water balance, in 2015, the total volume of water pumped into the system (SIV) equalled 258 00 m^3, which gives an average daily demand of 706,85 m^3/d. In the same year, real water losses (RL) equalled 18 083 m^3, which gives an average 49,54 m^3 of water lost each day. Such amount can be generated by approx. three leaking DN 90 water pipes in low pressure zone or approx. two leaking DN 160 water pipes in high pressure zone. On the other hand, these real water losses can be generated by leaking household connections. As was examined in paper [20], a single leaking household connection can cause over 7 m^3 of water lost per day. Therefore, the daily real water loss in the analysed system can be as well generated by seven leaking household connection (assuming similar diameter and pressure in a pipe).

Table 3. The results of possible volume of water lost through a single pipe break.

Pressure zone	Volume of lost water (m^3/d)			
	DN 90	DN 125	DN160	Average
Low pressure	17.82	18.85	19.00	18.59
Medium pressure	21.85	22.18	23.16	23.40
High pressure	24.18	24.56	25.73	24.83

The total base demand in all nodes in EPANET hydraulic model based on billing records equalled 567,92 m^3/d. In comparison with System Input Volume (SIV), that total daily amount of water equalled 706,85 m^3/d, which states that water losses in a system equals 19.65%. In order to include water losses in hydraulic conditions, all base demands were multiplied by 1.245 factor. The graph representing flow velocity in pipes at the maximum demand hour (19:00) in a model with equally distributed water losses for all nodes is presented in figure 4A. The flow velocity values in pies are relatively small (below 0,4 m/s) which is a typical situation for nearly all water supply networks in Poland. The graph representing flow velocity in pipes at maximum demand hour (19:00) in a model with water losses simulated as PDD outflows in two selected nodes (marked as green stars in upper right corner) in medium pressure zone is presented in figure 4B. The total additional outflow from two selected nodes equalled the average water loss in a system (nearly 50 m^3/d) and caused the flow velocity values increase (over 0.5 m/s) in the north part of the network. In the other part of a *WDS*, the hydraulic conditions (flow velocity) were similar to the model with equally distributed water losses.

4. Conclusions

Water losses analysis conducted on the basis of the IWA balance does not only give a comprehensive view of the losses' nature – neither leakages location nor amount of water lost through different pipes are identified this way. As revealed by the FEFLOW software analysis, daily amount of water losses (in this case 50 m^3/d) can be generated by two or three damaged pipes or a small number of household connections, further impeding to locate a leakage in reality. The assumption that water losses should be equally distributed in a DMA can lead to covering actual locations of leakages and can suggest the mistaken impression that a network operates properly. The analysis conducted in EPANET software proved that the hydraulic model reaction to leakages locations specified in nodes can be a tool supporting the process of leakages locating.

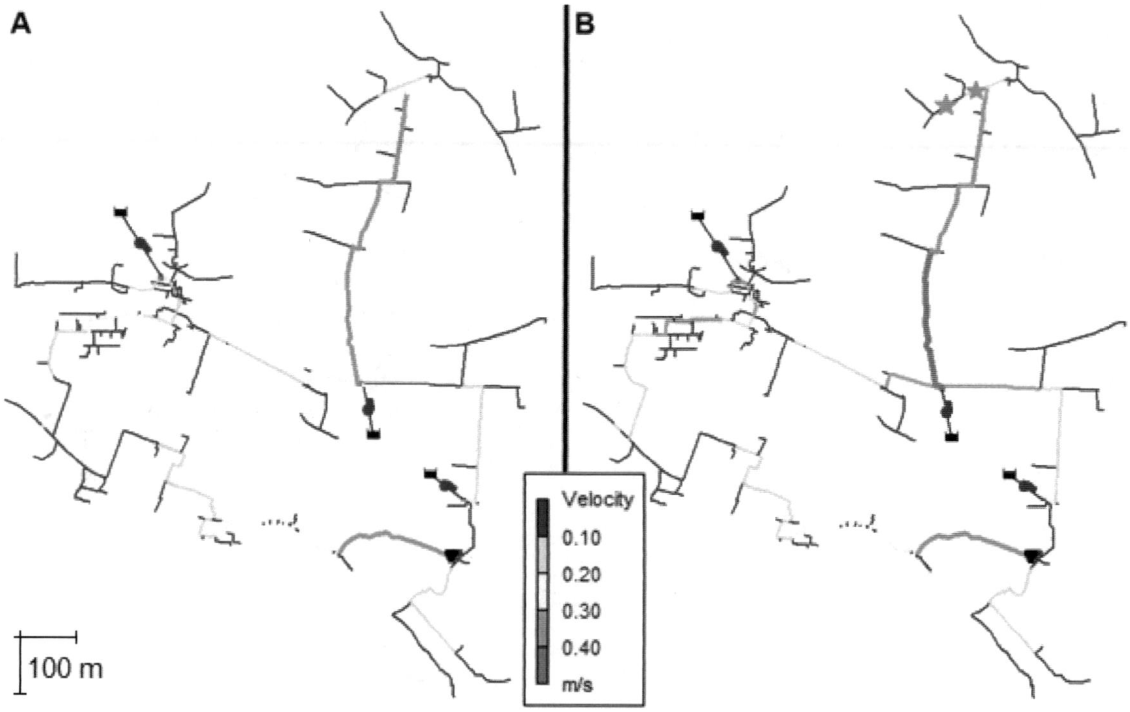

Figure 4. Comparison of flow velocity in pipes at maximum demand hour (19:00) between two approaches (4A: equally distributed water losses for all nodes, 4B: water losses simulated as PDD outflows in two selected nodes) of including water losses in hydraulic WDS model.

Acknowledgments
This article was founded by the statutory activity for young scientists (FNM/130-10/IŚ/2019) of the Faculty of Environmental Engineering, Lublin University of Technology.

5. References

[1] Farley M and Trow S 2003 *Losses in Water Distribution Networks. A Practitioner's Guide to Assessment, Monitoring and Control* (London:IWA Publishing)
[2] Lambert A and Hirner W 2000 *Losses from water supply systems. Standard terminology and recommended performance measures* (IWA Blue Pages)
[3] AWWA M32 2017 *Computer modeling of water distribution systems* (AWWA Denver)
[4] Zheng Y W, Farley M, Turtle D, Kapelan Z, Boxall J, Mounce S, Dahasahasra S, Mulay M, Kleiner Y 2011 Water loss reduction (Bentley Institute Press, Exton, Pensylvania)
[5] Bałut A 2019 Modelling of water distribution systems in low pressure condition *Gaz, Woda i Technika Sanitarna* **93**(6) 201–5 DOI: 10.15199/17.2019.6.4 (in Polish)
[6] Iwanek M, Suchorab P and Choma A 2018 Water Losses Analysis in Selected Group Water Supply Systems *ACEE* **11**(1) 133–40 DOI: 10.21307/ACEE-2018-013
[7] Głowny Urząd Statystyczny 2016 *Rocznik Statystyczny Rzeczypospolitej Polskiej (Statistical Yearbook of the Republic of Poland)* (Warszawa) (in Polish)
[8] Widomski M K, Broichsitter B, Zink S, Fleife A, Horn H R and Stępniewski W 2015 Numerical modelling of water balance for temporary landfill cover in North Germany *J Plant Nutr Soil Sci* **178**(3) 401–12 DOI: 10.1002/jpln.201400045
[9] Iwanek M, Kowalski D and Kwietniewski M 2015 Model Studies of a Water Outflow From an Underground Pipeline Upon Its Failure *Ochr Sr* **37**(4) 13–7 (in Polish)

[10] Suchorab P, Kowalska B and Kowalski D 2016 Numerical investigation of water outflow after the water pipe breakage *Rocz Ochr Środ* **18**(2) 416–27

[11] Widomski M K, Majerek D and Iwanek M 2018 Numerical modeling of water and heat transport in porous building material *AIP Conference Proceedings* **1988**(1)

[12] Diersch H J G 2005. *FEFLOW 5.2 Finite Element Subsurface Flow and Transport Simulation System, Reference Manual* (WASY Ltd. Berlin)

[13] www.epa.gov/water-research/epanet

[14] Haestad Methods Water Solutions 2013 *Computer applications in hydraulic engineering* (Bentley Institute Press, Exton, Pensylvania)

[15] Kowalski D 2009 Water Quality Management in a Water Distribution System *Ochr Sr* **31**(3) 37–40

[16] Świtnicka K, Suchorab P and Kowalska B 2017 The optimisation of water distribution system using Bentley WaterGEMS software *ITM Web of Conferences* **15** DOI:10.1051/ itmconf/ 20171503009

[17] Walski T M, Bezts W, Posluszny E T, Weir M H and Whitman B E 2016 Modeling leakage reduction through pressure control *American Water Works Association Journal* **98**(4) 147–55. DOI: 10.1002/j.1551-8833.2006.tb07642.x

[18] Lambert A 2001 What Do We Know About Pressure/Leakage Relationships in Distribution Systems. Proc. IWA Specialised Conference: System Approach to Leakage Control and Water Distribution Systems Management, Brno, Czech Republic 89–96

[19] Walski T M, Chase D V, Savic D A, Grayman W, Beckwith S, Koelle E 2007 *Advanced water distribution modeling and management* (Exton, Pensylvania: Bentley Institute Press)

[20] Iwanek M, Suchorab P and Suchorab Z 2019 FEM simulation of water lost through damaged household water connection *MATEC Web of Conference* **252** 05008(2019)CMES'18 DOI: 10.1051/matecconf/201925205008.

IOP Conf. Series: Materials Science and Engineering **710** (2019) 012003 doi:10.1088/1757-899X/710/1/012003

Designing and FEM simulation of the helicopter rotor and hub

K Skiba

Lublin University of Technology, Faculty of Mechanical Engineering, Department of Thermodynamics, Fluid Mechanics and Aviation Propulsion Systems, Nadbystrzycka 36 Str., 20-618 Lublin, Poland

k.skiba@pollub.pl

Abstract. The paper presents a preliminary design of the three-blade rotor. The geometric model was prepared in the Siemens NX 12.0 programme. After the preparation of the model and kinematic analysis, the structure was checked for strength by means of NX Nastran software. In addition, structural solutions of bearing rotors currently used in the construction of helicopters were discussed and a comparative analysis of the developed rotor design with other rotors was carried out in order to define the basic advantages and disadvantages of each solution. The main rotor of a helicopter is considered the most important component of the rotorcraft, which determines its performance. Rotating blades generate the resultant forces and moments in the rotor hub that allow the helicopter to fly. In the case of a single-rotor helicopter design, a tail rotor must be added to obtain directional control. Cyclic and collective control of blade inclination influences the generation of aerodynamic and inertial loads, which depend on the azimuthal position of the blade on the rotor disc. The number of rotor blades and the properties of the blade-hub connections strongly influence the level of load transfer from the blades through the hub to the helicopter fuselage. By incorporating a set of hinges, it is possible to reduce the component loads of the rotor, tilt and rolling moments transferred to the hull. A conventional articulated rotor head with built-in hinges requires frequent maintenance, which increases helicopter operating costs. Furthermore, a complex structure with a large number of components increases the weight of the articulated rotor.

1. Introduction

The aviation industry is the subject of numerous scientific studies, both in terms of aircraft propulsion and particular systems. Initiatives are underway to improve the rotorcraft [1, 2], and simulation studies of propulsion systems are also being undertaken [3, 4, 5, 6, 7]. Numerical fluid mechanics and other simulation methods allow advanced research to be completed in a much shorter time than required by the analytical method [8]. The possibilities of applying innovative technologies to increase the performance of rotorcraft, with particular focus on morphing technology [9, 10] or [11] are also being investigated. In recent years, the segment of unmanned aerial vehicles has been developing rapidly, as evidenced by market analyses predicting an intensive increase in the popularity of unmanned aerial vehicles [12, 13, 14] or updates of legal acts concerning UAVs [15].

This paper presents the results of numerical analyses carried out to test different blade geometries on a three-blade rotor designed for use on a research stand. The blades of the helicopter's rotor are loaded with aerodynamic and inertia forces. During the flight of the aircraft, due to aerodynamic and mass

IOP Conf. Series: Materials Science and Engineering **710** (2019) 012003 doi:10.1088/1757-899X/710/1/012003

forces, the blades make significant movements and undergo bending and torsional deformations, which in turn may cause high stresses and bending moments, especially in the blade base. Research and development projects to develop new aviation technologies typically require both the model of the test object and the equipment needed to accomplish the research.

In the scope of the implemented project aimed at developing a new concept of variable-geometry blades requires the use of a new rotor to test blades with intelligent materials. The prototype blade structure is different from the classic unmanned aerial vehicle blades. It is characterised by reduced stiffness and increased torsional susceptibility. The design incorporates systems causing additional forces generating torsional moments in the construction. Intelligent materials can also have the effect of increasing the temperature in their area of operation. These phenomena affect the performance of the rotor and other components. The design of the rotor must, therefore, curb these effects to the maximum extent.

The research analysis carried out in the field of helicopter rotor construction and maintenance shows that a rotor with a fully hinged head could reduce the interaction between the blades and the rotor to the minimum [16]. This kind of rotor consists of a head with a joint of the "Flap-hinge" type, which allows the blade to move freely in the plane containing the blade and shaft axes. The connection reduces the impact of blade interactions with each other, especially in triple and multi-blade systems. The other joint "Lag-hinge" type is a hinge that allows the blades to move forwards and backwards in the plane of rotation. The helicopter rotor is exposed to centrifugal forces that cause the Coriolis Effect in the system. The movement around this hinge reduces the effects of these phenomena. The last hinge is a mechanism for changing the blade's angle of attack. It consists of a hinge mounted along the blade axis and a pusher lever that limits its movement. Figure 1 shows a classic rotor system with articulated joints.

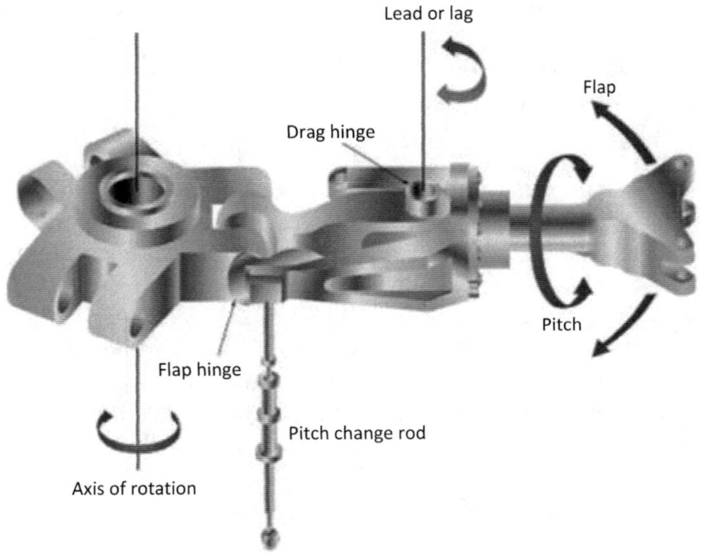

Figure 1. View of a typical helicopter rotor joint system [17].

2. Methodology

The rotor construction design has imposed certain restrictions resulting from the geometric dimensions of the blade. The diameter of the rotor must not exceed 2 m and the beginning of the blade base must be of the diameter smaller than 0.3 m. Moreover, smart material modules installed in the blades require an electricity supply. This requires the design and manufacture of a hollow shaft with a diameter that allows the electric wires to supply the actuators in the blades to be placed inside. The basic design assumptions of the rotor and swash plate/control disc resulting from the blade design and unmanned aerial vehicle performance are presented in Table 1.

IOP Conf. Series: Materials Science and Engineering **710** (2019) 012003 doi:10.1088/1757-899X/710/1/012003

Table 1. Design parameters of the rotor.

Parameter	Definition	Value	Unit
D_w	Rotor diameter	2	m
N_L	Number of rotor blades	3	pcs
ω_{max}	Maximum design rotor speed	180	rad/s
N_s	Engine power	30	kW
M_s	Engine torque	160	Nm
MTOW	Maximum take-off weight	150	kg
α	Overtaking control	30	deg
β	Longitudinal control range of motion	+/-7	deg
δ	Transverse control range of motion	+/-4	deg
s	General control range of motion	+/-12	mm

Designing the construction of a full-joint rotor of the helicopter maintaining all the required parameters is connected with the necessity of manufacturing and selecting components of the rotor with maximally reduced overall, volume and mass dimensions. Elements such as bearings, sliding rings or lock nuts exhibit specific strength properties. For the other structural elements of the rotor, these parameters must be determined.

Guidelines for fulfilling the certification requirements are written down in the certification specification. The designed rotor, due to its dimensions and performance, is subject to the certification specification of the European Union Aviation Safety Agency: Certification Specifications and Acceptable Means of Compliance for Small Rotorcraft CS-27 [18]. As stated in CS 27.303 Factor of safety "Unless otherwise provided, a factor of safety of 1.5 must be used. This factor applies to external and inertia loads unless its application to the resulting internal stresses is more conservative." The rotor construction must ensure a safety factor of at least k=1.5.

3. Research object

The strength calculations of the designed elements are carried out using the finite element method (FEM). In this paper, the results of calculations of key elements of the designed rotor during the UAV hover operation will be presented. Numerical operations were carried out with the use of NX Nastran module, which is a part of Siemens NX 12 software. The designed solid model of the rotor with blades and control disc is shown in figure 2.

Numerical investigations were carried out for the rotor components subjected to the highest loads during operation. For this purpose, the rotor head with the shaft and swing joints with blade anchors were isolated from the solid model. The blade structure was reduced to the base and replaced by connections that allow the imposition of aerodynamic loads and inertia forces generated by the blade. The connections of the moving components and their interactions were modelled as Surface-to-Surface contact. The rigidly connected elements were modified using a Surface-to-Surface adhesive.

The strength analysis was performed for the rotating rotor at full speed and with blade configurations to ensure maximum performance. The model was loaded with equivalent forces and moments applied to the blade base due to bearing, aerodynamic and centrifugal forces, and torsional moment (figure 4). The force values are shown in Table 2.

Table 2. Loads acting on the rotor.

Kind of load	Value	Unit
Aerodynamic lift force	1500	N
Centrifugal force of the blade	6156	N
Aerodynamic drag force of the blade	40	N
Torque	160	Nm
Rotational speed	180	rad/s

IOP Conf. Series: Materials Science and Engineering **710** (2019) 012003 doi:10.1088/1757-899X/710/1/012003

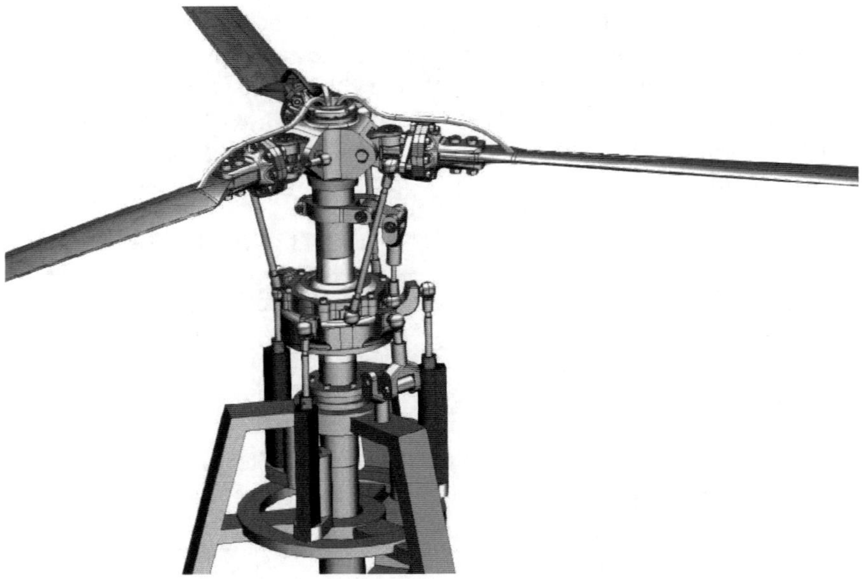

Figure 2. The three-dimensional model of the rotor prototype.

Figure 3 shows the force points and their vectors. The constraint points in the model were applied to the lower part of the shaft, restraining XY-plane displacements and Z-axis rotations. The Z-axis displacement was limited by restraining the contact surface of the shaft-locking nut and the position frame thrust bearing – the location where the rotor's aerodynamic lifting force is transferred to the stationary position frame. The constant angle of attack of the blades and thus the direction of force acting on the individual blades were achieved by limiting the movement of the pusher lever mounting holes in the Z-axis by simulating the operation of the control disc in the hovering position.

Figure 3. Force application locations and their vectors (red vector – loads, blue vector – constraint).

The material used in the construction of the rotor is high-quality structural steel C45 (EN). In its structure, apart from the carbon content of 0.42% to 0.5%, there are also small amounts of silicon, manganese, chromium, nickel, molybdenum, copper, sulphur and phosphorus admixtures, which all contribute to its mechanical, chemical and physical properties and strength. Due to the simplicity of machining, it is widely used in the industry. It is applied in the production of tools and machine elements that are subject to medium loads and at the same time require resistance to abrasion. This type of steel is ideal for the production of shafts and spindles, axles, unhardened gears, electric motor shafts, hubs for wheels, discs, rods, rollers and pump rotors [19]. The physical characteristics are shown in Table 3.

Table 3. Physical properties of C45 (EN) steel in standardised state.

Parameter	Value	Unit
Density	7850	kg/m³
Young's modules	205000	MPa
Shear modulus	78800	MPa
Poisson's number	0.30	–
Specific heat	470	J/kgK
Coefficient of thermal expansion	11.1	µm/mK
Thermal conductivity	48	W/mK

The blade yoke structure is designed using (AW-7075) aluminium alloy. It is an aluminium alloy with very high mechanical strength and high fatigue strength. It has medium corrosion resistance and is suitable for machining, polishing and grinding. Due to its properties, it is often used in machine elements and structures working under heavy loads. Given its physical and strength properties, it is widely used in the manufacture of elements in the aerospace industry [20]. Physical properties of the material are presented in Table 4.

Table 4. Physical properties of aluminium alloy AW-7075 (EN) [20].

Parameter	**Value**	**Unit**
Density	2810	kg/m³
Young's modules	72000	MPa
Shear modulus	27100	MPa
Poisson's number	0.33	–
Congealing temperature	475	°C
Melt flow temperature	635	°C
Specific heat	862	J/kgK
Coefficient of thermal expansion	23.5	µm/mK
Thermal conductivity	134	W/mK

The strength properties Ry (yield strength) and Ru (ultimate strength) of the applied materials are presented in Table 5.

Table 5. Strength properties of the applied materials.

Material identification	Ry [MPa]	Ru [MPa]	Permissible stresses [MPa]		
			k_r	k_g	k_s
C45 (EN)	600	355	170	205	110
AW-7075 (EN)	480	390	120	130	70

Bearings in the designed structure are mostly made of high carbon steel. Applied bearings have been selected from the catalogue in accordance with the manufacturer's guidelines concerning permissible forces.

IOP Conf. Series: Materials Science and Engineering **710** (2019) 012003 doi:10.1088/1757-899X/710/1/012003

4. Simulation results

The calculations were carried out for the full model assembly. This approach makes it possible to observe the behaviour of components and their mutual interaction resulting from the forces. The key element of each rotor is the shaft, hub and pins holding the joints and blade yokes. These components are exposed to the highest loads. Figure 4 shows the distribution of stresses on the rotor shaft. The reduced stresses according to this hypothesis reach a maximum value of 17.23 MPa concentrated in the area of the lower nut securing the shaft in the bearing housing of the stand frame. The distribution of the reduced stresses is shown in figure 5.

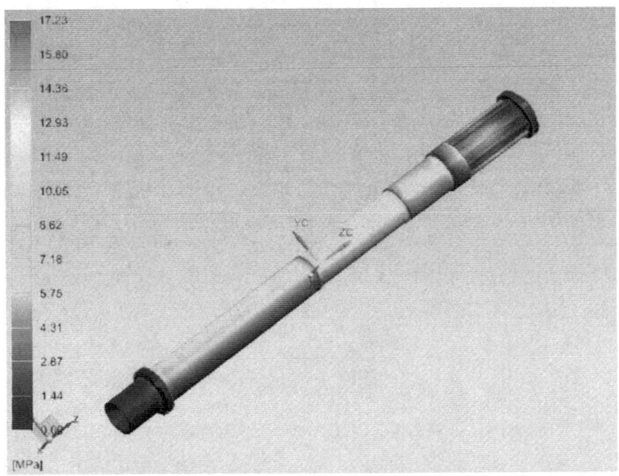

Figure 4. Reduced stress distribution of the shaft.

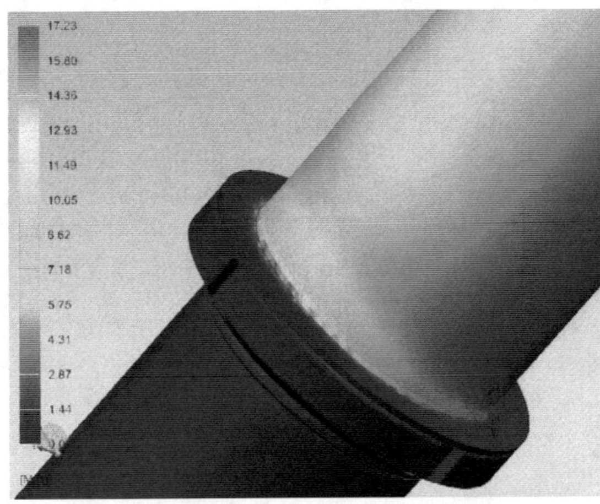

Figure 5. The area where the highest dike-induced stresses occur.

The lifting force generated by the blades with the intelligent material actuator on the rotor caused a stress of 9.72 MPa at the top of the shaft at the point of installation of the nut holding the rotor head on the multi-purge of the shaft (figure 5). Under these forces, the shaft model was also extended by 0.0668 mm along the direction of the axis of rotation (figure 6 and 7).

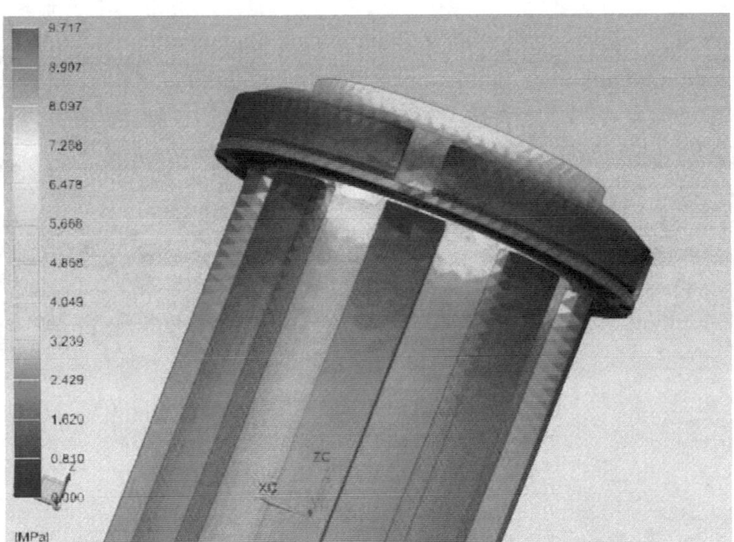

Figure 6. Distribution of stresses at the base of the head nut.

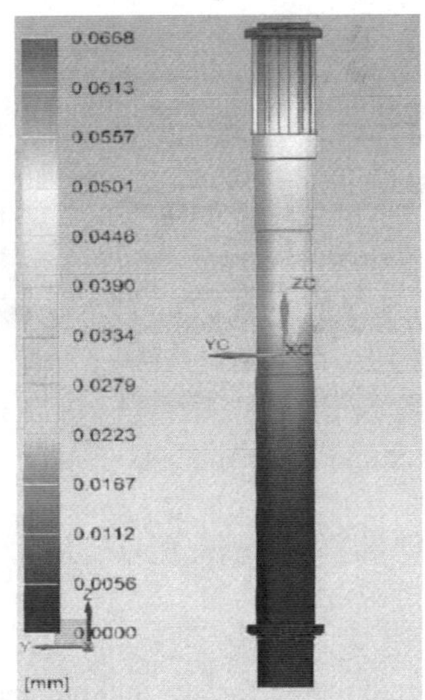

Figure 7. Elongation of the shaft along the Z-axis.

IOP Conf. Series: Materials Science and Engineering **710** (2019) 012003 doi:10.1088/1757-899X/710/1/012003

The stresses occurring in the design of the shaft are four times lower than the permissible stresses for the material from which it will be made. The element, therefore, fulfils the certification condition with a safety factor above k=3.

The rotor head is the connecting element between all the blades and thus the forces and interactions occurring in the rotor. This element transmits the torque from the shaft to the blades via a splined connection. Damage to this component is destructive to the entire rotor. The distribution of global stresses in the model is shown in figure 8. The numerical analysis showed that the highest stresses occurred in the area of the oscillating hinge pins and reach the value of 77.08 MPa. Figure 9 shows the value of stresses in the area of mounting holes.

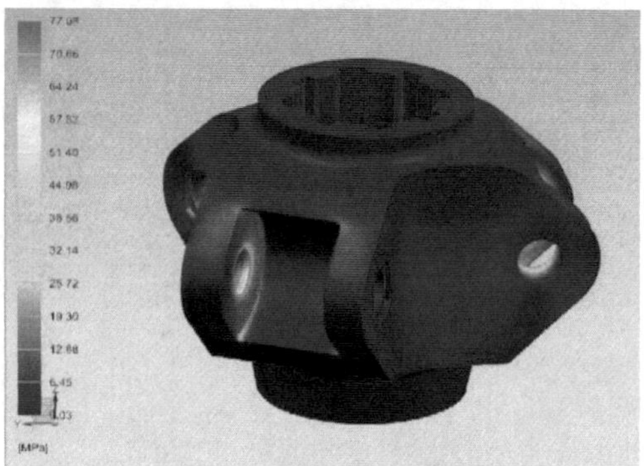

Figure 8. Stress distribution of the head.

Figure 9. Area of the highest reduced stresses.

The other important part of the rotor is the mounting yokes of the blades. In the designed rotor, due to the limited space, the construction consists of four key components: a seat element, the blade's base in the yoke, a modular bearing, a bearing housing and a pusher lever arm. The construction with a closed modular bearing allows the blade angle of attack to be changed independently of the blade force generated by the blade. The distribution of the reduced stresses is shown in figure 10. The stresses in the analysed model are concentrated mainly in the yoke part of the notch area resulting from the assembly pocket for the blade base. The highest value of these stresses is 44.40 MPa (figure 11).

Figure 10. Reduced stress distribution on the shackle of the blade.

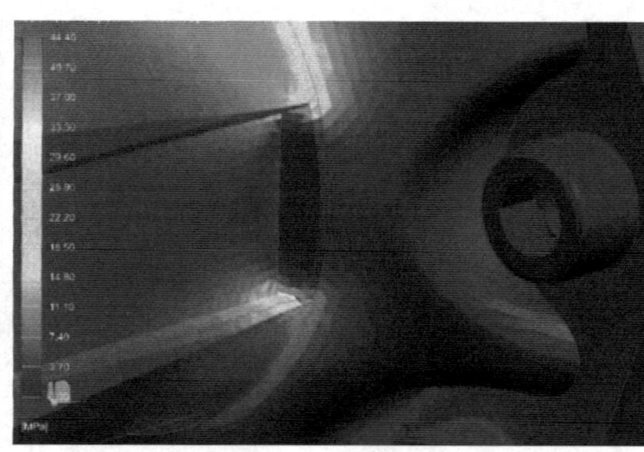

Figure 11. Tensions in the notch area.

Figure 12 illustrates the stress distribution in the axial section of the yoke. The tension at the limit of the change of the seating diameter of the modular bearing and the inner surface of the first hole for the blade-mounting bolt can be distinguished here. The highest acting force is the centrifugal force acting along the blade axis. The highest displacement in this direction is 0.0087 mm (figure 13).

IOP Conf. Series: Materials Science and Engineering **710** (2019) 012003 doi:10.1088/1757-899X/710/1/012003

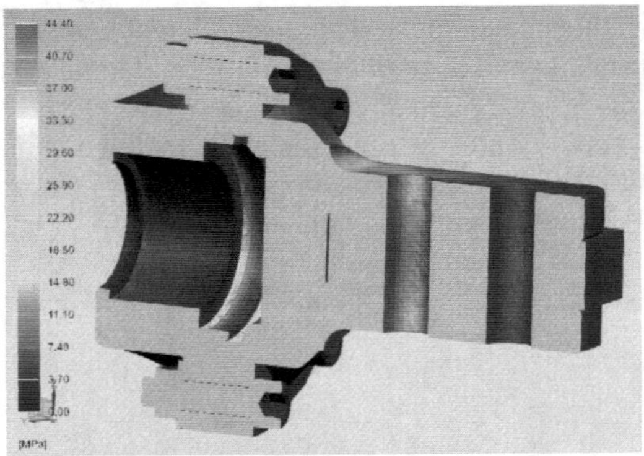

Figure 12. Stress distribution inside the shackle.

Figure 13. Extension of the shackle structure along the centrifugal force of the blade.

A pin is an element that acts as a link between the blade yoke and the rotor head. The numerical analysis showed that this particular element of the structure is characterised by the highest reduced stresses. Figure 14 shows the global stress distribution of the pin. The maximum values reached the value of 100.80 MPa and were focused in the radius connecting the bearing resistance surface with the cylindrical part of the pin (figure 15).

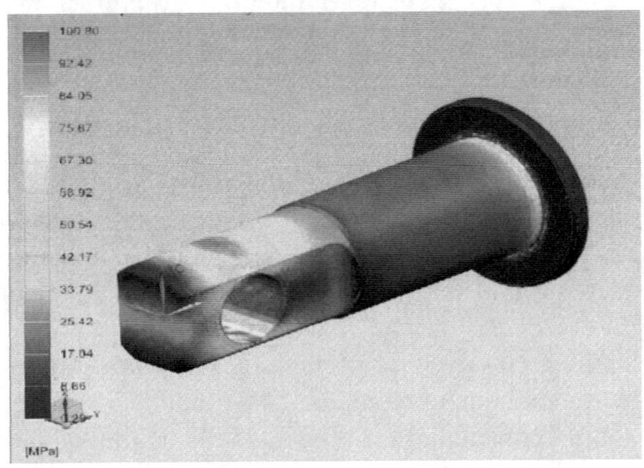

Figure 14. Stress distribution of the pin.

Figure 15. Highest stress area in the pin model.

Stress concentration with values close to 100 MPa also occurs around the pin seat hole in the element as a "Lag-hinge" (figure 16). Axial displacements – tension due to the centrifugal force of the rotor and blade elements caused a 0.0086 mm elongation of the element (figure 17).

Figure 16. Stress area of the pin model in the seat section.

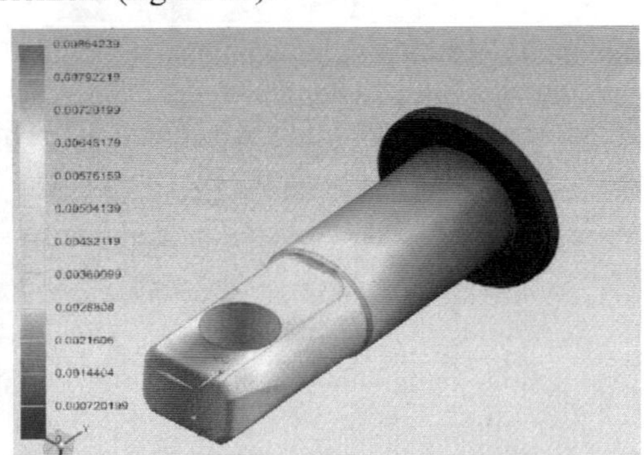

Figure 17. Extension of the pin design along the centrifugal force of the blade.

The last part that binds the shovel to the head is the swing joint. The steel element ensures independent blade and head displacement, eliminating unwanted stresses caused by the use of susceptible blades with intelligent materials. The distribution of reduced stresses is shown in figure 18. The value of the highest stress of 56.74 MPa was recorded within the seat of the "Lag-hinge" joint bearings. The forces transferred by the analysed element caused its deformation of 0.0050 mm (figure 19).

Figure 18. Reduced stress distribution of the articulated components of the blade swinging joint.

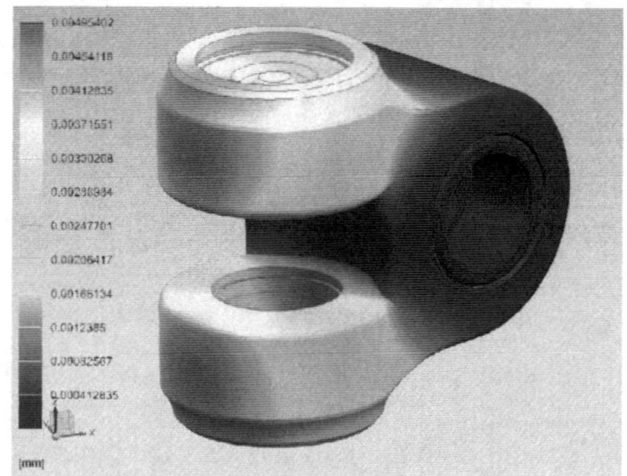

Figure 19. Extension of the joint structure along the action of the blade's centrifugal force.

5. Conclusions

This paper presents the main assumptions for the development of the new design of the rotor. The method of analysis and results of strength calculations of the prototype of the rotor designed for unmanned aerial vehicles using intelligent materials to modify the geometry of blades are presented. Such an application requires modification of the shaft in order to be able to supply power to the actuators installed in the blades. From the analysis, the safety factor was determined at the level of $k>1.5$, which gives grounds to obtaining certification for the strength of the design. The construction of the joint rotor has numerous moving elements in its structure. Despite the high safety factor, the places with the highest stresses, where the signs of wear and tear may appear, have been indicated. The numerical analysis showed that the element under the highest loading is the pin connecting the modular bearing of the blade yoke with the element of the horizontal oscillation joint. The stresses in this element result in a safety factor of $k=1.7$. The other analysed rotor components have a safety factor of more than 2 and in the case of the shaft; this factor is more than 3. However, it must be remembered that the structure is as strong as its weakest cell. The presented design of the rotor for unmanned aerial vehicles adapted to work with blades with intelligent materials meets the requirements for certification testing.

Acknowledgements

This work has been financed by the Polish National Centre for Research and Development under the LIDER programme. Grant Agreement No. LIDER/45/0177/L-9/17/NCBR/2018.

6. References

[1] Greaves M, Elasha F, Worskett J, Mba D, Rashid H and Keong R 2012 *Vibration Health or Alternative Monitoring Technologies for Helicopters* https://www.easa.europa.eu

[2] Goh GD, Agarwala S, Goh GL, Dikshit V, Sing SL, Yeong WY 2017 Additive manufacturing in unmanned aerial vehicles: Challenges and potential *Aerosp Sci Technol.* **63** 140-151

[3] Grabowski Ł, Czyż Z and Kruszczyński K 2014 Numerical Analysis of Cooling Effects of a Cylinders in Aircraft SI Engine *SAE Tech Pap.* 2014-01-28

[4] Wendeker M and Czyż Z 2016 Analysis of the bearing nodes loads of turbine engine at an unmanned helicopter during a jump up and jump down maneuver. *Ekspolatacja i Niezawodność - Maint Reliab.* **18** (1) 89-97

[5] Sochaczewski R and Szlachetka M 2019 Numerical Analysis of a Fuel Pump for an Aircraft Diesel Engine *MATEC Web Conf.* **252** 01003
[6] Czyż Z, Siadkowska K and Sochaczewski R 2019 CFD Analysis of Charge Exchange in an Aircraft Opposed-Piston Diesel Engine *MATEC Web Conf.* **252** 04002
[7] Grabowski Ł, Siadkowska K and Skiba K 2019 Simulation Research of Aircraft Piston Engine Rotax 912 *MATEC Web Conf.* **252** 05007
[8] Raczynski R. 2019 The use of co-simulation methodology in the project of PZL SW-4 helicopter adaptation to maritime version *45th European Rotorcraft Forum* Warsaw University of Technology
[9] Fincham J H S and Friswell M I 2015 Aerodynamic optimisation of a camber morphing aerofoil *Aerosp Sci Technol.* **43** 245-255
[10] Ajaj R M, Beaverstock C S and Friswell M I 2016 Morphing aircraft: The need for a new design philosophy *Aerosp Sci Technol.* **49** 154-166
[11] Sarkar P and Raczynski R 2017 Recent Progress in Flow Control for Practical Flows 2017 *Doerffer P, Barakos GN, Luczak MM* eds. Springer 126-135
[12] SESAR European Drones Outlook Study 2016 **93**
[13] Mazur M and McMillan J 2016 Global Market for Commercial Applications of Drone Technology Valued at over $127 bn *PwC*
[14] Joshi D. 2017 Commercial Unmanned Aerial Vehicle (UAV) Market Analysis – Industry trends, companies and what you should know *Bus Insid.*
[15] Siadkowska K 2017 Prawne aspekty eksploatacji dronów *Stud Iurid Lublinensia* **26** (3) 93-106
[16] Bramwell ARS, Done G and Balmford D 2001 Bramwell's Helicopter Dynamics Oxford: Butterworth-Heinemann
[17] Flight Mechanic 2017 Helicopter Structures and Airfoils – Main Rotor Systems
[18] European Union Aviation Safety Agency 2018 Certification Specifications and Acceptable Means of Compliance for Small Rotorcraft CS-27 *Amendmant 6*
[19] Kronos EDM 2019 Stal narzędziowa konstrukcyjna. Stal C45 (1.0503)
[20] Kronos EDM 2019 Metale kolorowe. Aluminium. PA9 (AW-7075).

IOP Conf. Series: Materials Science and Engineering **710** (2019) 012004 doi:10.1088/1757-899X/710/1/012004

Application of decision–making parametric structures in the analysis of a compound planetary gear

A Deptula[1], J Drewniak[2] and W Macek[3,4]

[1] Opole University of Technology, Faculty of Production Engineering and Logistics, Poland
[2] University of Bielsko-Biala, Faculty of Machine Building and Computer Science, Poland
[3] Energopiast sp. z o.o., Polna 12, 55-011 Siechnice, Poland
[4] Wroclaw School of Information Technology, ul. Ks. M. Lutra 4, 54-239 Wrocław, Poland

a.deptula@po.edu.pl

Abstract. The design process of planetary gears can be divided into two main stages, i.e. conceptual design and construction, that is the preparation of construction documentation. In the first stage, the constructor has a very limited number of project data. In the case of a reduction or multiplier transmission these are usually output parameters (torque and speed), characteristics of the driving and the driven machine, Therefore, the stage of gearbox design must be supported by preliminary calculations, which supports the selection of the correct technical solution, leading to the stage of the structural design preparation. In the method of systematic search, we used an algorithm for generating induction decision trees, based on entropy growth as a method related to machine learning. In the next step, the decision-making parametric structures were employed. In the future, the presented solution will allow further analyses and syntheses such as checking the isomorphism of the proposed solutions, determining the validity of constructions and/or operating parameters of the analysed gears.

1. Introduction

Graph-theoretical ideas are highly utilised by computer science applications. Especially in research areas of computer science data mining, clustering, images capturing, network etc; however, the scope of their applications includes studying and modelling in various areas such as, the traveling salesman problem, the shortest spanning tree in a weighted graph, obtaining an optimal match of jobs and men and locating the shortest path between two vertices in a graph. In the last dozen or so years, we can observe an extremely rapid trend for graph theory applications, decision trees and networks in various fields of science and technology: the theory of operational research, genetics, linguistics, econometrics, electronics to name a few. Thus, the need for further development of this theory and its methods has been growing. From a mathematical point of view, the theory of graphs and decision trees does not bring revolutionary innovation. The advantage of these methods, however, is the adaptation of mathematical methods and apparatus to the convenient modelling of phenomena of the contemporary world. Graphs, networks and trees are convenient as a type of a formal apparatus for machine system modelling, in hydraulic and transmission systems [1, 2, 3].

IOP Conf. Series: Materials Science and Engineering **710** (2019) 012004 doi:10.1088/1757-899X/710/1/012004

Other attempts have been made to model planetary gears by means of diagrams e.g. Wolf's pictograms [4], but the evolution of these methods has been halted by the lack of generalisation and lack of connections with other branches of mathematics [5, 6]. Also, the method based on signal flow graph theory for modelling of gears [7, 8] has not been quite popular. On the contrary, the graph-based methods have been independently and extensively developed for several recent years all over the world [7- 9]. Unlike graphs, dendrite-tree structures do not have cycles, but there may be a different number of initial vertices. Therefore, a different approach has been developed as a translation of a directed graph of dependence, among others, for parametrically acting structures. This approach is different from the established solutions described in the literature on parametric automation machines and their applications, related to control systems, operating systems, importance analysis of construction and/or operating parameters, and analysis of gears previously modelled using other types of graphs.

2. Analysis of planetary gears

The planetary gear has derived its name from the similarity to the planetary system, with which it shares the common feature: the "sun," the central component around which the other elements are orbiting like planets. It is made of two concentric gear wheels: a central (solar) wheel with external toothing, fixed permanently on the shaft, and a crown wheel (also called the central one) with internal toothing. Other elements are 3 to 5 small gears – satellites connected with a yoke (a jumper). The design and model diagram of the selected variant of the transmission [10- 11] is shown in figure 1.

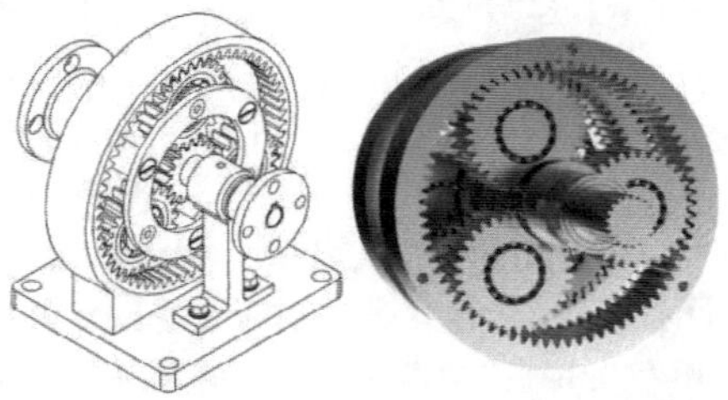

Figure. 1. Single-row gears - construction scheme and model diagram [10-11].

One of the most characteristic features of a planetary gear is the ability to transfer power based on three gear ratios. The system of internal and external wheels transfers the drive with high torque even from engines with high powers, while maintaining a small size. Thanks to this, the gear unit can be placed in a small drive medium, which does not burden the machine structure. The advantages of planetary gears include the following:
- one of the highest efficiencies (98- 99%),
- high reliability and low consumption,
- high power transfer,
- constant gear ratio (this is the only advantage of the chain transmission),
- other transmissions have slip gears that change the transmission ratio.

A simple gearing has a limited ratio. Typically, the absolute value of the base ratio does not exceed 10 [12]. When a larger gear ratio is needed, multi-stage systems or coupled systems are used. figure 2 shows a diagram of a two-stage ring gear [13].

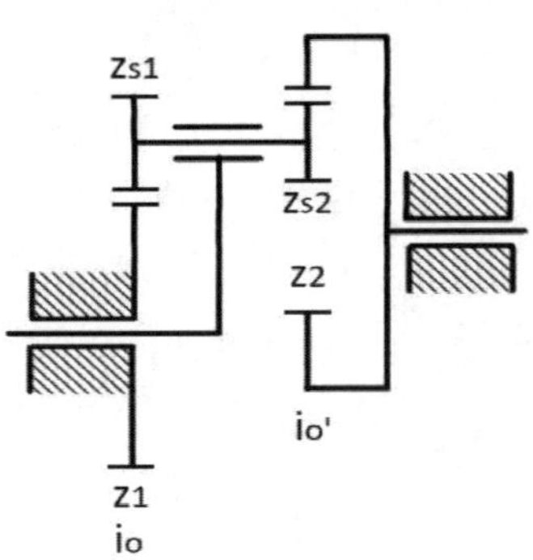

Figure. 2. Diagram of a two-stage planetary gearbox with a first i_o. and second i'_o degree ratio [13].

Figure 3 presents an exemplary constructional solution of the transmission from figure 2.

In circulating gears, as well as in other types of gear transmissions, not all parameters can be determined simultaneously at the very beginning of the design, particularly when the contour graph method is used.

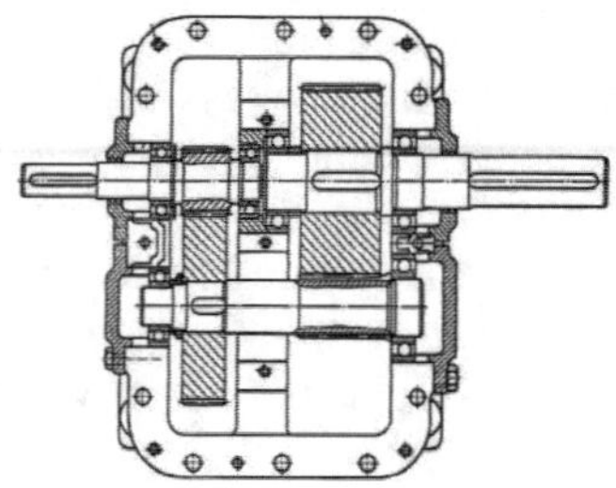

Figure. 3. Structural solution of a two-stage circular gear - bearing gears were used on both stages [10].

3. Subjects of investigations

3.1 2DOF planetary gear

The first considered gear is graphically presented as a functional scheme (Figure 4). It consists of three pseudo-stages, despite having 3 satellite wheels and 3 arms, referred to as carriers. In fact, there are not three classical stages because the system has a serial-parallel layout. In the classical meaning, the stages are serially arranged. The gear consists of eight movable parts, among them: three pinions, – i.e. sun gears denoted in Figure 4 – as 1, 4 and 7, three planet gears 2, 5 and 8, three-ring gears 3, 6 and 9 as well as three arms h_3, h_5 and h_8. However, it should be highlighted that gears 3 and 4 as well as wheels 6 and 7 are connected, creating the element that is simultaneously the sun and ring wheel. Similarly, all arms are united creating one arm h, which drives the output shaft II. The input shafts are these two – marked as I and III, respectively. Moreover, shaft I drives the sun gear 1, and independently shaft III drives the ring gear 9. Therefore, this is 2 DOF planetary gear, which is also calculated below, based upon an adequate formula. The kinematical scheme of the gear – grasping precisely all the above-listed relationships and descriptions – is shown in figure 1. It allows for execution of necessary structural and kinematic analyses. Moreover, all rotational pairs (full joints) are equipped in ball bearings (one for every kinematical pair) in accordance with recommendations of the classical mechanism and machine science.

$$M = 3 \cdot n - 2 \cdot c_5 - c_4 = 3 \cdot 8 - 2 \cdot 8 - 6 = 2,$$

where: $n = 8$ – the number of moving links, $c_5 = 8$ – the number of joints of class 5 (full joints), $c_4 = 6$ – the number of joints of class 4 (half joints).

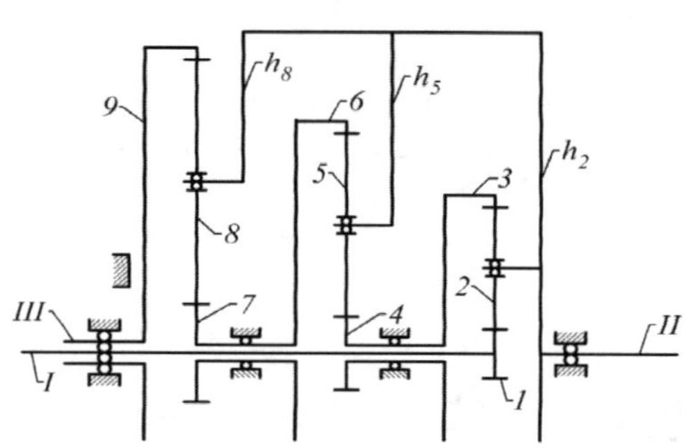

Figure. 4. Kinematic scheme of compound 2 DOF planetary gear [14].

In general, the considered system has two DOF gears, as in [14, 15]:

The geometrical and structural data are as follows:

a) tooth numbers for consecutive geared wheels:

$z_1 = z_4 = z_7 = 18$, $z_2 = z_5 = z_8 = 45$, $z_3 = z_6 = z_9 = -108$

b) moduli of geared wheels (the dimensions are given in appropriate standards):
$m_I = m_1 = m_2 = m_3 = 1\,\text{mm}$, $m_{II} = m_4 = m_5 = m_6 = 1.5\,\text{mm}$, $m_{III} = m_7 = m_8 = m_9 = 2\,\text{mm}$,
c) the number of satellite wheels in consecutive parallel planes: $s_2 = s_5 = s_8 = s = 3$.
For example, the pitch radiuses of adequate geared wheels have the following values:

$$r_1 = \frac{m_I \cdot z_1}{2} = 9\text{ mm},\ r_2 = \frac{m_I \cdot z_2}{2} = 22.5\text{ mm},\ r_3 = \frac{m_I \cdot z_3}{2} = -54\text{ mm},\ r_4 = \frac{m_{II} \cdot z_4}{2} = 13.5\text{ mm}$$

3.1 1 DOF planetary gear
The second object of kinematic analysis is identical to the previous planetary gear except for the ring gear 9 that here is braked (Figure 5). What follows is that the gear has now one input shaft I and one output shaft II, because the second input shaft III is immobilised [15].

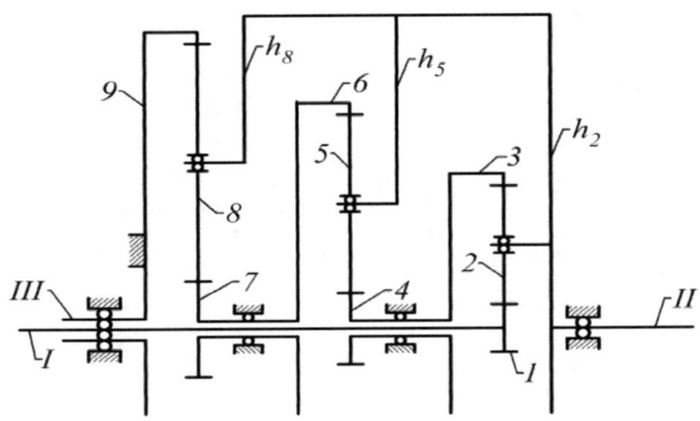

Figure. 5. Kinematic scheme of compound 1 DOF planetary gear.

The system possesses one DOF, because:

$$M = 3 \cdot n - 2 \cdot c_5 - c_4 = 3 \cdot 7 - 2 \cdot 7 - 6 = 1,$$

where: $n = 7$ – the number of moving links, $c_5 = 7$ – the number of class 5 joints (full joints), $c_4 = 6$ – the number of class 4joints (half joints).

Moreover, it should be noted that in the case of ring gear braked 9, the rotational kinematic pair (full joint) between this wheel and the base is inactive, and as such is not taken into account in structural considerations.

4. Method of contour equations (graphs)

Graphs play an important role in the kinematic analysis. Kinematic analysis is the study of relative motions associated with the links of a mechanism or machine and is a critical step towards the proper design of mechanisms. The problem can be formulated by the graphical, vector algebra, matrix, or other mathematical methods. The general idea of the graph-based modelling of mechanical systems consists of the following steps [5]:
- discretisation of a mechanical system. This means that appropriate simplifications have to be made. Some aspects are omitted. Some structural elements are considered essential and they are interpreted as graph vertices.
- assignment of the graph to the mechanism (especially the planetary gear) from special rules.
- derivation of special subgraphs, e.g. f-cycles or contours. These subgraphs can be singled out based upon the graph-theoretical algorithms, which causes that the approach is simple and algorithmic,
- listing the codes of these graph elements. The encoding rules are clearly defined, thus preventing mistakes,
- generation of a system of equations in an algorithmic way using codes.
- solving the obtained system in a chosen algebraic way to obtain needed angular velocities, ratios, forces, accelerations, etc.

4.1. Method of contour equations for the analysis of 2 DOF gear

Kinematic data for the first considered planetary gear have been established from or related to the structural layout of the considered planetary gear:

rotational velocity of shaft I: $\omega_I = \omega_1 = 150.08\,\mathrm{rad/s}$,

rotational velocity of shaft III: $\omega_{III} = \omega_9 = -0.314\,\mathrm{rad/s}$ additional data resulting from the gear structure:

$\omega_4 = \omega_3$, $\omega_{h_2} = \omega_{h_5} = \omega_{h_8} = \omega_h = \omega_{II}$, $\omega_7 = \omega_6$,

The following problems can be solved:

find the absolute output rotational velocity $\omega_{II} = \omega_{h_2} = \omega_{h5} = \omega_{h_8} = \omega_h$ of the output shaft II (that is also arms h_2, h_5, h_8 forming the main arm h connected to the shaft II - Figure 1), if the input shafts I (sun 1) and II (ring gear 9) rotate with input rotational velocities $\omega_I = \omega_1$ and ω_9, respectively,

find the transmission ratio $\left(i_{1,h}\right)_{\omega_9 \neq 0}$,

find the absolute rotational velocities ω_2, $\omega_3 = \omega_4$, ω_5, $\omega_6 = \omega_7$ and ω_8 of the other movable elements, i.e. gears 2, 3, 4, 5, 6, 7 and 8.

The formulated tasks will be first solved by means of the contour graph method. As mentioned, the basics of the method may be found in the book [16]. Therefore here, the equations are used without detailed explanations or derivations. The further considerations are related to the contour graph assigned to the analysed gear (Figure 6).

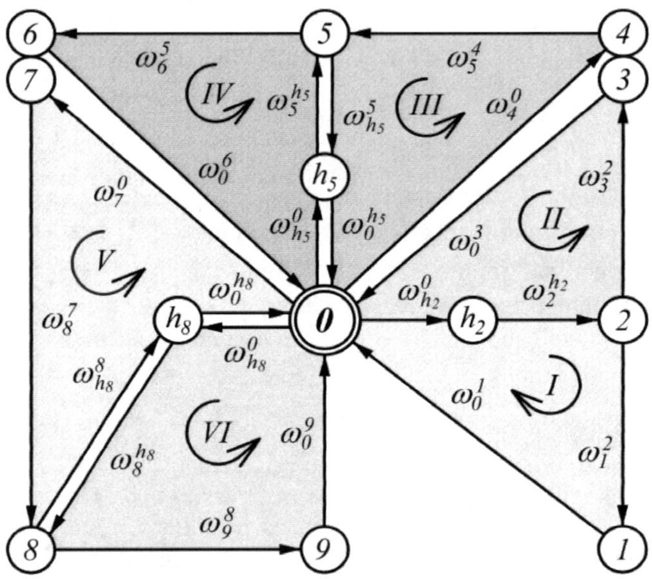

Figure. 6. Contour graph model of 2 DOF planetary gear.

Aiming for increasing the usefulness of graph modelling as well as to improve the understanding of the derived equations, the descriptions of relative rotational velocities were placed near the adequate arcs. For example, $\boldsymbol{\omega}_1^2$ denotes the vector of the relative rotational velocity of wheel 1 in relation to wheel 2. We use here notation rule that vector variables are written in bold instead of a traditional arrow above them

The number of independent contours is determined based on the following formula [16]:

$N = c - n = 14 - 8 = 6$,

where c – the number of joints ($c = c_5 + c_4 = 14$).

The algorithm for the method consists of contour equations for consecutive contours [16]:

for contour I:

$$\boldsymbol{\omega}^{0}_{h_2}+\boldsymbol{\omega}^{h_2}_{2}+\boldsymbol{\omega}^{2}_{1}+\boldsymbol{\omega}^{1}_{0}=\mathbf{0}, \tag{1}$$

$$\left(\mathbf{r}_1+\mathbf{r}_2\right)\times\boldsymbol{\omega}^{h_2}_{2}+\mathbf{r}_1\times\boldsymbol{\omega}^{2}_{1}=\mathbf{0}, \tag{2}$$

for contour II:

$$\boldsymbol{\omega}^{0}_{h_2}+\boldsymbol{\omega}^{h_2}_{2}+\boldsymbol{\omega}^{2}_{3}+\boldsymbol{\omega}^{3}_{0}=\mathbf{0}, \tag{3}$$

$$\left(\mathbf{r}_1+\mathbf{r}_2\right)\times\boldsymbol{\omega}^{h_2}_{2}+\left(\mathbf{r}_1+2\mathbf{r}_2\right)\times\boldsymbol{\omega}^{2}_{3}=\mathbf{0}, \tag{4}$$

for contour III:

$$\boldsymbol{\omega}^{0}_{4}+\boldsymbol{\omega}^{4}_{5}+\boldsymbol{\omega}^{5}_{h_5}+\boldsymbol{\omega}^{h_5}_{0}=\mathbf{0}, \tag{5}$$

$$\mathbf{r}_4\times\boldsymbol{\omega}^{4}_{5}+\left(\mathbf{r}_4+\mathbf{r}_5\right)\times\boldsymbol{\omega}^{5}_{h_5}=\mathbf{0}, \tag{6}$$

for contour IV:

$$\boldsymbol{\omega}^{0}_{h_5}+\boldsymbol{\omega}^{h_5}_{5}+\boldsymbol{\omega}^{5}_{6}+\boldsymbol{\omega}^{6}_{0}=\mathbf{0}, \tag{7}$$

$$\left(\mathbf{r}_4+\mathbf{r}_5\right)\times\boldsymbol{\omega}^{h_5}_{5}+\left(\mathbf{r}_4+2\mathbf{r}_5\right)\times\boldsymbol{\omega}^{5}_{6}=\mathbf{0}, \tag{8}$$

for contour V:

$$\boldsymbol{\omega}^{0}_{7}+\boldsymbol{\omega}^{7}_{8}+\boldsymbol{\omega}^{8}_{h_8}+\boldsymbol{\omega}^{h_8}_{0}=\mathbf{0}, \tag{9}$$

$$\mathbf{r}_7\times\boldsymbol{\omega}^{7}_{8}+\left(\mathbf{r}_7+\mathbf{r}_8\right)\times\boldsymbol{\omega}^{8}_{h_8}=\mathbf{0}, \tag{10}$$

for contour VI:

$$\boldsymbol{\omega}^{0}_{h_8}+\boldsymbol{\omega}^{h_8}_{8}+\boldsymbol{\omega}^{8}_{9}+\boldsymbol{\omega}^{9}_{0}=\mathbf{0}, \tag{11}$$

$$\left(\mathbf{r}_7+\mathbf{r}_8\right)\times\boldsymbol{\omega}^{h_8}_{8}+\left(\mathbf{r}_7+2\mathbf{r}_8\right)\times\boldsymbol{\omega}^{8}_{9}=\mathbf{0}. \tag{12}$$

Since the edges of the graph (digraph) have directions, they can be called arcs. The oriented cycles are in this case referred to as contours to distinguish them from others. The contours are marked by Roman numerals.

5. Application of decision–making parametric structures

In the graphs of dependencies parametrically analysed are the so-called 'bound' decisions. The results obtained in the subsequent decisions depend on the initial decisions, which enables the creation of dynamic models. The decision process and the space of the possible states of the analysed system are described parametrically from each of the graph tops.

A graph is an ordered pair G=(X, R), in which X is a finite set of elements called vertices of a graph, and R is a set of pairs $(x_i, x_j)(x_i, x_j \in X)$ called the edges of the graph. In the case of parametric graphs, the notation introduced by Deptuła [17, 18] defines the signs: G=(Q, Z), where Z is a set of pairs $(z_i, z_j)(z_i, z_j \in Z)$. Figure 7 shows an example of a parametric graph G.

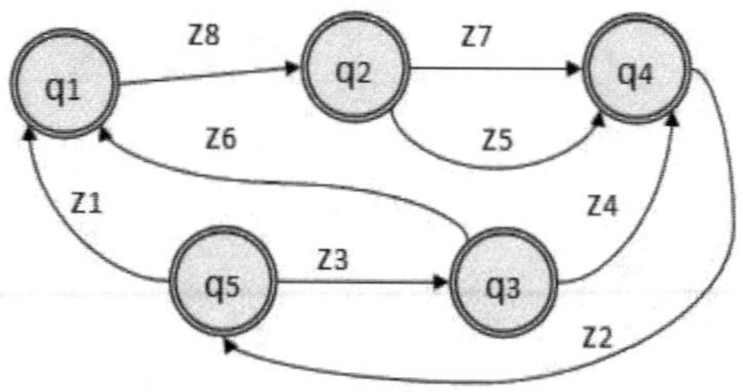

Figure. 7. An oriented dependency graph.

The oriented game graph is shown in Figure 7 is composed of a set of vertices Q:

$$Q = \{q_1, q_2, q_3, q_4, q_5\} \tag{13}$$

and of a set of edges Z, that is an ordered pair of vertices:

$$Z = \{z_1, z_2, z_3, z_4, z_5, z_6, z_7, z_8\} \tag{14}$$

The path in the G=(Q, Z) is the edge sequence $(z_{i_1}, z_{i_2}), (z_{i_2}, z_{i_3}), \ldots, (z_{i_{k-1}}, z_{i_k})$ in which for each $j \in \{2,3,\ldots,k\} (z_{i_{j-1}}, z_{i_j}) \in R$ and vertices $q_{i_1}, q_{i_2}, \ldots, q_{i_k}$ are different pairs. Vertex q_{i_1} is called the beginning of the path, and the top q_{i_k} – the end of the path. As a result of the graph distribution from the chosen vertex, a tree structure with cycles is obtained in the first step and, next, a general game tree structure is obtained. Each of them has an appropriate analytical formulation G_i^+ and G_i^{++}. A game tree structure is a part of the systematic searching method. A start vertex q_i is chosen in the first step. The algorithm of graph distribution of dependence on parametric structures can be found in the literature [19].

The graph in figure 7 is distributed into 5 structures forming the D set:

$$D = \{G_{\{q1\}}^{++}, G_{\{q2\}}^{++}, G_{\{q3\}}^{++}, G_{\{q4\}}^{++}, G_{\{q5\}}^{++}\} \tag{15}$$

Figures 8-9 show the structures from each of the vertices: q1, q2, q3, q4, q5 of the graph from figure 7.

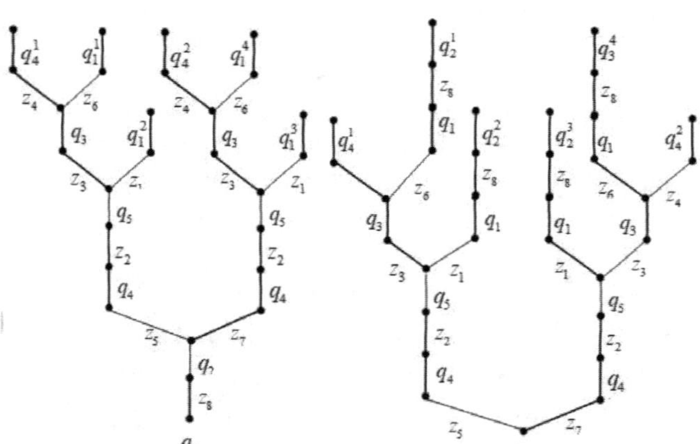

Figure. 8. Game-tree structures with initial vertices: *q_1, q_2,*.

IOP Conf. Series: Materials Science and Engineering **710** (2019) 012004 doi:10.1088/1757-899X/710/1/012004

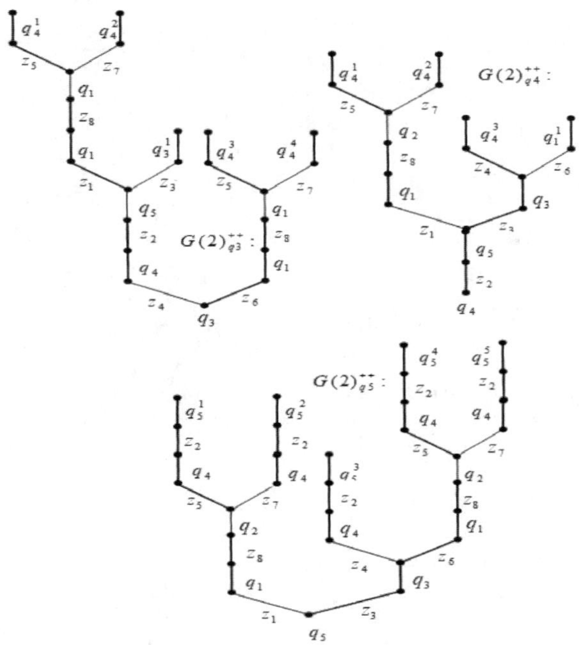

Figure. 9. Game-tree structures with initial vertices: q_3, q_4, q_5.

Each structure has its own analytical formula. For example, the structures in figure 4 are described by the following formulas:

$$\begin{cases} G^{++}_{q3} = ({}^{0}q_3({}^{1}z_4q_4({}^{2}z_2q_5({}^{3}z_1q_1({}^{4}z_8q_2({}^{5}z_5q_4^1, z_7q_4^2)^5)^4, z_3q_3^1)^3)^2, \\ z_6q_1({}^{2}z_8q_2({}^{3}z_5q_4^3, z_7q_4^4)^3)^2)^1)^0 \\ G^{++}_{q4} = ({}^{0}q_4({}^{1}z_2q_5({}^{2}z_1q_1({}^{3}z_8q_2({}^{4}z_5q_4^1, z_7q_4^2)^4)^3, z_3q_3({}^{3}z_4q_4^3, z_6q_1^1)^3)^2)^1)^0 \\ G^{++}_{q5} = ({}^{0}q_5({}^{1}z_1q_1({}^{2}z_8q_2({}^{3}z_5q_4({}^{4}z_2q_5^1)^4, z_7q_4({}^{4}z_2q_5^2)^4)^3)^2, \\ z_3q_3({}^{2}z_4q_4({}^{3}z_2q_5^3)^3, z_6q_1({}^{3}z_8q_2({}^{4}z_5q_4({}^{5}z_2q_5^4)^5, z_7q_4({}^{5}z_2q_5^5)^5)^4)^3)^2)^1)^0 \end{cases}$$

5.1. Structural representation including game-tree structures

Kinematic synthesis is the reverse problem of kinematic analysis. In this case, the designer is challenged to devise a new mechanism that satisfies certain desired motion characteristics of an output link. The kinematic synthesis problem can be further divided into three interrelated phases:

1. Type synthesis refers to the selection of a specific type of mechanism for product development. During the conceptual design phase, the designer considers many types of mechanism as possible and decides what type shows the highest potential of meeting the design objectives.

2. Number synthesis deals with the determination of the number of links, type of joint, and number of joints needed to achieve a given number of degrees of freedom of a desired mechanism.

Various methodologies have been developed for the systematic enumeration of kinematic structures. A thorough understanding of the structural characteristics of a given type of mechanism is critical for the development of an efficient algorithm. In the 2000s, certain approaches for structural synthesis of kinematic chains were presented [20-25]. Nelson et al. [26] the kinematic, power flow, and efficiency analysis of epicyclic gear trains. Wojnarowski et al. [27] presented a survey of works connected with the problem of the modelling of gears by means of versatile graph theory models. Zawiślak [4] reviewed selected graph applications for gear design from the AI perspective. A single atlas of graphs can be used to enumerate an enormous number of mechanisms. An exemplary transformation is shown in figure 10.

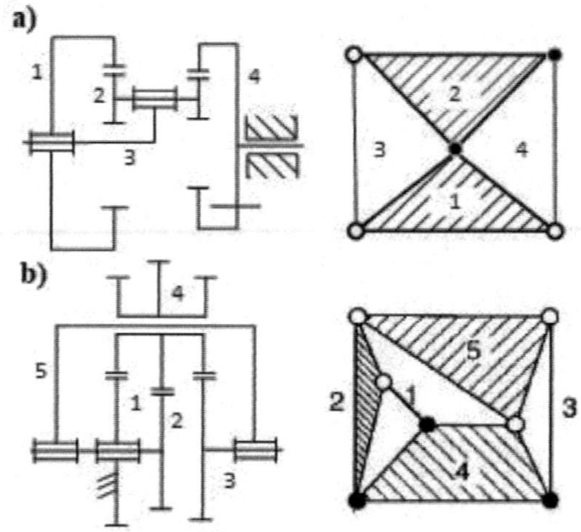

Figure. 10. Functional schematic and structural representation.

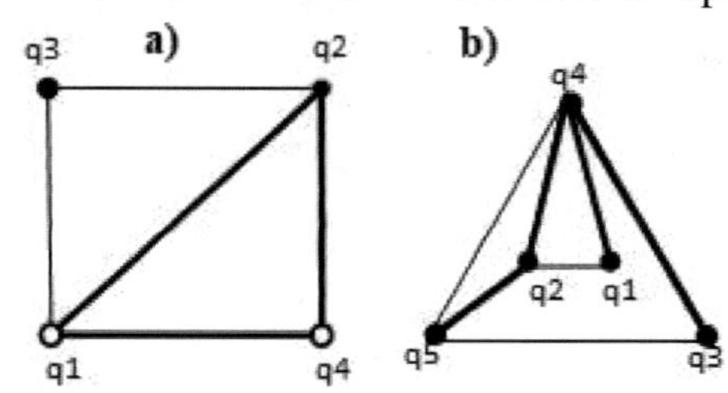

Figure. 11. Graph representation for structures from figure 10.

A special group of graphs are parametric graphs. The structural topology of a mechanism can be uniquely identified. Using graph representation, the similarity and difference between two different mechanism embodiments can be easily recognised. In particular, parametrically acting graphs can be used for systematic classification of mechanisms

For the contour graph model of 2 DOF planetary gear presented in figure 6, a dependency graph can be used (figure 12).

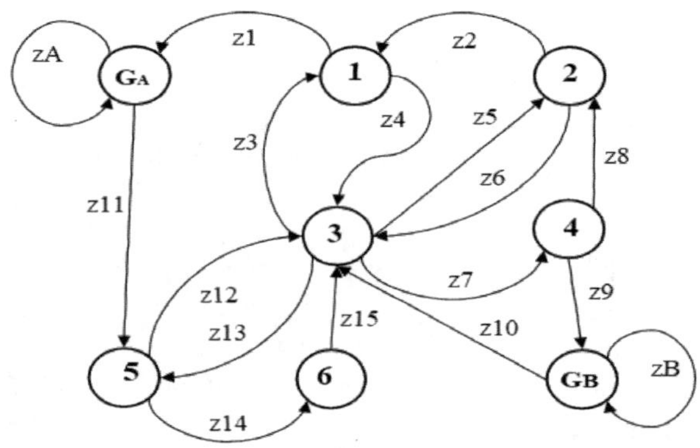

Figure. 12. Dependency graph for contour graph model of 2 DOF planetary gear.

For the graph with a path from the entrance to the output from Figure 12, it is possible to build a set D of parametrically acting structures:

$$D = \left\{ G_1^{++}, G_2^{++}, G_3^{++}, G_3^{++}, G_5^{++}, G_6^{++} \right\} \tag{16}$$

Each of the structures has an analytical formula describing the distribution of the dependency graph. Exemplary structures ($G_A^{++}, G_B^{++}, G_1^{++}, G_4^{++}$) have the following formulas

$$\begin{aligned} G_A^{+} = & ({}^{0}q_A({}^{1}z_{11}q_5({}^{2}z_{12}q_3({}^{3}z_5q_2({}^{4}z_6q_3, z_2q_1({}^{5}z_4q_3, z_1q_A)^{5})^{4}), \\ & z_7q_4({}^{4}z_8q_2, z_9q_B({}^{5}z_{10}q_3, z_Bq_B)^{5})^{4}, z_{13}q_5, z_3q_1)^{3}, z_{14}q_6({}^{3}z_{15}q_3)^{3})^{2}, \\ & z_Aq_A)^{1})^{0} \end{aligned} \tag{17}$$

$$\begin{aligned} G_B^{+} = & ({}^{0}q_B({}^{1}z_{10}q_3({}^{2}z_2q_2({}^{3}z_6q_3, z_2q_1({}^{4}z_4q_3, z_1q_A^{1}({}^{5}z_{11}q_5 \\ & ({}^{6}z_{12}q_3, z_{14}q_6({}^{7}z_{15}q_3)^{7})^{6}, z_Aq_A)^{5})^{4})^{3}, z_7q_4({}^{3}z_8q_2, z_9q_B)^{3}, \\ & z_{13}q_5, z_3q_1)^{2}, z_Bq_B)^{1})^{0} \end{aligned} \tag{18}$$

$$\begin{aligned} G_1^{+} = & ({}^{0}q_1({}^{1}z_4q_3({}^{2}z_5q_2({}^{3}z_6q_3, z_2q_1)^{3}, z_7q_4({}^{3}z_8q_2, z_9q_B \\ & ({}^{4}z_{10}q_3, z_B)^{4})^{3}, z_{13}q_5({}^{3}z_{12}q_3, z_{14}q_6({}^{4}z_{15}q_3)^{4})^{3}, z_3q_1)^{2}, \\ & z_1q_A(z_{11}q_5, z_Aq_A)^{2})^{1})^{0} \end{aligned} \tag{19}$$

$$\begin{aligned} G_4^{+} = & ({}^{0}q_4({}^{1}z_8q_2({}^{2}z_6q_3({}^{3}z_5q_2, z_7q_4, z_{13}q_5({}^{4}z_{12}q_3, \\ & z_{14}q_6({}^{5}z_{15}q_3)^{5})^{4}, z_3q_1({}^{4}z_4q_3, z_1q_A({}^{5}z_{11}q_5, z_Aq_A)^{5})^{4})^{3}, \\ & z_2q_1)^{2}, z_9q_B({}^{2}z_{10}q_3, z_Bq_B)^{2})^{1})^{0} \end{aligned} \tag{20}$$

In the next stage, we select the optimal structuring that is parametric. The following relationships were taken additionally into account: $\omega_3^0 = \omega_4^0 = \omega_3 = \omega_4$, $\omega_6^0 = \omega_7^0 = \omega_6 = \omega_7$, $\omega_{j,0} = -\omega_{0,j} = \omega_j$ for $j = 1, 2, \ldots, 9$:

$$\omega_h + \omega_2^h + \omega_1^2 = \omega_1, \tag{21}$$

$$\omega_h + \omega_2^h + \omega_3^2 - \omega_3 = 0, \tag{22}$$

$$-\omega_h + \omega_3 + \omega_5^4 - \omega_5^h = 0, \tag{23}$$

$$\omega_h + \omega_5^h + \omega_6^5 - \omega_6 = 0, \tag{24}$$

$$-\omega_h + \omega_6 + \omega_8^7 - \omega_8^h = 0, \tag{25}$$

$$\omega_h + \omega_8^h + \omega_9^8 = \omega_9, \tag{26}$$

$$\left(r_1 + r_2\right) \cdot \omega_2^h + r_1 \cdot \omega_1^2 = 0, \tag{27}$$

$$\left(r_1 + r_2\right) \cdot \omega_2^h + \left(r_1 + 2r_2\right) \cdot \omega_3^2 = 0, \tag{28}$$

$$r_4 \cdot \omega_5^4 - \left(r_4 + r_5\right) \cdot \omega_5^h = 0, \tag{29}$$

$$\left(r_4 + r_5\right) \cdot \omega_5^h + \left(r_4 + 2r_5\right) \cdot \omega_6^5 = 0, \tag{30}$$

$$r_7 \cdot \omega_8^7 - \left(r_7 + r_8\right) \cdot \omega_8^h = 0, \tag{31}$$

$$\left(r_7 + r_8\right) \cdot \omega_8^h + \left(r_7 + 2r_8\right) \cdot \omega_9^8 = 0, \tag{32}$$

The matrix-form equation representing the above system of equations is given below (where $\omega_I = \omega_1 = 150.08\,\text{rad/s}$ and $\omega_9 = -0.314\,\text{rad/s}$):

$$\{\mathbf{A}\} \cdot [\mathbf{x}] = [\mathbf{b}] \tag{33}$$

Where:

$[\mathbf{x}]^T = [\omega_h \;\; \omega_2^h \;\; \omega_1^2 \;\; \omega_3^2 \;\; \omega_3 \;\; \omega_5^4 \;\; \omega_5^h \;\; \omega_6^5 \;\; \omega_6 \;\; \omega_8^7 \;\; \omega_8^h \;\; \omega_9^8]$

$[\mathbf{b}]^T = [157.08 \;\; 0 \;\; 0 \;\; 0 \;\; 0 \;\; -0.314 \;\; 0 \;\; 0 \;\; 0 \;\; 0 \;\; 0 \;\; 0]$.

In the next stage, the Willis method should be used and the analysis carried out for gear with 2 DOF.

In the work on compound planetary gears [28] for the analysed gear, a special algorithm was developed to determine the optimal number of teeth. This method can also be successfully used to analyse failure modes and causes [29-32].

IOP Conf. Series: Materials Science and Engineering **710** (2019) 012004 doi:10.1088/1757-899X/710/1/012004

6. Conclusions

The game-tree structures method can be an excellent tool for gathering and transforming data. After encoding, the data can be repeatedly reproduced and globally distributed via the programme. In expert systems, the question of structuring the inference process (obtaining knowledge) consists of two separate parts:

- mapping of deductive activities specific to a given subject field (defining the structure of inference)
- conducting the deduction process by the requesting mechanism (determination of the inference function).

This approach to the deduction process makes it possible to represent the inference structure in the form of a parametrically directed graph with a defined span (i.e. the maximum or the minimum number of arcs derived from one node).

Graph nodes in the inference structure represent individual information states, and edges are logical transition operations that result in state transformation. Initial nodes are variants of the initial problem description, end nodes are the ending of the inference process. The graph-based methods of analysis and synthesis of planetary gears provide an alternative for the accomplishing of the tasks in question. The use of graph-based methods for the analysis of the presented gear type with a closed internal loop has been described and performed step by step giving a detailed explanation to all activities. The methods are relatively uncomplicated, algorithmic and general. This confirms the usefulness of these methods for verifying the correctness of gear analysis. The basic concept of modelling accounts for selected general properties or aspects of a mechanism, e.g. analysis of kinematics. Therefore, the main rotating elements of a mechanical system are represented by graph vertices and the relations among them are modelled via edges. The applied rules of assignment start from the idea: the singled out main gear elements such as e.g. gear wheels, planets and arms are considered as vertices. In particular, all elements rotating around the main physical axis of a planetary gear are represented by the vertices of a polygon, moreover the mutual relations among them: "the rotation around the same axis" are coded via the polygon edges and their diagonals, however, the latter are not drawn for simplicity of the picture. However, they are used for the determination of f-cycles.

7. References

[1] Harary F 1969 *Graph theory* Addison- Wesley. Reading Mass

[2] Korzan B 1978 *Elementy teorii grafów i sieci- metody i zastosowania* WNT Warszawa

[3] Pijls W and De Bruin A 2001 Game tree algorithms and solution trees *Theoretical Computer Science* **252**(1-2) 197-215

[4] Zawiślak S 2010 *The Graph-based Methodology as an Artificial Intelligence Aid for Mechanical Engineering* Design Bielsko- Biała

[5] Drewniak J and Zawiślak S 2010 Graph methods in kinematical analysis of multi-speed epicyclic gears *Inter. J. of Appl. Mech. and Eng.* **17** 791-798

[6] Drewniak J and Zawiślak S 2009 *Kinematical and dynamical analysis of closed kinematic chains using graphs and contour equations* PAMM 9,1 pp. 547-548

[7] Wojnarowski J 1986 *Grafy i liczby strukturalne jako modele układów mechanicznych* Wydaw. Polit. Śląsk., Gliwice

[8] Uematsu S 1997 An application of graph theory to the kinematic analysis of planetary gear trains *Inter. J. of the Japan Soc. For Precis. Eng* **31** 141-146

[9] Deptuła A 2014 Application of multi-valued weighting logical functions in the analysis of a degree of importance of construction parameters on the example of hydraulic valves *Inter. J. of Appl. Mech. and Eng.* **19**(3) 539-548

[10] http://home.agh.edu.pl/~kmtmipa/_10.05.2019

[11] https://pl-protech.com/produkt/przekladnie-planetarne/ _10.05.2019

[12] Muller L and Wilk A 1996 *Zębate przekładnie obiegowe* Wydawnictwo Naukowe PWN Warszawa

[13] Jonak J 2016 *Komputerowo wspomagane projektowanie maszyn* Cześć 1 Politechnika Lubelska Lublin

[14] Cervantes- Sánchez J, Rico-Martínez J.M and Panduro-Calvario C A 2012 General and systematic framework for the kinematic analysis of complex gear systems *Meccanica* **47**(1) 3–21

[15] Wojnarowski J, Drewniak J, Kądziołka T, Kopeć J, Stańco K and Zawiślak S 2019 *Application of Contour Equations to Kinematic Analysis of Complex and Compound Planetary Gear*, In: Uhl T. (eds) Advances in Mechanism and Machine Science. IFToMM WC 2019, Mechanisms and Machine Science, vol 73. Springer pp 987-996 Cham

[16] Marghitu D 2005 *Kinematic chains and machine components design* Elsevier Amsterdam San Diego. Academic Press London

[17] Deptula A 2014 *Game- tree structures with the complex complexity level as a tool in knowledge engineering* Systems in Management 2, Z. 2/2014 rok ISSN 2084-5537

[18] Deptuła A and Partyka M. A 2011 Application of dependence graphs and game trees for decision decomposition for machine systems *Journal of Automation Mobile Robotics & Intelligent Systems* **5** 17-26

[19] Deptula A, Drewniak J and Partyka M.A 2017 Analysis of a planetary gear modeled with a contour graph taking into account the method of parametric play structures *Mechanik* **7** 640- 642

[20] Del Castillo J M *2002* Enumeration of 1-dof planetary gear train graphs based on functional constraints *J Mech. Des* **124** 723–732

[21] Rao Y.V.D and Rao A.C 2008 Generation of epicyclic gear trains of one degree of freedom *J Mech Des* **130** 052604–1-052604-8

[22] Salgado D R and Del Castillo J M 2014 Analysis of the transmission ratio and efficiency ranges of the four-, five-, and six-link planetary gear train *Mech Mach Theory* **73** 218–243

[23] Ma M Y and Xu X Y A 2011 Novel algorithm for enumeration of the planetary gear train based on graph theory *Adv Mater Res.***199** 392–399

[24] Ding H F, Huang Z and Cao Y 2006 The topological graphs creation automatically of kinematic chains and atlas database establishment Chin *J Mech Eng*.**42** 32–36

[25] Hsieh C F,. Hwang Y W and Yan H S 2008 *Generation and sketching of generalized kinematic chains* In: 2008, ASME international design engineering technical conferences and computers and information in engineering conference New York: ASME

[26] Nelson C A and Cipra R J 2005 Simplified kinematic analysis of bevel epicyclic gear trains with application to powerflow and efficiency analyses *J Mech Des* **127** 278–286

[27] Wojnarowski J, Kopeć J and Zawiślak S 2004 Gears and graphs *J Theor Appl Mech* **44** 139–162

[28] Deptula A, Drewniak J and Partyka M A 2019 The Method of Searching Trees in Determining of the Optimal Number of Wheel Teeth for a Compound Planetary Gear In: Uhl T. (eds) Advances in Mechanism and Machine Science *IFToMM WC 2019 Mechanisms and Machine Science* 73 Springer Cham pp 2853-2862

[29] Macek W 2019 Fractal analysis of the bending-torsion fatigue fracture of aluminium alloy *Engineering Failure Analysis* **99** 97–107

[30] Macek W 2019 Post-failure fracture surface analysis of notched steel specimens after bending-torsion fatigue *Engineering Failure Analysis* **105** 1154–71

[31] Szala M and Łukasik D 2018 Pitting Corrosion of the Resistance Welding Joints of Stainless Steel Ventilation Grille Operated in Swimming Pool Environment *International Journal of Corrosion* **2018**

[32] Szala M, Beer-Lech K and Walczak M 2017 A study on the corrosion of stainless steel floor drains in an indoor swimming pool *Engineering Failure Analysis* **77** 31–8

Gas network improvement proposal using numerical simulation

T Jurek[1] and M Iwanek[2]

[1] Polska Spółka Gazownictwa Ltd. (PSG), Branch Gas Company in Lublin, Diamentowa 15, 20-471 Lublin, Poland
[2] Lublin University of Technology, Faculty of Environmental Engineering, Nadbystrzycka 40 B, 20-618 Lublin, Poland

m.iwanek@pollub.pl

Abstract. Simulation software packages are very useful tools for evaluating hydraulic conditions in gas systems, and therefore for supporting the design process, network maintenance and decisions about rerouting or modernisation of a network. In the range of this paper, a model of a real branched gas network covering 3 towns was created in the GAZNET software. At first, actual hydraulic conditions of gas flow in the pipes were simulated to evaluate distributions of flow velocity and pressure in the pipes. The obtained results indicated meaningful pressure drops and increase in flow velocity in some sections of the model, which not only constitutes a threat to provide required parameters of gas to consumers, but also forecloses connecting new consumers to the network. Thus, two variants of improvements were proposed: (1) expanding the network by building two additional sections or (2) increasing the number of gas sources by building additional pressure reduction and measuring unit. Conducted simulations indicated that changes of the network proposed in both variants clearly improve hydraulic conditions in the network model. However, the first variant was chosen as a better alternative, because distribution of hydraulic parameters in the model was more homogeneous than in the second variant, as well as because of economic considerations.

1. Introduction

Gas network is the most efficient and convenient logistics to deliver natural gas. Numerical simulations of gas systems operation enable verification of existing or intended solutions of gas distribution networks. Moreover, they give a broader view of investments related to gas systems modernisation, extension or building by considering different technical aspects [1–5]. Nowadays, computer simulations are widely applicable in each branch of engineering science, including environmental engineering. They enable not only the assessment of networks and installations maintenance [e.g. 6–10], but also visualisations and analyses of phenomena connected with the systems operation [e.g. 11–15].

The software packages for environmental engineering include programs enabling simulation of hydraulic conditions in gas, water and wastewater systems, and therefore supporting the design process, network maintenance and decisions about rerouting or modernisation of a network [16–19]. Programs of this kind base on mathematical models – systems of equations – that reflect actual network operation having regard to necessary simplifications [1, 20]. If variables in the model depends on time, simulation of the network operating is dynamic, otherwise it is in steady state [21].

Steady state analysis is common for models of low- and medium-pressure gas networks [1, 21]. Simulation of gas networks at steady state enables to determine pressure at nodes and gas flows in individual pipes for known source pressures, composition of natural gas and gas load demand. The determined hydraulic parameters must satisfy first and second Kirchhoff's laws. Methods, that are commonly used in this kind of simulation, covers Newton-nodal, Newton-loop and Hardy Cross methods [1, 22–23]. Theoretical information on a steady-state simulation is described in detail in the publication of Osiadacz [1].

Using numerical simulations at the design stage simplifies elaboration of several design variants and selection of one of them as an optimal proposition, and thereby limits the risk to bear inflated costs of investment. The aim of our paper is the analysis of two variants of modernisation of the selected gas network, using the GAZNET software simulation. Modernisation is necessary to improve hydraulic conditions in the network and guarantee the undisturbed network operation.

2. Materials and methods

Investigations were conducted for the selected real branched medium-pressure gas network for 1247 consumers in three towns – A, B and C. The network consists of PE-HD pipes (25–180 mm in outer diameter) and steel pipes (200 mm in inner diameter). The total length of the network equals to 75534.7 m. The network is supplied by the pressure reduction and measuring unit located in town C. The capacity of the unit equals to 667 m^3/h and pressure at the outlet – 292.5 kPa.

Investigations included:

- creating a model of the gas network described above, with the GAZNET software,
- computer simulation of actual hydraulic conditions during gas flow in the pipes,
- evaluation of the actual hydraulic conditions,
- creating models of the modified gas networks (2 variants),
- computer simulations of hydraulic conditions in the modified gas networks,
- analysis of obtained results and selection of a better alternative.

The software used to build the network model and to conduct the simulations was GAZNET – an application implemented by the module SONET [24]. The GAZNET is similar to other programs intended for calculations of hydraulic parameters and simulations of gas networks operation [e.g. 16, 25–27]. The model of the analysed network covers 781 links and 782 nodes (figure1). The model reflects actual conditions noted on 27th November 2017, at the moment of maximal gas off-take and maximal pressure drop in the network. Data including gas off-takes measured by consumers' gas meters were obtained from the PSG company.

The created model was calibrated and then, the steady state simulation was conducted. A steady-state simulation in the GAZNET software bases on Kirchhoff's laws. Creation of the model and steady-state simulations were repeated for two variants of the network modernisation.

3. Results

The simulation of hydraulic condition in the actual network model indicated existence of pipes, in which gas flow velocity was close to maximal value (20 m/s) or even exceeded this value. The highest value of gas flow velocity (24.5 m/s) was observed in town A. Distribution of gas flow velocity in the entire analysed network model is given in figure 2.

The values of gas pressure in all pipes in town B and partly in town A occurred also not satisfactory (figure 3). Although all simulated values were covered by the range of 10–500 kPa characteristic for medium-pressure gas networks, pressure clearly lower than 100 kPa observed in the large part of the network indicates a risk of failure to provide required parameters of gas to consumers, as well as indicates inability to connect new consumers to the network.

Unsatisfactory hydraulic conditions in the gas network are the result of large gas off-take from the network (430 m^3/h) by the new industrial consumer in town A (figure1). Such a large off-take (64% of the whole gas supplied to the network) located in the central section of the network disturbs the gas system operation.

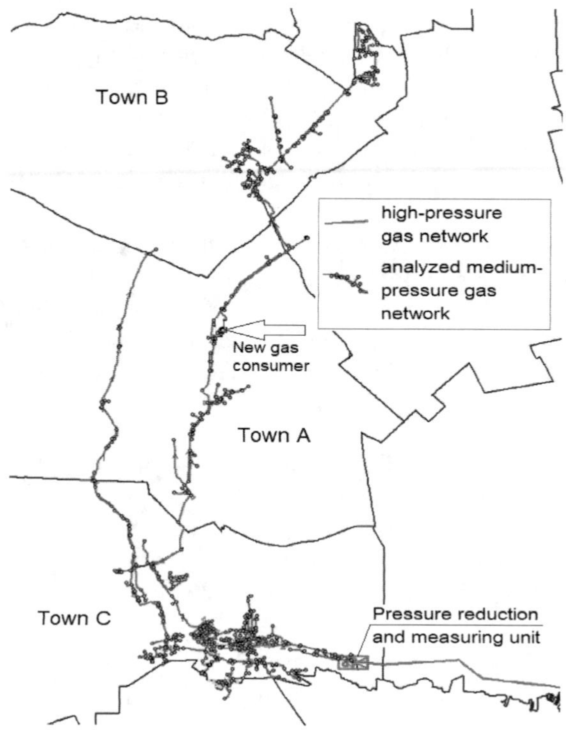

Figure 1. Analysed gas network model.

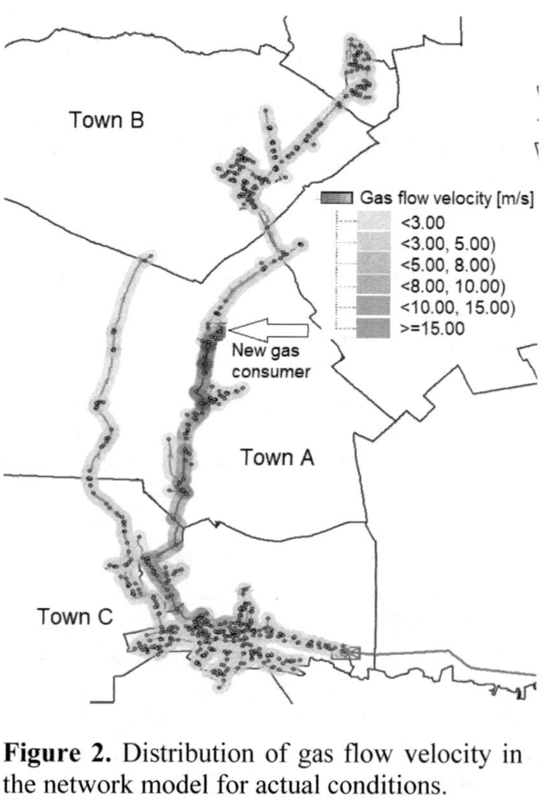

Figure 2. Distribution of gas flow velocity in the network model for actual conditions.

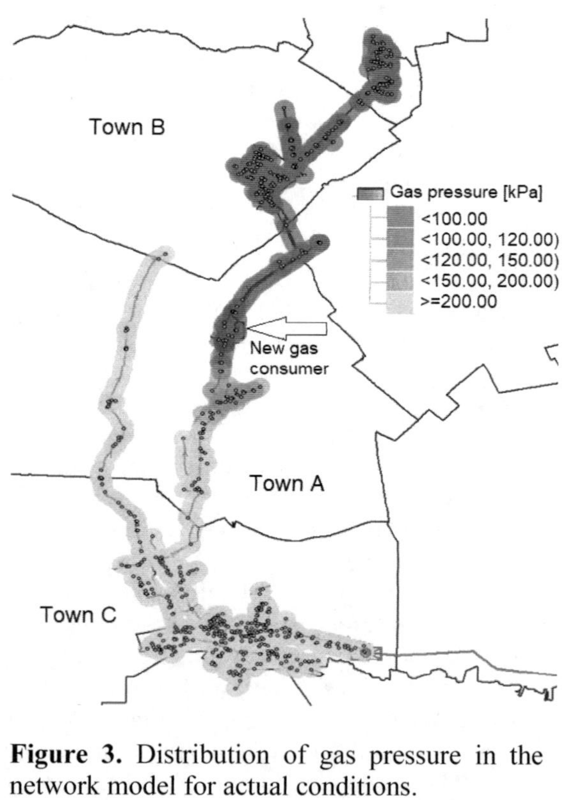

Figure 3. Distribution of gas pressure in the network model for actual conditions.

Two variants of gas network modernisation were proposed to improve the conditions of its operation. First one included expansion of the network and conversion of its structure from branched to branched and looped by building two additional sections (figure 4) :

- pipe PE-HD, SDR 11, *OD* = 110 mm, length 2226 m, in town B,
- pipe PE-HD, SDR 11, *OD* = 180 mm, length 3390 m, connected the chosen node in town A with the chosen node in town C.

The second variant included building the additional pressure reduction and measuring unit for town B (figure 4). The network would be connected with the unit by a steel pipe *DN* = 200 mm. The assumed capacity of the unit equalled to 167 m^3/h and pressure at the outlet to 220 kPa. All the parameters assumed in both variants (pipes dimensions in variant I, unit capacity and pressure in variant II were selected according to knowledge and competences of the experts from the PSG company).

The proposed variants of the network modernisation clearly improved the hydraulic conditions of gas flow. The conversion of the network structure according to variant I caused that velocity in all but one link in the model did not exceed 8 m/s (figure 5). Only two values higher than 8 m/s (12.5 m/s and 12.6 m/s) were observed in town A in two nodes 674 and 673 bounding the link supplying the new consumer, and they were the highest values of velocity observed in the model in variant I. Taking into account the rest of nodes, velocity was lower than 5 m/s in 99.00% of them (8 nodes with velocity 5–8 m/s).

Pressure in all gas pipes in the model for variant I exceeded 200 kPa with the lowest value 235.9 kPa in the node 673, to which the new consumer has been connected (figure 6). The highest value equalled 292.5 kPa for node 168 located in town C directly at the pressure reduction and measuring unit. Pressure was higher than 290 kPa in 51.85% of all nodes in the gas network model and lower than 240 kPa in 28.09%.

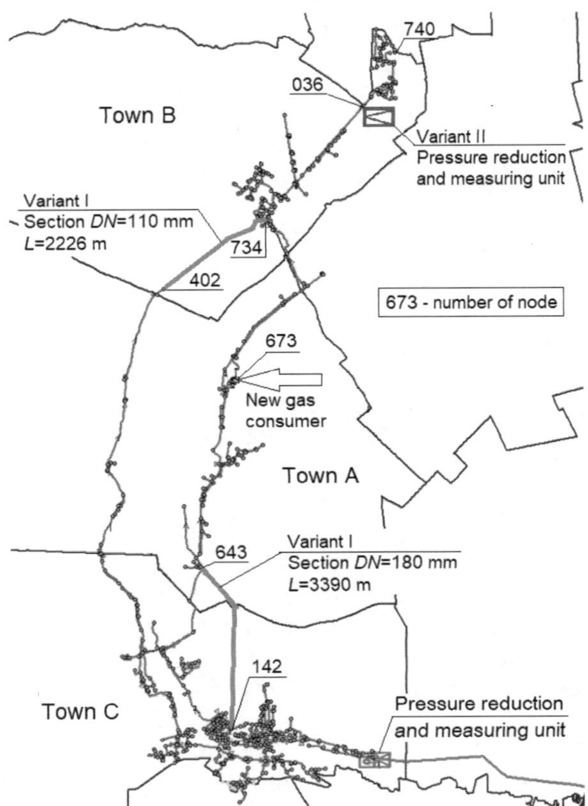

Figure 4. Model of gas network modernised according to variant I and II.

IOP Conf. Series: Materials Science and Engineering **710** (2019) 012005 doi:10.1088/1757-899X/710/1/012005

Similar results were observed for the model with the additional pressure reduction and measuring unit in town B according to variant II. Second source of gas supplying the network caused lowering velocity in the model (figure 7). Maximal values (13.6 and 13.7 m/s) were higher than for variant I, but they were observed in the same endpoints (nodes) of the link supplying the new consumer. Similarly as in variant I, these nodes were the only ones with velocity higher than 8 m/s. Velocity in the vast majority of the rest nodes (all but 5) was lower than 7 m/s.

Pressure determined for variant II covered greater range than for variant I: 209.6–292.5 kPa. The upper end of the range was the same, but the lower end was clearly lower – the difference equalled to 26.2 kPa (11.12%). Values higher than 290 kPa were obtained in 48.27% of all nodes and lower than 240 kPa in 34.44% (figure 8).

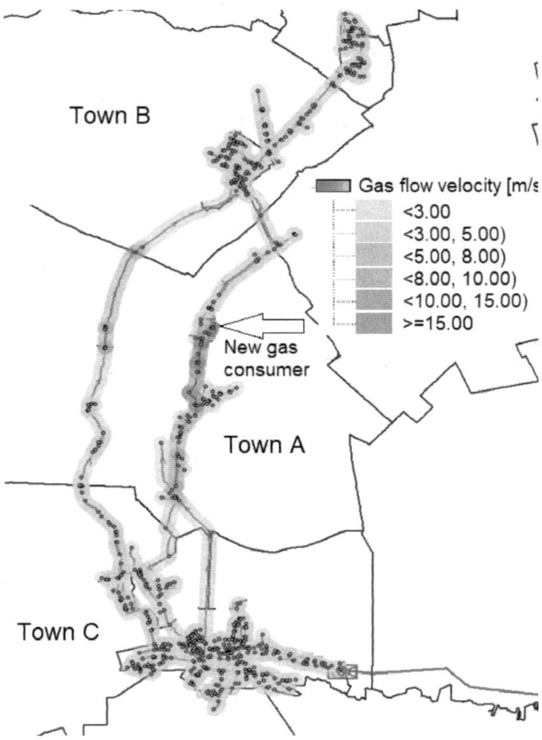

Figure 5. Distribution of gas flow velocity in the network model modernised according to variant I.

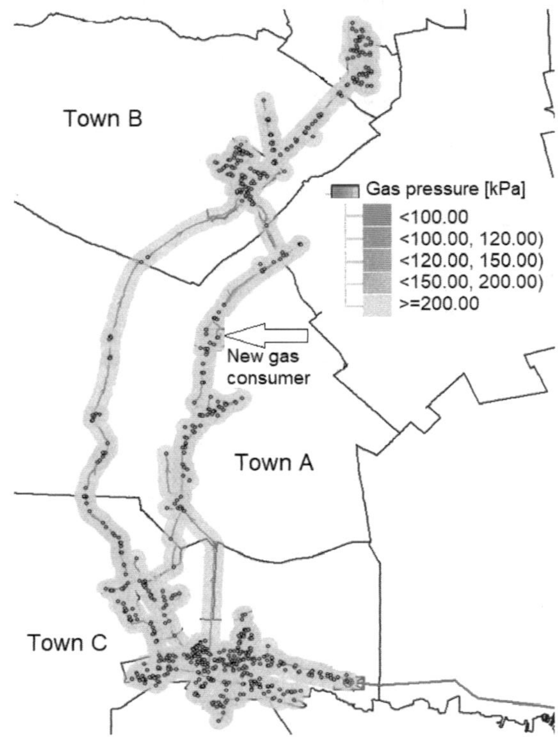

Figure 6. Distribution of gas pressure in the network model modernised according to variant I.

A more detailed analysis of calculated values indicated that distribution of both, velocity and pressure in the network model was more homogeneous for variant I than for variant II. Comparison of velocity and pressure for the actual conditions and for both of the proposed variants of the gas network modernisation is given in figures 9 and 10. The parameters are shown for selected nodes: 673 – a node to which the new consumer is connected, 740 – a node for which the lowest value of pressure was observed for actual conditions, 643 and 142 – nodes being endpoints of additional section $OD = 110$ mm in variant I, 402 and 734 – nodes being endpoints of additional section $OD = 180$ mm in variant I, 036 – a node to which the additional pressure reduction and measuring unit is connected in variant II, and nodes 407, 410 and 423 – selected nodes located near the node 673, characterised by adverse hydraulic conditions – relatively high gas flow velocity and low gas pressure.

IOP Conf. Series: Materials Science and Engineering **710** (2019) 012005 doi:10.1088/1757-899X/710/1/012005

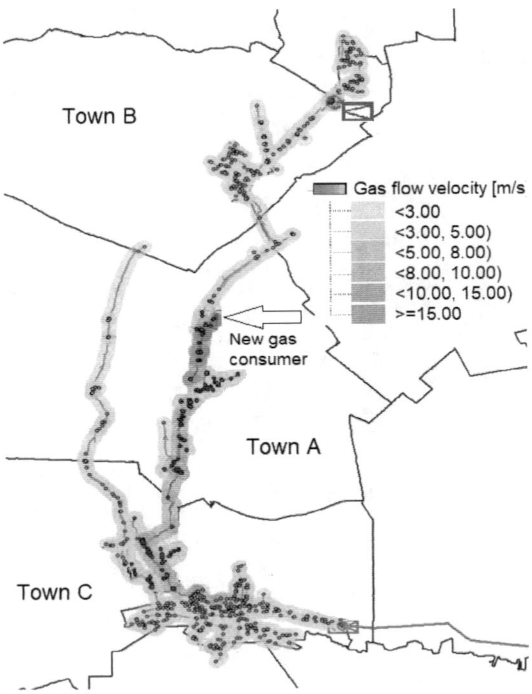

Figure 7. Distribution of gas flow velocity in the network model modernised according to variant II.

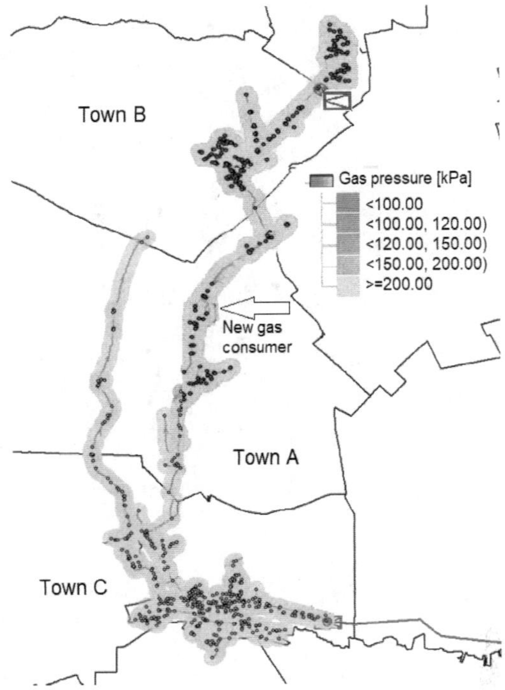

Figure 8. Distribution of gas pressure in the network model modernised according to variant II.

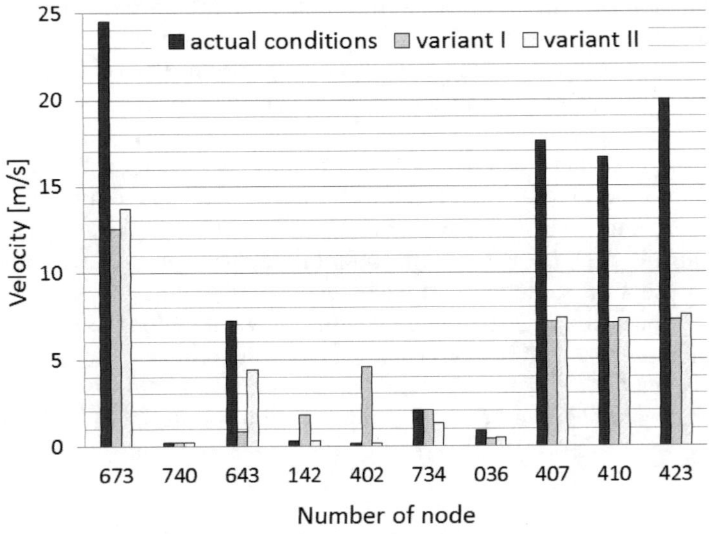

Figure 9. Comparison of gas flow velocity in selected nodes for actual conditions and two variants of network modernisation.

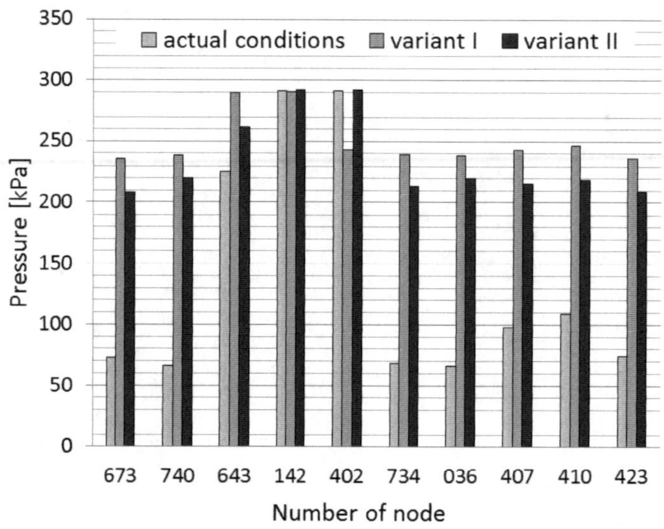

Figure 10. Comparison of gas pressure in selected nodes for actual conditions and two variants of network modernisation.

Conducted research indicates that both of the proposed variants clearly improve hydraulic conditions in the gas network supplying consumers in towns A, B and C. The obtained values of hydraulic parameters for both variants demonstrate that not only the gas system works properly in its current form, but also that system can be expanded to include new consumers from neighbouring communities. It is profitable for the PSG company, because of possible increase in volume of sold gas. The detailed analyses indicate a bit of an advantage of hydraulic conditions for variant I, but it should be underlined that differences between results obtained for variants I and II are minor. Although the results accuracy is enough to access the efficiency of the network modernisation, it is not enough to select a better alternative according to hydraulic conditions, with such small differences in simulated hydraulic parameters.

However, the choice of variant I is not only suggested by hydraulic analysis, but also supported by economic conditions. For variant I, the investment costs cover expenditure on designing and pipes installation only. For variant II the costs cover not only expenditure on designing and building the unit, but also on supplying the unit (designing and building a medium- or high-pressure gas network). Moreover, building a pressure reduction and measuring unit or a high-pressure gas network requires obtaining a decision concerning environmental conditions of the enterprise, which entails additional costs. Thus, the solution proposed in variant I is more profitable for the PSG company, because it enables proper operation of the network in both current and possibly expanded forms, at lower investment costs. Although it seems that variant I is a better alternative, a multi-criterial analysis is necessary to fully conclude.

4. Conclusion

Building the gas network model and computer simulation of its maintenance enabled to evaluate actual hydraulic conditions in the network supplying gas in three towns (A, B and C), and to locate the places where the gas system operation was disturbed. Moreover, computer simulation enabled to evaluate two proposals of gas system modernisation to improve hydraulic conditions in the system.

The conducted analyses indicated that both proposed methods for hydraulic conditions improvement were efficient and gave similar results – gas flow velocity did not exceed 20 m/s in network models and in vast majority of nodes was lower than 7 m/s, and gas pressure exceeded 200 kPa in all nodes. However, the first variant covering expansion of network and conversion of its structure from branched to branched and looped seemed to be slightly better because of less range of simulated values of gas flow velocity and pressure in the network model. Initial economical estimates also indicated the first variant as more beneficial, mainly for lower investment costs. However, a multi-criterial analysis is needed to make a full assessment of the proposed variants and select better alternative.

5. References

[1] Osiadacz A J 2001 *Statyczna symulacja sieci gazowych* (Warszawa: BIG, Fluid Systems) pp 1–407
[2] Fasihizadeh M, Sefti M V and Torbati H M 2014 Improving gas transmission networks operation using simulation algorithms: Case study of the National Iranian Gas Network *Journal of Natural Gas Science and Engineering* **20** 319–27
[3] Szoplik J 2017 Changes in gas flow in the pipeline depending on the network foundation in the area *Journal of Natural Gas Science and Engineering* **43** 1–12
[4] Hafsi Z, Elaoud S and Mishra M 2019 A computational modelling of natural gas flow in looped network: Effect of upstream hydrogen injection on the structural integrity of gas pipelines *Journal of Natural Gas Science and Engineering* **64** 107–17
[5] Lone S. Ramzan N, Saleemi A R and Karimi I A 2009 Modeling and Simulation of Gas Pipeline Network for Operational and Infrastructural Decisions *Computer Aided Chemical Engineering* **26** 453–458
[6] Janowska J, Widomski M K, Iwanek M and Musz A 2013 Numerical modeling of water flow through straight globe valve *Environmental Engineering IV*, ed A Pawłowski et al (Boca Raton: CRC Press-Taylor & Francis Group) pp. 41–9
[7] Pliżga O, Kowalska B and Musz-Pomorska A. 2016 Laboratory and numerical studies of water flow through selected fittings installed at copper pipelines *Rocznik Ochrona Środowiska* **18**(2) 873–84
[8] Suchorab P, Kowalska B and Kowalski D 2016 Numerical investigations of water outflow after the water pipe breakage *Rocznik Ochrona Środowiska* **18**(2) 416–27
[9] Iwanek M and Suchorab P 2017 The assessment of water loss from a damaged distribution pipe using the FEFLOW software *ITM Web of Conferences* **15** UNSP 03006
[10] Iwanek M, Suchorab P and Suchorab Z 2019 FEM simulation of water lost through damaged household water connection MATEC *Web of Conferences* **252** 05008
[11] Krukowski I, Widomski M, Iwanek M and Łagód G 2012 Exfiltration from sanitation pipes and transport of chosen pollutants–a model study *Ecological Chemistry and Engineering* A **19**(3) 311–20
[12] Iwanek M 2018 Numerical Investigations of moisture distribution in a selected anisotropic soil medium *Eurasian Soil Science* **51**(1) 73–80
[13] Raczkowski A, Suchorab Z, Połednik B, Zarzeka-Raczkowska E and Życzyńska A 2018 Computational fluid dynamics simulation of thermal comfort in naturally ventilated room by the fresh air valve installed in an external wall *AIP Conference Proceedings* **1988**(1), 020041
[14] Widomski M K, Majerek D and Iwanek M 2018 Numerical modeling of water and heat transport in porous building material *AIP Conference Proceedings* **1988**(1), 020048
[15] Raczkowski A, Suchorab Z and Brzyski P 2019 Computational fluid dynamics simulation of thermal comfort in naturally ventilated room *MATEC Web of Conferences* **252** 04007
[16] Osiadacz A, Chaczykowski M and Nowak A 2004 SimNetSSV 5 – pakiet oprogramowania do statycznej symulacji sieci gazowych *Nowoczesne Gazownictwo* **3**(IX) 37–40
[17] Woldeyohannes A D and Majid M A A 2011 Simulation model for natural gas transmission pipeline network system *Simulation Modelling Practice and Theory* **19**(1) 196–212
[18] Szoplik J 2016 Improving the natural gas transporting based on the steady state simulation results *Energy* **109** 105–16
[19] Świtnicka K, Suchorab P and Kowalska B 2017 The optimisation of a water distribution system using Bentley WaterGEMS software *ITM Web of Conferences* **15** 03009
[20] Taherinejad M, Hosseinalipour S M and Madoliat R 2017 Dynamic simulation of gas pipeline

IOP Conf. Series: Materials Science and Engineering **710** (2019) 012005 doi:10.1088/1757-899X/710/1/012005

networks with electrical analogy *Journal of the Brazilian Society of Mechanical Sciences and Engineering* **39**(11) 4431–41

[21] Gawron M 2011 Metody symulacji statycznej sieci gazowej *Zeszyty Naukowe Uniwersytetu Zielonogórskiego: Inżynieria Środowiska* **144**(24) 40–7

[22] Brkić D 2009 An improvement of Hardy Cross method applied on looped spatial natural gas distribution networks *Applied Energy* **86**(7-8) 1290–300

[23] Abeysekera M, Wu J, Jenkins N and Rees M 2016 Steady state analysis of gas networks with distributed injection of alternative gas *Applied Energy* **164** 991–1002

[24] Sygnity S. A. 2014 *Instrukcja użytkownika modułu SONET (User's Manual). Wersja dokumentacji 1.5*

[25] Bermúdez A, González-Díaz J, González-Diéguez F J, González-Rueda Á M and de Córdoba, M P F 2015 Simulation and optimization models of steady-state gas transmission networks *Energy Procedia* **64** 130–9

[26] Kogut K and Bytnar K 2007 *Obliczanie sieci gazowych. Przegląd programów komputerowych, tom II* (Kraków: Uczelniane Wydawnictwa Naukowo-Dydaktyczne AGH) pp 1–151

[27] Praks P, Kopustinskas V and Masera M 2015 Probabilistic modelling of security of supply in gas networks and evaluation of new infrastructure *Reliability Engineering & System Safety* **144** 254–64.

IOP Conf. Series: Materials Science and Engineering **710** (2019) 012006 doi:10.1088/1757-899X/710/1/012006

Nonlinear model and simulation of a rolling bearing

A Smagala[1] **and K Kecik**[2]

[1] R&D Department, Fabryka Lozysk Tocznych-Krasnik S.A., 6 Fabryczna Street, Krasnik, 23-210, Poland
[2] Lublin University of Technology, Faculty of Mechanical Engineering, 36 Nadbystrzycka Street, 20-618, Lublin, Poland

a.smagala@pollub.pl, k.kecik@pollub.pl

Abstract. This paper presents an analysis of a nonlinear model of a rolling bearing with two degrees of freedom. The acceleration response provides information regarding the relationship between vibration and a different number of balls, values of internal clearance and rotary shaft speeds. The acceleration response is higher when the radial internal clearance and the shaft speed increase. When the number of balls increases, the acceleration decreases for the entire range of radial internal clearance. These results are useful for understanding the vibration response mechanism of ball bearings under variable parameters.

1. Introduction

The rolling bearings are one of the most important components in the entire machine construction domain. They are widely used in various mechanical equipment across industries, including the aerospace, automotive, construction, mining, steel, paper, textile, railways, and renewable energy [1]. The bearings are frequently implemented due to their low friction characteristics and their load-carrying capacity. Rolling bearings work under different conditions and are frequently subject to heavy loads generated in the machinery, as well as to varying time and space dynamic loads, not to mention numerous other unknown factors that could affect the bearing life. Rolling bearings affect the running accuracy, reliability and service life of mechanical equipment. The complexity of bearing internal geometry and the loading mechanism will influence the vibration expression of its behaviour [2]. To prevent machine breakdown it is often necessary to diagnose and detect defects in rolling bearings so as to replace it before the fault reaches a certain degree. The dynamic behaviour of the bearing plays a pivotal role especially in high-rotation systems where large dynamical forces occur that can cause its deformation [3].

When the bearing is running and its defective component, either a rolling element, an outer raceway or an inner raceway, interacts with its corresponding mating component, either defective or non-defective, abrupt changes in the contact stresses occur. It can be a root cause of undesirable vibration, excited by the changes in the bearing structure, which can be also observed in other structural components connected to the bearing and consequently may lead to the emergence of acoustic signals. These signals can be used to detect the presence of a defect with the help of appropriate condition-based diagnostic techniques [4]. Theoretical ball bearing models help understand the mechanisms of vibration signal generation, thus enabling the study of the impact of different parameters, such as load and force transmission paths, to better understand the vibration generated at the beginning of an incipient defect. Despite the long history of the use of bearings the vast experience accumulated with regard to their fatigue-life prediction or load capacity calculation, relatively few models have been proposed to explain the dynamic behaviour of bearings [5]. Most of these models proposed by many researchers refer to the simulation of the ball bearing movement in the presence of a localised defect. As a first, Sunnersjo [6] proposed the mathematical model of bearing vibrations in which a two degree of freedom (2DOF) system was built, and which provided the load-deflection according to Hertzian contact theory. In the model, the mass and inertia of the rolling elements were neglected. A decade later, Sunnersjo [7] explained in detail the relation between surface irregularities and the vibration characteristics of the radial bearing with positive clearance and subjected to radial loads. Another study focused on the influence of inner ring waviness and the non-uniform diameter of the rolling elements on vibratory response [8]. At about the same time, McFadden and Smith [9] proposed a mathematical model of bearing vibration for a single-point and multi-point defect under radial load, where the defect-induced vibrations are modelled as a result of a series of impulses as the bearing ran at a constant frequency and radial load. In that model, it was assumed that the excitation was represented as a series of repeated impulses when the rolling element strikes the point of defect. These impulses may excite resonances in the bearing and machine. The bearing rotation exhibits a periodical characteristic of the frequency specified by the location of the defect, whether it is on the inner raceway, outer raceway or one of the rolling element [8],[10]. Feng et al. [11] were developing the model with 2DOF further considering the influence of bearing cage and ball slide and also the impact of vibration response for the defect in outer and inner raceway.

In the early 1990s, more refined bearing models were introduced. Lim and Singh [12] proposed the 5DOF model of bearing, which assumed the rigid inner and outer rings and deformable balls. Su et al. [13] extended the model proposed by McFadden for the determination of periodic characteristics of various loading and transmission path effects and their impact on vibration. These effects are associated with the axial and radial loads, misalignment and dynamic unbalance of the shaft [14]. In the model, the bearing vibration was modelled as the system output subjected to excitations from surface waviness and roughness through the lubrication film. It was shown that the vibration spectrum of a normal bearing under a preloaded condition has a pattern of equal frequency spacing distribution that is similar to that of a defective bearing [8]. In the late 1990s, Tandor and Choudry [15] proposed an analytical model for predicting the vibration frequencies of rolling bearings and the amplitudes of significant frequency bearing components caused by a defect on the inner/outer raceway or on the rolling elements. The model considered the effect of loading and pulse shape on the vibration amplitude [8]. Randal and Antoni proposed the vibration fault signal model in which the slip between the rolling elements was introduced. This slip may cause random fluctuation between the impulses due to the defect in the bearing. In the model, the vibration signal with random fluctuation will have a spectrum where the defect frequency components are smeared into each other [16]. Patil et al. [14] presented the 2DOF analytical model of ball bearing for predicting the effect of a localised defect on the bearing vibration and developed a computer program to simulate the defect on the raceway with the results presented in the time and frequency domains. In the model, contacts between the ball and the raceway are considered as non-linear springs and Hertzian contact deformation theory is used to calculate the contact forces. Grajale et al. presented in their work [8] the 2DOF mathematical ball bearing model with localised outer raceway defect. In the paper, the model of the bearing was considered as mass-spring-damper system considering each rolling element as a contact spring-damper pair, based on Hertzian contact deformation theory. The paper made use of Peak Value analysis to reveal the characteristic frequencies of the studied type of the defects. In Sadok's work [8], the dynamic behaviour of bearing as a coupled three-degree-of-freedom system was presented. In the paper, the generation of vibration by a point defect in bearing was modelled as a function of rotation, the distribution of the load, the bearing structure elasticity, the oil film characteristics and the transfer path between the bearing and the transducer. Cui et al. presented the 5DOF nonlinear vibration model for fault severity assessment of ball bearings [17]. In our study, the outer raceway defect size parameter was introduced into the dynamic model. The vibration response signal of rolling bearing was simulated under different fault size. The results of the study may help understand the vibration response mechanism of rolling bearing under a various degree of defects.

This work is focused on bearing behaviour with numerical 2DOF model. The vibration response signal is simulated under different number of balls, different radial internal clearance and different bearing rotary speed. The 2DOF model is implemented in MatLab-Simulink environment. This model accounts for the effect of normal force due to ball race contact and ignores the effects resulting from the traction force.

2. Model of bearing

In regular rolling bearing systems, the inner ring is seated on the shaft and rotates together with the rotating shaft, whereas the outer ring is mounted into the housing and the balls undergo pure rolling movement on the raceways. Usually, every ball bearing should be manufactured with more or less positive radial clearance. Hence, during the operation of the bearing, the raceway is divided into the loaded and unloaded zones, based on the range of radial load. The balls which enter the loaded zone are deformed and produce varying flexibility vibration. The rolling element bearing is the most critical element of rotating machines, therefore its proper description of the bearing is of high importance from the viewpoint of the dynamic behaviour of the entire machine. The key components of any vibratory system are the mass, stiffness, damping and the external forces. Therefore, the deep groove ball bearing with a single row of balls is simplified and modelled as a nonlinear spring-damper system (figure 1).

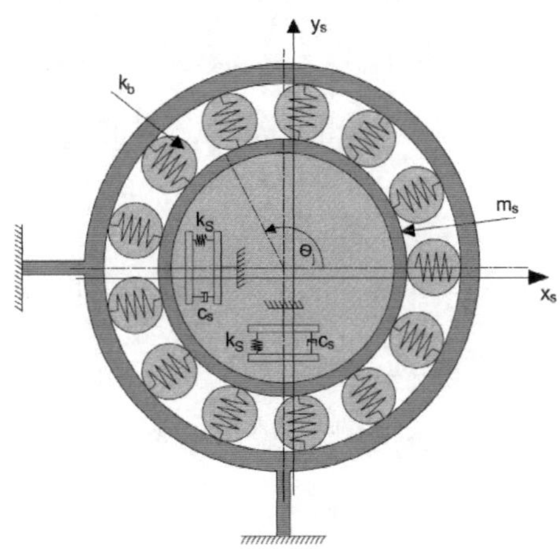

Figure 1. 2DOF model of the rolling bearing.

In the model, the outer ring was fixed in a rigid, non-deformable housing (the model considers the case of a non-rotating outer race), and the mass of the balls is neglected. The inner ring rotates together with the shaft. The elastic deformation between the raceways and each point of contact with the ball is assumed to be Hertzian. The assumptions and simplifications of the model are:

- the ball bearing model has equi-spaced balls rolling on the surface of the inner and outer raceway and there is no interaction between them,
- the slipping effect of the balls is neglected,
- the motion of race and balls occur in the bearing's plane,
- the deformations at the contacts are described by the Hertzian contact theory.

The inner raceway is fixed rigidly to the rotor shaft. Therefore, it rotates with a constant angular speed. The equations of motion of the inner race can be written using the Newton equations, and have the following form:

$$m_s\ddot{x}_s + c_s\dot{x}_s + k_s x_s + F_x = 0, \quad (1)$$

$$m_s\ddot{y}_s + c_s\dot{y}_s + k_s y_s + F_y = F, \quad (2)$$

where m_s is the mass of the shaft together with the inner ring, c_s, k_s are the damping and stiffness of the shaft with inner ring, the F_x, F_y are nonlinear forces that describe the contact between the inner raceway and the rolling elements. The complete deformation of any rolling element (j) is the result of relative displacement between the inner and outer rings (x_s, y_s), the angular position (φ_j) of the rolling element, and the radial internal clearance (*RIC*). The *RIC* parameter can be used for modelling of the lubricant film thickness. The deformation δ_j is obtained from [18]:

$$\delta_j = (x_s)\cos\varphi_j + (y_s)\sin\varphi_j - RIC, \tag{3}$$

The nonlinear contact forces of any rolling element with the raceway are described by the Hertzian theory. These forces have the following formulations [17]:

$$F_X = k_b \sum_{j=1}^{Z} H(\delta_j)\delta_j^{1.5}\cos\phi_j, \quad F_Y = k_b \sum_{j=1}^{Z} H(\delta_j)\delta_j^{1.5}\sin\phi_j, \tag{4}$$

$H(\delta_j)$ is the Heaviside function that defines the condition of deformability (the rolling element is in contact with the raceway or lack of contact – switch function), k_b means the stiffness of the rolling elements, and Z denotes the number of the rolling elements (balls).

The angular position of the *j-th* ball ϕ_j is a function of the period dt, the initial angular position of the bearing cage ϕ_0 and the angular speed of the bearing cage ω_c. The ϕ_j and ω_c are defined as follows:

$$\phi_j = \frac{2\pi(j-1)}{Z} + \omega_c dt + \phi_0, \; \omega_c = \left(1 - \frac{D}{d_m}\right)\frac{\omega_s}{2} \tag{5}$$

where ω_s is the angular speed of the shaft, D is the diameter of the ball, and d_m is the pitch diameter of the bearing.

3. Numerical results and discussion

The equations of the motion are solved numerically in MatLab-Simulink environment using ODE45 Runge-Kutta solver with the fixed time step 1·10-5. The parameters of the 2DOF model of a bearing, based on the commercial bearing 6009, are given in table 1.

Table 1. Geometrical and physical properties of the bearing model.

Parameter	Value
Number of balls, Z	11, 13, 15
Ball diameter, D	8.731 mm
Pitch diameter, d_m	60 mm
Contact stiffness, k_b	$1.8978 \cdot 10^{10}$
Stiffness of the inner race, k	4.241·104 N/m
Damping of the inner race, c	1376.8 Ns/m
Mass of inner race, m_s	0.48 kg

The calculations were conducted for different numbers of balls (Z), various values of the radial internal clearance (*RIC*) and different rotary speeds of the shaft (n). The paper presents an analysis of the vibration for different combinations of these parameters.

Figure 2 and Figure 3 show acceleration time response of the simulated bearing with thirteen balls corresponding to the x-direction and the y-direction. The *RIC* was fixed and equalled 10 μm. The rotary shaft speed was n=1500 rpm (blue line) and n=6000 rpm (orange line).

The amplitude of acceleration response in the y-direction is twice higher than the amplitude in the x-direction. This result is probably caused by the gravity force. Of course, the amplitude of acceleration has a higher level for the high rotary speed. A fourfold increase in the rotary speed causes an increase in the acceleration of even eight times (in the y-direction). Moreover, the rotary speed influences the time period vibration, which is clearly visible in figure 2 and 3.

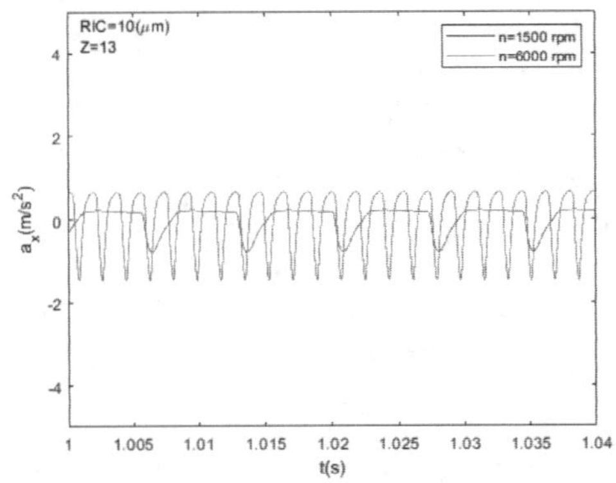

Figure 2. Simulation signal of acceleration in time domain in the x-direction for different rotary speed and fixed *RIC* level.

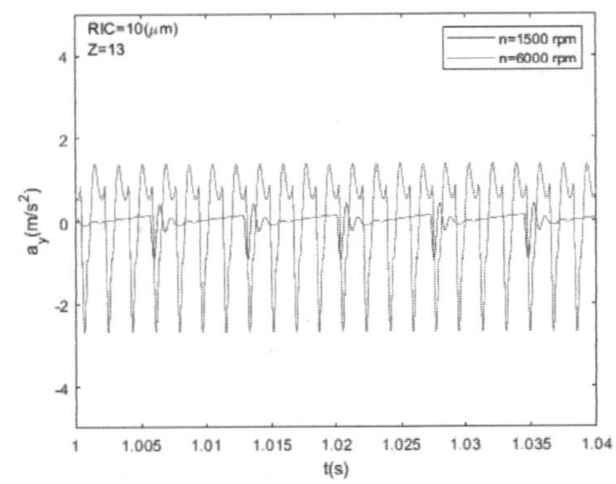

Figure 3. Simulation signal of acceleration in time domain in the y-direction for different rotary speed and fixed *RIC* level.

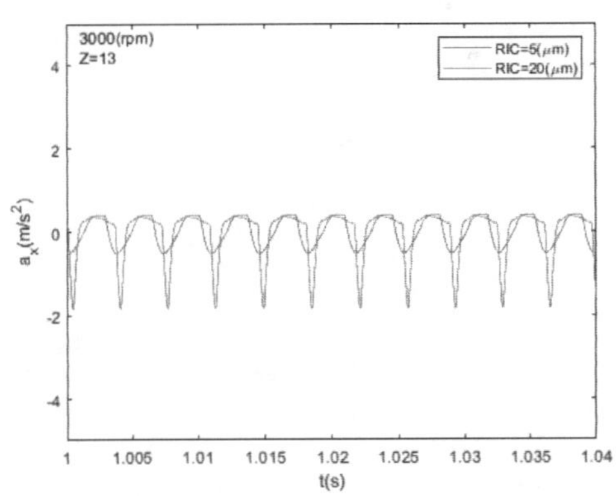

Figure 4. Simulation signal of acceleration in time domain in the x-direction for different *RIC* level.

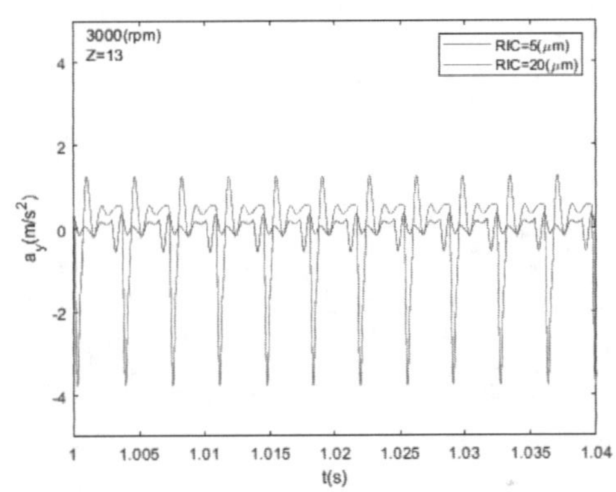

Figure 5. Simulation signal of acceleration in time domain in the y-direction for different *RIC* level.

Figure 4 and figure 5 show acceleration time responses for the bearing with thirteen balls and the rotary speed of shaft n=3000 rpm. The two values of the *RIC* parameter were compared. The small radial clearance was assumed *RIC*=5 μm (blue line), and high-level *RIC*=20 μm (orange line). The greater *RIC* causes an increase in the acceleration values. This effect can be explained by lower stiffness of the bearing system. The maximal level of acceleration obtained in the y-direction was about a_y=4 m/s^2 (figure 5).

Figures 6-9 present the root mean square (RMS) acceleration v the number of balls for the different rotary speed n and a different *RIC* value. The specified plot shows RMS acceleration in the x and y-direction with the y-axes on both left and right sides. The continuous line presents results for n=1500 rpm, the dotted line for n=3000 rpm, the dot for n=4500 rpm, and the dashed-dot line for n=6000 rpm. For all rotary speeds, the acceleration is smaller for the bearing with the larger number of balls. This is particularly evident for higher shaft speeds. Moreover, this relationship can be described by a simple linear function, which is very useful from the practical point of view.

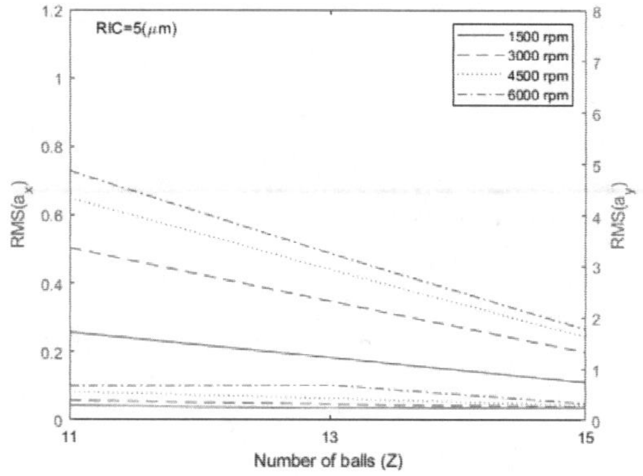

Figure 6. RMS value of acceleration in the x and y-directions v the number of balls, for *RIC*=5μm.

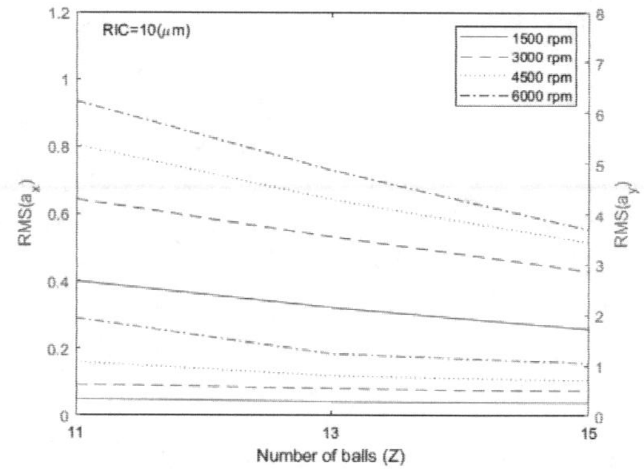

Figure 7. RMS value of acceleration in the x and y-directions v the number of balls, for *RIC*=10μm.

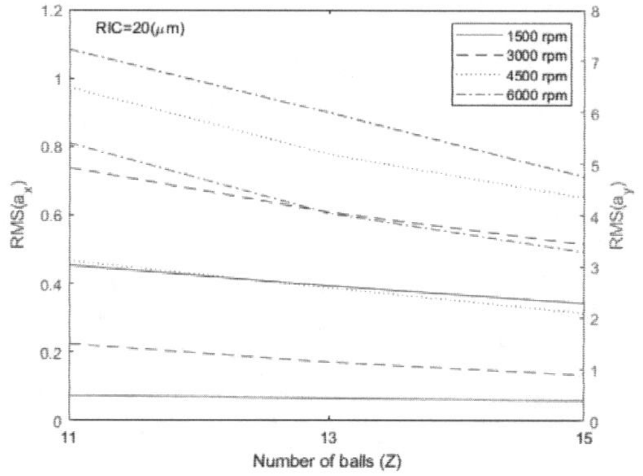

Figure 8. RMS value of acceleration in the x and y-directions v the number of balls, for *RIC*=20μm.

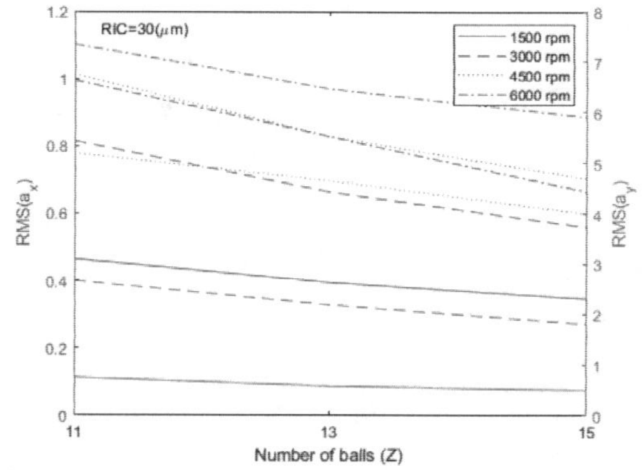

Figure 9. RMS value of acceleration in the x and y-directions v the number of balls, for *RIC*=30μm.

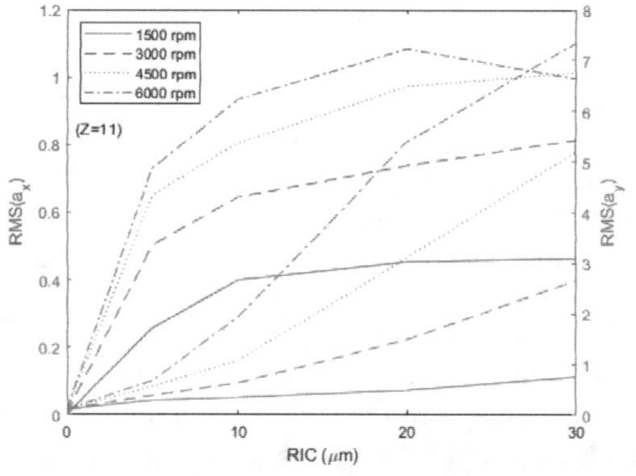

Figure 10. RMS value of acceleration in the x and y-directions v radial internal clearance *RIC*, for *Z*=11.

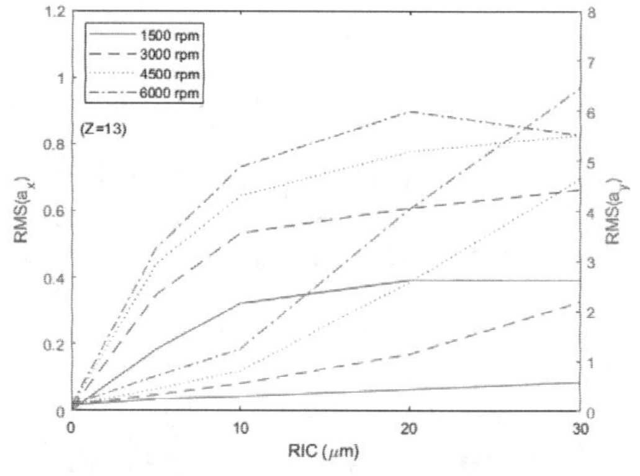

Figure 11. RMS value of acceleration in the x and y-directions v radial internal clearance *RIC*, for *Z*=13.

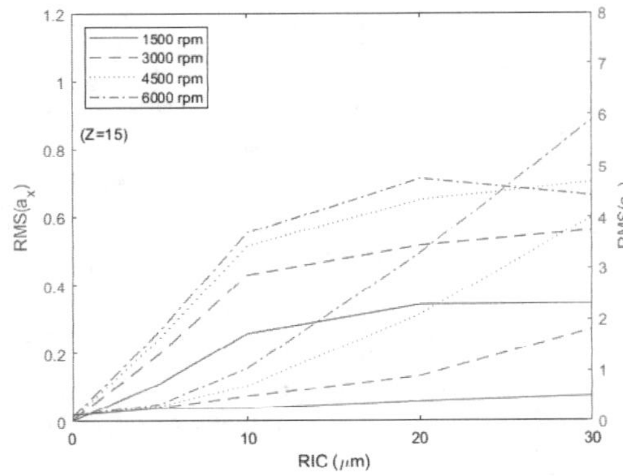

Figure 12. RMS value of acceleration in the x and y-directions v radial internal clearance *RIC*, for *Z*=15.

Figures 10, 11 and 12 show the influence of *RIC* parameter on the accelerations for different values of the shaft speed. The investigations have been performed for bearings with the number of balls: *Z*=11 (figure 10), *Z*=13 (figure 11) and *Z*=15 (figure 12). For shaft speeds *n*=1500 rpm, the acceleration responses are almost on the same level for the different number of balls. When the shaft speed increases the acceleration responses also increase in both directions. The increase of radial internal clearance has more impact on the acceleration response for the *y*-direction. The acceleration responds strongly in the function of radial internal clearance, especially at higher shaft speeds. The number of balls is a crucial parameter from the practical point of view.

4. Conclusions

In this paper, the 2DOF nonlinear numerical model of deep groove ball bearings was studied. The bearing vibration response for a different number of balls, value of radial internal clearance and shaft speed were analysed.

The numerical investigations show that the radial internal clearance and the shaft speed have a significant influence on the bearing vibration response in both directions. The bearing dynamic behaviour is similar in both directions. However, higher acceleration was obtained in the vertical direction. When the number of balls increases, the acceleration values are lower. The radial clearance is of great importance: its increase causes a marked rise in acceleration.

The next step of our research will be to include the temperature in the bearing model. Moreover, the radial internal clearance model will be modified.

5. References

[1] Harris T A and Kotzalas M N 2007 *Rolling Bearing Analysis - essential concepts of bearing technology*, 5th ed. (Florida: CRC Press, Boca Raton)

[2] Cong F, Chen J, Dong G and Pecht M 2013 Vibration model of rolling element bearings in a rotor-bearing system for fault diagnosis *Journal of Sound and Vibration* **332(8)** pp 208-97

[3] Cui L, Zhang Y, Zhang F, Zhang J and Lee S 2016. Vibration response mechanism of faulty outer race rolling element bearings for quantitative analysis *Journal of Sound and Vibration* **364** pp 67-76

[4] Singh S, Howard C Q and Hansen C H 2015 An extensive review of vibration modelling of rolling element bearings with localized and extended defects *Journal of Sound and Vibration* **357** pp 300-30

[5] Sadok S, Bechir B and Marc T 2007 A Numerical Model to Predict Damage Bearing Vibrations

Journal of Vibration and Control **13(11)** pp 1603-28

[6] Sunnersjo C S 1978 Varying compliance vibrations of rolling bearing *Journal of Sound and Vibration* **58(3)** pp 363-73

[7] Sunnersjo C S 1985 Rolling bearing vibrations: The effects of geometrical imperfections and wear *Journal of Sound and Vibration* **98(4)** pp 455-74

[8] Grajales J A, Lopez J F and Quintero H F 2014 Ball bearing vibrations model: Development and experimental validation. *Ingeniería y Competitividad* **16(2)** pp 279-88

[9] McFadden P D and Smith J D 1984 Model for the vibration produced by a single point defect in a rolling element bearing *Journal of Sound and Vibration* **96(1)** pp 69-82

[10] Liu J, Shao Y and Lim T C 2012 Vibration analysis of ball bearings with a localized defect applying piecewise response function *Mechanism and Machine Theory* **56** pp 156-69

[11] Feng N S, Hahn E J and Randall R B 2002 Using transient analysis software to simulate vibration signal due to rolling element bearing defect *Proc. of the 3rd Australian Congress on Applied Mechanics* (Sydney)

[12] Lim T C and Singh R 1990 Vibration transmission through rolling element bearing Part 1: Bearing stiffness formulation *Journal of Sound and Vibration* **139(2)** pp 179-99

[13] Su Y T Lin M H and Lee M S 1993 The effect of surface irregularities on the roller bearing vibrations *Journal of Sound and Vibration* **163(3)** pp 455-66

[14] Patil M S, Mathew J, Rajendrakumar P K and Desai S 2010 A theoretical model to predict the effect of the localized defect on vibrations associated with ball bearing *International Journal of Mechanical Sciences* **52(9)** pp 1193-201

[15] Tandon N and Choudhury A 1999 A review of vibration and acoustic measurement methods for the detection of defects in rolling element bearings *Tribology International* **32(8)** pp 469-480

[16] Randall R B and Antoni J 2011 Rolling element bearing diagnostics – A tutorial *Mechanical Systems and Signal Processing* **25(2)** pp 485-520

[17] Cui L, Chen X and Chen Sh 2014 Dynamics Modelling and Analysis of Local Fault of Rolling Element Bearing. Advance in Mechanical Engineering doi: 10.1155/2014/262351

[18] Mishra C, Samantaray A K, Chakraborty G, Zhang J, and Lee S 2017 Ball bearing defect models: A study of simulated and experimental fault signatures. *Journal of Sound and Vibration* **400** pp 86-112.

IOP Conf. Series: Materials Science and Engineering **710** (2019) 012007 doi:10.1088/1757-899X/710/1/012007

Computer modelling of automobile engine performance as the source of implications for automobile technology management

O Orynycz[1], K Tucki[2], A Wasiak[1] and R Mruk[2]

[1] Bialystok University of Technology, Department of Production Management, Wiejska Street 45A, 15-351 Bialystok, Poland
[2] University of Life Sciences, Institute of Mechanical Engineering, Department of Production Engineering Warsaw, Nowoursynowska Street 164, 02-787 Warsaw, Poland

o.orynycz@pb.edu.pl; karol_tucki@sggw.pl

Abstract. The need for finding the replacement for fossil fuels causes several technical and technological problems in the engine construction as well as the fuel production and its standardisation. This particularly concerns biofuels. The variety of biological resources and possible conversion routes leads to the multiplication of fuel properties. The regional dependence of biomass resources might cause interregional incompatibility of fuels, which should be avoided. Our previous studies have shown that computer modelling is capable of providing information on the dependence of engine performance upon the type of the fuel used, also under conditions of simulation of road tests. Based on computer simulations of the given engine cycle, the impact of the use of biofuels on carbon dioxide emissions from a diesel and gasoline engine was analysed. The tested fuels were: gasoline 95, diesel, ethanol, DME dimethyl ether, and FAME fatty acid methyl esters. The differences in CO_2 emissions of the tested fuels were large (up to 17%), but none of the analysed fuels exceeded the maximum emission limits allowed by legal regulations. The results of such simulations might be used for planning the development of various types of engines - possibly adjustable to fuels of various origin, as well as for planning the fuel composition assuring the best performance in a wide spectrum of engine constructions.

1. Introduction

Nowadays increasing attention is paid to the problems of environmental protection [1, 2]. The preservation of clean air is essential for the comfort of human life and health [3, 4]. The economic development, including transportation, forces the increase in energy consumption [5]. The side effect of burning fuel necessary for motor vehicle movement is the emission of pollutants, in particular CO_2 [6]. A huge and increasing number of passenger cars and the wide use of lorries and other means of transport causes that the attempts to mitigate air pollution become an enormous challenge [7, 8]. It should be emphasised that the environmental challenges are not only connected with the technological development but also with the very complex ways of technology management. Road transport, which is one of the key sectors of the European economy, has been considered for many years to be one of the major emitters of substances that impair the quality of atmospheric air [9, 10]. About 30% of total CO_2 emissions in the EU come from the transport sector, 72% of which – from road transport. There are two ways to reduce CO_2 emissions from cars: increasing vehicle efficiency or changing the fuel. Most of the energy used in transport comes from conventional fuels, but refineries are and will be forced through legal regulations to produce more and more environmentally friendly fuels [11]. The binding European directives oblige member states to implement biofuels and bio-components on the fuel market. In the directive's sense, alternative fuels are fuels or energy sources that serve, at least in part, as a substitute for petroleum-based energy sources in transport and which can potentially help to reduce the dependence of EU Member States on oil imports and support the decarbonisation of transport, consequently improving the environmental performance of this sector [12]. The fuels, alternative to diesel oil and gasoline have already been used on a large scale. Derivatives of plant materials such as vegetable oils or their esters are used for diesel engines while ethanol – as a replacement of petrol or its additive for spark ignition engines [13]. Alternative fuels and also electro-mobility are promoted and introduced to the market [14, 15]. Increasingly restrictive driving tests are being introduced, intensifying the race towards reduction of pollutant emissions [16, 17]. The designing and use of computer simulation programmes and models leading to determination and comparison the parameters of engine performance become significant tools in research and development works in automobile testing and evaluation [18, 19].

2. Materials and methods

The aim of this work was to determine the impact of biofuels on the emission of carbon dioxide from an internal combustion engine. For this purpose, computer simulations were used to model the given engine work cycle.

The actual technical specifications of the Fiat Panda 1.3 MultiJet passenger car, equipped with a diesel engine, and Fiat Seicento 1.1 with a gasoline engine [20, 21] were implemented in the simulation. The choice of the vehicles was dictated by their typically urban character (small displacement of the engine, small dimensions) and various fuels that they can be supplied with. The basic technical data for the 1.1 Fire and 1.3 MultiJet II engines are listed below in Table 1.

Table 1. Basic technical data for engines: 1.1 Fire and 1.3 MultiJet II.

Parameter	Units	1.1 Fire	MultiJet II
The number of cylinders	–	4	4
Cylinder layout	–	In-line	In-line
Injection type	–	Direct, multipoint	Direct, multistage
Compression ratio	–	9.6: 1	16.8 : 1
The diameter of the cylinder	mm	70	69.6
Piston stroke	mm	72	82
Displacement	cm^3	1108	1251
Maximum power	kW	40	55
Maximum torque	N·m	88	190
Rotational velocity at maximum power	rpm	5000	4000
Rotational velocity at maximum torque	rpm	2750	1500
Idle speed	rpm	850±50	850±20

The computer program enabled simulations for fuels with different properties and compositions: gasoline 95, diesel fuel; and for alternative fuels (100% content): ethanol, dimethyl ether DME, fatty acid methyl esters FAME. The basic properties of the fuels are listed in Table 2.

Table 2. The basic properties of the fuels used for simulations.

Parameter	Gasoline 95	Ethanol	DME	Diesel oil	FAME
Carbon content [%]	86.4	52.1	52.1	86.5	78.0
Hydrogen content [%]	13.6	13.1	13.1	13.4	12.0
Oxygen content [%]	0	34.7	34.7	0	10.0
Heating Value [MJ/kg]	45	30.4	28.8	44.0	37.0
Air to fuel ratio [g_{air}/g_{fuel}]	14.7	9.06	9.06	14.5	12.5

Scilab software [22] was used to build simulation models. It is an open source package for scientific applications offering advanced calculation functions. The software allows creating dynamic models based on complicated mathematical calculations and algorithms. Its scope of use includes design, simulation, recording and compilation of developed solutions. In addition to choosing the suitable software, it was also important to choose the right work cycle for each vehicle for a reliable verification of the obtained results. Each newly manufactured vehicle, before it is allowed to be used on the road, must undergo the vehicle approval procedure to determine the values of various parameters (including exhaust emissions) and compare them with the applicable standards. The NEDC homologation test (New European Driving Cycle) has been in force since 1990 and it has been the basis for the decision approving new cars for road use. A new test procedure has been in force since September 2017, however, it applies to new cars only [23, 24, 25, 26]. For those, there are completely different emission limits for both carbon dioxide and other exhaust components. Both Fiat Seicento and Fiat Panda were examined based on the former NEDC procedure and, therefore, this one was chosen for simulation in the present study.

The scheme of the simulation model developed as part of the conducted tests is shown in figure 1.

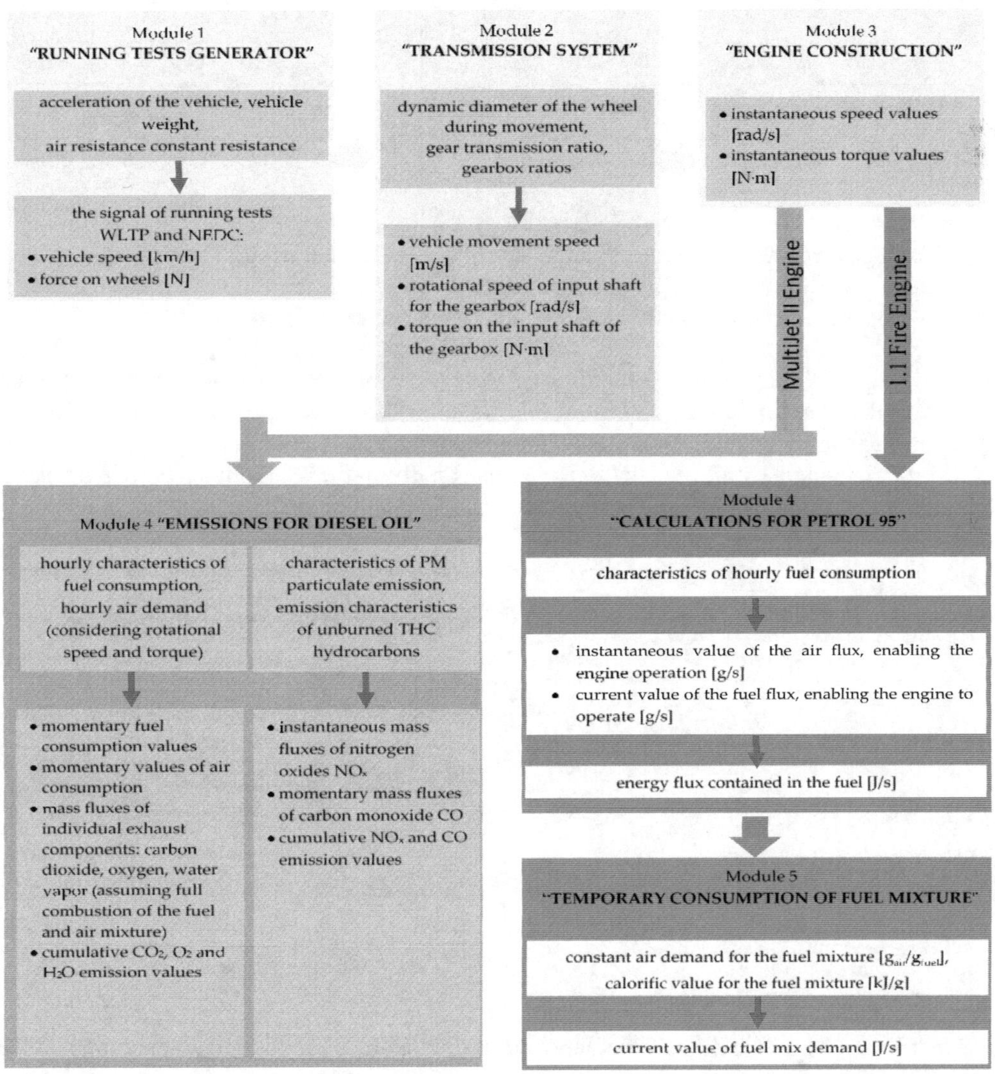

Figure 1. Diagram of the simulation model for calculation of the mass exhaust values.

3. Results

As mentioned, the simulation model was used to analyse the operation of two different engine types fed with different types of fuels according to the NEDC test procedure. Figure 2 shows the results generated by the "NEDC generator" block, showing the following waveforms: vehicle speed, the force acting on the wheels and the gear number in the gearbox.

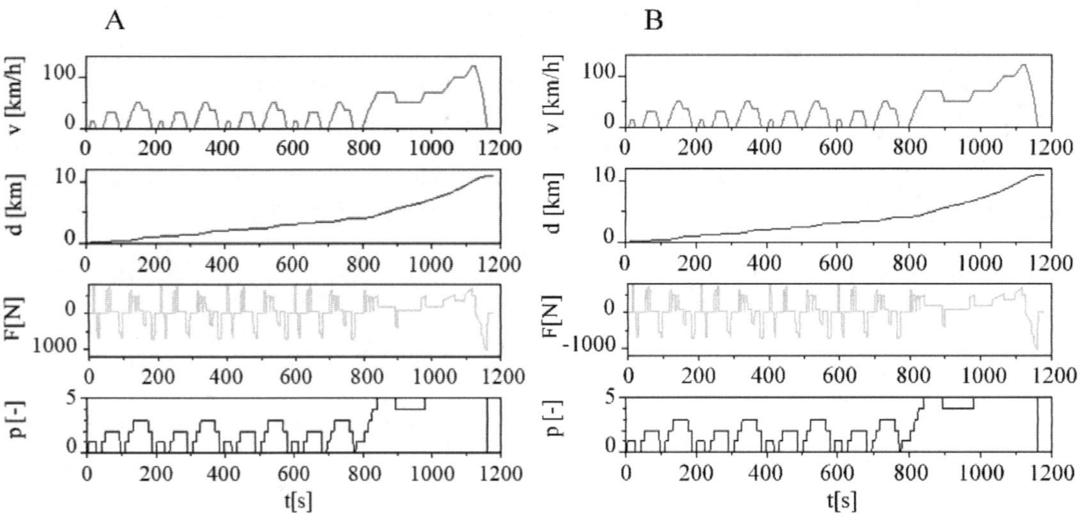

Figure 2. Signals generated by the "NEDC generator" module during the simulation (A – Fiat Seicento, B – Fiat Panda). v – vehicle speed, d – distance travelled by the vehicle during the test, F – force exerted on the wheels and p – gear number in the gearbox.

The above signals indicate the correctness of the entire model. It was found that the courses are consistent with the literature data, and that at this stage no deviations occur that would depend on the simulated engine type.

The results generated by the "Calculation emission from fuels" module, which determined the emission values of carbon dioxide and water vapour, are presented in figure 3.

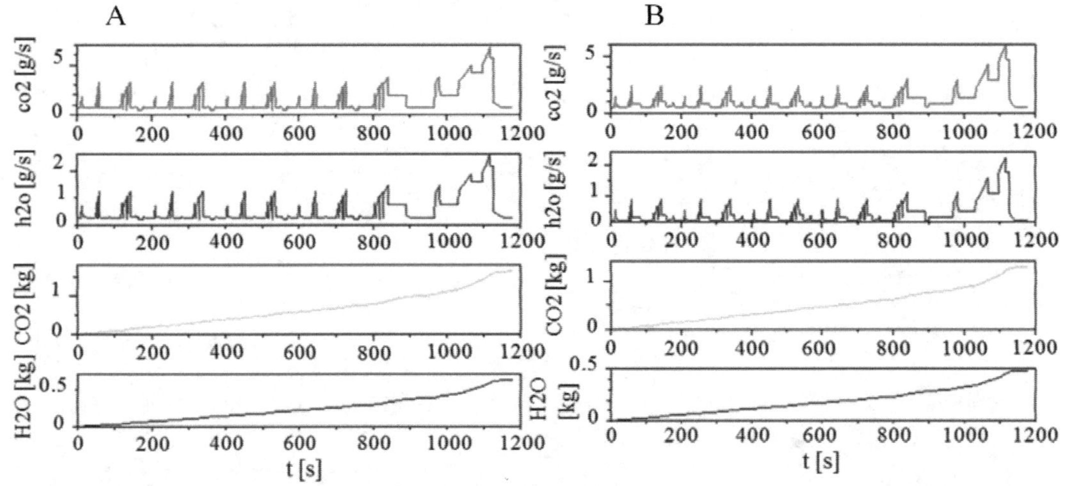

Figure 3. Signals generated by the "Calculation emission from fuels" module during the simulation (A – Fiat Seicento, B – Fiat Panda).

The graphs show time dependencies of: the mass flux of carbon dioxide emission during the test, mass flux of water vapour emission, total carbon dioxide emission and total water vapour emission from 95 gasoline and diesel oil.

On the basis of the above graphs, slight differences in the mass emission of carbon dioxide and water vapour for the considered fuels can be observed. However, the "total carbon dioxide emission," chart indicates higher emission from gasoline engine than from diesel.

The detailed results are presented in Table 3. The table gives the list of total carbon dioxide and water vapour emissions for individual fuels.

Table 3. Results of total carbon dioxide and water vapour emissions for the used fuels.

Emitted substance	Gasoline 95	Ethanol	DME	Diesel oil	FAME
CO_2 [kg]	1.51	1.41	1.54	1.29	1.38
H_2O [kg]	0.750	0.898	0.948	0.488	0.517

The differences in the emissions of the indicated fuels are in favour of diesel and its alternatives. During the test, the diesel engine emitted more than 17% less carbon dioxide than the gasoline engine. However, FAME fuel turned out to be more emissive in relation to conventional fuel in terms of both carbon dioxide and water vapour. The difference amounts to approx. 7% CO_2 and 6% H_2O respectively. In the case of gasoline and its substitutes, conventional fuel has shown carbon dioxide emissions in the middle between its own alternative fuels. Ethanol showed lower emissivity (more than 7% less CO_2 in the test), while dimethyl ether DME about 2% more. In the case of water vapour, the emission value compared to conventional gasoline was higher by about 20% for ethanol and 26% for DME, respectively.

The total carbon dioxide and water vapour emissions that discharged during the NEDC test by the tested engines are presented in Table 4. The results obtained are given in g/km, and compared with the permissible levels set out in relevant standards. On this basis, it was estimated whether a given propulsion unit could be allowed to operate on the basis of this very criterion.

Table 4. List of CO_2 emission results for individual fuels with values from regulations.

Emitted substance	Gasoline 95	Ethanol	DME	Diesel oil	FAME
CO_2 [kg]	1.51	1.41	1.54	1.29	1.38
CO_2 [g/km]	137.27	128.18	140.00	117.27	125.45
CO_2 [g/km] limit from the regulation	170	170	170	130	130

The limiting values were accepted accordingly to production year of the given automobile, i.e. 170 g/km for FIAT Seicento manufactured in 2010 and 130 g/km for FIAT Panda with diesel engine, produced in 2015. Despite the differences in carbon dioxide emissions from individual fuels, each of them is within the permissible range.

The car equipped with a diesel engine has shown lower CO_2 emission than the car fuelled with gasoline. Such a result can be attributed to the better fuel economy of the car with diesel engine in road conditions corresponding to the NEDC tests. It needs to be noted that the simulation programme does account for the formation of substances (environmentally rather harmful, e.g. tars) other than CO_2 and water vapour. Consequently, conclusions concerning the ecological impact of the cars under analysis cannot be considered as complete.

4. Conclusions

Based on the conducted simulations, the following conclusions can be made:

- In the conditions of NEDC tests and the use of dedicated fuels, the car equipped with a diesel engine has shown smaller CO_2 emission than the car with spark-ignition engine. This difference seems to be attributable to differences in the behaviour of cars as whole units;

- The comparison of diesel fuel with its alternative FAME fuel, showed a higher emission of carbon dioxide by about 7% when the FAME was used. This corresponds to a higher content of carbon in the chemical composition of FAME. It has to be recognised, however, that in the case of diesel fuel the emitted carbon is derived from fossils and, therefore, introduces additional carbon to the atmosphere, while the biofuel contains the carbon that was already in the atmospheric carbon cycle. Consequently, the use of biofuel has a positive ecological impact, despite higher emission of CO_2;
- With respect to gasoline and its substitutes, the lowest value of emitted carbon dioxide was obtained for ethanol, while the dimethyl ether, DME, emitted more CO_2 than conventional gasoline 95. While the first observation could be associated with smaller carbon content in ethanol than in other fuels, the result observed for DME appears to be in contrast with the chemical composition, and probably indicate some problems with the combustion of this fuel in the conditions required by the test. The problem requires further studies for deeper understanding;
- Despite observed differences in emissivity, in the case of both cars, each fuel analysed in the test procedure gave the result within the CO_2 emission limits allowed by the regulations;
- The developed simulation programme reflects the work cycle of real engines powered by different fuels and can be a helpful tool in the process of planning experiments or a valuable source of data for further consideration. The actual state of the programme is dedicated only to the determination of the main components of exhaust gases. In spite of the compliance with NEDC tests, it cannot serve as a tool for deeper ecological assessment of the car's performance.

Funding
The research was carried out under financial support obtained from the research subsidy of the Faculty of Engineering Management (WIZ) of Bialystok University of Technology. From the grant No. WZ/WIZ/4/2019 (Olga Orynycz, Andrzej Wasiak).

5. References

[1] Herran DS, Tachiiri K and Matsumoto K 2019 *Appl. Energ.* **243** 119-131 https://doi.org/10.1016/j.apenergy.2019.03.069
[2] Anifantis AS, Colantoni A and Pascuzzi S 2017 *Renew. Energy* **103** 115–127 https://doi.org/10.1016/j.renene.2016.11.031
[3] Pisoni E, Christidis P, Thunis P and Trombetti M 2019 *J. Environ. Manage.* **231** 249-255 doi number: 10.1016/j.jenvman.2018.10.039 (online)
[4] Wasiak A 2018 Modeling Energetic Efficiency of Biofuels Production. Green Energy and Technology. Cham, Switzerland: Springer Nature Switzerland. ISBN 978-3-319-98430-8. https://doi.org/10.1007/978-3-319-98431-5
[5] Mikielewicz D, Kosowski K, Tucki K, Piwowarski M, Stepien R, Orynycz O and Wlodarski R 2019 *Energies* **12** 1-21 https://doi.org/10.3390/en12163173
[6] Kuklinska K, Wolska L, and Namiesnik J 2015 *Atmos. Pollut. Res.* **6** 129-137
[7] Liu D, Guo X, and Xiao B 2019 *Sci. Total Environ.* **661** 750-766
[8] Sarkan B, Semanova S, Harantova V, Stopka O, Chovancova M. and Szala M 2019 *Proc. III International Conference of Computational Methods in Engineering Science CMES'18,* (Kazimierz Dolny/Poland) 06009 DOI: https://doi.org/10.1051/matecconf/201925206009
[9] Clean transport, Urban transport. Alternative fuels for sustainable mobility in Europe. Available online: https://ec.europa.eu/transport/themes/urban/cpt_en (accessed on 15 June 2019).
[10] Alalwan HA, Alminshid AH, and Aljaafari HAS 2019 *Renew Energ* **28** 127-139 https://doi.org/10.1016/j.ref.2018.12.006
[11] Tucki K, Orynycz O, Swic A and Mitoraj-Wojtanek M 2019 *Energies* **12** 1-22 https://doi.org/10.3390/en12152942
[12] Kupczyk A, Mączyńska J, Redlarski G, Tucki K, Bączyk A and Rutkowski D 2019 *Appl Sci* **9** 1-

13 https://doi.org/10.3390/app9020254
[13] Mączyńska J, Krzywonos M, Kupczyk A, Tucki K, Sikora M, Pińkowska H, Bączyk A, and Wielewska I 2019 *Fuel* **241** 989-996 https://doi.org/10.1016/j.fuel.2018.12.116
[14] Rietmann N and Lieven T 2019 *J. Clean. Prod.* **206** 66-75
[15] Krzywonos M, Tucki K, Wojdalski J, Kupczyk A, and Sikora M 2017 *Rocz Ochr Środ.* **19** 394-410 ISSN 1506-218X
[16] Mendoza-Villafuerte P, Suarez-Bertoa R, Giechaskiel B, Riccobono F, Bulgheroni C, Astorga C, and Perujo A 2017 *Sci. Total Environ.* **609** 546-555
[17] Triantafyllopoulos G, Dimaratos A, Ntziachristos L, Bernard Y, Dornoff J and Samaras Z 2019 *Sci. Total Environ.* **666** 337-346 https://doi.org/10.1016/j.scitotenv.2019.02.144
[18] Tucki K, Mruk M, Orynycz O, Botwińska K and Gola A 2019 *Sustainability* **2019** 11 1-15 https://doi.org/10.3390/su11082188
[19] Ambrozik A, Kurczyński D, Łagowski P and Warianek M 2016 *Proceedings of the Institute of Vehicles* **1** 23-36 ISSN 1642-347X
[20] Tucki K, Mruk R, Orynycz O, Wasiak A and Świć A 2019 *Sustainability* **2019** 11 1-19 https://doi.org/10.3390/su11164449
[21] Zembowicz J 2010 *Fiat Seicento* (Warszawa, Poland: Wydawnictwo Komunikacji i Łączności)
[22] Scilab Enterprises. Available online: https://www.scilab.org/ (accessed on 15 June 2019)
[23] Marotta A, Pavlovic J, Ciuffo B, Serra S and Fontaras G 2015 *Environ Sci Technol* **49** 8315-8322 doi number: 10.1021/acs.est.5b01364
[24] Dimaratos A, Tsokolis D, Fontaras G, Tsiakmakis S, Ciuffo B and Samaras Z 2016 *Transportation Research Procedia* **14** 3169-3178 https://doi.org/10.1016/j.trpro.2016.05.257
[25] Kikolski M and Ko Chien-Ho 2018 *Engineering Management in Production and Services* **2018** 10 70-79 doi number: 10.2478/emj-2018-0018
[26] Nazarko J, Czerewacz-Filipowicz K and Kuźmicz AK 2017 *Journal of Business Economics and Management* **2017** 18 1212-1227 doi number: 10.3846/16111699.2017.1404488

IT support for optimisation of abrasive water cutting process using the TOPSIS method

A Radomska-Zalas, A Perec and A Fajdek-Bieda

Jacob of Paradies University, Department of Technology, Teatralna 25, Gorzow Wlkp., Poland

aperec@ajp.edu.pl

Abstract. Modern production processes are becoming increasingly complex. At the same time, customers require high-quality products at the lowest price. This situation increases the importance of optimisation of production processes from the preparation of production to its implementation. Various methods of process optimisation have been devised to respond to the emerging phenomena including organisational aspects, device design, quality engineering and process automation. The TOPSIS method is used to make multi-criteria decisions. In this paper, a methodological study of this method was carried out, indicating conditions of applicability, assumptions and limitations, and the use of the TOPSIS method in the IT system supporting the optimisation of industrial processes on the example of waterjet cutting optimisation.

1. Introduction

In the production company, many activities are performed as a result of which input elements, such as materials, are transformed into output elements, i.e. the finished product. Production processes in the industrial environment are more and more complex and require the correlation of high quality of the product and the lowest possible price. The development of production processes led to the moment when optimisation took on a meaning, which denoted the drive towards lower production time and costs while maintaining the high quality of production. The work on optimisation methods considers not only organisational aspects but also the design of devices, their quality and automation of production processes. One of the forms integrating the mentioned factors is the IT system based on MCDM (Multiple criteria decision analysis) methods and algorithms of Decision Support Systems (DSS). There are also expert systems which seek the solution to the decision problem in logical reasoning with the use of an expert knowledge base. The source of knowledge is the analysis of processes. Such knowledge is important in intelligent systems, which use representation, understanding, design and implementation of knowledge. The dynamic development of technologies, products and requirements, however, shifts the emphasis away from the analysis of processes, and knowledge is often not used after making optimal decisions. It is the use of information systems that should facilitate the acquisition and processing of knowledge to optimise industrial processes.

Among the emerging production technologies, high-pressure water jet cutting is particularly attractive due to its environmental soundness [1] [2]. High-pressure water jet machining is one of the fastest developing advanced production technologies and is a preferred alternative to conventional material separation methods. This is largely due to the wide cutting possibilities of various materials [3], including multilayer materials of heterogeneous properties, as well as the precise cutting of a complex contour [4].

The optimisation of the cutting process with an abrasive water jet is supported by IT systems, which stems from the incapacity of the current models of prediction of abrasive water cutting effects to give satisfactory results in a wide range of parameter changes, in particular regarding new materials. This means that extensive research is needed to extend the empirical database, which requires the use of IT tools along with methods known as Design of Experiment. One of the methods supporting the selection of optimal parameters for various technological processes is the TOPSIS approach. The article presents the general construction of an IT system and an example of using the incorporated TOPSIS method for the selection of best parameters for optimising abrasive waterjet cutting. The TOPSIS method (Technique for Order Preference by Similarity to Ideal Solution) was proposed in 1981 by Hwang and Yoon [5] and assumes identifying the best solution from a complete set of solutions – alternatives [6]. According to the TOPSIS methodology, the best solution is the one that has the shortest distance from the best, and at the same time the farthest from the worst solution [8]. In other words, it consists in comparing an analysed decision's options with reference solutions: ideal solution non-ideal solution and assessing the distance between the variants and these solutions. As optimal, the chosen variant at the shortest distance from the ideal solution and the longest from the non-ideal one is chosen. The TOPSIS method algorithm consists of the following steps [7] [8]:

1. Establishment of a normalised decision matrix.
2. Determination of weighted normalised decision matrix.
3. Determining the evaluation of the weighted "ideal" solution and the weighty "non-ideal" solution.
4. Determining the distance from the weighted "ideal" and "non-ideal" solution.
5. Determination of relative distances from the weighted ideal solution.
6. Selection of the best variant.

The subsequent section of this work presents the results from analyses on the selection of optimal aluminium cutting parameters for the abrasive water jet technology using the TOPSIS.

2. Construction of an IT system to support process optimisation

Optimisation of production processes is the choice of the most advantageous solution, taking into account the key factors for an industrial enterprise, i.e. cost, time, product quality [9]. When designing an IT system supporting the optimisation of production processes, it is necessary to take into account the cost, time, quality of the product, but also such activities related to the production process and related factors as employees, machinery and equipment, storage or means of transport.

The design of the production optimisation support system was created based on the experience of companies gathered in the Lubuski Metal Cluster. Lubuski Metal Cluster needs a system that facilitates decision-making poorly structured at the strategic and tactical level of management. The main requirements for an optimisation decision support system are unit data retrieval, i.e. data extraction from collections. It was also necessary to ensure free access to and analysis of data, as well as to provide previously defined aggregate data. The result of the system's operation should be the preparation of projects of possible decisions and presenting the consequences of proposed decisions using calculation and simulation models.

To build the system, the quantitative and qualitative data was used, which made it possible to build a database and a database of methods that, thanks to artificial intelligence, allow the system to learn and adapt to changing data and changes in the organisation's environment. The main components of the system are:

- the user interface,
- a database and a database of methods that bring together the knowledge of the system,
- decision support module, which contains a description of the problem to be solved and
- enables the selection of artificial intelligence methods,
- a module for collecting data and transferring it to the database.

The project also assumes the need to consider the most important functions required by decision support systems, such as:

- a function that allows obtaining selected information from a database, spreadsheets, statistical data, historical data and other data sources available in the company. Acquiring information is the source of further analysis and is the basis for making decisions in the field of optimisation,

- a function identifying the problem, i.e. based on the specified data and specific input conditions indicating the place or area in the organisation that requires optimisation,
- conditional analysis function for the estimation of the probability of meeting the input conditions by the entered data and the suitability of this data for the process of optimisation of a given place or area in the enterprise,
- function that calculates the effects of certain decision options. This function makes it possible to estimate the effects of decisions made, considering the possibility of changing data or changes in the company's environment. Thanks to the use of such a function, the system user can work on various data and can make decisions supported by the analysis of these data,
- an optimisation function, which was recognised as the most important function, and which allows making decisions maximising or minimising the objective function without disturbing the set of input conditions. The result of the executed function is the presentation of several decision variants from which the user chooses the most advantageous variant.

Considering all input data and after analysing the problems of companies, the system project was created. The designed system supporting the optimisation of production processes is presented in Figure 1.

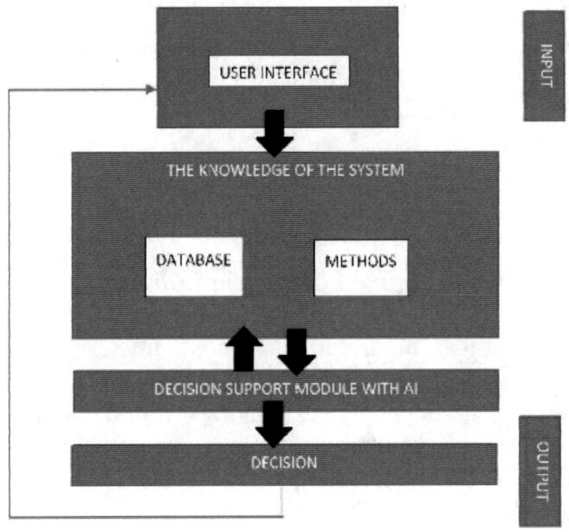

Figure. 1. Design of the support system for the selection of optimisation process parameters (own study).

At the entrance to the system there is a user interface, by means of which data is put into the system, as well as by means of which the user is familiarised with the decision options to choose from. Another element of the system is its knowledge, which consists of a database and a database of methods. System knowledge is data entered into the system, but also those that the system collects as a result of learning. The system supports decision-making in a module using artificial intelligence methods (decision trees, neural networks, generic algorithms). This module uses the knowledge of the system. The output of the system is a decision to help the user in the optimisation process. The output data is passed on to users and at the same time, they are input data read into the database and method database as a base for system learning.

The designed system enables the integration of data from various levels and areas of the company, remembers the effects of decisions taken, as well as adapts to changing data and the environment. The system makes it possible to identify general problems that are easy to specify, as well as problems, including smaller-scale ones, that are difficult to determine. The system collects data from various levels and finds decision-makers at different levels in the organisation hierarchy.

It should be mentioned that the system supports decision-making, but the decision-maker supervises and controls each stage of the process. The decision-maker can finish work with the system at any stage. Thanks to the possibility of supporting all stages of decision-making, the system also gives the opportunity to make many decisions often dependent on each other, and difficult to take because of different styles of decision-making by many decision-makers. In addition, the function of adapting to the changing environment of the system enables effective support of decision making. The decision is made faster, and at the same time exhibits high quality and accuracy. The proposed system adapts and learns, but also gives the opportunity to teach the decision-maker. System users learn the approach to planning and decision-making. Models that are part of the system give the possibility to apply different approaches and, consequently, different variants of the decision.

3. Investigation of the impact of the TOPSIS method on the selection of optimal parameters

Machining materials with a high-pressure abrasive stream is more complex than conventional machining methods. Water under high pressure, produced in a special high-pressure pump, is transformed inside the nozzle into a high-speed stream (amounting to 800 m/s), grabs and accelerates abrasive grains in the mixing chamber [10].

The addition of dry abrasive to the nozzle increases the cutting efficiency and, as a result, empowers the stream to cut any material [11] [12]. The most commonly used abrasive is garnet [13], but there are other natural and synthetic abrasives [14], such as crushed glass, olivine and aluminium [12] [15] [16]. The abrasive should be selected considering the relationship between nozzle life and workpiece performance [17] [18]. The results presented in the article refer to the less common choice – aluminium.

The article presents the process of testing the IT system while selecting the optimal parameters of the abrasive water cutting process [19] [20]. Two dependent output parameters of the process were introduced:

- **Width** – width of kerf. It is important that it is as long as possible, although the width is of minimal importance during a single cutting. It is acquired only in the case of a pattern/cutting of many elements from a material board. Width is expressed in [mm].
- **Rz** – one of the surface roughness parameters (quality). The coefficient should be as low as possible because then the quality of surface finish is high. Rz is expressed in [mm].
- In addition, three independent control parameters for the cutting process were introduced:
- **Traverse speed** – feed. The speed of movement of the stream with respect to the workpiece material expressed in [mm/s].
- **Pressure** – working pressure given in [MPa]
- **Abrasive flow rate** – abrasive flow rate expressed in [g/s].

The TOPSIS described above was chosen as the method for selecting the optimal parameters. The TOPSIS method was employed to verify its suitability for the selection of water cutting process parameters. As the TOPSIS method is crucial in assigning weights to individual parameters, three options were considered:

- highest weight for Width, - highest weight for Rz, - equivalent Width and Rz.

The reason for the adoption of three possibilities is the will to examine whether and how the assignment of the above-mentioned weight for one criterion will affect the selection of optimal parameters. For each of the possible selections of weights, the study was conducted in six stages, the results of which are given below:

- **Establishing a normalised decision matrix.**

Nine variants and two dependent criteria were adopted for the study. For each variant, independent criteria were assigned, the application of which produced a dependent criterion. The dependent criteria have been normalised by a quotient that follows the formula (1):

$$X_{ij} = \frac{x_{ij}}{\sqrt{\sum_{i=1}^{m} x_{ij}^2}} \ (1) \qquad (1)$$

where:

Xij – normalised j-criterion value in i-th option;
xij – real j-criterion value in i-th option;
m – number of possibilities.

A normalised decision matrix is provided in Table 1.

IOP Conf. Series: Materials Science and Engineering **710** (2019) 012008 doi:10.1088/1757-899X/710/1/012008

Table 1. Normalised decision matrix (own study).

Traverse speed	Pressure	Abrasive flow rate	Rz	Width	Rz	Width
[mm/s]	[MPa]	[g/s]	[mm]	[mm]		
0.5	350	2	13.61	1.05	0.27690763	0.40459123
0.5	300	1.5	14.43	0.95	0.29359127	0.36605874
0.5	250	1	18.33	0.845	0.37294026	0.32559961
0.75	350	2	12.76	0.94	0.25961362	0.36220549
0.75	300	1.5	15.67	0.85	0.31882018	0.32752624
0.75	250	1	14.49	0.75	0.29481202	0.28899374
1	350	1.5	19.52	0.85	0.39715187	0.32752624
1	300	1	20.46	0.75	0.41627701	0.28899374
1	250	2	16.38	0.75	0.33326576	0.28899374

- **Determination of weighted normalised decision matrix.**

Building a decision matrix, a probe was carried out for three possibilities:

- Greater importance of Width (0.75), smaller Rz (0.25); - Greater importance of Rz (0.75), lower Width (0.25); - The equal weight of Width (0.5) and Rz (0.5).

The normalised criteria values were multiplied by the weights. Table 2 presents the results for the three weighting allocation possibilities, obtained by means of calculations compliant with the formula (2):

$$WX_{ij} = X_{ij} * w_j \tag{2}$$

where:

WX_{ij} – normalised weighted j-criterion value in i-th option;

X_{ij} – normalised j-criterion value in i-th option;

w_j – the weight given to the jth criterion.

Table 2. Weighted normalised decision matrix (own study).

Traverse speed	Pressure	Abrasive flow rate	Rz	Width	Rz	Width	Rz	Width	Rz	Width
[mm/s]	[MPa]	[g/s]	[mm]	[mm]	0,75	0,25	0,25	0,75	0,5	0,5
0.5	350	2	13.61	1.05	0.36921018	1.61836494	1.107631	0.539455	0.553815	0.809182
0.5	300	1.5	14.43	0.95	0.39145502	1.46423494	1.174365	0.488078	0.587183	0.732117
0.5	250	1	18.33	0.845	0.49725368	1.30239845	1.491761	0.434133	0.745881	0.651199
0.75	350	2	12.76	0.94	0.3461515	1.44882194	1.038454	0.482941	0.519227	0.724411
0.75	300	1.5	15.67	0.85	0.42509357	1.31010495	1.275281	0.436702	0.63764	0.655052
0.75	250	1	14.49	0.75	0.39308269	1.15597495	1.179248	0.385325	0.589624	0.577987
1	350	1.5	19.52	0.85	0.52953583	1.31010495	1.588607	0.436702	0.794304	0.655052
1	300	1	20.46	0.75	0.55503602	1.15597495	1.665108	0.385325	0.832554	0.577987
1	250	2	16.38	0.75	0.44435435	1.15597495	1.333063	0.385325	0.666532	0.577987

- **Determining the evaluation of the weighted "ideal" solution and the weighty "non-ideal" solution.**

In the next step, PIS (Positive Ideal Solution) and NIS (Negative Ideal Solution) were calculated. PIS, or the "perfect" solution, contains the highest values of the maximised and the lowest minimised variables, NIS vice versa. The formulas (3) and (4) were used for calculations:

$$PIS^{+} = WX_1^{+}, WX_2^{+}, \dots, WX_n^{+} = \{max_i WX_{ij}, min_i WX_{ij}\} \quad (3)$$

$$NIS^{-} = WX_1^{-}, WX_2^{-}, \dots, WX_n^{-} = \{min_i WX_{ij}, max_i WX_{ij}\} \quad (4)$$

In Table 3, PIS and NIS values for the tested criteria are shown.

Table 3. The weighted "ideal" solution and the weighty "non-ideal" solution (own study).

	Rz	**Width**	**Rz**	**Width**	**Rz**	**Width**
	0,75	**0,25**	**0,25**	**0,75**	**0,5**	**0,5**
Ideal	0.555036	1.618365	1.665108	0.539455	0.83255403	0.80918247
Non-ideal	0.346151	1.155975	1.038454	0.385325	0.51922724	0.57798748

- **Determining the distance from the weighted "ideal" and "non-ideal" solution**

Subsequently, the distance of considered possibilities from the "ideal" and "non-ideal" solution was calculated. For this purpose, formulas (5) and (6) were used:

$$d_i^{+} = \sqrt{\sum_{j=1}^{n}(WX_{ij} - WX_j^{+})^2} \quad (5)$$

where:

d_i^{+} - distance from the "ideal" solution

$$d_i^{-} = \sqrt{\sum_{j=1}^{n}(WX_{ij} - WX_j^{-})^2} \quad (6)$$

where:

d_i^{-} - distance from the "non-ideal" solution

The distance of the variants from the "ideal" solution is presented in Table 4, and from "non-ideal" in Table 5.

Table 4. The distance from the weighted "ideal" solution (own study).

Traverse speed	**Pressure**	**Abrasive flow rate**	**Rz**	**Width**	**Rz**	**Width**	**Rz**	**Width**	**Rz**	**Width**
[mm/s]	**[MPa]**	**[g/s]**	**[mm]**	**[mm]**	**0,75**	**0,25**	**0,25**	**0,75**	**0,5**	**0,5**
0.5	350	2	13.61	1.05	0.185825842		0.557477526		0.278738763	
0.5	300	1.5	14.43	0.95	0.224754972		0.493425014		0.257189005	
0.5	250	1	18.33	0.845	0.321206507		0.202834784		0.180197123	
0.75	350	2	12.76	0.94	0.269030799		0.629196763		0.324591868	
0.75	300	1.5	15.67	0.85	0.334528415		0.403142171		0.248490233	
0.75	250	1	14.49	0.75	0.48993201		0.509721464		0.335359662	
1	350	1.5	19.52	0.85	0.309312915		0.128103799		0.15880535	
1	300	1	20.46	0.75	0.462389982		0.154129994		0.231194991	
1	250	2	16.38	0.75	0.475452339		0.366073685		0.28463063	

Table 5. The distance from the weighted "non-ideal" solution (own study).

Traverse speed	Pressure	Abrasive flow rate	Rz	Width	Rz	Width	Rz	Width	Rz	Width
[mm/s]	[MPa]	[g/s]	[mm]	[mm]	0,75	0,25	0,25	0,75	0,5	0,5
0.5	350	2	13.61	1.05	0.462964575		0.168941943		0.233767951	
0.5	300	1.5	14.43	0.95	0.311571227		0.170381725		0.168445767	
0.5	250	1	18.33	0.845	0.210408433		0.455926558		0.2381841	
0.75	350	2	12.76	0.94	0.292846988		0.097615663		0.146423494	
0.75	300	1.5	15.67	0.85	0.173170165		0.242334933		0.141282265	
0.75	250	1	14.49	0.75	0.046931198		0.140793594		0.070396797	
1	350	1.5	19.52	0.85	0.239553479		0.552546731		0.285667807	
1	300	1	20.46	0.75	0.208884523		0.62665357		0.313326785	
1	250	2	16.38	0.75	0.098202854		0.294608561		0.147304281	

- **Determination of relative distances from the weighted ideal solution and selection of the best variant.**

In the last step, the relative proximity coefficient of the decision variants to the ideal solution was determined; formula (7) was used for this purpose:

$$RD_i = \frac{d_i^-}{d_i^+ + d_j^-} \tag{7}$$

On the basis of the obtained values of the relative closeness coefficient of decision variants, assuming that the higher the RDi value the better the decision-making variant, the system generated a ranking of variants, which is presented in Table 6.

Table 6. Relative distances from the weighted ideal solution and selection of the best variant (own study).

Traverse speed	Pressure	Abrasive flow rate	Rz	Width	Rz 0,75 / Width 0,25	Position	Rz 0,25 / Width 0,75	Position	Rz 0,5 / Width 0,5	Position
[mm/s]	[MPa]	[g/s]	[mm]	[mm]						
0.5	350	2	13.61	1.05	0.286418907	9	**0.767431973**	**3**	0.543873388	6
0.5	300	1.5	14.43	0.95	0.419063944	8	0.74332631	4	0.604248107	5
0.5	250	1	18.33	0.845	0.604208954	5	0.307903289	7	0.43070079	7
0.75	350	2	12.76	0.94	0.47880661	7	**0.865693459**	**1**	**0.689132232**	**2**
0.75	300	1.5	15.67	0.85	0.658911465	4	0.624564634	5	0.637526338	4
0.75	250	1	14.49	0.75	**0.912582577**	**1**	**0.783565972**	**2**	**0.826504802**	**1**
1	350	1.5	19.52	0.85	0.56354865	6	0.188207888	8	0.357288956	9
1	300	1	20.46	0.75	**0.688823988**	**3**	0.19740425	9	0.424583554	8
1	250	2	16.38	0.75	**0.828812055**	**2**	0.554084338	6	**0.658966486**	**3**

4. Results and discussion

The article presents the results that the IT system returned for the nine possible combinations of input parameters introduced. Two dependent and three independent criteria for process optimisation of abrasive water jet cutting were introduced into the system. Because the method is important for the weights assigned to a particular criterion, a test was carried out for three possibilities, i.e. more weight was assumed for Width, then higher for Rz, and the last possibility was equal weights. Table 6 shows the results for given possibilities. Figure 2 presents the values of dependent parameters and position results for individual options.

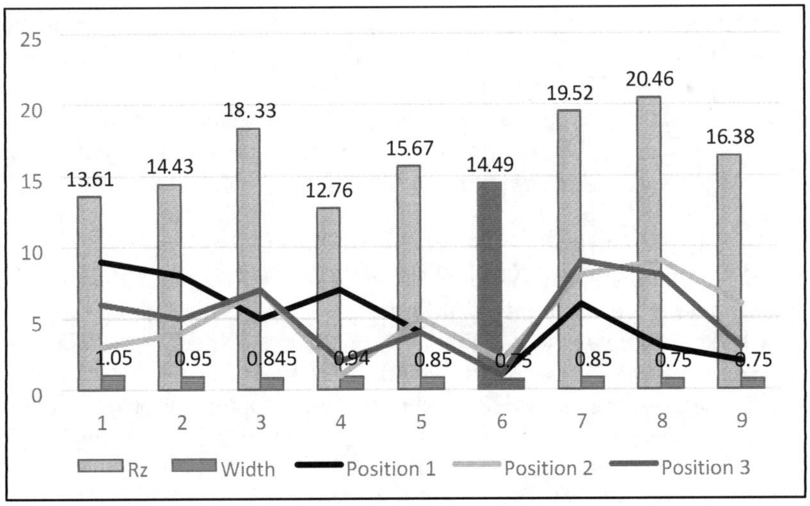

Figure 2. The values of dependent parameters and position results for individual options (own study).

When analysing the three highest classified parameters, one option is particularly worth the attention, i.e. 0.75 for traverse speed, 250 for pressure and 1 for abrasive flow rate, which gives 14.49 for Rz and 0.75 for Width. The indicated result, regardless of the assigned weights, takes into account the recommendation that Width and Rz should be as small as possible. Despite the large dispersion for the remaining results, this one is highly indicated for three options.

5. Summary and conclusion

Although the TOPSIS method works best for multiple criteria, even with two gives satisfactory results. The general assessment of the results leads to the conclusion that the use of the TOPSIS method enables the selection of optimal process parameters that significantly affect the test process, and at the same time allows the user to omit those factors that have a negligible impact on the results. It can simplify the test process very much, reducing the time to obtain reliable results and lowering the cost of testing, because it reduces the number of tests required. However, it should be remembered that one of the limitations of this method is the need to give weight, which is usually subjective.

The TOPSIS method does not require IT support, but the use of an IT system as a tool supporting the calculation process is important. It enables the creation of a knowledge base, which in the context of repeatability of research will allow the decision-makers to increase the correctness of the final results, as well as to compare the results achieved for various multi-criteria analysis methods. It is also important to use artificial intelligence mechanisms in the system. An IT system equipped with the AI mechanisms gives the possibility of repeated use in making a concrete decision, considering different input states each time. Such a system thus becomes a universal tool with a simulation character. It should be added that the use of an expert system, particularly in the optimisation of production processes, may have an impact on shortening the entire process cycle, which gives the opportunity to analyse a larger number of cases, and in combination with the possibility of using historical data increases the efficiency and accuracy of the decisions taken.

6. References

[1] Patyk R et al. 2018 Experimental and numerical researches of duplex burnishing process in aspect of achieved productive quality of the product *AIP Conference Proceedings* **1960** 070021 doi:10.1063/1.5034917

[2] Kukiełka L et al. 2016 Analysis of the States of Deformation and Stress in the Surface Layer of the Product after the Burnishing Cold Rolling Operation *Materials Science Forum* **862** 278-87

[3] Wessels V, Grigoryev A, Dold C and Wyen C-F 2012 Abrasive waterjet machining of three-dimensional structures from bulk metallic glasses and comparison with other techniques *J. Mater. Res.* **27**(8) 1187–92 doi: 10.1557/jmr.2012.36
[4] Ćojbašić Ž, Petković D, Shamshirband S, Tong Ch W, Sudheer Ch, Janković P, Dučić N and Baralić J 2016 Surface roughness prediction by extreme learning machine constructed with abrasive water jet *Precision Engineering* **43** 86–92 doi:10.1016/j.precisioneng.2015.06.013
[5] Hwang C L and Yoon K 1981 *Multiple Attribute Decision Making Methods and Applications* (Springer-Verlag) doi:10.1007/978-3-642-48318-9
[6] Decui L, Zeshui X, Dun L and Yao W 2018 Method for three-way decisions using ideal TOPSIS solutions at Pythagorean fuzzy information *Information Sciences* **435** 282-95
[7] Kasprzak D 2018 Przedziałowa metoda TOPSIS dla grupowego podejmowania decyzji *Optimum. Economic Studies* **4**(94) 256-73 doi:10.15290/oes.2018.04.94.19
[8] Yuvaraj N and Pradeep Kumar M 2014 Multiresponse Optimization of Abrasive Water Jet Cuuting Process Parameters Using TOPSIS Approac *Materials and Manufacturing Processes* **30**(7) 882-89 doi:10.1080/10426914.2014.994763
[9] Ťavodova M, Kalincova D and Slovakova I 2018 Evaluation of Some Parameters of Hard Surfacing Treatment of the Functional Surfaces of Forestry Tools *Management Systems in Production Engineering* **26**(4) 222-26
[10] Valíček J, Hloch S and Kozak D 2009 Surface geometric parameters proposal for the advanced control of abrasive waterjet technology *The International Journal of Advanced Manufacturing Technology* **41**(3–4) 323–28
[11] Producing GMA in Australia *GMA Garnet Australia* Accessed: 18-Mar-2019 available: https://www.gmagarnet.com/en-gb/about-gma/producing-gma-australia
[12] Cenac F, Zitoune R, Collombet F and Deleris M 2015 Abrasive water-jet milling of aeronautic aluminium 2024-T3 *Proceedings of the Institution of Mechanical Engineers Part L: Journal of Materials Design and Applications* **229**(1) 29–37 doi:10.1177/1464420713499288
[13] Aydin G, Kaya S and Karakurt I 2017 Utilization of solid-cutting waste of granite as an alternative abrasive in abrasive waterjet cutting of marble *Journal of Cleaner Production* **159** 241–47
[14] Sriragen A K and Sathiya P 2017 Optimisation of Process Parameters for Gas Tungsten Arc Welding of Incoloy 800HT Using TOPSIS *Materials Today: Proceeding* **4** 2031-2039
[15] Perec A 2016 Abrasive Suspension Water Jet Cutting optimization using orthogonal array design *Procedia Engineering* **149** 366–73
[16] Supriya S B and Srinivas S 2018 Machinability Studies on Stainless Steel by Abrasive Water Jet – Review *Materials Today: Proceedings* **5** 2871–76
[17] Perec A Pude F Kaufeld M and Wegener K 2017 Obtaining the selected surface roughness by means of mathematical model based parameter optimization in Abrasive Waterjet Cutting *J of Mech. Eng.* **63**(10) 606–13
[18] Kinik D, Gánovská B, Hloch S, Monka P, Monková K and Hutyrová Z 2015 On-line monitoring of technological process of material abrasive water jet cutting. On-line praćenje tehnološkog postupka rezanja materijala abrazivnim vodenim mlazom *Tehnički vjesnik* **22**(2) 351–57
[19] Perec A and Radomska-Zalas A 2019 Modeling of abrasive water suspension jet cutting process using response surface method *AIP Conference Proceedings* **2078** 020051 doi:10.1063/1.5092054
[20] Perec A, Radomska-Zalas A, Bieda A, Musiał W, Praźmo J, Sobczak R and Nagnajewicz S Study into abrasive water jet machining of corundum based ceramics. Lecture Notes in Mechanical Engineering *Advances in Manufacturing Engineering and Materials* (to be published)

IOP Conf. Series: Materials Science and Engineering **710** (2019) 012009 doi:10.1088/1757-899X/710/1/012009

Numerical analysis of the rotor in the co-simulation methodology

K Siadkowska, R Raczyński and M Wendeker

Lublin University of Technology, Faculty of Mechanical Engineering, Department of Thermodynamics, Fluid Mechanics and Aviation Propulsion Systems, Nadbystrzycka 36, 20-618 Lublin, Poland

k.siadkowska@pollub.pl

Abstract. The co-simulation method in the composite analysis of the rotor enables connection between Multi Body Dynamics (MBD), Finite Element Method (FEM) and Computational Fluid Dynamics (CFD) by using MATLAB code control algorithms. The idea of unidirectional analysis assumes the use of a rigid rotor model with implemented flexible elements of specific mass-rigidity properties, a special Modal Natural Frequency matrix. In this paper, the MBD model complemented with flexible elements was loaded with inertial forces and additional aerodynamic forces. These aerodynamic forces were determined from CFD analyses considering the aeroelasticity of the rotor blade (static and dynamic characteristics appropriate for a given aerodynamic airfoil). The validation of the computational model was performed on the basis of Virtual Blade Model analyses of the rotor, completed with additional algorithms based on the Blade Element Theory method.

1. Introduction

Co-simulation is a methodology for integrating various algorithms and numerical environments to create continuous data coupling transmission in the computational loop (for bidirectional analyses) and to enable data transmission without feedback (for unidirectional analyses). The co-simulation methodology is capable of analysing physical phenomena based on different discretisation methodologies using a different mathematical description, e.g. Lagrange, Euler. The co-simulation-based approach has the following objectives: reducing numerical computing time, which has a direct impact on product implementation costs; reducing expensive stand tests, mainly fatigue and endurance tests; improving computational convergence compared to conventional numerical computational combinations; analysing multiple physical phenomena in a single step (computational loop).

However, this approach shows many disadvantages such as: advanced hardware, complicated integration of boundary conditions between numerical environments, complicated schemes of element discretisation, difficult selection of the time step size for transient analyses and a long-term process of preparing a computational model.

In the world aviation industry, there is a lot of research connected with the development of rotorcraft [1, 2, 3, 4] and application of innovative technologies improving their performance [5, 6, 7]. In the paper [8] numerical investigations of aircraft components were made. In work [9] numerical research with Ansys Fluent was examined in order to check the static longitudinal stability for the given ranges of horizontal stabiliser angles and angles of attack. Due to the necessity to meet numerous safety requirements resulting from aviation regulations, R&D on passenger aircraft is very

IOP Conf. Series: Materials Science and Engineering **710** (2019) 012009 doi:10.1088/1757-899X/710/1/012009

expensive. Unmanned aircraft [10, 11, 12] have recently intensively developed so technologies used in manned aircraft naturally migrate to unmanned aircraft. The Brandenburg Institute for Society and Security report [13] on unmanned aerial systems used in civil aviation claims that the number of unmanned aircraft vehicles (UAV) projects and manufacturers more than doubled only between 2005 and 2011. The strongest growth was observed in civil and commercial applications. The leading manufacturers of UAV are the USA (30.33%), France (6.42%), Great Britain, Israel and Russia. The costs of innovative research conducted in unmanned aircraft are much lower, which also affects the development of technologies used in passenger aircraft and the need to integrate unmanned airborne UAV systems with passenger aviation [14, 15].

This publication focuses on performing integrated numerical simulations of a prototype test articulated main rotor, dedicated to rotorcraft with a maximum starting mass of MTOW=150 kg. The main rotor is an object on which both, mass (inertial) loads and complex states of load resulting from aerodynamics and dynamics of rotational motion of a system with many degrees of freedom, act. For this purpose, a series of numerical analyses based on the co-simulation methodology has been used to achieve a unidirectional coupling of issues covering:

— aerodynamics of the main rotor: loads resulting from the generated thrust force and resistance on a particular blade, phasicity (variability) of the thrust force resulting from the cyclic change of the blade setting angle, variable loads resulting from the generated control torque (acting along the blade axis of geometric rotation);
— inertia of the main rotor: centrifugal force loads, phasicity (variability) of the thrust force resulting from the cyclic change of the blade setting angle, loads resulting from the interaction of gravitational forces (static blade deflection, gap over the beam);
— dynamics of the main rotor: issues of loads resulting from structural dynamics and body dynamics in rotational motion (resonance, frequency of natural vibrations of the rotor blades), loads of assemblies located in a heterogeneous field of both tangential and axial accelerations, loads resulting from kinematic couplings of the control system, loads resulting from the dynamics of the control system, loads resulting from fluctuations of the blade in the plane of thrust and rotation of the main rotor.

2. Methodology

The main method of scientific research used in this paper was computer simulation which has not been widely used at the beginning of the 20th century [16]. Currently, the method of computer simulation is considered to be one of the most important methods dedicated to the study of phenomena for which the real process would be too expensive or dangerous or the object of research does not exist yet.

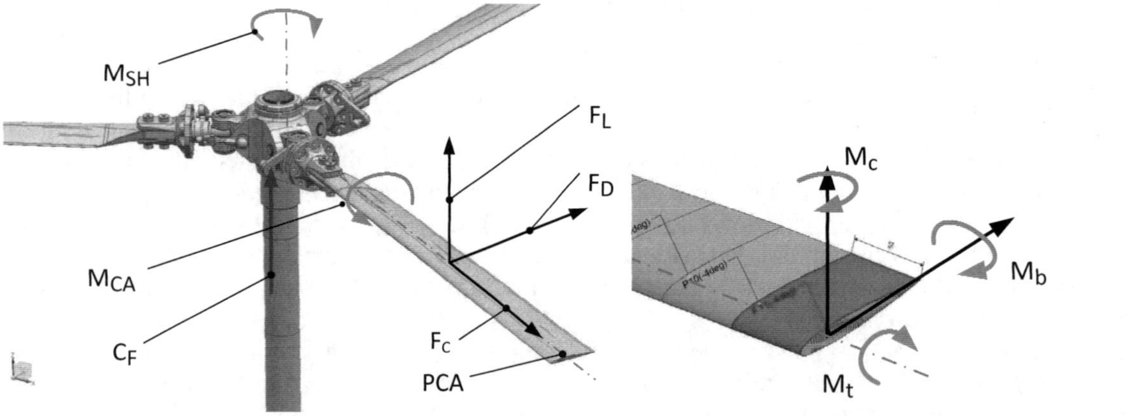

Figure 1. Global load system a) on the rotor, b) on the blade, where: Mt – twist moment, Mb – beam wise moment, Mc – chordwise moment, PCA – pitch control axis, FC – centrifugal force, FD – drag force, FL – lift force, CF – control force, MCA – control moment, MSH – shaft moment.

The use of coupled analyses enables integration of calculation algorithms with different definitions of boundary conditions, as well as a continuous data exchange between individual calculation blocks. The figure 1 shows a typical simplified load diagram working on the rotor.

As can be seen from the above diagram, loads in the rotor are associated with both aerodynamic loads (FL, FD, Mc, Mt) and dynamic loads (MSH, Mb, Mc, FC). It should be noted that the Fc, Mb and Mc loads are both aerodynamically and structurally related.

An application of Co-Simulation methodology to combined analyses allows to integrate CFD and MBD computational issues. As indicated in the paper [17], currently, there is no simulation tool to simulate the whole system with a necessary complexity, level of details and a sufficient performance. For this reason, a combination of many simulation environments is applied using the so-called cooperative simulation (co-simulation, cooperative simulation, coupled simulation) [18, 19, 20]. The methodology of co-simulation and its application have been described in detail, among others, in the publications [17] and [21].

Co-simulation is a methodology to integrate different algorithms, numerical environments to create continuous data transmission coupling over the time of the computational loop (for bidirectional analyses) and to enable data transmission without feedback (for one-way analyses). A characteristic feature of the co-simulation methodology is the ability to analyse physical phenomena based on different discretisation methodologies, characterised by a different mathematical description, e.g. Lagrange, Euler. The above assumptions serve as a foundation for basic blocks of Co-Simulation couplings used within the framework of numerical analyses of the carrier rotor:

a) a numerical analysis of the MBD model considering the susceptibility to the component element of the rotor by means of linear discretisation of the element (figure 2). Such a numerical analysis is performed in the time domain, which means that the loads are transferred on an ongoing basis under constant boundary conditions of the restraint (figure 3).

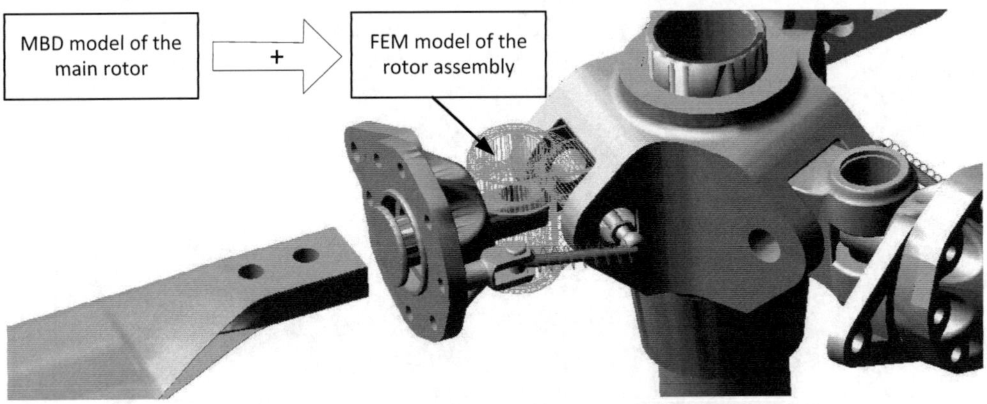

Figure 2. The use of elements susceptible the MBD analysis.

Figure 3. Visualisation of Von Mises Stress.

IOP Conf. Series: Materials Science and Engineering **710** (2019) 012009 doi:10.1088/1757-899X/710/1/012009

b) a numerical analysis of the MBD model including external aerodynamic forces (figure 4). External loads can be discretely adapted based on the aerodynamic characteristics obtained from CFD analyses. Inside the ADAMS environment, the resultant force is determined on the basis of the current setting angle and the current rotational speed of the supporting rotor shaft (figure 5).

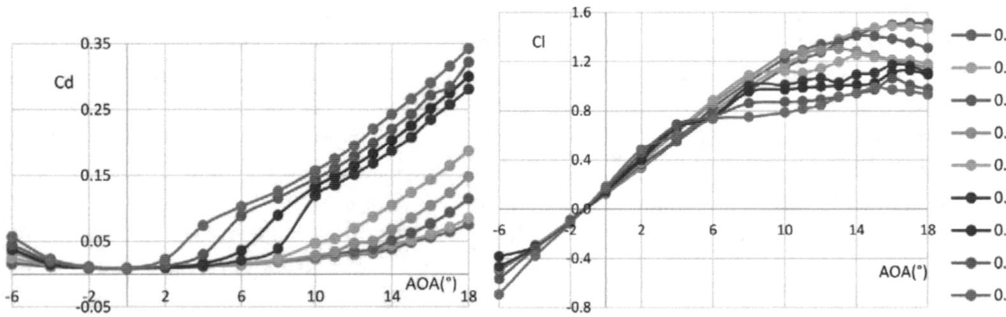

Figure 4. Aerodynamic characteristics drag force (on the left) and lift force coefficient (on the right) for NACA 23012 clean airfoil [22].

In order to supplement the numerical MBD model of the supporting rotor, a MATLAB Simulink or MSC EASY5 integrator is used, which allows to determine the value of aerodynamic loads acting on a particular blade section. The integrated value of the partial forces determines the total aerodynamic force of thrust and resistance according to the diagram below (figure 6).

$dD = \frac{1}{2}\rho U^2 c * dr * cd$ – aerodynamic drag force

$dL = \frac{1}{2}\rho U^2 c * dr * cl$ – lift force of the blade element section

$dT = \mathrm{dLcos}\emptyset - dDsin\emptyset$ – total thrust force of the rotor

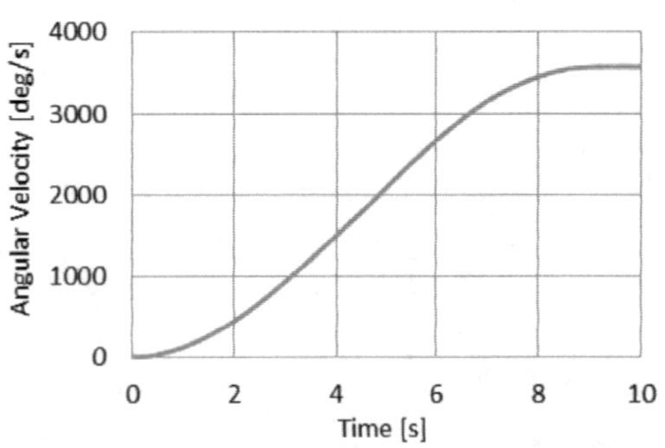

Figure 5. Current rotor speed value for co-simulation methodology for unidirectional coupling.

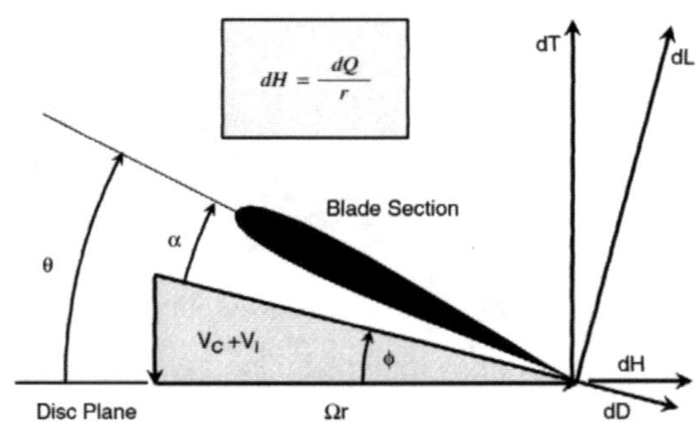

Figure 6. Schematic diagram for determining aerodynamic loads.

The visualisation of the coupling in the MATLAB is shown in this diagram below (figure 7). The diagram below allows for coupling between the mechanical part responsible for the rotational speed and the aerodynamic part considering the aerodynamic characteristics and atmospheric parameters in accordance with the current ISA standards.

IOP Conf. Series: Materials Science and Engineering **710** (2019) 012009 doi:10.1088/1757-899X/710/1/012009

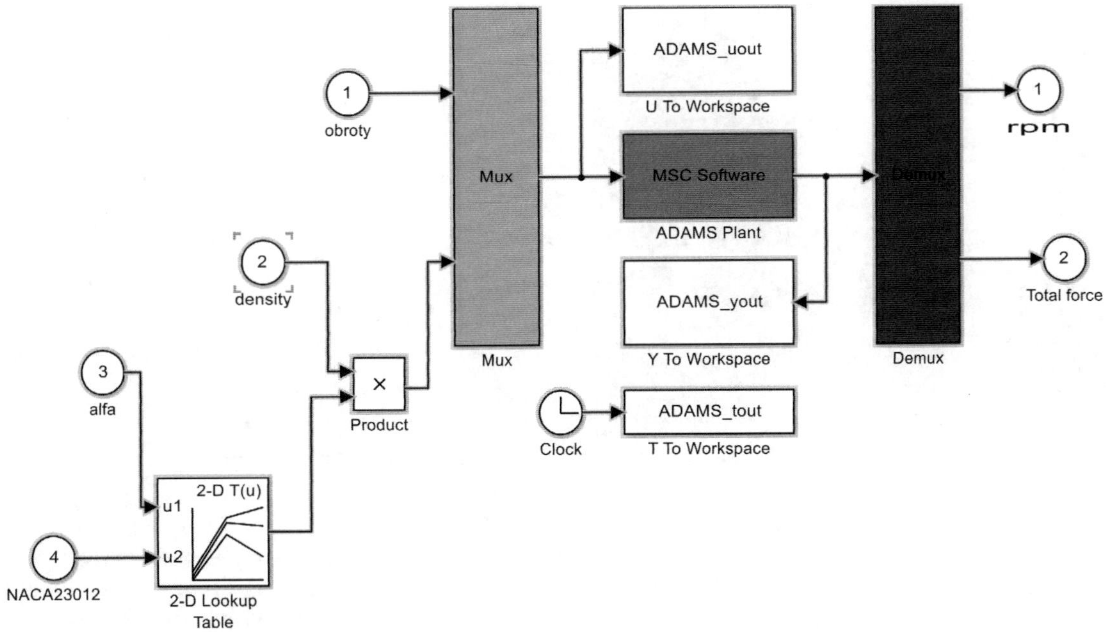

Figure 7. Visualisation of the aerodynamic internal coupling.

3. Research object

The research object is an articulated load-bearing rotor in which the horizontal and vertical pivot joints are mounted on a common axis. The angle is changed by means of an axial joint to which the rotor blades are attached. The diagram below shows the schematic diagram of the supporting rotor and its numerical interpretation in the MSC ADAMS environment (figure 8).

Figure 8. Visualisation of the numerical model schematic diagram of the CAD model (on the left) and MBD numerical model (on the right).

The MSC ADAMS numeric model is based on a geometric CAD model imported into the MSC ADAMS environment as a Parasolid including mass parameters and a linear definition of the material. The ADAMS environment represents MBD issues within which to every geometric figure are assigned appropriate mass geometrical properties and kinematic pairs that unambiguously define degrees of freedom of an object in computational space. Additionally, the computational model enables a simulation of elastic damping elements simulating the hydraulic damper with variable compartment characteristics. Figure 9 shows the geometric interpretation with the typical characteristics of the viscous silencer used in the calculations.

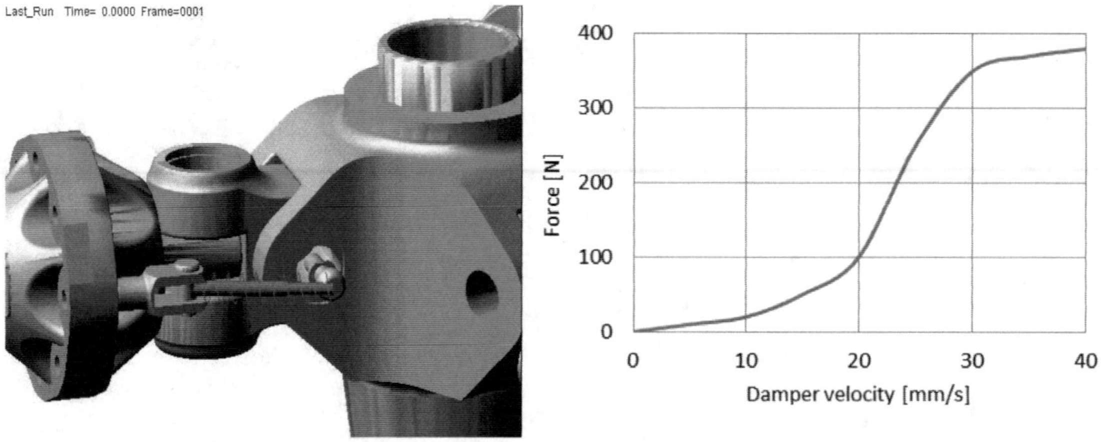

Figure 9. Characteristics of the horizontal fluctuation damper.

The following schemes are adopted for the blade modelling:

a) an indirect method consisting in segmenting the blade (figure 10). The numerical MES analyses enabled us to specify different types of stiffness in each direction, i.e. flexural stiffness in the plane of higher stiffness, flexural stiffness in the plane of lower stiffness and torsional stiffness. In the environment of the MBD model, the synthesis of individual sections takes place by means of a bushing type intermediate element or an element of stiffness matrix M [6x6].

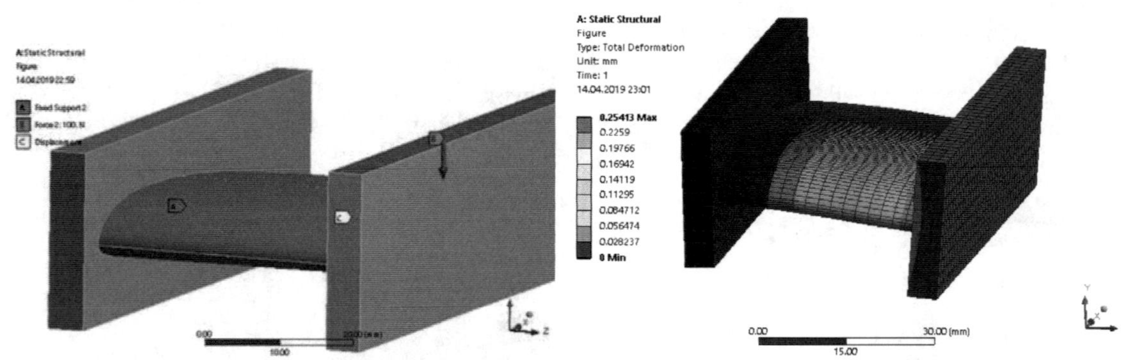

Figure 10. MES model of the blades section.

The mass and inertia values for the individual sections are determined according to the analysis of the CAD model parameterised by the material properties (figure 11).

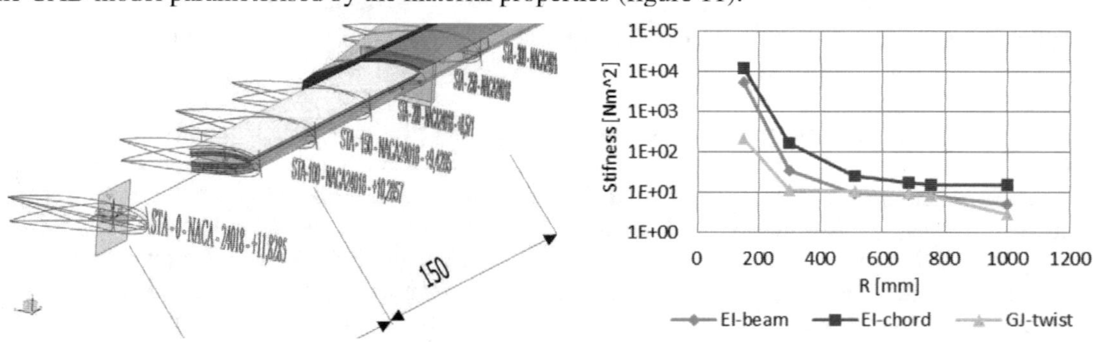

Figure 11. Airfoils distribution along blade (on the left) and stiffness distribution (on the right).

IOP Conf. Series: Materials Science and Engineering **710** (2019) 012009 doi:10.1088/1757-899X/710/1/012009

b) a direct method by replacing a rigid model with an FEM model and a complete database with information on the form-related deformations relevant to individual forms of natural vibration (Modal Natural Frequency database).

The coupled numerical MBD model includes a geometric interpretation of the geometric model of the rotor hub with the basic assemblies of the supporting rotor. The following figure shows the interpretation of the calculation model (figure 12).

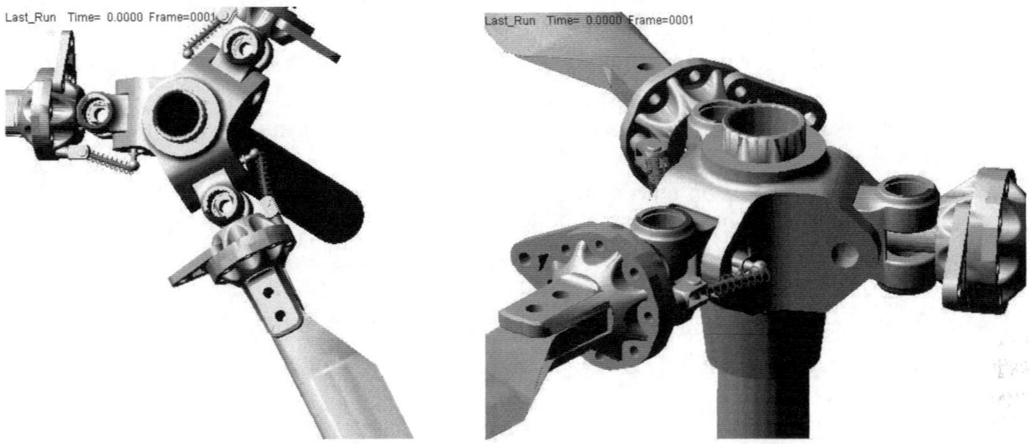

Figure 12. Visualisation of MBD interpretation of the calculation model.

The MBD model, supplemented with the susceptible elements, was loaded with inertial forces and additional aerodynamic forces determined from the CFD analyses considering the aeroelasticity of the carrier rotor blade (static and dynamic characteristics appropriate for a given aerodynamic airfoil). The calculation model will be validated in line with the Virtual Blade Model analyses of the carrier rotor, supplemented with additional algorithms based on the Blade Element Theory method.

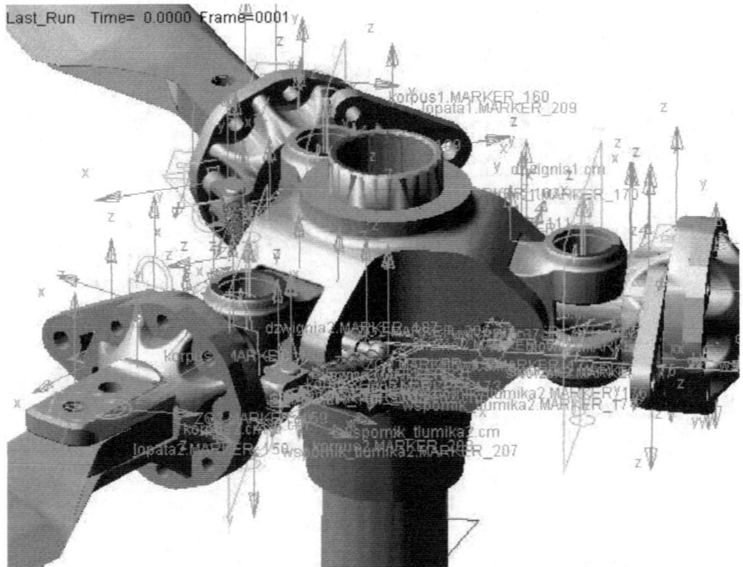

Figure. 13. Visualisation of the MBD model together with the system of ties determining the degrees of freedom

IOP Conf. Series: Materials Science and Engineering **710** (2019) 012009 doi:10.1088/1757-899X/710/1/012009

The application of the co-simulation method in the combined analysis of the carrier rotor enables the connection between the MBD, FEM and CFD environments via control algorithms using the MATLAB code. The idea of one-way connection assumes the use of a rigid rotor model to which susceptible elements with specific mass-rigidity properties and the individual matrix of Modal Natural Frequency are entered. The idea behind the MBD model is to use appropriate kinematic pairs reflecting the physics of the motion of the articulated supporting rotor (figure 13).

The following connections are important for the model of the supporting rotor:

- a sliding kinematic pair with a single degree of freedom to provide an unambiguous linear transformation of the system corresponding to the action of the general stroke;
- a kinematic pair of rotation with a single degree of freedom corresponding to the rotation of the rotor within the set speed range;
- a kinematic pair of rotation with a single degree of freedom to provide rotation corresponding to the fluctuations of the blade in the plane of thrust;
- a kinematic pair of rotation with a single degree of freedom to provide rotation corresponding to the fluctuations of the blade in the plane of rotation;
- a kinematic pair of rotation with a single degree of freedom to provide rotation corresponding to the rotation of the axial joint;
- a three-degree freedom spherical kinematic pair corresponding to the operation of a control disc correlated with the cyclic control action of the rotor.

4. Simulation results

Numerical analyses cover complex issues of correlation between particular groups of analyses. One of the basic goals is to provide necessary information on the mechanics and dynamics of rotor operation. The diagrams below show the results of the dynamic analysis of the rotor based on the Co-simulation methodology including the coupling between the rigid MBD model and susceptible elements, as well as aerodynamic forces and moments (figure 14 and figure 15). Combined analyses allow to evaluate only kinematics of motion to determine possible collisions and, above all, dynamic loads, including stability of work. The following figure shows the visualisation of the change in the cyclic angle at the comb joint caused by transverse restraint. A large area of instability due to system vibration is clearly visible.

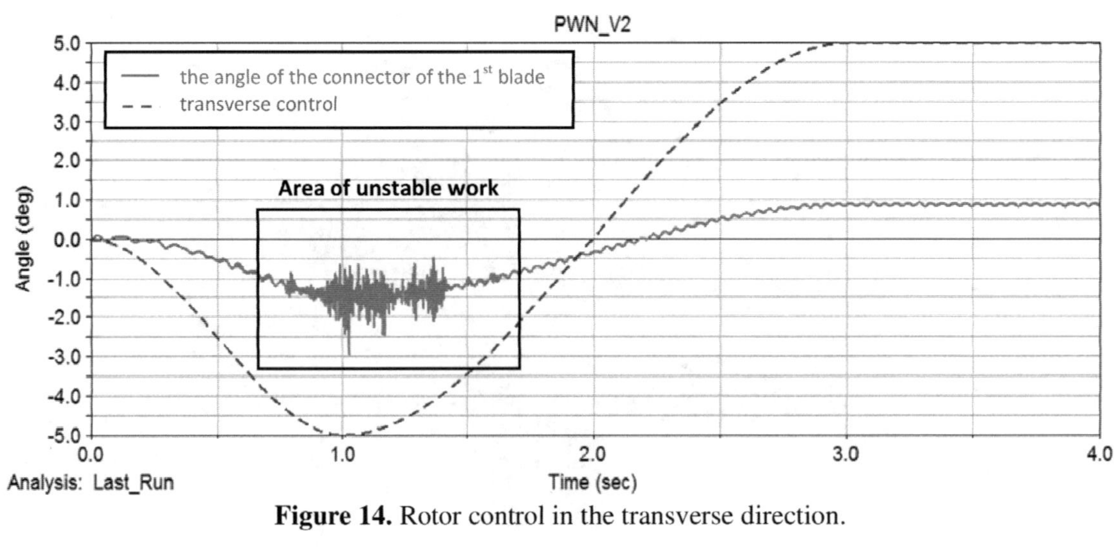

Figure 14. Rotor control in the transverse direction.

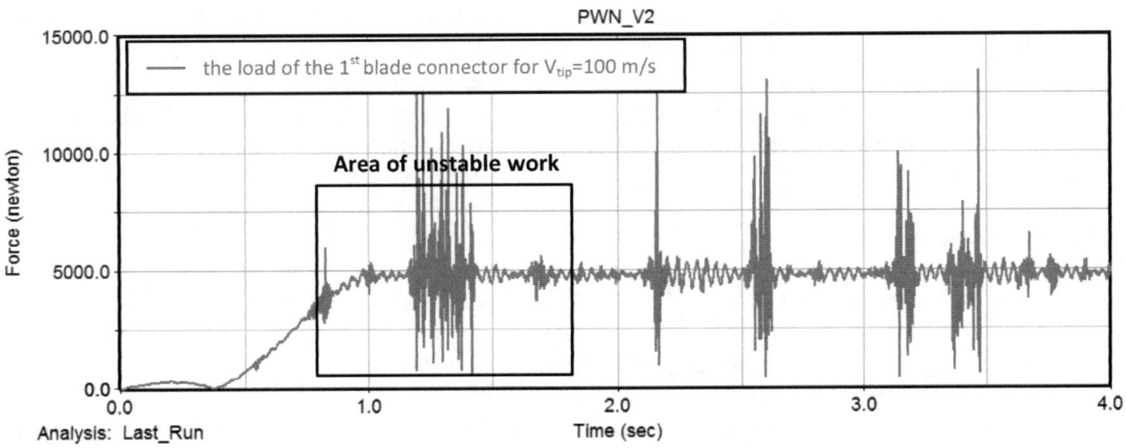

Figure 15. Visualisation of the load flow, with a clear area of high instability.

An additional advantage of the Co-simulation approach is using feedback on the borderline between MBD and the rigid body and implementation of susceptible elements by the possibility of integration with algorithms of structural dynamics. The analyses of frequencies and forms of vibrations of a system with many degrees of freedom (such as the load-bearing rotor) enable easy validation of the model using the results of numerical analyses based on the Galerkin method, etc. The diagram below shows the visualisation of the Campbell diagram of the model rotor. Comparing the results presented in figure 11 with numerical results, a correlation in the nominal speed range (RPM relative value = 1) and the form of single-node vibration in h1 in the plane of rotation with the area of unstable operation is visible (figure 16).

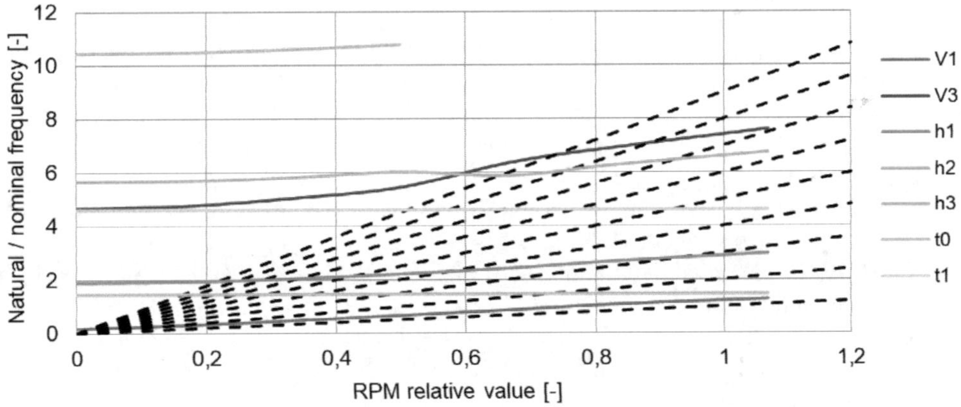

Figure 16. Campbell diagram corresponding to the common form of the vibration of the main rotor.

5. Conclusions

This co-simulation approach aims to: reduce the time required for numerical computing, which has a direct impact on product implementation costs; reduce the impact of costly, mainly fatigue and endurance, bench tests; improve computational convergence compared to conventional numerical computational combinations; and analyse multiple physical phenomena in a single step (computational loop). However, it also shows many drawbacks such as high hardware requirements, a complex integration of boundary conditions between numerical environments, complex component discretisation schemes, a problematic selection of time step sizes for transient analysis, and a long process of building the computational model.

This paper discusses the research using the method of computer simulation so the goal of creating a virtual laboratory to perform static and dynamic tests was achieved. Summing up, the Co-simulation analyses easily and transparently allow for an evaluation and optimisation of the rotor at the level of the computational model, which has a significant impact on the reduction of product implementation costs. The ultimate plan is to conduct a laboratory experiment to verify the adopted design assumptions.

Acknowledgements
This work has been financed by the Polish National Centre for Research and Development under the LIDER program. Grant Agreement No. LIDER/45/0177/L-9/17/NCBR/2018.

6. References

[1] Decoret J-M, Surin J, Merlin M and Sahuc C 2009 Investigation of the Technical Feasibility and Safety Benefit of an Integrated Engine / Rotor / Transmission Health Monitoring and HOMP System for Part 27 Piston Engine Small Helicopters *EASA*
[2] Greaves M, Elasha F, Worskett J, Mba D, Rashid H, Keong R 2012 Vibration Health or Alternative Monitoring Technologies for Helicopters *EASA*
[3] Mba D, Place S, Rashid H, Keong R 2012 Research Project EASA.2011/5 HELMGOP - Helicopter Main Gearbox Loss of Oil Performance Optimisation *EASA*
[4] Grabowski Ł, Siadkowska K and Skiba K 2019 Simulation Research of Aircraft Piston Engine Rotax 912 *MATEC Web Conf.* **252** 05007
[5] Fincham J H S and Friswell M I 2015 Aerodynamic optimisation of a camber morphing aerofoil *Aerosp Sci Technol.* **43** 245-255
[6] Ajaj R M, Beaverstock C S and Friswell M I 2016 Morphing aircraft: The need for a new design philosophy *Aerosp Sci Technol.* **49** 154-166
[7] Goh G D, Agarwala S, Goh GL, Dikshit V, Sing SL and Yeong W Y 2017 Additive manufacturing in unmanned aerial vehicles: Challenges and potential. *Aerosp Sci Technol.* **63** 140-151
[8] Czyż Z, Siadkowska K and Sochaczewski R 2019 CFD Analysis of Charge Exchange in an Aircraft Opposed-Piston Diesel Engine *MATEC Web Conf.* **252** 04002
[9] Czyż Z, Łusiak T, Karpiński P and Czarnigowski J 2018 Numerical investigation of the gyroplane longitudinal static stability for the selected stabilizer angles *J Phys Conf Ser.* **1101**(1) 012003
[10] Mazur M and McMillan J 2016 Global Market for Commercial Applications of Drone Technology Valued at over $127 bn *PwC*
[11] Joshi D. 2017 Commercial Unmanned Aerial Vehicle (UAV). Market Analysis – Industry trends, companies and what you should know *Bus Insid.*
[12] Duszczyk M 2017 Eksplozja popularności bezzałogowców dopiero nadciąga *Rzeczpospolita*
[13] Skrzypietz B T 2012 Unmanned Aircraft Systems for Civilian Missions. *Brandenburg Institute for Society and Security* **1** 1-28
[14] Ainley M and Clough A 2009 Interim Report of the Preliminary Impact Assessment on the Safety of Communications for Unmanned Aerial Systems (UAS) *EASA*
[15] Clough A, Ainley M and Hunt S 2009 Preliminary Impact Assessment on the Safety of Communications for Unmanned Aircraft Systems *EASA*
[16] Apanowicz J 2002 Metodologia Ogolna
[17] Bleicher F, Duer F, Leobner I, Kovacic I, Heinzl B and Kastner W 2014 Co-simulation environment for optimizing energy efficiency in production systems *CIRP Ann - Manuf Technol.* **63**(1) 441-444
[18] Zheng W, Yan J, Chen X, Xu S and Li Y 2018 Technology and engineering application of cross area HVDC interconnection system high-precision simulation modeling based on ADPSS *Glob Energy Interconnect.* **1**(5) 627-635

[19] Bartel A, Brunk M and Schöps S 2013 On the convergence rate of dynamic iteration for coupled problems with multiple subsystems *J Comput Appl Math.* **262** 14-24

[20] Xie F, Wang J and Wang Y 2011 Modelling and co-simulation based on AMESim and Simulink for light passenger car with dual state CVT *Procedia Eng.* **16**(Apee) 363-368

[21] Brecher C, Esser M and Witt S 2009 Interaction of manufacturing process and machine tool *CIRP Ann - Manuf Technol.* **58**(2) 588-607

[22] Sarkar P and Raczynski R 2017 Recent Progress in Flow Control for Practical Flows. *Doerffer P, Barakos GN, Luczak MM*, eds. Springer 126-135.

IOP Conf. Series: Materials Science and Engineering **710** (2019) 012010 doi:10.1088/1757-899X/710/1/012010

Modelling of transverse shrinkage of injection-moulded parts using experimental methods and fuzzy logic theory

P Stączek

Lublin University of Technology, Faculty of Mechanical Engineering, Department of Automation, Nadbystrzycka 36, 20-618 Lublin, Poland

p.staczek@pollub.pl

Abstract. The method of building a mathematical model of fuzzy structure and the set of its rules, on the basis of experimental results is presented. The developed method is illustrated on the example of modelling the relationship of the transverse shrinkage of HDPE injection–moulded part (moulding) versus the shape of the moulding and the conditions of the injection moulding process. The developed shrinkage model can be helpful at the design stage of the moulding geometry, as well as mould cavity and the runner system. Due to the fuzzy structure, the model can be easily modified and generalised into similar moulding injection processes. It is also a good starting point for the optimisation of the process, aimed at compensating the shrinkage value in the entire volume of the moulding by adjusting the pressure profile in the holding phase of moulding injection cycle.

1. The purpose of mathematical modelling

In principle, two types of mathematical models are distinguished: phenomenological and experimental (behavioural). The purpose of creating the first one is to learn about the internal structure and mechanisms governing the modelled system. In contrast, experimental models are primarily created to give formal description of relationships between causal (input) variables and the effects of their interaction on the system (output variables). Experimental models can be used to approximate the system's response to input variables with values other than the values used to build the model [1, 2]. Therefore, experimental models are called equivalent: "approximation models" or "cause-and-effect models".

Approximation models are usually built on the basis of the results of a series of experiments (previously planned or not) in which the values of input variables and the corresponding values of the output signals from the system are measured. Next, the mathematical function is matched to the collected data sets, which approximates the experimental data with the smallest error. For practical reasons, the most popular approximation is a polynomial of 2nd or 3rd degree (mostly). An alternative to the approximation with an explicit equation is the construction of artificial neural network. Models of this form allow approximation of very complex relationships, among others: interactions of input variables and strong nonlinearities inside the modelled system.

A slightly different approach to system modelling may be noticed in case of applying the theory of fuzzy logic. Models with fuzzy structure are something intermediate between phenomenological and experimental models. They allow to aggregate in one model the impartial (but often incomplete) knowledge of the mechanisms governing the modelled system and the results of observing behaviour of the system, i.e. purely experimental knowledge; thus, so–called “expert knowledge” [3, 4]. The advantage of this approach is the ability to model very complex relationships, and to include in the model many partial relations between variables in the form of the so-called “rules”. The disadvantage of the model constructed in this way is its “fuzziness”, resulting in an approximate estimation of the values of output variables from the system [5]. The reason of this is, that both the knowledge base (rules) of the modelled system and the mechanism of inference from rules operate on approximate values, just like: “very small”, “small”, “close to zero” etc.

The paper presents the original method of building a fuzzy model (its set of rules) on the basis of experimental results. The developed method is illustrated on the example of modelling the relationship of the transverse shrinkage value of HDPE injection–moulded part (moulding) against the shape of the moulding and the conditions of the injection moulding process.

2. Polymer shrinkage phenomenon

The shrinkage (equivalent: after–shrinkage) is a phenomenon of reduction of moulding dimensions due to changes occurring during the moulding injection process and later. The primary shrinkage of thermoplastics occurs in the mould cavity [6, 7]. Changes in the specific volume of the material during cooling and phase transitions (solidification, crystallisation) are the reason in this case. The secondary shrinkage is caused by stress relaxation processes inside the solidified moulding which may go on for dozens of days after injection [8].

Shrinkage is usually an undesirable phenomenon and is the cause of dimensional imprecision of mouldings, surface collapse, distortion and deformation [10, 11]. Therefore, it must be considered already at the stage of designing the geometry of the part produced and the injection mould. The FEM computer simulations of filling the mould cavity with the liquefied polymer are widely used on this stage [12, 13, 14, 15, 16].

The value of shrinkage may be intentionally controlled, mainly during the holding phase of moulding injection cycle [17, 18]. Then, the shrinkage changes are compensated by feeding an additional volume of plastic to the molten core of the moulding. It is possible only until the whole moulding or plastic in the runner or in the gate is completely solidified [7, 18, 19, 20].

3. The general form of shrinkage model

The developed model of shrinkage was necessary for the development of decision algorithms (rules defining the adjusting of the injection process settings) designed for the equalisation of the shrinkage value in the entire volume of the injection moulding [21]. A necessary condition for effective shaping of shrinkage is the knowledge of its dependence on the main control variables of the injection process and the geometry of moulding. It was necessary to develop a model of transverse shrinkage S_p of an injection moulded part of HDPE in the function of three independent variables: nominal wall thickness d_n, distance from the gate x and the pressure in the holding phase p_d:

$$S_p = f(d_n, x, p_d) \tag{1}$$

where transverse shrinkage S_p is defined as the ratio of the difference in the dimension of the cavity d_n and the actual wall thickness d of the moulding to d_n:

$$S_p = (d_n - d) / d_n * 100\% \tag{2}$$

Due to the complexity of the physical phenomena that determine the shrinkage, as well as their dynamic and multidimensional character, it was decided that the model (1) will be built using experimental techniques.

4. Moulding characteristics

The shrinkage model (1) was developed for a series of thin–walled beams with graduated thickness (table 1, figure 1) produced from HDPE. The beams consisted of up to three sections with intentionally varying thickness along their axis. Five different thickness profiles were chosen:

"constant", "ascending", "descending", "thin–thick–thin" and "thick–thin–thick" according to direction of polymer flow. The mould cavity was filled with plastic through the slit–shape gate located at the shorter side of the beam (figure 1). Two identical symmetrically located cavities and runners were made in the mould.

5. Experimental results

For each of the five types of beams (table 1), a series of injection cycles was made with three different holding pressure values p_d: 15, 30 and 60 MPa. The value of transverse shrinkage (2) was measured at three points located on the longitudinal axis of the beam (bottom of table 1), corresponding to the centres of its three segments of different thickness, spaced from the gate by x – 25, 75 and 125 mm, respectively. The results of the measurements are shown graphically in figure 2. Numbers in the table grids lying at the intersections of columns and rows (d_n and x) are the mean values of transverse shrinkage S_p measured in the respective beam segments. The location of the number in the table cell indicates which type of moulding (of which thickness profile) the result is from. Each of three tables contains the results obtained for three different holding pressures. Negative values of shrinkage mean that the thickness of the moulding was greater than the corresponding dimension of the cavity (effect of high holding pressure).

Table 1. Nominal dimensions (in millimeters) of injection moulded beams

Mark	Dimensions	Thickness profile
4–4–4	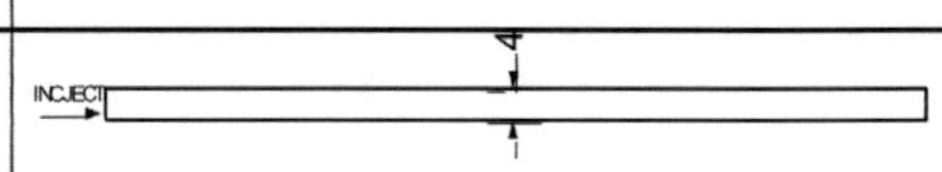	constant
2–4–6	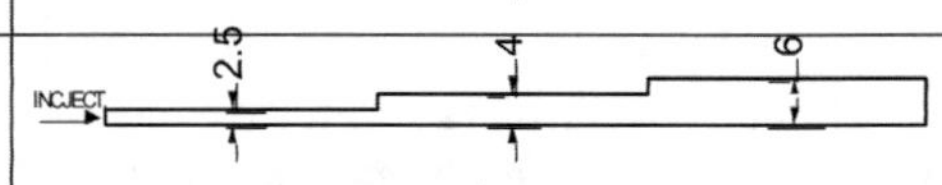	ascending
6–4–2	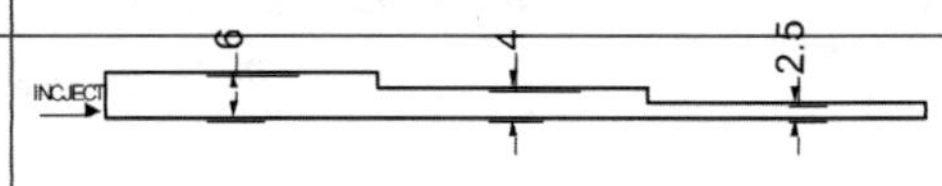	descending
2–6–2	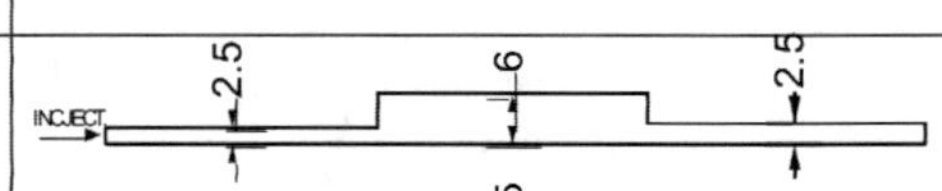	thin–thick–thin
6–2–6	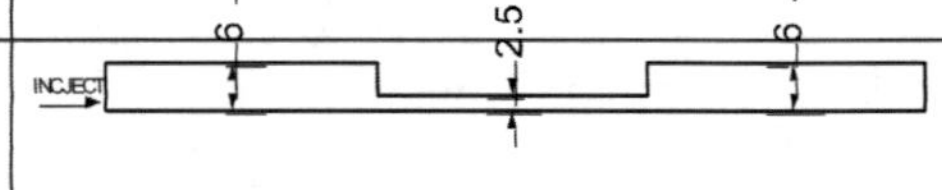	thick–thin–thick
	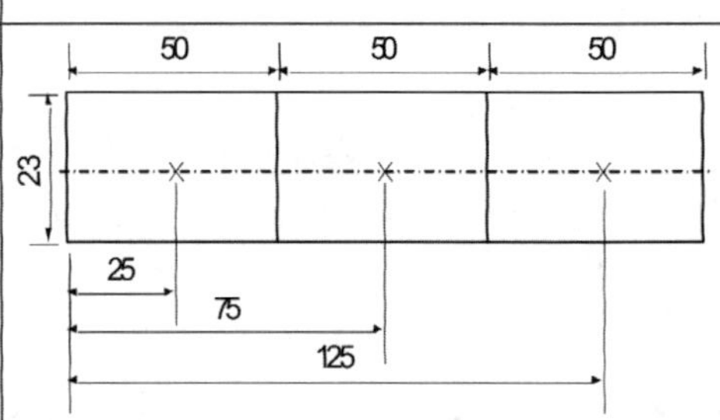	X marks indicate thickness measurement points

Comparing the shrinkage values S_p of beams segments with the same nominal thickness d_n and distance from the gate x, but belonging to mouldings of different thickness profiles (i.e. within one table cell of figure 2), it can be concluded that the values of transverse shrinkage are similar to each other, except for the segments of beams with a 2–4–6 profile (with a grey background). Hence, shrinkage values of respective beam segments for all four (2–6–2, 4–4–4, 6–2–6, 6–4–2) thickness profiles, i.e. excluding results for 2–4–6 profile, were averaged and shown in figure 3.

Figure 1. 3D model of injection moulded mould for shrinkage study.

Variation in transverse shrinkage along the axis of the beam with a thickness profile of 2–4–6 for three holding pressure p_d values is shown in figure 4. Only for this shape of the moulding an untypical tendency of the shrinkage increase versus the increase of the holding pressure was observed (in the middle beam segment of thickness $d_n = 4$ mm).

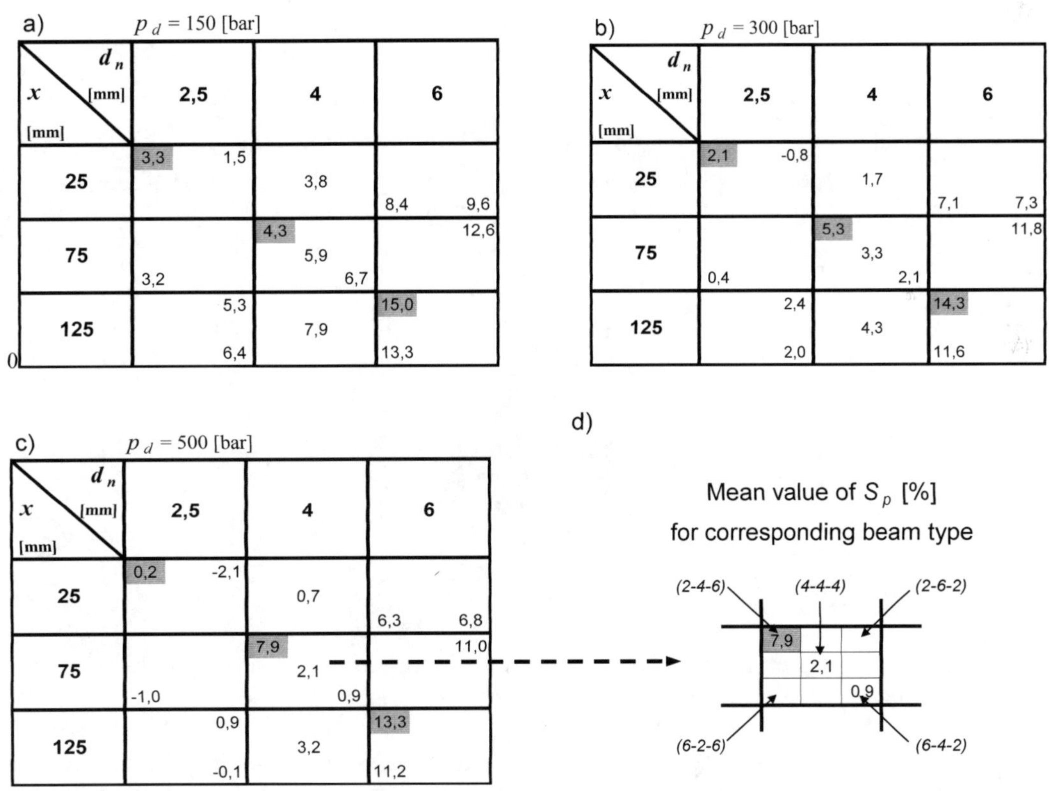

Figure 2. (a)-(c) transverse mould shrinkage (mean values in %) against nominal wall thickness d_n and distance from the gate x for three values of packing pressure p_d ; (d) explanation of results presentation.

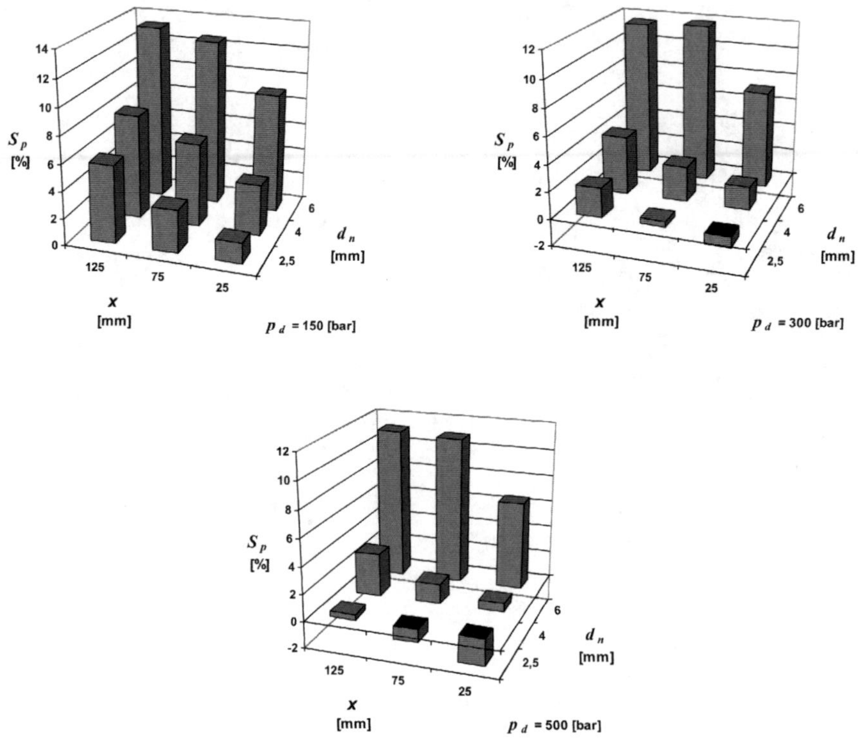

Figure 3. Mean value of transverse mould shrinkage S_p for beams segments of following thickness profiles: 2–6–2, 4–4–4, 6–2–6 and 6–4–2, i.e. excluding results for 2–4–6 beam; d_n – nominal thickness of segment, x – distance from the gate, p_d – packing pressure.

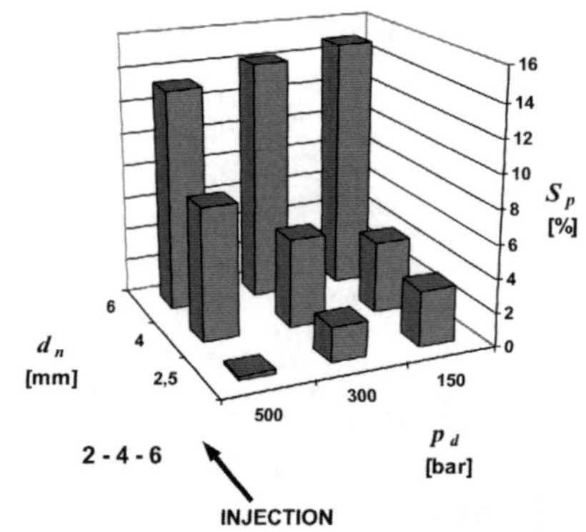

Figure 4. Mean value of transverse mould shrinkage S_p for beam of ascending thickness profile 2–4–6; d_n – nominal thickness of segment, p_d – packing pressure.

The results of the experiment showed that the value of transverse shrinkage in beams segments of the same thickness d_n is similar, regardless of the beam thickness profile. The exception is the case of a beam with increasing thickness along polymer flow direction, marked as 2–4–6 (figure 2, numbers on grey background). Therefore, it was decided that the model of shrinkage (1) will be extended by a new independent variable – beam thickness profile Δd_n along the flow direction of the polymer in the cavity (i.e. along the beam axis):

$$S_p\,(d_n\,, x\,, p_d\,, \Delta d_n) \tag{3}$$

where Δd_n is the difference in the nominal thickness of the segment located farthest from the gate and the thickness of segment nearest the gate (e.g. for the moulding marked 2–4–6, the thickness profile is Δd_n = 6mm–2.5mm = 3.5mm).

6. Fuzzy model of transverse shrinkage

The theory of fuzzy logic was used to approximate the experimental results. The model with a fuzzy structure was constructed and its basic elements shown in figure 5 and figure 6. Three linguistic values (LV) were defined for the first three independent variables of model (3): **S**mall, **M**edium, **B**ig. For the fourth independent variable Δd_n (thickness profile), the following LVs were created: **D**escending, **C**onst., **A**scending. The triangular membership functions were defined – figure 5a and 5b. Figure 5c presents LV and their membership functions for the model output variable S_p.

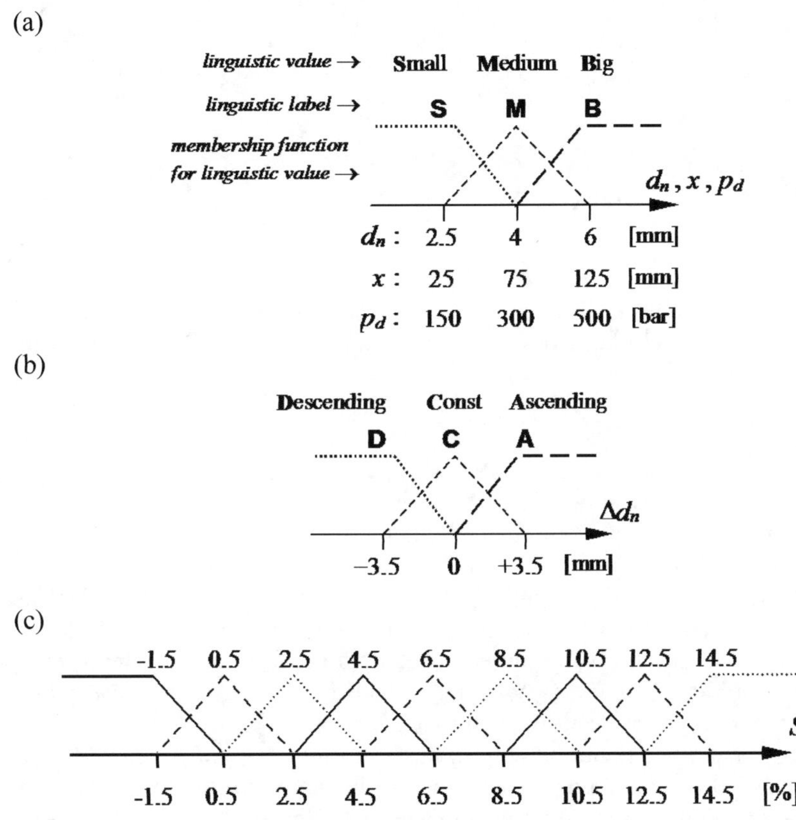

Figure 5. Fuzzy variables of transverse shrinkage model $S_p\,(d_n\,, x\,, p_d\,, \Delta d_n)$; (a) and (b) independent variables, (c) output variable.

The model's fuzzy rules have the following general form:

$$\textbf{if}\ \ d_n \text{ is ...}\ \ \textbf{and}\ \ x \text{ is ...}\ \ \textbf{and}\ \ p_d \text{ is ...}\ \ \textbf{and}\ \ \Delta d_n \text{ is ...}\ \ \textbf{then}\ \ S_p \text{ is ...} \tag{4}$$

where dots in predicates (before "than") should be replaced with linguistic variables of corresponding independent variables (figure 5a and 5b) and "S_p is" (the successor) should be followed by LV for a shrinkage (figure 5c). The complete set of model rules (3×3×3×3 = 81 rules) is presented graphically in figure 6. The linguistic value of the shrinkage S_p in the successor of each rule (4) was selected on the basis of experimental results presented in figure 2. The LV of S_p was chosen for which position of the maximum of membership function was the closest to the measured shrinkage value.

The Mamdani method of reasoning (a typical FLC controller) was applied [3,4]. Centre of area (COA) defuzzification algorithms was used because of its good approximation ability.

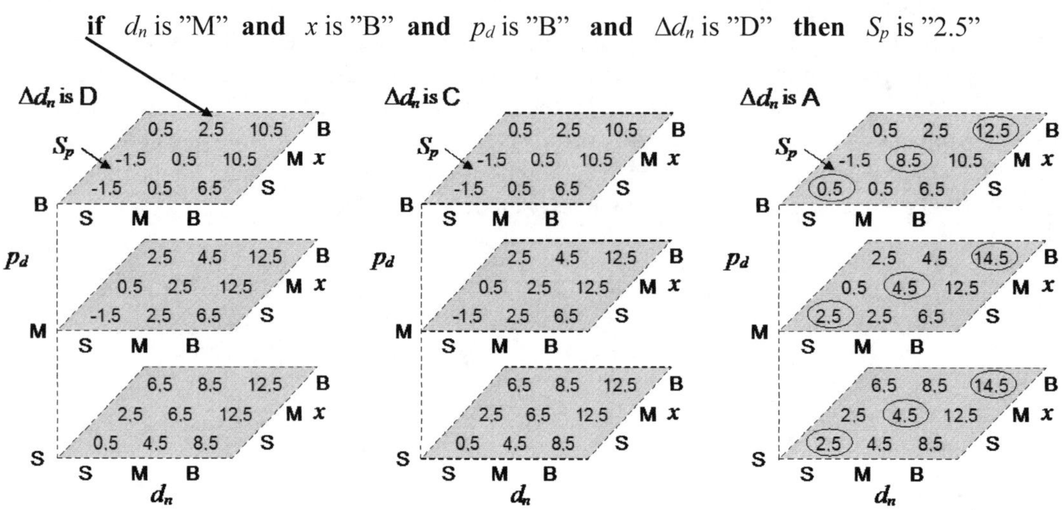

Figure 6. Graphical presentation of 81 fuzzy rules for model $S_p(d_n, x, p_d, \Delta d_n)$. Circles indicate LV in successors of rules which were built based on empirical results for beam of ascending thickness profile 2–4–6 (Δd_n is Ascending).

7. Conclusions

The method of building a fuzzy logic model of multiple input variables was presented. The experimental data of modelled relationship was used to build a set of rules for fuzzy model. An advantage of the described approach is simplicity of including partial expert knowledge into model by modifying chosen rules, which reduces the error of model approximation.

The developed method was used to build a fuzzy logic model of transverse shrinkage of HDPE mouldings. The model includes following variables: nominal dimensions of the mould cavity, profile of the wall thickness of the moulding considered along with the flow path of the plastic, the distance from the gate and the holding pressure (single-stage). The fuzzy structure of the constructed model allows its easy modification (generalisation) to similar injection processes.

The developed shrinkage model can be helpful at the design stage of the moulding geometry and the mould runner system. It is also a good starting point for the optimisation of the process, aimed at compensating the shrinkage in the entire volume of the moulding by adjusting the pressure profile in the holding phase of moulding injection cycle.

8. References

[1] Montgomery D 2013 *Design and analysis of experiments* (New Jersey: John Wiley & Sons Inc.)
[2] Dean A, Morris M, Stufken J and Bingham D 2015 *Handbook of design and analysis of experiments* (Washington: CRC Press)
[3] Pedrycz W 1993 *Fuzzy control and fuzzy systems* (Taunton: Research Studies Press Ltd.)
[4] Ross T J 2010 *Fuzzy logic with engineering applications* (Chichester: John Wiley & Sons Inc.)
[5] Yager R R and Zadeh L A 2012 *An introduction to fuzzy logic applications in intelligent systems* (Berlin: Springer Science & Business Media)
[6] Osswald T A, Turng T and Gramann P J 1996 *Injection molding handbook* (Munich: Hanser

Publishers)
[7] Isayev A I and Kwon K 2009 *Injection molding. Technology and fundamentals* (Munich: Carl Hanser Verlag GmbH & Co. KG) pp 779–808
[8] Osswald T A and Menges G 1996 *Materials science of polymers for engineers* (Munich: Hanser Publishers)
[9] Guevara–Morales A and Figueroa–López U 2014 *J. Mater. Sci.* **49** pp 4399–415
[10] Mohan M, Ansari M N M and Shanks R A 2017 *Polym.–Plast. Technol. Eng.* **56** pp 1–12
[11] Kurt M, Kaynak Y, Kamber O S, Mutlu B, Bakir B and Koklu U 2010 *Int. J. Adv. Manuf. Technol.* **46** pp 571–78
[12] Rosli M U, Ishak M I, Jamalludin M R, Khor C Y, Nawi M A M and Syafiq M A K 2019 *IOP Conference Series: Materials Science and Engineering* **551** 012108
[13] Wilczyński K and Narowski P 2019 *Polym.* **11** 639
[14] Huang M, Liu C T, Zhao Z F and Shen C Y 2011 *Adv. Mater. Res.* **189–193** pp 2121–4
[15] Raz K, Zahalka M and Chval Z 2017 *Manuf. Technol.* **17** pp 79–83
[16] Monka P, Hloch S, Andrej A, Somsak M and Murgas F 2016 *Manuf. Technol.* 2016 **16** pp 561–569
[17] Jansen K M B 1998 *Int. Polym. Process.* **13** pp 309–17
[18] Jansen K M B, Van Dijk D J and Keizer K P 1998 *Int. Polym. Process.* **13** pp 417–24
[19] He W, Zhang Y F, Lee S, Fuj J Y H and Nee A Y C 1998 *J. Intell. Manuf.* **9** pp 17–27
[20] Stączek P, Bogucki M and Płaska S 2003 *Polim.* **48** pp 714–19
[21] Stączek P 2009 *Polim.* **54** pp 42–50

IOP Conf. Series: Materials Science and Engineering **710** (2019) 012011 doi:10.1088/1757-899X/710/1/012011

Predictive modelling for air temperature and humidity in a mushroom production process

M Awtoniuk[1], M Daniun[1], D Komarchuk[2] and S Syrotyuk[3]

[1] Warsaw University of Life Sciences, Institute of Mechanical Engineering, Nowoursynowska 164, 02-787 Warsaw, Poland
[2] National University of Life and Environmental Sciences of Ukraine, Education and Research Institute of Energetics, Automation and Energy Efficiency, 12 Heroyiv Oborony, 03041 Kiev, Ukraine
[3] Lviv National Agricultural University, Faculty of Mechanization and Power Engineering, Volodymyra Velykoho 1, 80381 Dubliany, Ukraine

michal_awtoniuk@sggw.pl

Abstract. Mushrooms (Agaricus bisporus) are produced in a growing room with a heating, ventilation and air conditioning (HVAC) system. Cultivation is cyclical and consists of several successive phases. Every production phase has different environmental requirements. Microclimate is a system with multiple input signals (e.g. control signals of the solenoid valves of HVAC system) and multiple output signals (compost temperature, air temperature, air relative humidity and carbon dioxide concentration inside the growing room). This paper presents model for prediction for air temperature and air relative humidity inside the mushroom growing room. To determine the dynamic behaviour of the microclimate, the parametric identification method was used. Several models with ARX structures were created. A fit index was chosen to evaluate the quality of models. Finally, optimal prediction horizon for each model was determined.

1. Introduction

Poland is the European leader in the white bottom mushroom production. Polish producers export the most mushrooms in the world. It is estimated that every third imported mushroom in the world comes from Poland [1, 2]. Concurrently, growth of the mushroom industry is noticeable. Over the years 2003-2013, mushroom harvest in Poland increased by nearly 62% (from 165 thousand tonnes in 2003 to 267 thousand tonnes in 2013) with a simultaneous increase in a cultivated area only by 25% (from 190 ha in 2003 to 239 ha in 2013) [3]. It suggests an increase in professional experience of technologists, as well as an improvement in technology itself. Not without significance is the involvement of technology increasing the level of automation. Expectations of producers in the field of production process automation are getting higher, the aim of which is to improve product quality and reduce energy consumption of the process. Optimisation of energy consumption requires development of prediction models for controlled processes [4]. In production of mushrooms, a control object is the microclimate of the growing room. The microclimate of the mushroom-growing plant is described by several physical quantities, including compost temperature, air temperature, relative air humidity and carbon dioxide concentration in the air [5-7].

Many research projects have been described in literature with regard to greenhouse microclimate modelling [8-11]. However, it is not possible to transfer these models directly to the field of mushroom cultivation. The main reason is different plant physiology. The signals measured in the greenhouse and the mushroom growing chamber are different. Examples include a role of artificial light and protection against intense sunlight [12]. Due to the fact that mushrooms do not perform photosynthesis, these signals in the model can be omitted. Substantial differences can also be identified with regard to the common microclimatic factors [13]. An example is carbon dioxide concentration. Greenhouse plants require an additional CO_2 supply, while in mushroom cultivation, the concentration of CO_2 is excessive, and it has to be removed from production chamber. The mushroom microclimate model available in literature is usually based on laws of physics [7]. Models of this kind are very accurate, what is very important in simulation studies; however, their complexity makes them impractical for control systems. This is the reason why we decided to build the parametrical models.

Model development for prediction of temperature and relative humidity inside a growing room is the main goal of the paper. An additional goal is to determine the optimal prediction horizon. The model may be used as an element of a predictive control system of the microclimate in the growing room. Additionally, thanks to the prediction of air parameters it will be possible to more precisely schedule technological operations, e.g. mushroom harvest.

2. Materials and methods

The measurement data was obtained from HVAC system of the mushroom growing room. Figure 1 shows diagram of the HVAC system of the mushroom growing room with all measurement points marked. The signals were sampled every 10 minutes. The mushroom farm was located in the city of Nowosielec in Mazowieckie Voivodship, Poland. The parameters describing the mushroom farm are shown in Table 1.

Due to specificity of mushroom production, the cultivation process has been divided into three stages (Figure 2). Each of the stages have different microclimate requirements. Stages were determined in such a way, that they directly preceded the period of mushroom harvesting. The process of harvesting mushrooms is repeated three times during the production cycle, and harvesting periods are called flushes [14]. For each of the designated stages, a local model of the microclimate has been built. Describing the process using several local models is more accurate than using one global model. Determining a global model very often requires many simplifications [15].

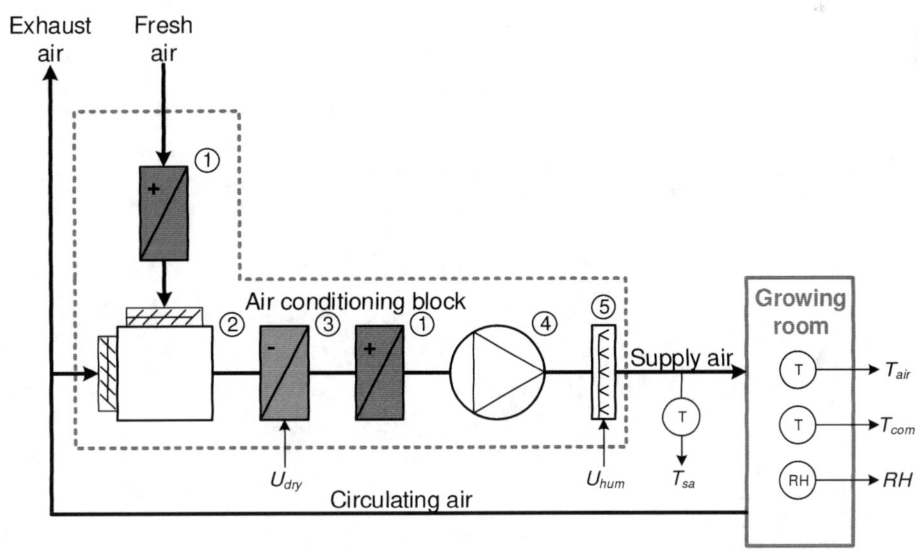

Figure 1. Diagram of HVAC system: 1 – heater; 2– mixing chamber; 3 – cooler (used as dehumidifier also); 4 – fan; 5 – humidifier; U_{dry} – dehumidifier control signal; U_{hum} – humidifier control signal; T_{sa} – supply air temperature; T_{air} – air temperature inside growing room; T_{com} – compost temperature; RH – air relative humidity inside growing room.

Table 1. Cultivation characteristic parameters.

Parameter	Value
Number of growing rooms	6
Growing room length [m]	29
Growing room width [m]	7
Growing room height [m]	4
Growing room volume [m^3]	812
Number of shelves	4
Number of rows	2
Shelves width [m]	1.34
Shelves length [m]	25.2
Crop area [m^2]	270
Growing room volume to crop area ratio	3.007

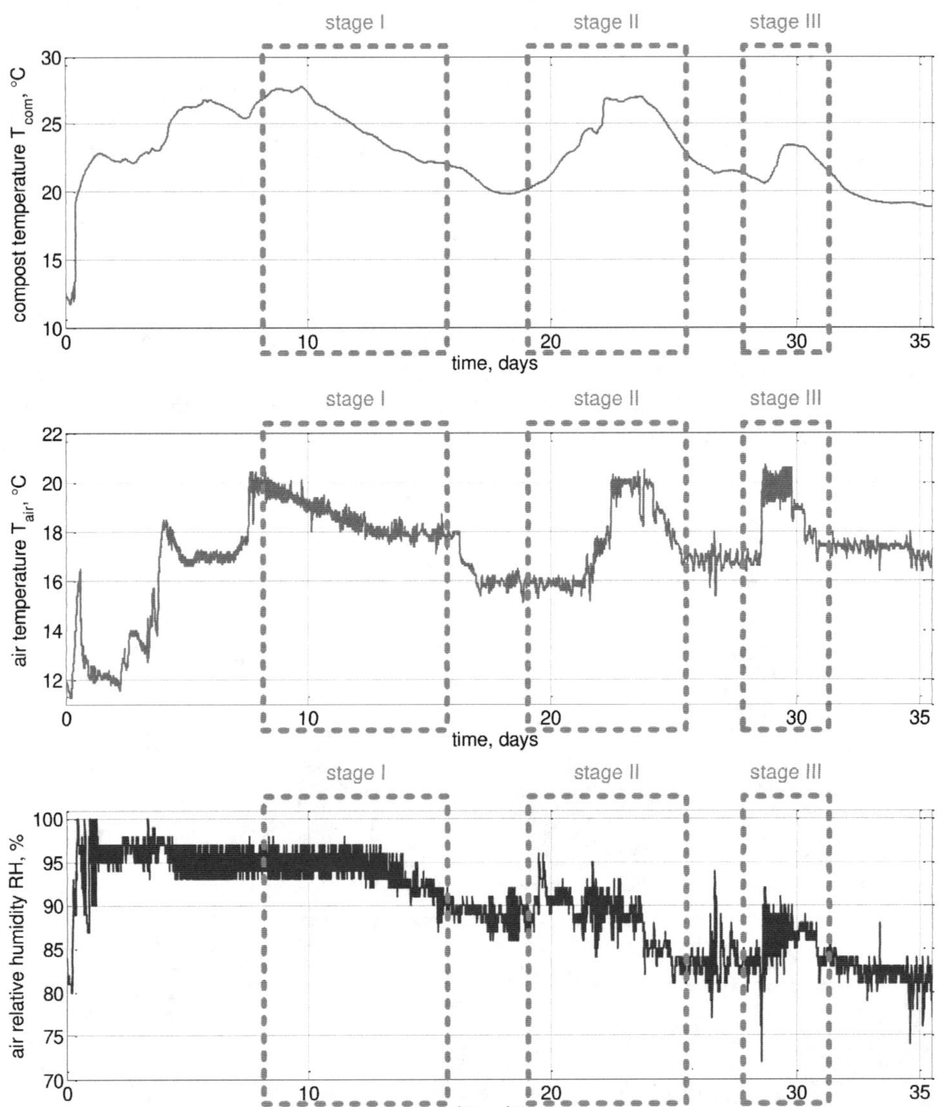

Figure 2. Changes of microclimate parameters during mushroom production process.

3. Model development

The model describing air temperature and air relative humidity inside mushroom growing room has a MIMO structure (many inputs and outputs). There are following input signals: supply air temperature T_{sa}, compost temperature T_{com}, dehumidifier control signal U_{dry} and humidifier control U_{hum}. Air temperature T_{air} and air relative humidity RH signals are coupled, therefore they are also present among the input signals. Every MIMO model can be divided into two MISO models (multiple inputs, one output). We marked MISO models as M_{Tair} and M_{RH} (Figure 3).

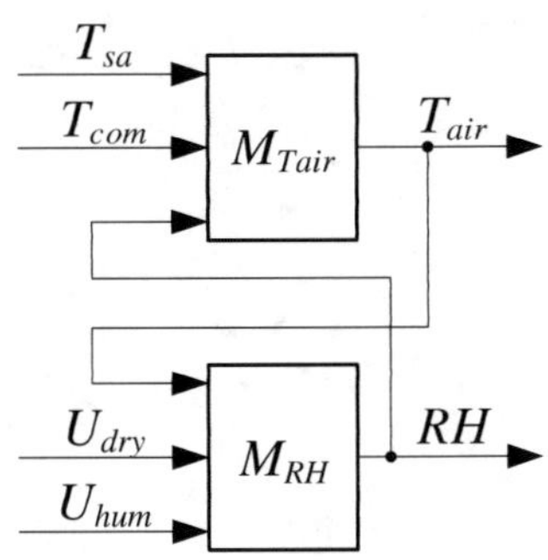

Figure 3. Model block diagram.

ARX model structure was used for each of the models. ARX models are widely used to predict thermal behaviour of microclimate in different buildings, such as greenhouses and offices [9, 11, 16, 17]. The following equations describe models' structure:

$$\begin{aligned} M_{Tair} &: A(z)\cdot T_{air}(t) = B_1(z)\cdot T_{sa}(t-n_{k1}) + B_2(z)\cdot T_{com}(t-n_{k2}) + B_3(z)\cdot RH(t-n_{k3}) + e(t) \\ M_{RH} &: A(z)\cdot RH(t) = B_1(z)\cdot T_{air}(t-n_{k1}) + B_2(z)\cdot U_{dry}(t-n_{k2}) + B_3(z)\cdot U_{hum}(t-n_{k3}) + e(t) \end{aligned} \quad (1)$$

$$A(z) = 1 + \sum_{i=1}^{n_a} a_i z^{-i}, \quad B_1(z) = \sum_{i=1}^{n_{b1}} b_{1i} z^{-i+1}$$

$$B_2(z) = \sum_{i=1}^{n_{b2}} b_{2i} z^{-i+1}, \quad B_3(z) = \sum_{i=1}^{n_{b3}} b_{3i} z^{-i+1}$$

where: $T_{sa}(t-n_{k1})$, $T_{com}(t-n_{k2})$, $RH(t-n_{k3})$ – input signals of M_{Tair} model at time t-n_{k1}, t-n_{k2}, t-n_{k3}; $T_{air}(t)$– output signal of M_{Tair} model at time t; $T_{air}(t-n_{k1})$, $U_{dry}(t-n_{k2})$, $U_{hum}(t-n_{k3})$ – input signals of M_{RH} model at time t-n_{k1}, t-n_{k2}, t-n_{k3}; $RH(t)$ – output signal of M_{RH} model at time t; $e(t)$ – noise signal at time t; z – discrete operator; a_i – coefficients of polynomial A; b_{1i}, b_{2i}, b_{3i} – coefficients of polynomials B_1, B_2, B_3; n_a – degree of polynomial A; n_{b1}, n_{b2}, n_{b3} – degrees of polynomials B_1, B_2, B_3; n_{k1}, n_{k2}, n_{k3} – delays of input signals.

Values of delays n_k were selected by means of the Model–Based Estimation method. In this method, the delay for which the value of the loss function V is the lowest should be searched out [18, 19]. The loss function is equal to

$$V(\boldsymbol{\theta}) = \frac{1}{N}\sum_{t=1}^{N} \varepsilon^2(t) \quad (2)$$

where: $\boldsymbol{\theta}$ – model parameters vector, N – number of measurement data, ε – residuals.

Figure 4 presents selection of delays n_k for the three stages of the M_{Tair} model. Selection was limited to a range of 1-10 samples. Values of delays equal to $n_{k1} = [1 \;\; 3 \;\; 1]$ for stage I, $n_{k2} = [1 \;\; 2 \;\; 1]$ for stage II, and $n_{k3} = [1 \;\; 1 \;\; 1]$ for stage III have been selected.

IOP Conf. Series: Materials Science and Engineering **710** (2019) 012011 doi:10.1088/1757-899X/710/1/012011

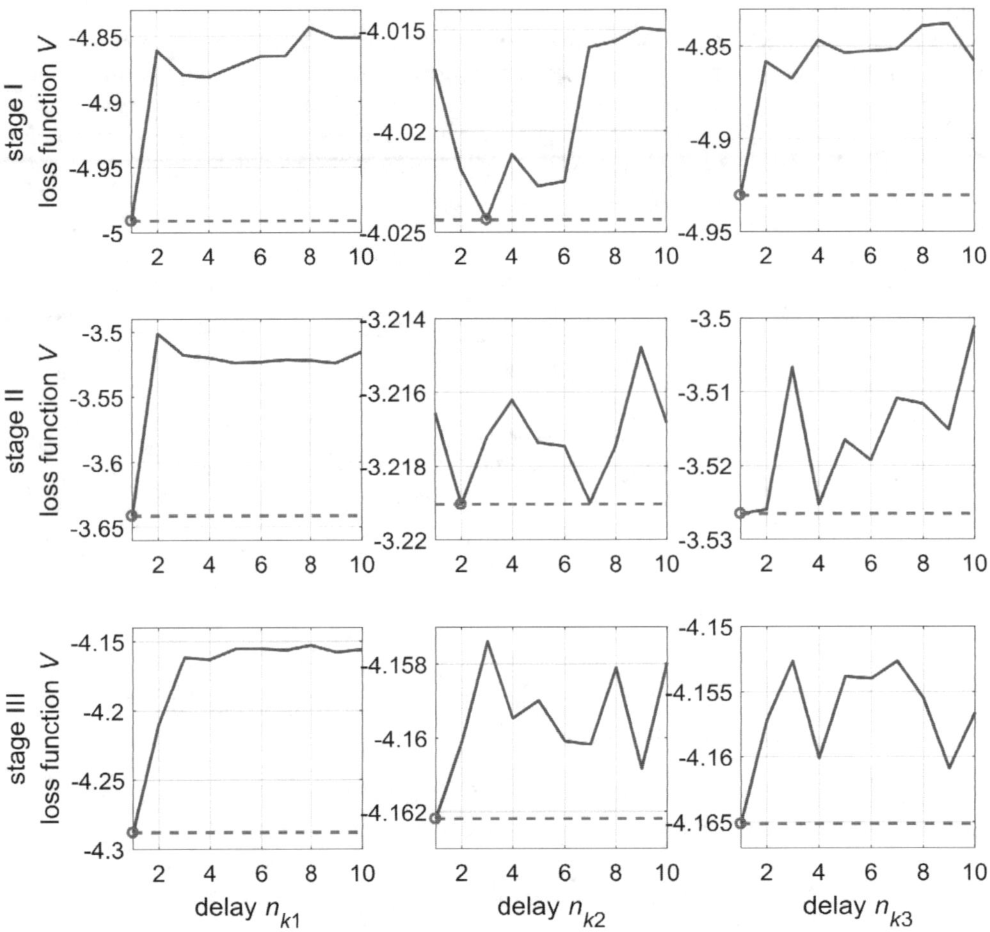

Figure 4. Estimation of delays n_k for the M_{Tair} model.

Selection of delays for the M_{RH} model has been made accordingly to the same procedure. The delays were equal to $n_{k1} = [2 \quad 1 \quad 2]$ for stage I, $n_{k2} = [1 \quad 1 \quad 1]$ for stage II, and $n_{k3} = [3 \quad 1 \quad 1]$ for stage III.

The next stage of model development was to determine its order. The order of the ARX model is determined by the degree of polynomial A, i.e. n_a. Additionally, for each n_a different combinations of models are possible due to the degrees of polynomials B, i.e. n_{b1}, n_{b2}, n_{b3}. A model within a range of $n_a, n_{b1}, n_{b2}, n_{b3} \in \{1,...,10\}$ was the most desired. An additional limitation $n_a \geq \max(n_{b1}, n_{b2}, n_{b3})$ was assumed to make the developed ARX model a proper transfer function. Thanks to this, the number of combinations of possible models was limited from 10 000 to 3 025. Goodness of fit was evaluated by means of *fit* index, which can be calculated as follows

$$fit = \left(1 - \frac{\left\| y(t) - \hat{y}(t) \right\|_2}{\left\| y(t) - \bar{y} \right\|_2}\right) \cdot 100\% \tag{3}$$

where: $y(t)$ – output signal (measured), $\hat{y}(t)$ – predicted output signal, $\bar{y}$ – mean of output (measured) signal.

For each stage, a learning and testing dataset was determined. The *fit* index was calculated for the one-step-ahead prediction mode for both learning and testing dataset. The model order has been selected by analysing the fitness of many models with different orders (Figure 3). The conducted analysis was based on the principle of parsimony, i.e. fewer complex models were preferred (the lowest possible n_a value), with the best possible *fit* (the highest possible *fit* index value) [20]. High *fit* index value for a testing dataset was a priority. To improve readability of the figure, only the models with the best *fit* for each order n_a were presented (Figure 5).

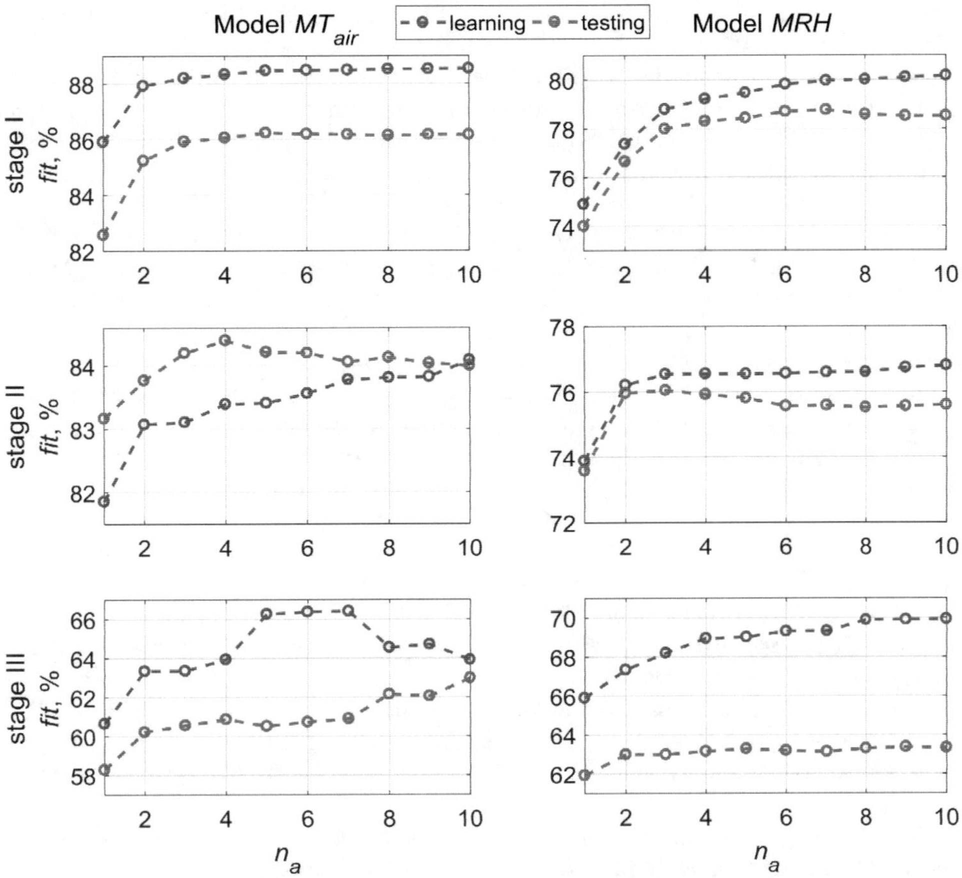

Figure 5. Influence of the model order n_a on the *fit* value for learning and testing dataset.

Table 2 presents detailed information about the structure of selected models. It also includes values of the *fit* index calculated for the learning dataset (one-step-ahead prediction), as well as for the testing dataset (*k*-step-ahead prediction).

Table 2. Model structure and *fit* index values for M_{Tair} and M_{RH} models.

Model structure $[n_a\ n_{b1}\ n_{b2}\ n_{k1}\ n_{k2}\ n_{k3}]$	*fit* index [%]						
	learning dataset	testing dataset (*k*-step-ahead prediction)					
		k=1	*k*=2	*k*=3	*k*=4	*k*=5	*k*=6
model M_{Tair}							
stage 1: [3 3 1 1 1 3 1]	88,2	85,9	84,6	83,2	81,5	79,8	78,3
stage 2: [2 2 2 2 1 2 1]	83,1	83,8	78,9	76,1	73,5	70,0	67,2
stage 3: [2 2 2 1 1 1 1]	63,3	60,2	49,1	44,7	42,6	40,5	38,9
model M_{RH}							
stage 1: [3 1 3 3 2 1 2]	78,9	78,0	76,8	76,4	75,1	74,6	73,9
stage 2: [2 1 2 1 1 1 1]	76,2	76,0	69,4	64,8	60,4	56,5	53,3
stage 3: [2 2 2 1 3 1 1]	67,3	63,0	51,7	45,0	38,8	33,3	28,9

IOP Conf. Series: Materials Science and Engineering **710** (2019) 012011 doi:10.1088/1757-899X/710/1/012011

4. Conclusions

The aim of this paper, i.e. development of a model for prediction of air temperature and relative humidity has been achieved. The values of *fit* index for stages I and II are at a similar level and are much higher than for stage III. This suggests difficulties with the development of a model describing the microclimate in the final stage of production. The reason for these difficulties may be the changing over time force of crops impact on the microclimate in the growing room, e.g. decreasing biological potential or decreasing compost activity.

In most cases, *fit* index values for the M_{RH} model were lower. This may be influenced by the selection of input signals of the model in the form of control signals of humidifier U_{hum} and dehumidifier U_{dry}. These signals can be replaced by an additional measurement of humidity of the air supplied to the growing room. It could also improve the M_{RH} model *fit* index. Unfortunately, this measurement is not currently carried out in production practice.

The model for stage I retained acceptable predictive properties for the whole tested predictive horizon (up to 6 samples, which for 10 minutes sampling time corresponds to an hour predictive signal value). For stage II, the prediction horizon should be shortened to 4 and 2 samples for M_{Tair} and M_{RH} respectively.

The results indicate that the developed model can be used in the design of automatic control systems in practice. Models can be a part of predictive control systems. The next stage of research will be the development of an optimiser. The task of the optimiser will be to calculate the value of control signal of the actuators in such a way as to ensure the optimum value of objective function. The objective function should be multi-criteria, so as to combine in itself to maximise the control quality and minimise the energy consumption of the process.

5. References

[1] Gaze R 2005 *The Mushroom Journal* **670** 11–13
[2] McCarty T 2010 *Mushrooms: Industry and Trade Summary* (Washington: United States International Trade Commission)
[3] Nosecka B 2015 *Rynek owoców i warzyw* **53** (in Polish)
[4] Orynycz O and Wasiak A 2019 *MATEC Web of Conferences* **252** 06013
[5] Han J-H, Kwon H-J, Yoon J-Y, Kim K, Nam S-W and Son J E 2009 *Biosystems Engineering* **104** 417–424
[6] Hayes R 1991 *Science and Cultivation of Edible Fungi*, ed M Maher (Rotterdam: Balkema) p 307
[7] Wachowicz E and Woroncow L 2010 *Technical Sciences* **13** 30–39
[8] Cooman A and Schrevens E 2006 *Biosystems Engineering* **94** 517–524
[9] Patil S L, Tantau H J and Salokhe V M 2008 *Biosystems Engineering* **99** 423–431
[10] Uchida F H and Pieters J G 2004 *Neurocomputing* **56** 423–428
[11] Uchida F H, Pieters J G and Deltour J M 2003 *Biosystems Engineering* **84** 147–157
[12] Cooman A and Schrevens E 2007 *Biosystems Engineering* **96** 249–255
[13] Rodríguez F, Berenguel M, Guzmán J and Ramírez-Arias A 2015 *Modeling and Control of Greenhouse Crop Growth* (Cham: Springer International Publishing)
[14] Sakson N 2013 *Pieczarka Uprawa Intensywna* (Poznań: PWRiL) (in Polish)
[15] Janiszowski K 2015 *Identyfikacja modeli parametrycznych w przykładach* (Warszawa: Akademicka Oficyna Wydawnicza EXIT (in Polish)
[16] Obstawski P, Bakoń T and Czekalski D 2019 *Applied Thermal Engineering* **148** 420–429
[17] Mustafaraj G, Chen J and Lowry G 2010 *Energy and Buildings* **42** 348–356
[18] Ljung L 1999 *System Identification: Theory for the User* (Englewood Cliffs: Prentice Hall)
[19] Tangirala A K 2015 *Principles of system identification: theory and practice* (Boca Raton: CRCPress)
[20] Söderström T and Stoica P 1997 *Identyfikacja systemów* (Warszawa: PWN) (in Polish)

IOP Conf. Series: Materials Science and Engineering **710** (2019) 012012 doi:10.1088/1757-899X/710/1/012012

Modelling hydrodynamic characteristics of the underwater glider based on Computational Fluid Dynamics

K Strycznieiwcz[1], W Strycznieiwcz[2] and R Szczepaniak[3]

[1] Warsaw University of Technology, Department of Aerodynamics, Institute of Aeronautics and Applied Mechanics, Nowowiejska 24, 00-665 Warsaw, Poland
[2] Łukasiewicz Research Network - Institute of Aviation, Department of Aerodynamics, Al. Krakowska 110/114, 02-256 Warsaw, Poland
[3] Polish Air Force University, Dywizjonu 303 35, 08-521, Dęblin, Poland

kamila.stryczniewicz@meil.pw.edu.pl

Abstract. The underwater glider is buoyancy-propelled Autonomous Underwater Vehicle (AUV). Their propulsion relies upon changing their buoyancy with internal pumping systems enabling them upward and downward motion. In the presented paper, the forward gliding motions are generated by hydrodynamic lift forces exerted on a pair of hydroplanes attached to a glider hull. The hydrodynamic characteristics of a glider were determined using Computational Fluid Dynamics (CFD). A 3D model was created for the simulation of flow behaviour in the Baltic Sea. The lift and drag forces distribution at different angles of attack was studied for Reynolds number 10^5. The flow velocity was 0.5 m·s-1 and the angle of the attack varied from −30° to 30° in steps of 2°. Results from the CFD constitute the basis for the calculation the equations of motions of the glider in 6 degrees of freedom. The vehicle motion simulation will be achieved through numeric integration of the equations of motion. This work will contribute to dynamic modelling and three-dimensional motion simulation of the torpedo-shaped underwater glider.

1. Introduction

The inspiration to this work has come from intensive research in the field of modelling the dynamics of underwater vehicles, their manoeuvrability and performance, driven by their increasing applications in oceanography, military, natural environment safety. The research objective is to study the general dynamics model of an underwater vehicle with the aid of methods of the CFD for accurate assessment of dynamic effects between the vehicle and the external seawater environment. The vehicles are inertia propelled, i.e. move due to mass or configuration of mass change, or hybrid propelled, i.e. using inertia and engines to move. The steady-state gliding motion is defined by Zhang et al., 2012, as: "a particular change in buoyancy and fixed position of moving mass, the state variables of the glider remain unchanged and angular velocity remains zero for its sawtooth gliding motion".

The paper focuses on inertia-based propelled AUV, i.e. not equipped with thrusters. AUVs constitute, then, a class of underwater vehicles not equipped with external propulsion systems. There are numerous types of such AUV currently available, such as Slocum [1], Spray [2] or Sea-glider [3]. They are designed for long-range missions, even 9 months on a single mission, consume small energy, are cost-efficient and exhibit significant endurance. They are applied in oceanography, in rescue and cleaning sea missions. Their propulsion relies upon changing the buoyancy with internal pumping systems enabling them upward and downward motions, and their forward gliding motions are generated by the hydrodynamic lift forces exerted on a pair of wings attached to a glider hull. The resulting AUV motion is a "saw-tooth motion" pattern in the vertical plane and a straight line motion in the horizontal plane. Low energy and easy propulsion have enabled for the involvement of various AUV in real applications, where either a single glider or a group [4, 5]. Another propulsion method for AUV is by using a movable internal mass [6, 7]. Other kind of motion of an AUV described in literature is along a spiral [8]. The spiral motion can be obtained by rotating an internal mass, e.g., the battery pack, or a rudder attached to the tail section of the glider.

2. Methods

An adequate model of hydrodynamic interactions is utterly important for a mathematical description of an AUV dynamics. In the late 1960s, Gertler and Hagen [9] proposed a fully analytical yet quite complex set of formulas for hydrodynamic force and moment components in the body-fixed reference frame, which are reminiscent of flight dynamics. Most recently, the CFD methods have become increasingly popular in hydromechanical modelling of AUVs. RANS-based simulations have been used to help calibrate the coefficient-based models [10, 11] or to provide steady-state hydrodynamic characteristics for such models [12-14]. Different turbulence models applied to RANS simulations of flows past a drifting submarine have been investigated and compared [15]. In a similar work [16], the RANS simulations have been used to calculate flow and forces past a submarine performing steady turns.

One of the most advanced applications of the CFD method consists in solving the full set of equations for unsteady turbulent flow (URANS model) simultaneously, i.e. fully-coupled with the equations of motion. It means that either the flow equations need to be solved in a domain which changes in time in *a priori* unknown way or the flow problem needs to be posed and solved in the body-fixed non-inertial reference frame. In the latter case, the far-field conditions become time-dependent and non-uniform. Yet another approach to the CFD-based determination of the hydrodynamic loads is presented in the study by Xiaocui et al. [17]. In this work, the computational domain is divided into two parts: the internal part surrounding the submarine and the "far-field" part around. Each part has been meshed independently. The entire domain moves linearly at an instantaneous velocity of the submarine but only the internal mesh performs rotary motion following the spatial orientation of the vehicle. This way, the problem of deforming meshes is replaced with the interpolation problem at the interface of sliding meshes. The accuracy of force prediction is exceptionally high. In this study Commercial CFD solver ANSYS Fluent 18.2 has been used to solve the steady-state Reynolds averaged Navier Stokes (RANS) equations. The Shear Stress Transport (SST) $k - \omega$ model was chosen for the turbulence model, which is widely used in flow separation problems and likely to be found at the aft of the AUV.

2.1. Geometry

The geometry of the AUV was created in SolidWorks 2018, commercial CAD software, and was designed based on the gliders commonly known in the literature. Our glider has a torpedo-shaped hull with tail shapes and fixed hydroplanes at the rear part. Torpedo-type hull is a cylindrical body with a large ratio between the length and diameter, L/D = 8. This shape provides good properties and is widely used by all major manufacturers of AUVs. The vehicle presented in figure 1 is symmetrical along the XZ and XY planes, for this reason, the calculations were performed for the half of the body to save the calculation costs.

Figure 1. The studied geometry.

IOP Conf. Series: Materials Science and Engineering **710** (2019) 012012 doi:10.1088/1757-899X/710/1/012012

To model the profiles of the bow and stern of the vehicle, *Myring Equations* are commonly used. These theoretical equations describe curves for bow and stern of torpedo bodies generating the smallest possible drag coefficient [21]. The equations are as follows:

- Bow,

$$r_1(x) = \frac{1}{2}D\left[1 - \left(\frac{x-a}{a}\right)^2\right]^{\frac{1}{n}} \tag{1}$$

- Stern,

$$r_2(x) = \frac{1}{2}D - \left[\frac{3D}{2c^2} - \frac{tg\theta}{c}\right](x-a-b)^2 + \left[\frac{D}{c^3} - \frac{tg\theta}{c^2}\right](x-a-b)^3 \tag{2}$$

where all parameters of the equations are geometric (except for parameter *n*) and are shown in figure 2. These equations are used with the purpose of validation the studied geometry according to the optimal hydrodynamic shape.

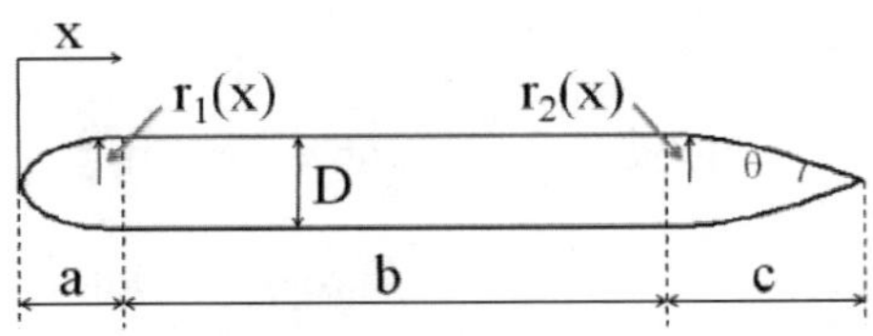

Figure 2. Schematic figure of the AUV hull and geometric parameters of Myring Equations [21].

Table 1. Formatting sections, subsections and subsubsections.

Parameter	Dimension, m
a	0.31
b	1.1
c	0.6
D	0.25

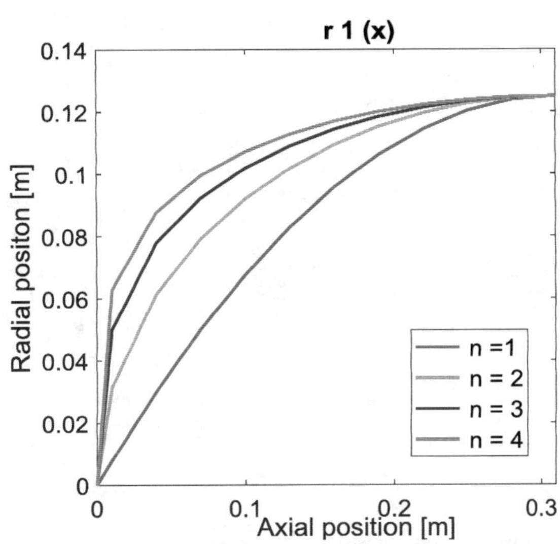

Figure 3. Bow profiles analyzed for different values of parameter n.

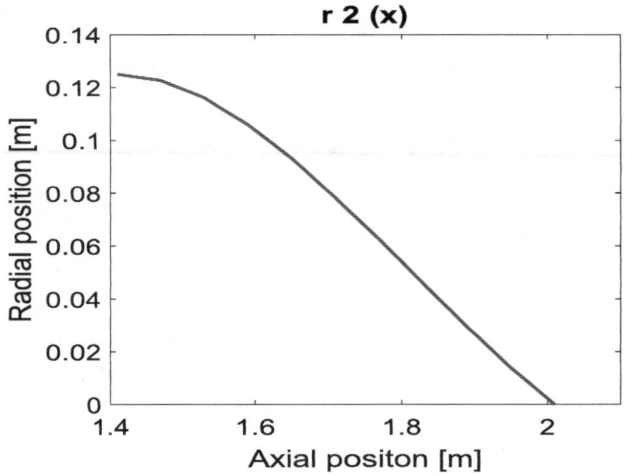

Figure 4. Stern profile for θ=13°.

Myring equations show that the bow of the studied glider is well-fitted if n=2. A higher length-to-diameter ratio (L/D) is desirable for a glider. The predicted value of $L / D = 8$ is sufficient.

2.2. CFD Analysis

The glider under analysis is assigned for the Baltic Sea mission; therefore, the maximum operational depth is 25 m, where the absolute pressure is around 3500 hPa. According to the [22], the density of the water in the Baltic Sea at the 10°C is ρ=1005 kg·m^{-3} and the dynamic viscosity is μ=1.33e-3 kg·m^{-1}·s^{-1}. The kinematic viscosity was specified at ν=1.3e-6 m^2·s^{-1}.

ITTC [21] recommendations for marine CFD applications to specify that the computational domain around the experimental glider, shown in figure 5, should extend La (1-2 L) in the upstream of the leading edge of the body (L_h = 1.3 L) in the radial direction (from the centreline of the body) and L_f=3-5 L in the downstream of the trailing edge of the body. These dimensions of the external domain were examined.

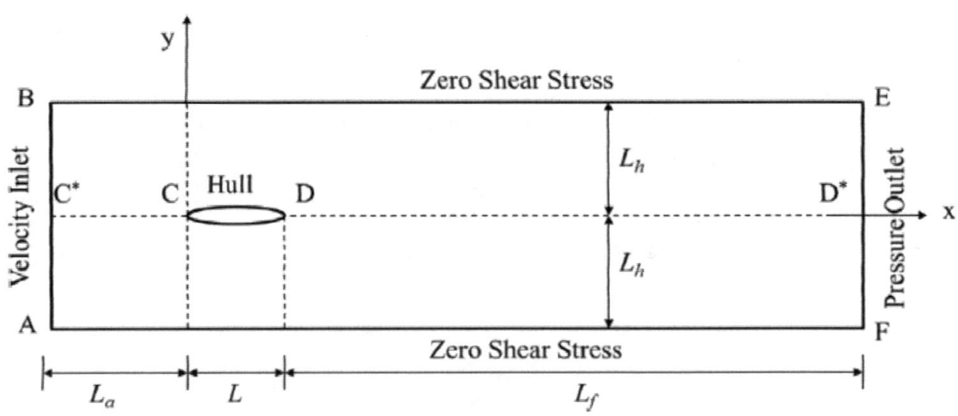

Figure 5. Computational domain for glider with boundary conditions [18].

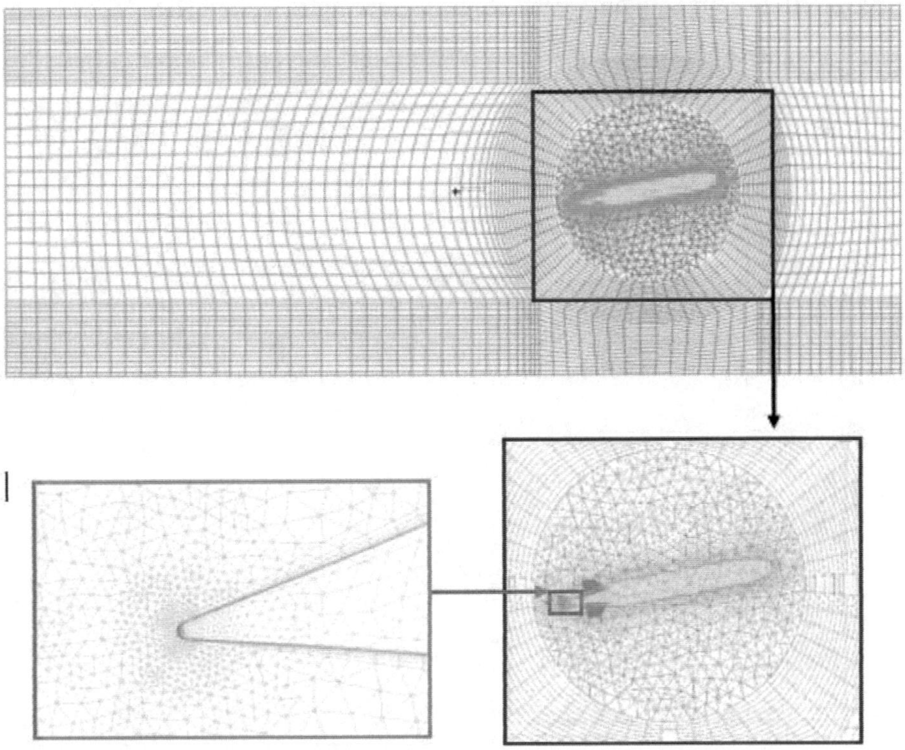

(Total elements = 1373174)

Figure 6. Distribution of the cell around the 3D model of the underwater glider.

The inner domain (cylinder shape) around the body was discretized with the unstructured mesh. The outer domain was the structured mesh. Mesh interface between the two domains was established. The volume tetra mesh was created with the Robust (Octree) method. The mesh should be dense in areas where the flow velocities are sensitive to grid spacing and coarse in other areas. Due to that reason the region around the hull was concentrated. 10 prism layers proved enough to perform flow characteristics in the boundary layer, figure 6. The thickness of the boundary layer was set according to the ITTC, CFD marine recommendation [19]:

$$\delta = 0.035\, L \cdot Re^{-\frac{1}{7}} \tag{3}$$

where: L – the length of the glider [m], Re – Reynold number $= \frac{v \cdot L}{\vartheta}$, (v- velocity of the glider [$m \cdot s^{-1}$]; υ – kinematic viscosity [$m^2 \cdot s^{-1}$]). Relying on [19], the non-dimensional wall distance can be defined in terms of Reynolds number in the following way:

$$\frac{y}{Lpp} = \frac{y^+}{Re\sqrt{\frac{Cf}{2}}} \tag{4}$$

where: y is the first required cell height and C_f is an estimate of the skin friction coefficient, based on the ITTC standard method.

$$C_f = \frac{0.075}{(log_{10}Re-2)^2} \tag{5}$$

IOP Conf. Series: Materials Science and Engineering **710** (2019) 012012 doi:10.1088/1757-899X/710/1/012012

Table 2. Flow parameters.

Parameter	Dimension, m
V, ms^{-1}	0.5
Re	$7.7 \cdot 10^5$
C_f	0.004966

The values of wall distance parameter distribution are presented in figure 7.

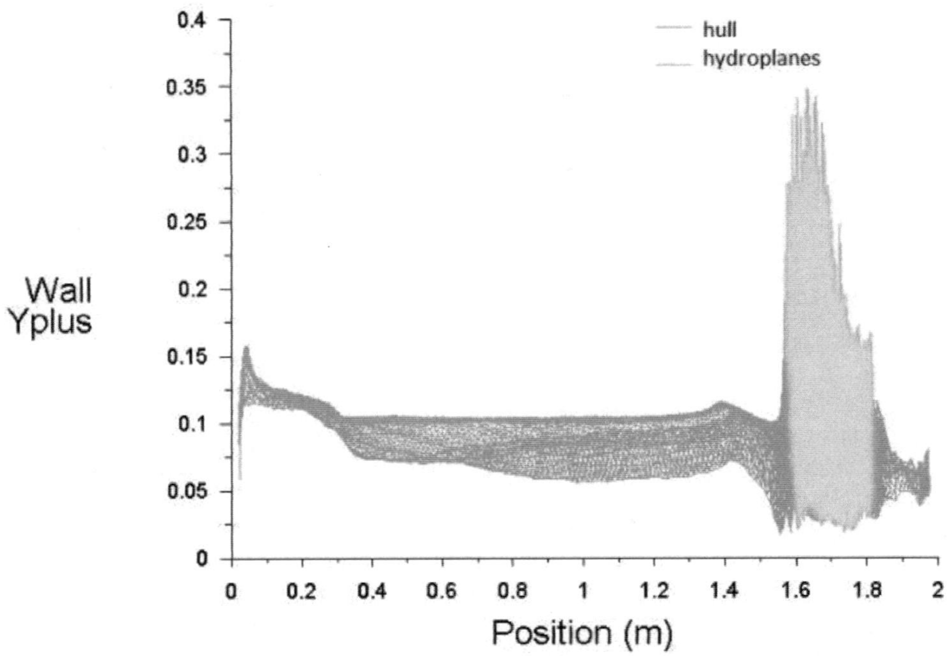

Figure 7. Wall distance parameter distribution.

Commercial CFD solver ANSYS Fluent 18.2 was used to solve the steady-state Reynolds averaged Navier Stokes (RANS) equations. The Shear Stress Transport (SST) k – ω model is chosen for the turbulence model, which is widely used in flow separation problems and likely to be found at the aft of the AUV. The convergence is decided by the standard deviation of the viscous pressure resistance coefficient and frictional resistance coefficient displayed in per cent of the average force. The convergence criterion in the present study is set as 1% for the viscous pressure resistance coefficient and frictional resistance coefficient. Computations are carried out until the steady state is reached [20]. The glider was tested for the constant velocity 0.5 m·s-1 and range of pitch angle ($\alpha = -30°$ to $30°$ in steps of 2°), the drag and lift coefficients are computed for the front reference surface.

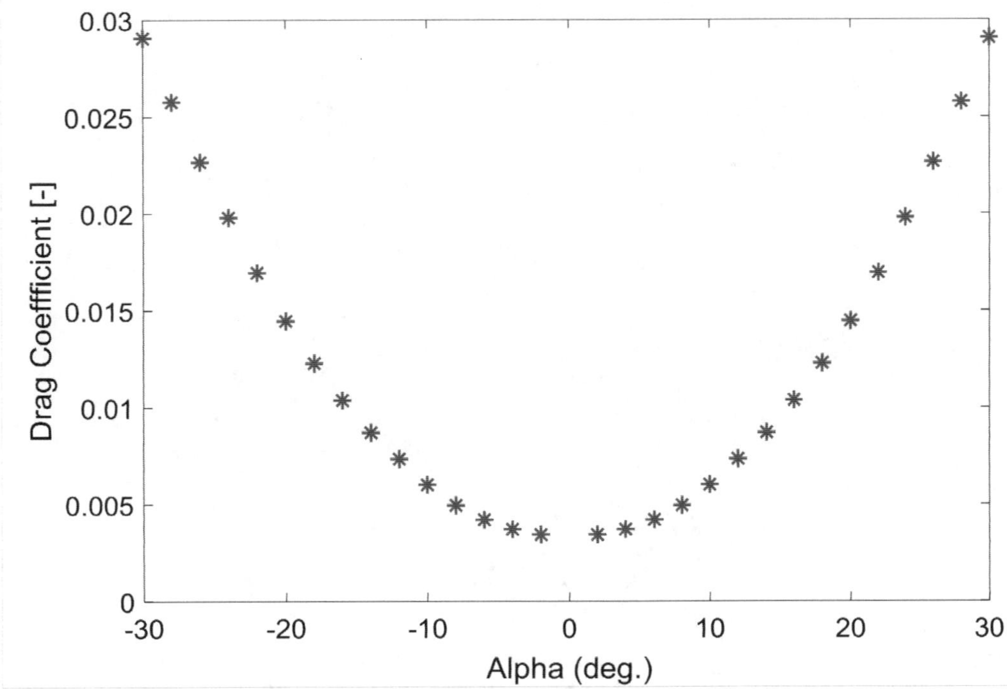

Figure 8. Drag coefficient as a function of pitch angle α.

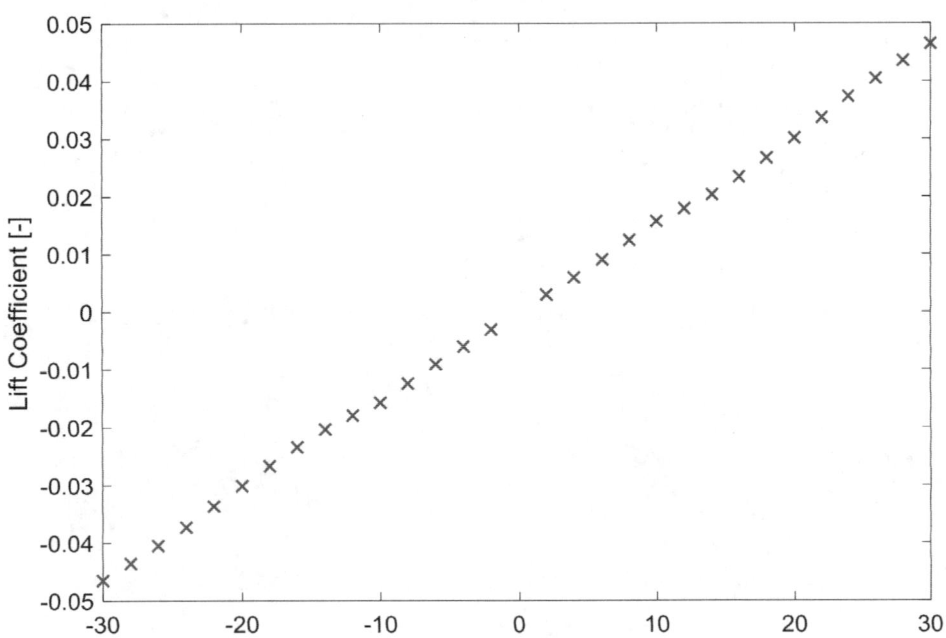

Figure 9. Lift coefficient (Cl) as a function of pitch angle α.

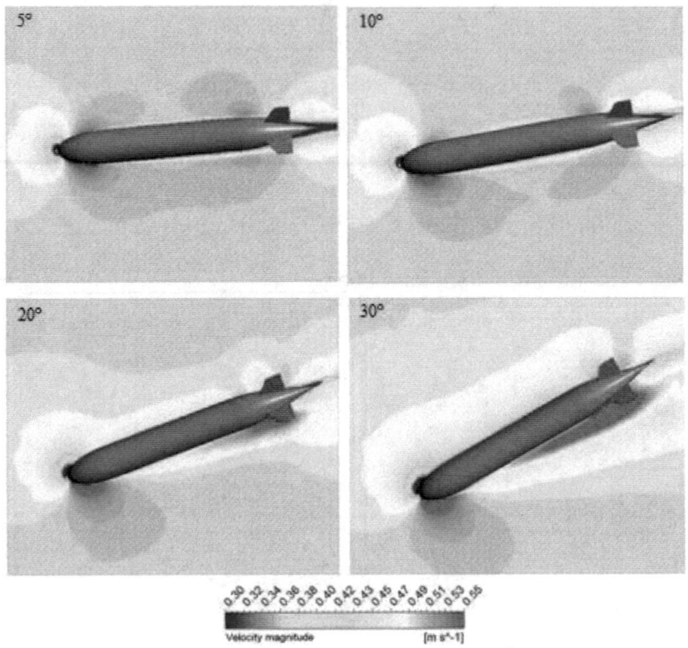

Figure 10. Velocity magnitude around the glider body: V=0.5 m·s-1, pitch angle from 5° to 30°. X0Z geometry plane.

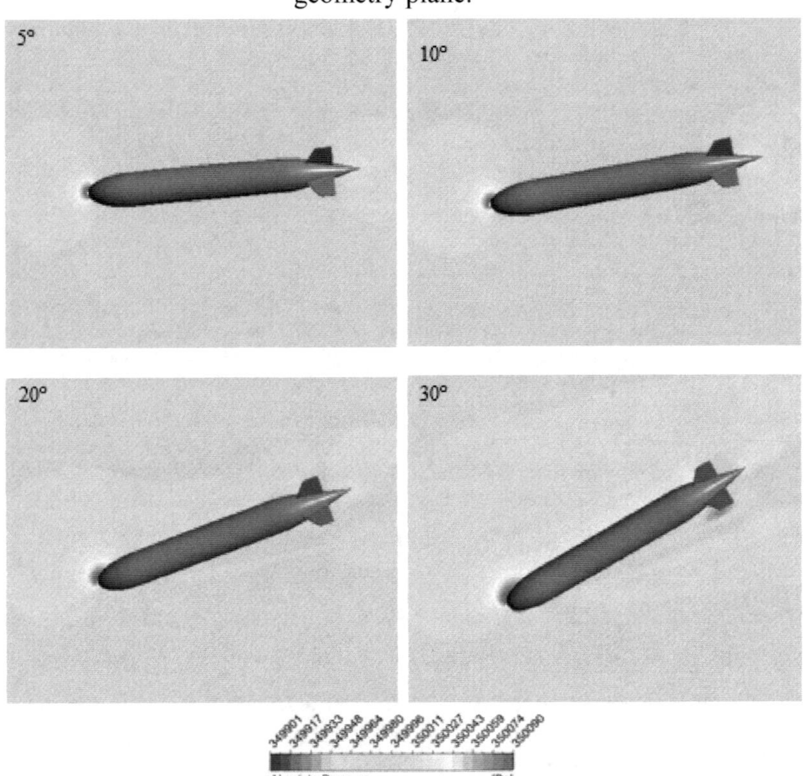

Figure 11. Absolute pressure around the glider body: V=0.5 m·s-1, pitch angle from 5° to 30°. X0Z geometry plane.

3. Conclusions

In this paper, the hydrodynamic single-phase flow around the underwater glider was discussed. The study was related to seawater flow in the turbulent regime by using the ANSYS Fluent R18.2 commercial software. The results show the proper tendency in the hydrodynamic characteristic as a function of pitch angle α. Furthermore, the results were compared with the results presented in the literature [18]. The other force, such as added mass, needs to be considered in the future. The results from the CFD constitute the basis for the calculation of equations of motions of the glider in 6 degrees of freedom. Therefore, vehicle motion simulation will be achieved through numeric integration of the equations of motion. The equations of motion are formulated as Ordinary Differential Equations. The ODE problem will be solved by Runge-Kutta 4th order method in MatLab. This work will contribute to dynamic modelling and three-dimensional motion simulation of a torpedo-shaped underwater glider.

4. References

[1] Webb D C, Simonetti P J and Jones C P SLOCUM, an underwater glider propelled by environmental energy 2001 *IEEE J. Oceanic Eng.* **26** 447–52

[2] Sherman J, Davis R E, Owens W B and Valdes J The autonomous underwater glider ''spray'' 2001 *IEEE J. Oceanic Eng.* **26** 437–46

[3] Eriksen C C, Osse T J, Light R D, Wen T, Lehman T W, Sabin P L, Ballard J W and Chiodi A M 2001 Sea-glider: a long range autonomous underwater vehicle for oceanographic research *IEEE J. Oceanic Eng.* **26**, 424–36

[4] Zhang F, Fratantoni D M, Paley D, Lund J and Leonard N E Control of coordinated patterns for ocean sampling 2001 *Int. J. Control* **80** 1186–99

[5] Leonard N E, Paley D A, Davis R E, Fratantoni D M, Lekien F and Zhang F Coordinated control of an underwater glider fleet in an adaptive ocean sampling field experiment in Monterey bay 2001 *J. Field Robot.* **27** 718–40

[6] Graver J, Leonard N E Underwater glider dynamics and control. In: 12th International Symposium on Unmanned Untethered Submersible Technology, Durham, 2001, 1–14.

[7] Bhatta P Leonard N E Nonlinear gliding stability and control for vehicles with hydrodynamic forcing 2008 *Automatica* **44** 1240–50.

[8] Zhang S, Yu J, Zhang A and Zhang F Spiraling motion of underwater gliders: Modeling,analysis, and experimental results 2013 *Ocean Engineering* **60** 1–13

[9] Gertler M, Hagen G R Standard equations of motion for submarine simulation *Report* 2510 Naval Ship Research and Development Center, June 1967.

[10] Su Y, Zhao J, Cao J and Zhang G Dynamics modeling and simulation of autonomous underwater vehicles with appendages 2013 *J. Marine Sci. Appl.* **12** 45-51.

[11] de Barros E A, Pascoal A, de Sa E Investigation of a method for predicting AUV derivatives 2008 *Ocean Engineering* **35**, 1627–36.

[12] Isa K, Arshad M R and Ishak S A hybrid-driven underwater glider model, hydrodynamics estimation, and an analysis of the motion control 2014 *Ocean Engineering* **81**, 111-29.

[13] Liu F, Wang Y, Niu W, Ma Z and Liu Y Hydrodynamic performance analysis and experiments of a hybrid underwater glider with different layout of wings 2014 *IEEE*, 978-1-4799-3646-5/14

[14] Singh Y, Bhattacharyya S K and Idichandy V G CFD approach to steady-state analysis of an underwater glider 2014 *IEEE*, 978-1-4799-4918-2/14

[15] Phillips A B, Turnock S R and Furlong M Influence of turbulence closure models on the vortical flow field around a submarine body undergoing steady drift 2010 *J. Marine Sci. Techn.* **15** 201-17

[16] Zhang J T Jordan A M, Gerber A G, Gordon A, Holloway L and Watt G D Simulation of the flow over axisymmetric submarine hulls in steady turning 2013 *Ocean Engineering* **57** 180–196

[17] Xiaocui W, Yiwei W, Chenguang H, Zhiqiang H and Ruiwen Y An effective CFD approach for marine-vehicle maneuvering simulation based on the hybrid reference frames method 2015 *Ocean Engineering* **109** 83–92
[18] Singh Y, Bhattacharyya S K, Idichandy V G CFD approach to modelling, hydrodynamic analysis and motion characteristics of a laboratory underwater glider with experimental results 2017 *J. Ocean Eng Sci* **2** 90–119
[19] ITTC, Recommended procedures and guidelines: practical guidelines for ship CFD applications, 7.5, *ITTC*, 2011, pp. 1–18.
[20] Ichihashi N, Ikebuchi T and Arima M *Proc. 18th ISOPE conf* 2008 Canada 156–61. ISBN: 1-880653-68-0
[21] Myring D F A Theoretical Study of Body Drag in Subcritical Axisymmetric Flow 1976 *AEQ* **27** 186-94
[22] Lepparanta M and Myrberg K 2009 *Physical oceanography of the Baltic Sea* (Chichester UK Praxis Publishing Ltd) ISBN 978-3-540-79702-9

Simulation of unsteady flow with cavitation in plastic pipes using the discrete bubble cavity and Adamkowski models

K Urbanowicz[1], A Bergant[2,3] and H F Duan[4]

[1] West Pomeranian University of Technology, Szczecin, Poland
[2] Litostroj Power d.o.o., Ljubljana, Slovenia
[3] University of Ljubljana, Ljubljana, Slovenia
[4] The Hong Kong Polytechnic University, Hong Kong, PR China

kamil.urbanowicz@zut.edu.pl

Abstract. The work presents two modified cavitation models for the simulation of transient flow in pressure plastic pipes. The first model is a discrete bubble cavitation model, the prototype of which was presented by Shu, and the second one is the Adamkowski model. In the latter model, the problem encountered in the classical model (DVCM - discrete vapour cavitation model) related to artificial damping of pulsation, which results from approximate timing of cavity opening and collapse. In both models, the corrected efficient calculation of Zielke convolution integral was used to simulate the unsteady wall shear stresses. The numerical results from the two models agree well with the results of measurement in the literature.

1. Introduction

Transient cavitating flow often occurs in hydraulic piping systems that were not thoroughly analysed at the design stage. This is often the result of abandoning the detailed numerical analysis of possible consequences resulting from the dynamics of flows. Designers focus mostly on stationary flows and the only dynamic equation applied is the simplified Joukowsky formula for the determination of appropriate wall thickness. In fact, not only high pressures but also low pressures resulting from water hammer are a threat. In conditions of vapour pressure, the liquid stream column separation occurs. Two types of transient cavitation can be distinguished in pressure pipelines [1]. The first one is gaseous cavitation, which is a slow-changing phenomenon, the air bubbles emitted from the gas are dissolved in the liquid as a result of the pressure drop (the secretion lasts for a second, and the possible dissolution takes even longer). Gaseous cavitation results in an average reduction of the pressure wave speed, therefore, it helps increase the damping of pressure waves. The above features are conducive to securing the system, hence, this type of cavitation is beneficial even if it does not occur in a steady state in the analysed system. The second type of cavitation, the vaporous cavitation, takes place when the pressure drops to the vapour pressure. This is a fast-changing phenomenon, i.e. the evaporation requires only microseconds to form and to disappear given that there is a sudden increase in pressure. This type of cavitation is responsible for the appearance (in analysed hydraulic systems) of secondary pressure waves, which result from the implosion of vapour areas created in the pipe. In contrast, the interference between the primary (resulting from a fast closure of the valve) and these secondary pressure waves can result in pressure surges exceeding even twice the maximal pressure calculated from the Joukowsky formula. In practice, strengthening and wave suppression depend on the geometry and material properties of the analysed hydraulic systems, initial conditions prevailing in the system before the occurrence of unsteadiness as

IOP Conf. Series: Materials Science and Engineering **710** (2019) 012013 doi:10.1088/1757-899X/710/1/012013

well as boundary conditions taking place after the occurrence of unsteady flow. When implosion of vapour bubbles occurs at the wall of the pipe, cavitation erosion occurs: when the bubbles collapse (the vapour areas decrease until they completely disappear) the local velocity of the separating phase edges increases rapidly. The collision of the stream surrounding these edges in the last stage of closure is a source of large local pressures, which may be much higher than the pressure calculated from the Joukowsky formula.

Due to the above-mentioned impacts of cavitation on hydraulic systems, there is a great need to analyse them at the design stage. In this paper, a one-dimensional flow modelling was undertaken in a simplified hydraulic system consisting of a pressure vessel, a pipeline and a valve. From the literature on unsteady flows occurring in elastic pipes [2-8], a number of models can be distinguished, including discrete vapour cavity model (DVCM), discrete gas cavity model (DGCM), discrete Adamkowski cavity model (DACM) and discrete bubble cavity model (DBCM). DVCM [3] and DGCM [2,8] have already been modified and used to model transient cavitating flows occurring in plastic pipes [9-12]. In this paper, the modified DACM and DBCM models will be developed and presented. Unsteady hydraulic resistance (skin friction loss) will be modelled in accordance with the improved solution presented in [13,14]. The calculated results obtained from the modified DBCM and DACM models will be compared with the experimental results found in the literature.

2. Modelling transient cavitating pipe flow

The unsteady flow of liquids in plastic pipes can be described by the system of equations [15]:

$$\begin{cases} \frac{1}{\rho_l c_0^2}\frac{\partial p}{\partial t} + \frac{\partial v}{\partial x} + \Xi \int_0^t \frac{\partial p(u)}{\partial t} \cdot w_J(t-u)du = 0, \\ \rho_l \frac{\partial v}{\partial t} + \frac{\partial p}{\partial x} + \frac{4\rho_l}{D}\left(\frac{v|v|}{8} f + \frac{4\nu_l}{D}\int_0^t \frac{\partial v}{\partial t}(u) \cdot w(t-u)du\right) = 0, \end{cases} \tag{1}$$

where p – pressure $[Pa]$, v – mean cross-section velocity $[m/s]$, t – time $[s]$, x – distance along the pipe $[m]$, ρ_l – liquid density $[kg/m^3]$, c_0 – pressure wave speed $[m/s]$, D – inner pipe diameter [m], f – Darcy-Weisbach friction coefficient $[-]$, ν_l – liquid kinematic viscosity $[m^2/s]$, $w(t-u)$ – weighting function $[-]$, $w_J(t-u)$ – time derivative of pipe material creep compliance function $[s^{-1}Pa^{-1}]$, $\Xi = \frac{D}{e}\xi$ – enhanced ξ parameter $[-]$, e – thickness of pipe wall $[m]$, ξ – parameter describing support condition of pipe $[-]$.

Comparing the above system of equations and the system describing the flow in metal conduits [14], the only difference between the systems appears in the equation of continuity (the first of the equations (1)) with an additional convolution integral term describing a retarded strain rate in the plastic pipe wall. The solution of the above system for an idealised case of flow in which the gas areas were not able to form can be found in the literature [15,16].

In this paper, we extend to plastic pipes the following two models: Shu's bubble cavitation model (discrete bubble cavitation model – DBCM)[4], and the Adamkowski model [17,18] (discrete Adamkowski cavitation model – DACM). TDACM is based on the above system of equations (1) assuming only a specific solution defining the average value of flow velocity at the nodes in which cavitation occurs. The DBCM needs a more detailed explanation.

2.1. Discrete bubble cavity model (DBCM)

The equation of motion of two-phase vaporous cavitating flow in the horizontal pipeline is:

$$\frac{\partial}{\partial t}(\rho_m v_m A) + \frac{\partial}{\partial x}(\rho_m v_m^2 A) + A\frac{\partial p}{\partial x} + \pi D \tau_m = 0 \tag{2}$$

where A – pipe cross-sectional area $[m^2]$, v_m – mixture velocity $[m/s]$, τ_m – wall shear stress of two-phase mixture $[Pa]$.

The equation that describes the mixture density ρ_m is:

$$\rho_m = \alpha\rho_v + (1-\alpha)\rho_l. \tag{3}$$

where α – volumetric fraction of liquid phase $[-]$, ρ_v – density of vapour phase $[kg/m^3]$, ρ_l – density of liquid phase $[kg/m^3]$.
After differentiation of equation (2) it follows:

$$\frac{\rho_m v_m}{A}\frac{dA}{dt} + \rho_m \frac{dv_m}{dt} + \rho_m v_m \frac{\partial v_m}{\partial x} + v_m \frac{d\rho_m}{dt} + \frac{\partial p}{\partial x} + \frac{4}{D}\tau_m = 0 \tag{4}$$

Writing the continuity equation separately for the vaporous and liquid phase

$$\begin{cases} \frac{\partial}{\partial t}(\rho_v(1-\alpha)A) + \frac{\partial}{\partial x}(\rho_v(1-\alpha)Av_v) = 0 \\ \frac{\partial}{\partial t}(\rho_l \alpha A) + \frac{\partial}{\partial x}(\rho_l \alpha A v_l) = 0 \end{cases} \tag{5}$$

where v_v – average velocity of vapour flow [m/s] and v_l – average velocity of liquid flow [m/s].
By adding the above equations to each other, one gets:

$$\frac{\partial}{\partial t}(\rho_v(1-\alpha)A + \rho_l \alpha A) + \frac{\partial}{\partial x}(\rho_v(1-\alpha)Av_v + \rho_l \alpha A v_l) = 0 \tag{6}$$

Next assuming the non-slip flow, in which the vapour phase has the same velocity as the liquid one $v_v = v_l = v_m$:

$$\frac{\partial}{\partial t}A\rho_m + \frac{\partial}{\partial x}(Av_m\rho_m) = 0 \tag{7}$$

After differentiation:

$$A\frac{d\rho_m}{dt} + \rho_m \frac{dA}{dt} + A\rho_m \frac{\partial v_m}{\partial x} = 0 \tag{8}$$

From the above equation two useful forms of this equation can be derived:

$$\frac{1}{\rho_m}\frac{d\rho_m}{dt} + \frac{1}{A}\frac{dA}{dt} + \frac{\partial v_m}{\partial x} = 0 \tag{9.1}$$

and

$$v_m \frac{d\rho_m}{dt} + \frac{\rho_m v_m}{A}\frac{dA}{dt} + v_m \rho_m \frac{\partial v_m}{\partial x} = 0 \tag{9.2}$$

Equation (9.2) makes three terms disappear in equation (3). The resulting final form of the equation of motion is similar to the single-phase flow one:

$$\rho_m \frac{dv_m}{dt} + \frac{\partial p}{\partial x} + \frac{2}{R}\tau_m = 0 \tag{10}$$

In plastic pipes [19]:

$$\frac{d\rho_l}{dt} = \frac{\rho_l}{K_l}\frac{dp}{dt}, \quad \frac{d\rho_v}{dt} = \frac{\rho_v}{K_v}\frac{dp}{dt} \quad \text{and} \quad \frac{1}{A}\frac{dA}{dt} = \frac{\Xi}{E}\frac{dp}{dt} + 2\frac{d\varepsilon_r}{dt} \tag{11}$$

where: E – Young modulus [Pa], K_l – bulk modulus of liquid phase [Pa] and K_v – bulk modulus of vapour phase [Pa].

The total derivative of equation (3) is:

$$\frac{d\rho_m}{dt} = \alpha\frac{d\rho_l}{dt} + (\rho_l - \rho_v)\frac{d\alpha}{dt} + (1-\alpha)\frac{d\rho_v}{dt} \tag{12}$$

Using equations (11) and (12) in equation (9.1) yields:

$$\left[\rho_m \frac{\Xi}{E} + \frac{\alpha\rho_l}{K_l} + \frac{(1-\alpha)\rho_v}{K_v}\right]\frac{dp}{dt} + (\rho_l - \rho_v)\frac{d\alpha}{dt} + 2\rho_m \frac{d\varepsilon_r}{dt} + \rho_m \frac{\partial v_m}{\partial x} = 0 \tag{13}$$

The first term on the left-hand side of the above equation, in square brackets, is a formula for the pressure wave speed. Due to the fact that in this paper the numerical solution will be based on a constant rectangular grid, the value of the pressure wave propagation velocity will be determined for the flow occurring in the initial time. Subsequently, we deal with a single-phase liquid flow, which determines the value of the liquid phase concentration coefficient $\alpha_0 = 1$. The same square bracket is reduced to the form:

$$c_0^{-2} = \left[\rho_l\left(\frac{\Xi}{E} + \frac{1}{K_l}\right)\right] \tag{14}$$

Therefore, the equation of continuity can be finally described as:

$$\frac{1}{c_0^2}\frac{dp}{dt} + (\rho_l - \rho_v)\frac{d\alpha}{dt} + 2\rho_m\frac{d\varepsilon_r}{dt} + \rho_m\frac{\partial v_m}{\partial x} = 0 \tag{15}$$

In non-slip flows, the share of the dispersed phase is of the statistical nature, i.e. the volume α and mass concentration c_m are equal to the relevant proportions of dynamic nature, i.e. transport concentration and the degree of dryness [20]; then the following relationship applies: $v_m = \frac{v}{\alpha}$, where v is superficial velocity of the liquid phase.

2.2. Discrete Adamkowski Cavity Model (DACM)

The main assumption of the Adamkowski model [17,18] is that a segment with few vapour zones may be replaced with only one vaporous zone, which is equivalent with regard to the mass and energy conservation laws. When the pressure falls to a vapour pressure p_v in a given cross-section, the conservation law requires the equality between the cavity volume in the corresponding pipe segment and the sum of the cavity volumes in the original pipe segment. This law also requires the wave period along the pipe length to be preserved. In the intermediate cross-sections "j" of the pipeline, two streams of the flowing liquid with velocities v_j^+ and v_j^- are replaced with one continuous fluid stream, represented with the velocity v_j. The energy conservation law requires that the sum of the kinetic and potential energies of elementary liquid masses separated by a vaporous zone has to be equal to the total energy following the reconnection of the two separate liquid streams, which is why:

$$dm\left[\frac{(v_j^+)^2}{2} + gz_j^+ + \frac{(v_j^-)^2}{2} + gz_j^-\right] = dm\left[\frac{(v_j)^2}{2} + gz_j^- + \frac{(v_j)^2}{2} + gz_j^-\right] \tag{16}$$

From the above equation the formula for mean velocity in intermediate cross-section is given by:

$$v_j = \sqrt{0.5\left[(v_j^+)^2 + (v_j^-)^2\right] + g(z_j^+ - z_j^-)} \tag{17}$$

In the case of the horizontal pipeline, the difference $(z_j^+ - z_j^-)$ is equal to 0. The z_j^+ and z_j^- are altitudes of the centres of cross-sectional surfaces limiting the cavity within the computational cross-section "j" from the right and left sides, respectively.

As in the water hammer case, the mean flow direction changes numerous times until final flow suppression occurs, the Adamkowski equation for mean velocity is expressed by:

$$v_j = sgn[0.5(v_j^+|v_j^+| + v_j^-|v_j^-|)]\sqrt{\left|0.5\left[(v_j^+)^2 + (v_j^-)^2\right]\right|} \tag{18}$$

This equation (18) is used when the pressure falls to the vapour pressure $p_j = p_v$, i.e. when the column separation of liquid stream takes place.

2.3. Wall shear stress model

In DACM the wall shear stress is modelled as for a single-phase flow. The equation used in this work which represents wall shear stress as a sum of quasi-steady component τ_q and an unsteady component τ_u is:

$$\tau = \tau_q + \tau_u = \frac{\rho v|v|}{8}f + \frac{4\mu}{D}\int_0^t \frac{\partial v(t-u)}{\partial t}\cdot w(u)du \tag{19}$$

and its effective numerical solution, employed in this work is [14]:

$$\tau(t+\Delta t) \approx \frac{\rho f v_{(t)}|v_{(t)}|}{8} + \frac{4\mu}{D}\sum_{i=1}^{3} \underbrace{\left[A_i y_i(t) + \eta B_i\left[v_{(t)} - v_{(t-\Delta t)}\right] + [1-\eta]C_i\left[v_{(t-\Delta t)} - v_{(t-2\Delta t)}\right]\right]}_{y_i(t+\Delta t)} \tag{20}$$

where

$$\eta = \frac{\int_0^{\Delta\hat{t}} w_{class.}(u)du}{\int_0^{\Delta\hat{t}} w_{eff.}(u)du},\ A_i = e^{-n_i\Delta\hat{t}},\ B_i = \frac{m_i}{\Delta\hat{t}n_i}[1-A_i],\ C_i = A_i B_i, \tag{21}$$

A_i, B_i, C_i are constant during simulation and are calculated prior to the simulation with the coefficient describing the effective weighting function; $\Delta\hat{t}$ – is a dimensionless time; $w_{class.}(u)$ and $w_{eff.}(u)du$ are respectively the classic inefficient weighting function (laminar or turbulent depending on the type of flow before the transients) and efficient weighting function (are the approximation of the classic weighting function). In this paper, the three-term efficient weighting function, presented in a former study [13] was used. These coefficients as the simulation was in the turbulent regime were properly scaled using the universal weighting function procedure.
The second analysed DBCM is a two-phase flow model, which required the following modification in the effective numerical solution:

$$\tau(t+\Delta t) \approx \frac{\rho_m f v_{(t)}|v_{(t)}|}{8\alpha_{(t)}^2} + \frac{4\mu_m}{D}\sum_{i=1}^{3} \underbrace{\left[A_i y_i(t) + \eta B_i\left[\frac{v_{(t)}}{\alpha_{(t)}} - \frac{v_{(t-\Delta t)}}{\alpha_{(t-\Delta t)}}\right] + [1-\eta]C_i\left[\frac{v_{(t-\Delta t)}}{\alpha_{(t-\Delta t)}} - \frac{v_{(t-2\Delta t)}}{\alpha_{(t-2\Delta t)}}\right]\right]}_{y_i(t+\Delta t)} \tag{22}$$

The μ_m is the mixture viscosity calculated in present work from a simple Dukler's solution [4]:

$$\mu_m = \alpha\mu_l + (1-\alpha)\mu_v \tag{23}$$

2.4. Numerical scheme
With the formulation of cavitation and shear stress model above, the method of characteristics (MOC) based on a rectangular grid is adopted in this study to solve the continuity and motion equations (e.g. equations (10) and (15) when the first cavitation model is used). The principle of this MOC is to convert the original set of partial differential equations (PDE) into an ordinary differential equation (ODE) by considering the wave evolution process along the characteristic lines ($v \pm c_0$), at which the solution can be integrated from known information at last-time step. The obtained ODE is then solved by the central finite difference (FD) scheme to determine the numerical solutions at the current time step along the pipeline. To ensure the stability and convergence, a Courant number for the simulation process is less than or equal to 1 for the reservoir-pipeline-valve system used in this study. More details about the mathematical formulation and application of MOC can be found in a former work [16]. The final computer programmes were written in MatLab, which is a multi-paradigm numerical computing environment and proprietary programming language developed by MathWorks (plain MatLab, no extra toolboxes were needed). The identification of the calculation algorithm forced to filter the creep compliance function according to the procedure presented in former studies [15,16].

3. Simulation examples
The obtained mathematical formulas described in detail in the previous section were used to write computer programmes for modelling unsteady runs in a hydraulic system built from a reservoir-pipe-valve. The number of experimental studies on unsteady flows in this type of systems during which cavitation areas were created is limited, with only Güney [9,10] and Soares [12] being known to the authors. In this paper, the studies obtained by Güney will be used for comparison. The test stand was composed of a 43.1 meters long (R=0.0208 m and e= 0.0042 m) LDPE pipe. Two cases of turbulent flow were analysed, in which the main difference was a significant difference in the temperature of the flowing liquid – in Case 1 the temperature was $T_1 = 13.8$°C and in Case 2 – $T_2 = 38.5$°C. The temperature significantly affects the mechanical properties of the pipe material: a drastic change in the

creep function occurs, the derivative of which is the main function affecting the modelled damping of pressure oscillation.

Values of the necessary input parameters for computer programmes are summarised in Table 1, while coefficients describing the creep function for the two analysed temperatures are given in Table 2. The method of characteristics was used with a constant number of reaches N=64. Such a selected number of reaches the following time steps $\Delta t_{case\,1} = 0.0022\ [s]$ and $\Delta t_{case\,2} = 0.0031\ [s]$. The results of the simulation tests obtained in relation to the experimental data are-presented in Figure 1.

Table 1. Details of Güney's research.

Case	T	μ_v	μ_l	c_0	p_v	ρ_v	ρ_l	v_0	p_{Rr}
1	13.8	9.56e-6	0.0012	300	1.55e3	0.012	999.3	1.28	1.2955e5
2	38.5	1.026e-5	6.7e-4	220	6.81e3	0.048	992.6	1.33	1.2985e5

where: T – temperature [°C]; μ_v – dynamic viscosity of vapour [Pa·s]; μ_l – dynamic viscosity of water [Pa·s]; p_v – vapour pressure [Pa]; ρ_v – density of water vapour [kg/m³]; v_0 – initial velocity [m/s]; p_{Rr} – reservoir pressure [Pa]

Table 2. Creep compliance function coefficients.

Case	T	J_0	J_1	J_2	τ_1	τ_2
1	13.8	$1.11\cdot10^{-9}$	$0.637\cdot10^{-9}$	$0.871\cdot10^{-9}$	0.0166	1.747
2	38.5	$2.12\cdot10^{-9}$	$2.097\cdot10^{-9}$	$3.570\cdot10^{-9}$	0.0347	3.077

where: J_i – creep-compliance coefficients [Pa⁻¹] and τ_i – retardation times [s]

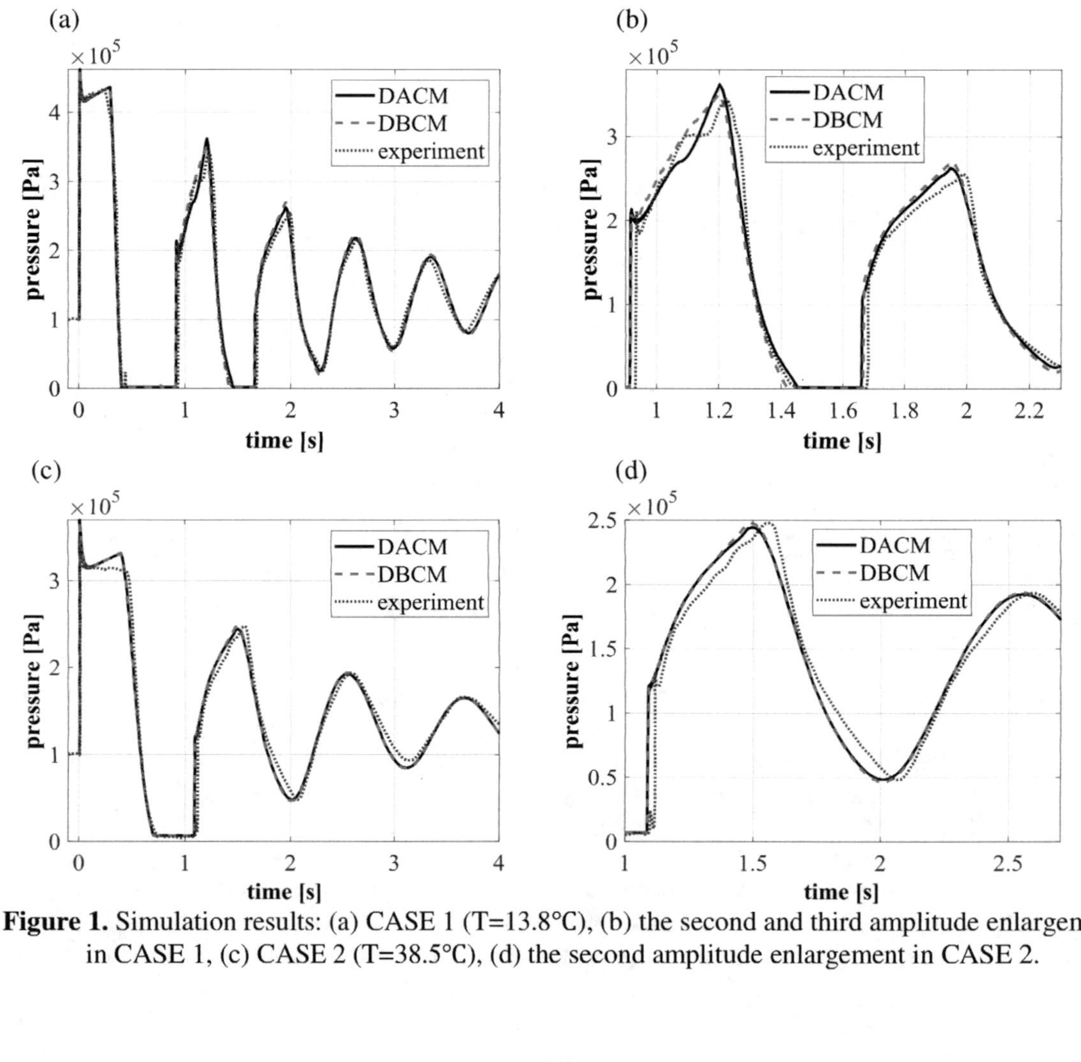

Figure 1. Simulation results: (a) CASE 1 (T=13.8°C), (b) the second and third amplitude enlargement in CASE 1, (c) CASE 2 (T=38.5°C), (d) the second amplitude enlargement in CASE 2.

From the presented results, it can be seen that in the analysed Case 1 (Figures 1a and 1b) both flow models produce accurate results. The detailed remarks regarding Case 1 are as follows:

a) the visible peak of pressure at the top of the first amplitude (Figure 1a) as well as the subsequent drop followed by a gentle experimentally observed increase in pressure was modelled with both DBCM and DACM;
b) the slight peak observed in experimental studies at the beginning of the second amplitude ($t = 0.96\ s$) was modelled correctly by both models (Figure 1b);
c) in the unsteady flow without the cavitational liquid stream column ($t > 1.65\ s$), it can be seen that the phase shift of the simulated amplitudes increases over time. The result of the above behaviour should be sought in the change of the pressure wave speed c during the water hammer or measurement and approximation errors at the stage of defining the experimental creep function by Güney.

Detailed comments on analysed Case 2 were similar, although it was noted that:

a) in the first experimental amplitude, as in Case 1 after the original peak, a gentle pressure increase. Which, remembering given by Güney, the pressure drops on the length of the pipe (Case I: $h = 2.88\ m$ and Case II: $h = 2.93\ m$) is unusual. For similar initial parameters, one expects a similar course of pressure changes, as modelled by the analysed numerical models (Figure 1c);
b) both numerical models (DBCM and DACM) starting from the third amplitude overestimate the minimum pressures occurring between successive amplitudes (see the results for $t = 3.1\ s$);
c) in the analysed case, it was harder to find simulation differences between the analysed numerical models (Figure 1d).

4. Conclusions

The paper presents the modelling of unsteady flows of liquids in pressure polymer pipes. During pressure drop to the saturated vapour pressure, there were areas of cavitation that were modelled with two modified models, namely: DBCM – based on two-phase flow equations and DACM – an improved version of the classical DVCM model. The usefulness of the presented models for flow analysis in plastic pipes was examined for the first time. From the research, it becomes clear that the presented models provided high-accuracy simulation of the unsteady flows with the cavitational liquid column separation, and that both similarly modelled cavitation areas. It was hard to see simulation differences between the results obtained from the DBCM and DACM models.

Subsequently, a precise quantitative analysis of the results obtained will be carried out in relation to other experimental results known in the literature in which cavitation areas appeared in plastic pipes. Applied experimental creep functions (developed by Güney, and filtered using method described in a former paper [16]) showed high efficiency in modelling the analysed flows. This further indicates that the exact "maps" of the creep function in a wide range of times and temperatures will allow in the future a complete exclusion of the calibration method (estimation of creep compliance coefficients) commonly used today.

5. References

[1] Bergant A, Simpson A R, Tijsseling A S 2006 *J. Fluids Struct.* **22** 135
[2] Kranenburg C 1972 The effect of free gas on cavitation in pipelines induced by water hammer *Proc. of the 1th Int. Conf. on Pressure Surges* (BHRA) 41
[3] Safwat H H, Van Den Polder J 1973 *J. Fluids Eng.-Trans. ASME* **95** 91
[4] Shu J J 2003 *Int. J. Pressure Vessels Pip.* **80** 187
[5] Simpson A R, Bergant A 1994 *J. Hydraul. Eng.-ASCE* **120** 361
[6] Soares A K, Martins N M C, Covas D I C 2017 *J. Hydraul. Res.* **55** 731
[7] Streeter V L 1983 *J. Hydraul. Eng.-ASCE* **109** 1408
[8] Wylie E B 1984 *J. Fluids Eng.-Trans. ASME* **106** 307
[9] Gally M, Güney M, Rieutord E 1979 *J. Fluids Eng.-Trans. ASME* **101** 495
[10] Güney M S 1977 *Contribution a l'étude du phénomène de coup de bélier en conduite viscoélastique* (PhD thesis, Lyon) (in french)

[11] Hadj-Taïeb L, Hadj-Taïeb E 2009 *Int. J. Model. Simul.* **29** 206
[12] Soares A K, Covas D I C, Carriço N J G 2012 *J. Hydraul. Res.* **50** 228
[13] Urbanowicz K 2017 *J. Theor. Appl. Mech.* **55** 1029
[14] Urbanowicz K 2018 *ZAMM-Z. Angew. Math. Mech.* **98** 802
[15] Urbanowicz K, Firkowski M 2018 *Springer Proceedings in Mathematics & Statistics* **249** 385
[16] Urbanowicz K, Firkowski M 2018 Effect of creep compliance derivative in modeling water hammer in viscoelastic pipes *Proc. of the 13th Int. Conf. on Pressure Surges* (BHRA) 305
[17] Adamkowski A 2004 *Transient states in the vortex systems of water machines* (Scientific papers of Institute of Fluid-Flow Machinery, Polish Academy of Sciences, 534/1493/2004) (in polish)
[18] Adamkowski A, Lewandowski M 2009 *J. Fluids Eng.-Trans. ASME* **131** 071302
[19] Covas D 2003 *Inverse transient analysis for leak detection and calibration of water pipe systems modelling special dynamic effects* (PhD thesis, Imperial College London (University of London))
[20] Hsu Y Y, Graham R W 1976 *Transport Processes in Boiling and Two-Phase Systems: Including Near-Critical Fluids* (Washington D.C., Hemisphere Publishing Corp.; New York, McGraw-Hill Book Co.).

IOP Conf. Series: Materials Science and Engineering **710** (2019) 012014 doi:10.1088/1757-899X/710/1/012014

Recurrence-based diagnostics of rotary systems

B Ambrożkiewicz[1], N Meier[2], Y Guo[3], G Litak[1] and A Georgiadis[2]

[1] Lublin University of Technology, Faculty of Mechanical Engineering, Department of Automation, Nadbystrzycka 36, p.727, 20-618 Lublin, Poland,
[2] Leuphana University, Institute of Product and Process Innovation (PPI), Volgershall 1, 21339 Lüneburg, Germany,
[3] Kunming University of Science and Technology, Faculty of Mechanical and Electrical Engineering, Kunming City, Yunnan Province, China

b.ambrozkiewicz@pollub.pl

Abstract Rotary mechanisms are commonly used for transferring rotational movement in diverse industrial applications in mechanical engineering. Components such as gears and rolling bearings have found their purpose in various automotive, machinery or agriculture systems. During operation in mechanisms, they are subjected to defects or changes of their key parameters. This paper considers the application of recurrence plots (RP) and recurrence quantification analysis (RQA) in the detection of teeth crack in a planetary gear system and radial internal clearance (RIC) in a double row self-aligning ball bearing. Raw signals are obtained from accelerometers installed in test rigs. The analysis consists of a statistical analysis approach and calculations of basic RQA parameters i.e. recurrence rate, determinism and length of the longest diagonal. In the paper, we extract information about the fault detection in one of the rotary systems and about bearing operational parameters with nonlinear dynamics identification.

1. Introduction

Powerful tools applied to analyse nonlinear systems include recurrence plots and a recurrence quantification analysis. They have been used in a variety of areas of nonlinear science e.g. physiology, engineering, physics, astronomy. In general, the recurrence plot (RP) is an advanced technique of signal analysis. For the first time, it was proposed by Eckmann [1] as a graphical representation of a dynamical process. The RP represents a square matrix of time events, plotted with points. Black dots correspond to those time events at which a dynamical system recurs. It shows all time events when the dynamical trajectory of the system meets the same area in the phase space [2] with the assumed conditions.

The efforts of diagnostics on rotary systems are mainly focused on digital signal processing (DSP) technologies, which can be divided into the time (statistical) and frequency (spectral) domain analysis [3]. Furthermore, numerous studies consider the application of extended standard DSP methods, such as Fourier and Wavelet Transforms [4,5] or Hilbert Transform (HT) in the detection of defects and operational parameters in rotary systems [6,7]. One of the advantages of the abovementioned methods is their computation time, however, they do not allow for easy acquisition of information buried in noised or non-stationary time series. The application of recurrence plots analysis seems to be adequate to obtain additional information coming from measurements.

In this work, we present a promising approach to distinguish the particular nonlinear behaviours in two rotary systems, the planetary gear system and in the self-align ball bearing. Note that nonlinearities can be introduced to the system by defects and other operational conditions represented by a friction level, damping, and load. Their share in spectra is hard to determine unambiguously with standard methods, such as the fast Fourier transform (FFT) or signal envelope; by the application of recurrence plot analysis, however, we can obtain specific patterns in the phase space trajectory. Moreover, by application of the recurrence quantification analysis (RQA), it is possible to assign a quantificator value to describe each feature. The main advantage of the RQA is using fairly short data lengths. Besides, RQA has also a good anti-noise ability.

Up to now, several articles have considered diagnostics of rotary systems with the abovementioned method, for example, in the evaluation of a degradation level in bearings [8,9] or a fault detection of servo valves [10]. Our work investigates and compares both analyses of the fault detection in the planetary gear system with a cracked tooth [11] and the recognition of variable radial internal clearance in the self-align ball bearing [12].

2. Recurrence plot (RP) and recurrence quantification analysis (RQA)

Many dynamical systems are characterised by their property to come back to the previous state, referred to as recurrence. This technique was first proposed in seminal work of Henri Poincare in 1890 [13] and almost one hundred years later, developed by Eckmann et al. [1] formulating the recurrence plot method. In order to obtain a quantitative interpretation of the RP method, recurrence quantification analysis was introduced to obtain numeral information hidden in recurrence plots. Nowadays, the mentioned analyses are used for the detection of quantitative changes in the behaviour of dynamical systems and give satisfying results in various applications.

2.1. Recurrence plot (RP) method

Recurrence analysis is a graphical method designed to locate hidden patterns, non-stationarity and structural changes occurring in the dynamical system [14], by revealing all the times when the phase space trajectory meets the same area in the phase space. We obtain the recurrence when a state x at time i at a different time j is indicated within a two-dimensional square matrix [R] with ones (recurrence state) and zeros (no recurrence):

$$[R_{i,j}] = \Theta\left(\varepsilon - \left\|\{x_i\} - \{x_j\}\right\|\right), \qquad i,j = 1,\dots,N \tag{1}$$

where, N denotes the number of considered states x_i, ε_i states for the threshold and $\Theta(\ )$ is the Heaviside function. Recognised states on recurrence plot are represented in the following way:

$$[R_{i,j}] = \begin{cases} 1: & \{x_i\} \approx \{x_j\} \\ 0: & \{x_i\} \neq \{x_j\} \end{cases}, \qquad i,j,\dots,N \tag{2}$$

$\{x_i\}\approx\{x_j\}$ means equality up to an error (or distance) ε. To sum up, the matrix $[R_{i,j}]$ compares states of the system at both states and if they are similar, they are marked by one in the matrix $R_{i,j}=1$ or when they are different the matrix is $R_{i,j}=0$. Obtained matrix creating recurrence plot, shows us when similar states of the considered dynamical system occur.

In our research for both considered systems, we have found the parameter of the embedding dimension m determined with the value of the first minimum of the false nearest–neighbours (FNN) algorithm [13]. On the other hand, the time delay τ was found by the first minimum in the mutual information (MI) function.

2.2. Recurrence quantification analysis (RQA)

In order to support the interpretation of recurrence plots, recurrence quantification analysis (RQA) was introduced, which quantifies the number and the duration of recurrences for the considered dynamical system by its state space trajectory. At first, it was used for exploration of a non-stationary cardiac signal [15], but with development of recurrence techniques, they began to be applied also in various engineering systems, such as a diesel engine [16], a bevel gearbox [17] or a car's braking system [18]. In our work, we focus on three main quantificators i.e. recurrence rate (RR), determinism (DET) and the average length of the diagonal (L) [19,20].

- Recurrence rate (RR)

  The recurrence rate is a basic RQA parameter, which describes the density of recurrence points on the recurrence plot obtained by different input parameters.

$$RR = \frac{1}{N^2}\sum_{i,j=1}^{N} R_{i,j} \quad \text{for } |i-j|\geq 1 \tag{3}$$

- Determinism (DET)

  By determinism, the percentage of recurrence points creating diagonal lines in the recurrence plot of minimal length lmin is expressed.

$$DET = \frac{\sum_{l=l_{min}}^{N} lP(l)}{\sum_{l=1}^{N} lP(l)} \tag{4}$$

  P(l) in formula (4) states for the frequency distribution of the lengths l of the diagonal lines.

- Average length of the diagonal lines (L)

  This measure refers to the predictability time of the investigated system.

$$L = \frac{\sum_{l=l_{min}}^{N} lP(l)}{\sum_{l=l_{min}}^{N} P(l)} \tag{5}$$

3. Test rigs and measurements

The first of analysed systems was the single-stage planetary gearbox transmission (type: 2K-H), figure 1 concerns two cases, namely, when the system works properly and with planet gear defect, figure 2. For measurements of the system's response, we used an accelerometer mounted on the ring gear presented in figure 1 (3). In the test, only one constant rotational velocity of the sun gear was considered, in order to obtain spectra for both cases with the same characteristic rotational frequencies. We focused on the part of spectra where we got the difference between the proper and the defective system's work.

Figure 1. Planetary gear system used for tests – numbers indicate eddy probe (1) and piezoelectric accelerometers (2-4).

Figure 2. One of planetary gears with a damaged tooth.

Moreover, in our research, we decided to determine if there is the possibility to detect the change of the radial internal clearance (RIC) in the double-row self-aligning ball bearing using RP and RQA. For this, we used the experimental system, figure 3, consisting of the three-phase motor, which was propelling a shaft with two above mentioned ball bearings mounted on the conical adapter sleeves and housed in the adapter with accelerometers mounted into it. We obtained the system's response from one MEMS sensor only. In figure 3 markings L and F on the housings state for locating and fixed bearing, correspondingly. On the first one, we were changing its RIC by pressing or pulling it onto the conical adapter sleeve. The measurement procedure of bearing's RIC is described in the following articles [21, 2]. The test was performed for three different clearance values and by one rotational velocity as in planetary gear system.

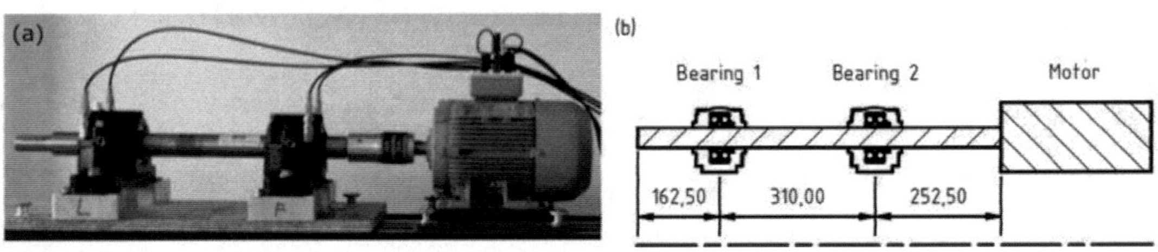

Figure 3. Test rig used for the experiment (a) and its cross-sectional view (b) with indicated distances.

4. Results and discussion

One of the proper approaches to start recurrence analysis is the application of basic statistical tools like a mean or a standard deviation for obtained signals. This is the first step in which we could identify numerically the difference between each system's operational state. In our analysis, we calculated the standard deviation for signals and their values are aggregated in Table 1. The discrepancies of SD values in each system are the result of applying different accelerometers and their montage.

Table 1. Standard deviation values for time courses of all considered cases.

Planetary Gear System	Healthy		Planet Gear Crack
SD (m/s^2)	0.1392		0.2918
RIC (μm)	7	34	46
SD (m/s^2)	2.8727	3.9073	4.6406

The next step is to find values of three crucial input parameters for drawing recurrence plot i.e. embedding dimension – m, time delay – τ and threshold – ε. Two first parameters were found with the approach described in the last paragraph of sub-section 2.1, but there is still no unambiguous method for the threshold determination. Because two different rotary systems are compared, we decided to use the same threshold value ε=0.5 (in standard deviation units) for all considered cases, in order to ensure the consistency of obtained data. Our approach we can name as the empirical one and all the values are collected in Table. 2. That value we assumed based on recurrence rate results, figure 4a, where we can observe the clear separation between each recurrence rate line for both systems.

Table 2. Values of recurrence plots input parameters, τ is represented in sampling intervals.

	m	τ	ε
Planetary Gear System	3	2	0.5
RIC	2	3	

The last step of analysis is RQA, which we performed for three recurrence quantificators described in subsection 2.2. All results are presented for adopted signal intervals in figure 4. It needs to be emphasised that with the increase in standard deviation value we obtain higher values of each quantificators in ε domain. In both systems, each considered state is recognisable on all plots, that what makes the systems different from each other is the sensitivity on gear damage and clearance value in different ε range.

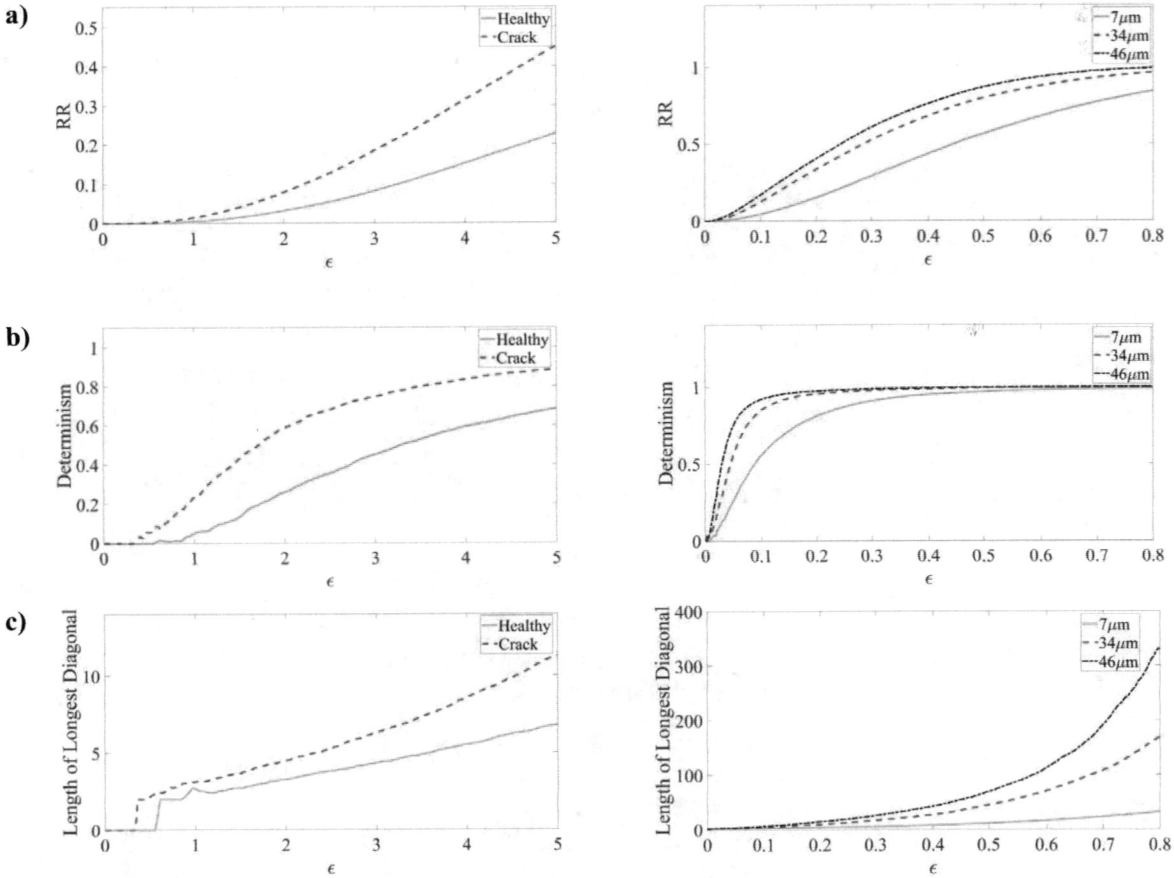

Figure 4. RQA results for the planetary gear system (in the left panel) and the radial internal clearance (in the right panel). The order of quantificators is the following: a) Recurrence rate, b) Determinism, c) Length of the longest diagonal.

Representative examples of studied RP are presented in figure 5 and figure 6. For all considered cases they are reproducing dynamic system trajectories returning to the same area in the embedded space, thus giving a recurrence point marked with a black dot. Analysing results obtained for planetary gear system, we can determine that RP during planetary gear crack is characterised with small fluctuations indicating periodicity in the system. The fluctuations in question are strongly connected with increased nonlinearity, including phase couplings between main and super- and sub-harmonics during rotation of the cracked gear, which brings a more regular pattern of the RP. In the studied bearing system, we noticed that the clearance, which causes a nonlinear effect can lead to the observation of lines splitting in the monotonic pattern.

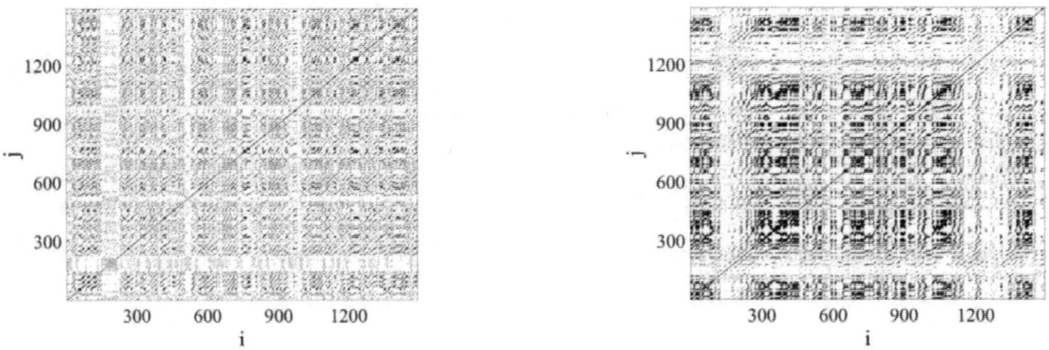

Figure 5. Recurrence plots for the state with planet gear crack (left) and for the proper operation (right).

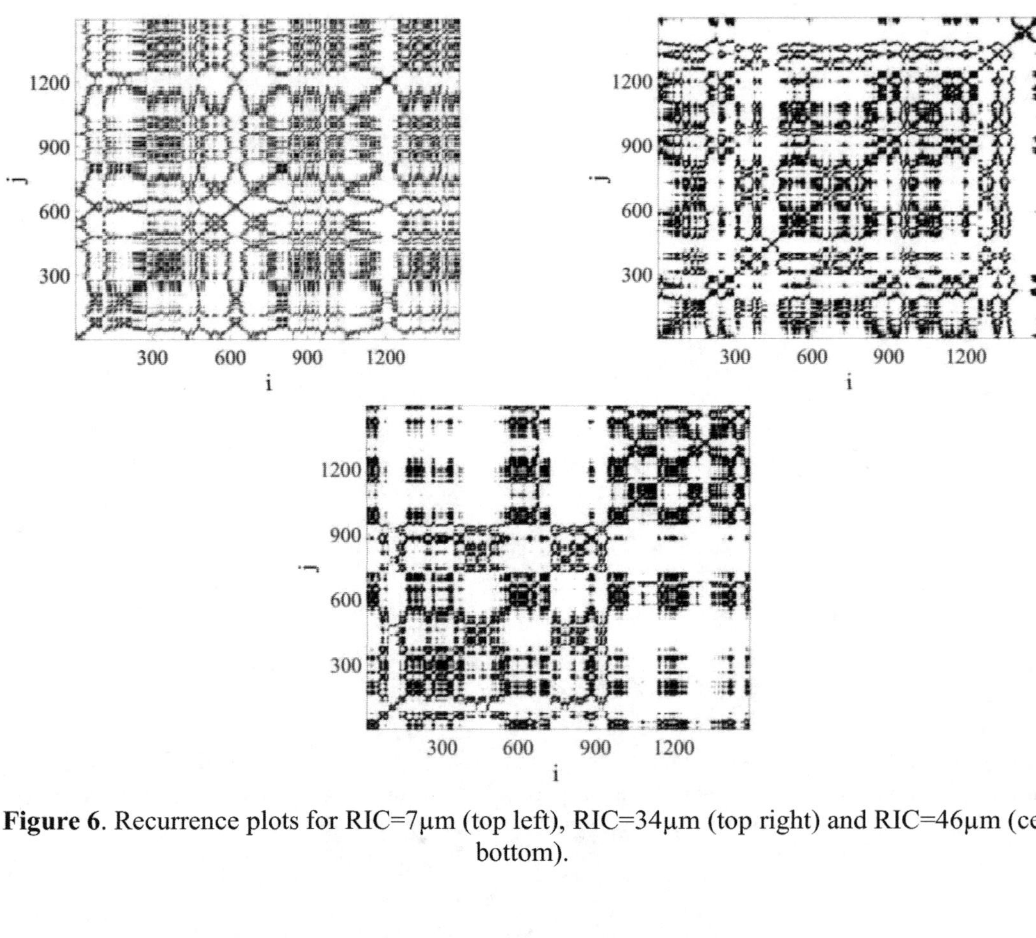

Figure 6. Recurrence plots for RIC=7μm (top left), RIC=34μm (top right) and RIC=46μm (centre bottom).

IOP Conf. Series: Materials Science and Engineering **710** (2019) 012014 doi:10.1088/1757-899X/710/1/012014

5. Conclusions

Based on RPs and various RQAs, we detected planetary gear faults and clearance level effects. In the considered systems, we focused on the increasing nonlinearities reflected in the system responses. This response can become non-harmonic, non-periodic or intermittent by large nonlinearity. It is proved by isolated areas on the RP consisting of recurrence points localised closely to each other. Additionally, non-harmonicity causes RP's pattern to be more compact (improving contrast between black and white regions) with increasing RIC. In light of this study, we are convinced that RP and RQA statistics are useful tools for faults and operational parameter identification.

6. References

[1] Eckmann J P, Olifsson Kamphorst S and Ruelle D 1987 Recurrence plots of dynamical systems *Europhysics Letters* **4** pp 973-977

[2] Marwan N 2008 A historical review of recurrence plots *The European Physical Journal Special Topics* **164** pp 3-12

[3] Ruqiang Y, Yuning Q, Zhoudi H and Gao R X 2011 Rolling bearing defect severity evaluation using recurrence plot entropy *IEEE International Instrumentation and Measurement Technology Conference* doi: 10.1109/IMTC.2011.5944161

[4] Lou X and Loparo K A 2004 Bearing fault diagnosis based on wavelet transform and fuzzy interference, *Mechanical Systems and Signal Processing* **18** pp 1077-1095

[5] Peng Z K and Chu F L 2004 Application of the wavelet transform in machine condition monitoring and fault diagnostics: A review with bibliography *Mechanical Systems and Signal Processing* **18** pp 199-221

[6] Karioja K, Lahdelma S, Litak G and Ambrożkiewicz B 2018 Extracting periodically repeating shocks in a gearbox from simultaneously occurring random vibration *15th International Conference on Condition Monitoring and Machinery Failure Prevention Technologies, CM/MFPT 2018*, pp 456-464

[7] Soualhi A, Medjaher K and Zerhouni N 2015 Bearing health monitoring based on Hilbert-Huang transform, support vector machine and regression *IEEE Transactions on Instrumentation and Measurement* **64** pp 52-62

[8] Qian Y, Yan R and Hu S 2014 Bearing Degradation Evaluation Using Recurrence Quantification Analysis and Kalman Filter *IEEE Transactions on Instrumentation and Measurement* **63** pp 2599-2610

[9] Yan R, Qian Y, Huang Z and Gao R X 2011 Rolling bearing defect severity evaluation using recurrence plot entropy *IEEE International Instrumentation and Measurement Technology Conference 2011*

[10] Samadani M, Kitio Kwuimy C A and Nataraj C 2014 Fault detection and severity analysis of servo valves using recurrence quantification analysis *Annual Conference of the Prognostics and Health Management 2014*

[11] Ambrożkiewicz B, Guo Y, Litak G and Wolszczak P 2019 Dynamical response of a planetary gear system with faults using recurrence statistics *Springer Proceedings in Physics* **228** chapter 9 pp 177-185

[12] Meier N, Ambrożkiewicz B, Georgiadis A and Litak G 2019 Verification of measuring the clearance using kurtosis, recurrences and neural networks and comparison of these approaches *2019 IEEE Sensors, Montreal*, pp 1-4

[13] Marwan N, Romano M C, Thiel M and Kurths J 2007 Recurrence plots for the analysis of complex systems *Physics Reports* **438** pp 237-329

[14] Franch J B and Contreras D 2002 Recurrence plots in nonlinear time series analysis *Journal of Statistical Software* **7**

[15] Zbilut J, Thomasson N and Webber C L 2002 Recurrence quantification analysis as a tool for nonlinear exploration of nonstationary cardiac signals *Medical Engineering & Physics* **24** pp 53-60

[16] Sen A K, Longwic R, Litak G and Górski K 2008 Analysis of cycle-to-cycle pressure oscillations in a diesel engine *Mechanical Systems and Signal Processing* **22** pp 362-373
[17] Syta A, Jonak J, Jedliński Ł and Litak G 2012 Failure Diagnosis of a Gear Box by Recurrences *Journal of Vibration and Acoustics* **134**
[18] Oberst S and Lai J C S 2011 Statistical analysis of brake squeal noise *Journal of Sound and Vibration* **330**
[19] Marwan N 2011 How to avoid potential pitfalls in recurrence plot based data analysis *International Journal of Bifurcation and Chaos* **21** pp 1003-1017
[20] Schinkel S, Dimigen O and Marwan N, 2008 Selection of recurrence threshold for signal detection *The European Physical Journal* **164** pp 45-53
[21] Georgiadis A, Gong X and Meier N 2018 Vibration analysis based on the spectrum kurtosis for adjustment and monitoring of ball bearing radial clearance *MATEC Web of Conferences* **211**
[22] Meier N, Biyani Y and Georgiadis A 2018 Determination of bearing clearance by the application of neural networks *2018 IEEE Sensors, New Delhi*, pp 1-4

Design and optimisation methods for interactive mobile VR visualisation

S Skulimowski, M Badurowicz, M Barszcz and J Montusiewicz

Lublin University of Technology, Department of Computer Science, Nadbystrzycka 38, Lublin, Poland

s.skulimowski@pollub.pl

Abstract. The article describes the process of creating a VR mobile application for Google VR platform. The application presents a virtual museum exhibition as part of cross-cultural competence and user interface usability research. The process of creating the application requires taking into account the hardware and software limitations and the specificity of mobile devices. In addition, a higher level of optimisation helps build positive user experience, which is essential in VR applications. The process includes, among others: planning activities, preparing assets, implementing VR vision and interaction modules. The paper includes descriptions of individual steps, problems and other non-standard situations that occurred during the creation and testing of the VR application. In addition the results of the initial performance tests and the methods used to optimise and maintain the stability of the application are presented. By taking into account all of the mentioned factors and the fact that many solutions are yet not well described, the description of the process is expected to provide a point of reference and help for future developers of mobile VR applications.

1. Introduction

In recent years, VR (Virtual Reality) is developing dynamically, becoming a very promising technology. Computer games and art are its primary scope of application [1]. However, VR is equally important in medicine [2], education (training of pilots, doctors, surgeons [2], and other specialist [3,4,5]), engineering (design, simulations, repairs) or business communication (where VR combined with telepresence can replace expensive employee journeys in many cases) [6]. More and more often VR technology is used to prepare virtual tours of historic buildings [7].

Companies producing VR equipment (Steam and HTC, Oculus or Sony) are competing to outdo the parameters of the rival products. These are devices that offer rising opportunities each year, however, they are still very expensive. That is why new low-cost solutions are being sought. The market offers relatively cheaper mobile solutions compared to available professional ones (i.e. VR glasses - Oculus Rift, HTC Vive, as well as PlayStation VR). Their equivalent on the mobile market is, among others, Samsung Gear VR and Google Cardboard. Instead of the built-in screen, these solutions use only the user's phone, which takes over all the work and allows the image to be processed and displayed [5]. These solutions are, therefore, much more accessible financially than their prices established high-end competitors. In addition, the entire system is completely wireless and can be used anywhere. No wiring allows free head rotation. These solutions are the most forward-looking, nevertheless, today they offer far worse parameters.

According to literature analysis, many VR applications are currently being developed for mobile platforms [8,9]. Some of them serve to broaden the experience of visiting the museum [10,11], while others support the cross-culture competency education process [12,13,14]. These are areas similar to the one intended to explore in the present paper. In no other case are methods of proper development and optimisation of application operation described. This confirms the authors' belief that the prepared article will be useful for future developers.

In the process of creating proprietary mobile applications in VR technology, the Unity environment is often used. Although it is usually used to create games, it may be often found in scientific experiments, such as: testing the behaviour of a homogeneous robotic hive in a space with defined obstacles [15], generating digital landscapes built of trees with variable structures for VR [16], or simulation of nuclear decommissioning devices in VR [17].

The purpose of this research is:

- to develop an optimised application in the context of operational stability (fluency) and minimisation of resource consumption (energy and operational memory) for mobile devices,
- to organise knowledge in the field of procedures for designing and producing interactive VR presentations for selected mobile devices.

The authors have developed the application and have optimised it to become a low-cost mobile VR platform, due to the need to reduce the costs of using VR, to make it possible to use low-cost VR goggles [18], instead of stationary VR platforms such as Oculus Rift or HTC Vive.

2. The subject of study

The subject of the research is the VR application created within the framework of the work [18], along with the equipment of Mirza Ulugbek's madrasah located in Samarkand, Uzbekistan. The room model and its equipment were developed through the usage of reverse engineering technology, i.e. 3D scanning and classical modelling based on photographic documentation. Several models have been developed: among others, a hexagonal table typical for Uzbek culture, a book stand - *lauch* and a book placed on the stand. In addition, textures from photographs were used, representing objects such as turban and *chapan* - an Uzbek coat, carpet or *kilims*. The developed application was made as part of a project implemented by the Institute of Computer Science of the Lublin University of Technology "3D Digital Silk Road" [19]. One of the objectives of the application was to present museum objects without the necessity of going on an actual trip to a real museum, and as a way to improve the attractiveness of museums.

In addition, the application is used for:

- testing user interfaces in virtual space,
- educating students of the Lublin University of Technology in the field of cross-culture competencies [20].

A stereoscopic view from the user's perspective of the presented application is presented in Figure 1.

The application allows the user to interact with objects through the VR interface. After interacting with the exhibit, its description is presented. Depending on the situation for which the digital room will be used, its content is supplemented with additional objects. For this purpose, other objects belonging to different cultural circles have been made, such as the Roman Catholic cross, a *Torah* in the form of a scroll or the Star of David. These objects were entirely made using classic 3D modelling in the 3ds Max programme and incorporated in the application. In addition, a model of a virtual sculpture belonging to the Gallery of the Art of Socialist Realism, which was obtained using the 3D scanning technique using the Artec Eva scanner, was placed in the student's room.

IOP Conf. Series: Materials Science and Engineering **710** (2019) 012015 doi:10.1088/1757-899X/710/1/012015

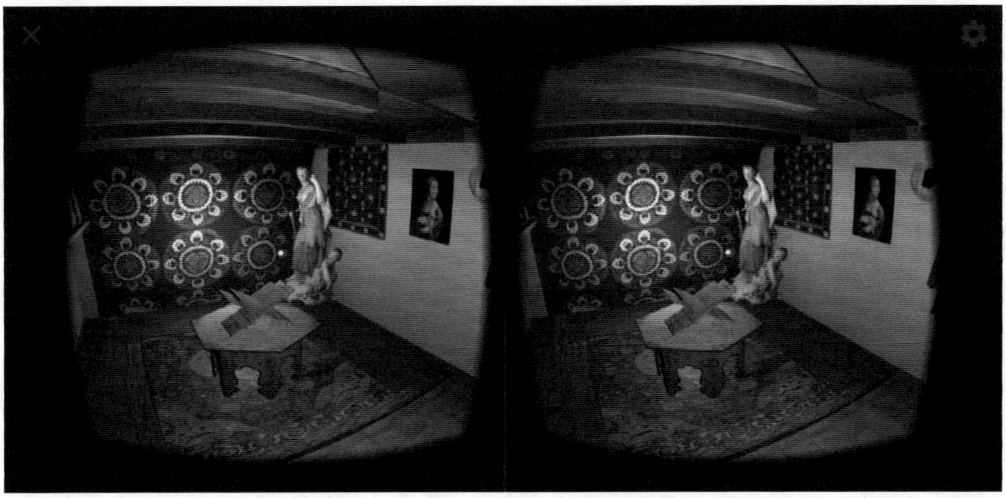

Figure 1. Stereoscopic view of the scene from the perspective of the user (the first-person perspective).

From the dominant mobile platforms, i.e. Android and iOS, the Android platform was selected for the purposes of the presented application as more common and available. Among mobile platforms in Uzbekistan, up to 87% of the market belongs to Android [21], whose additional advantage is the availability of free tools for programmers.

The choice of mobile technology as a base for VR applications is, however, associated with the lower computing power of mobile devices [22]. For the best results of VR usage and to prevent motion sickness occurrence it is necessary to achieve the proper application frame rate [16] - the minimum value is 24 frames per second. This caused initial problems with application performance that had to be resolved through proper application optimisation.

3. Application development process

The application development process included the use of many Information Technology tools (IT tools) - Figure 2. These were tools in the field of visualisation design and testing (Figure 2 L1), 3D modelling (Figure 2 L2), and communication with external devices (Figure 2 L3).

The Unity engine was used to integrate the interaction mechanisms and the virtual environment. The choice of the environment was justified by the possibility of porting the created project to many distribution platforms, which fits in with the mission of increasing accessibility. It soon turned out, however, that although the ability to port project means that it is possible, it by no means makes any implication of its simplicity. The steps necessary to complete the VR application in the Unity environment and the problems that arose during the work on the project are described in Sections 3.1 and 3.2.

3.1. Programming environment configuration

In order to assume control over and perform stereoscopic projection on Android mobile devices, it was necessary to use the Google VR SDK - GVR SDK (E1) Unity extension. This required additional configuration determining the exact version of the Android API that is used in the project and launching support for VR Cardboard and Daydream (W0) technologies.

Further problems with the Unity and GVR SDK environments were:

- the need to separately install the Android Debug Bridge (ADB) tool (E2),
- the need to use the ADB tool only in a specific version compatible with Unity (E3),
- the inability to use two versions of the ADB tool in parallel on one development machine (E3),

- failure to detect Android SDK and Android Tools on the Unity side (E6, E7),
- an incorrect version of the installed JDK environment, i.e. JDK9 (E5) – unsupported by the Unity community, which resulted in an enigmatic message "Resource compilation failed! Failed to recompile Android resource files" (E4).

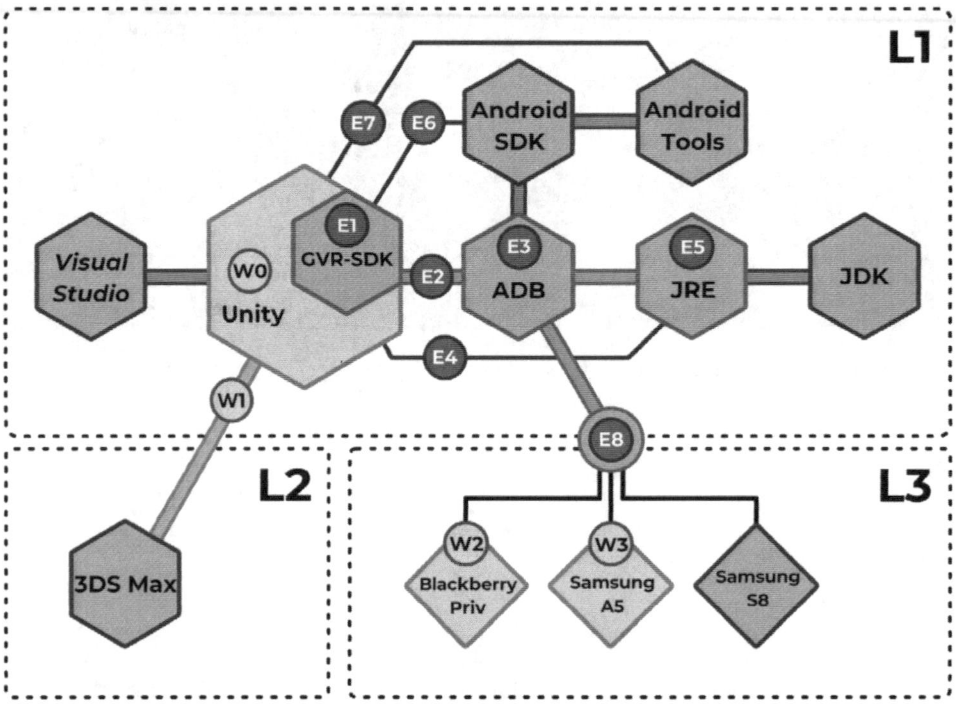

Figure 2. Diagram of dependencies between tools used in the project, with division into layers (Ls) and considering the areas where errors (Es) and warnings (Ws) occurred.

To sum up, every further issue related to the application building itself revealed a defect in programming, resulting in a loss of integrity with tools operating at ever-lower levels of abstraction (E1, E2, E4, E6, E7). The information contained in the official Unity documentation did not take into account the scenario that was described above, and most of the suggestions regarding potential solutions came from online forums and blogs of people who faced similar obstacles in the past. For example, the lack of detection of Android SDK and Android Tools has been solved by downloading custom tools for Android and a separate manager for Android SDK components (E6, E7).

3.2. Mobile VR application development

After dealing with the problems described above, the authors proceeded to the proper development of the application. The process included:

- creating C# scripts describing all planned types of moving in virtual space and interacting with selected objects on the stage,
- creating shaders, describing the way of displaying objects in key moments of the projection, such as: the moment of focusing the participant's attention on the object, the initiation of interaction, the interaction effect, the termination of interaction and ending focus on the object,
- the arrangement of developed 3D models on the stage and assigning methods and interaction effects to them.

Due to the specificity of the Unity environment and the application, it was impossible to introduce automatic tests. This significantly extended the application development time – searching for potential errors and verifying the operation of scripts during the time of further modifications.

Based on the created prototype, manual tests were carried out on a group of IT students of the Lublin University of Technology in order to identify the initial, repeated errors. The prototype contained only the basic mechanisms of interaction and imported models.

Most of the participants had clear problems with concentration or balance. This was due to the low frame rate. That is why work on the optimisation of applications was required to be carried out later, to improve the performance by increasing the number of frames per second of the displayed VR scene.

4. Application optimisation procedure

4.1. Tools for testing application performance and evaluation criteria

The basic element that assisted programmers in optimising the presented application were the tools for profiling the performance of the Unity platform (Figure 3).

Performance profiling tools present primarily the time that the application allocates to the execution of individual stages of its operation, such as the time of rendering a single frame, the duration of scripts (Figure 3 C), the order of rendering of the elements of the scene in each frame (Figure 3 D) and similar.

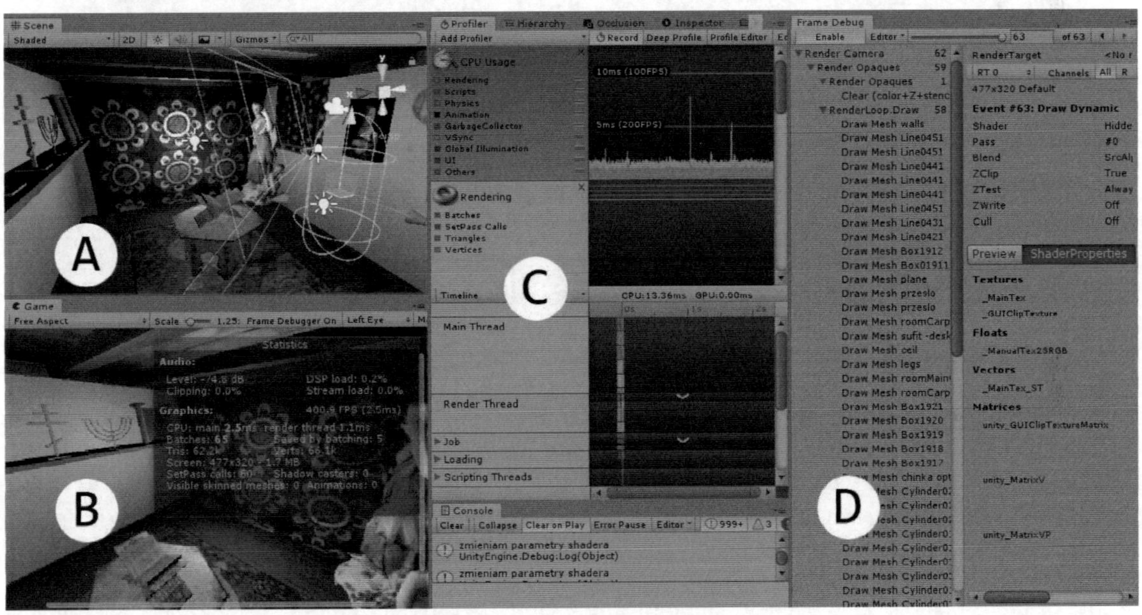

Figure 3. Unity environment during local testing of application operation: A – scene editor with preview of the user avatar position in space, B – FPP view simulation with a list of basic projection statistics, C – profiler window with time-varying computer resource allocations, D –a debugger frame window with a view of the order of rendering objects on stage in each frame.

The use of profiling tools has shown that the basic operation that takes up the application's operating time is the rendering of the scene itself, not the interaction script. Hence, there was no need to modify the scripts' code; however, it was necessary to focus on the stage itself. That is why the authors used the tool provided by Unity, Stats (Figure 3 B), which shows the number of triangles and vertices in the current view of the scene, the number of operations on the graphics card – dynamic and cached batches and occupancy of VRAM.

This mechanism proved to be effective in searching for 3D objects that had too complex a grid. According to preliminary experiments, it emerged that the stage must contain less than 100,000 triangles and this value was set as a base value for subsequent stages of the optimisation process.

In addition, a script developed by the authors presenting the number of presented frames per second displayed in the application based on the Time.unscaledDeltaTime property of the Unity platform, responsible for determining the time elapsed between successive frames generated in the application was used to assess the optimisation process. Displaying the frame rate indicator (FPS) was placed outside the user's field of view and served only for the development purposes. However, this mechanism has been abandoned in favour of mechanisms provided to programmers directly by the Google VR platform presenting overlay on the application displaying the current frame rate along with the graph of history, which is presented in Figure 4. This allows programmers to evaluate the decrease in the number of frames generated depending on the user's movement in the virtual world, which allows spots inefficient system components.

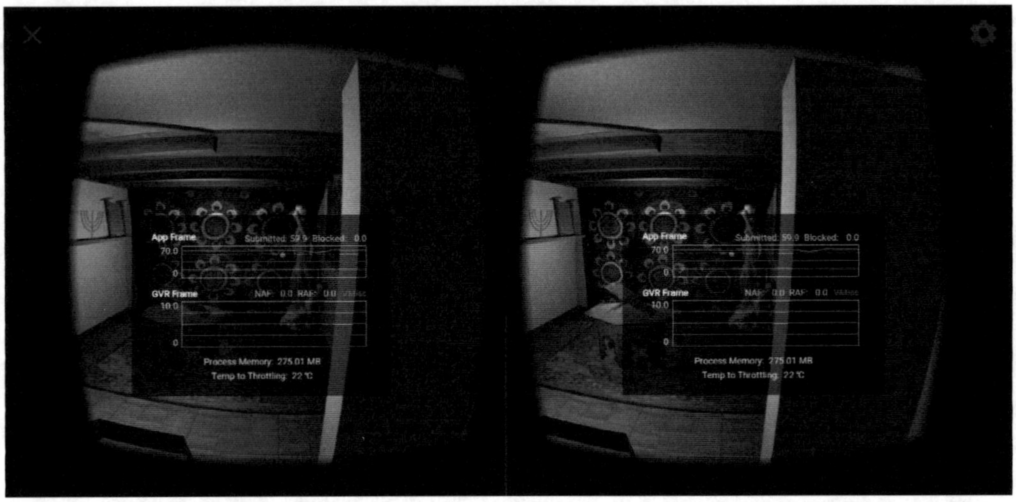

Figure 4. Application statistics screen provided by the Google VR platform.

An important element presented by the tools of the Google VR platform is also the temperature remaining until reaching the threshold value when lowering the processor clock speed of the mobile device occurs (thermal throttling), which can further affect the decrease in the number of frames generated per second.

4.2. Optimisation methods implementation

Some of the problems have been solved by using a suitably efficient mobile device. The performance increase was achieved thanks to the transition during testing for Samsung A5 (W2) and later Samsung S8 instead of the initially used Blackberry Priv (W1).

Some of the optimisation methods have been developed in accordance with the recommendations of the Unity platform manufacturer [23], i.e. such steps as:

- baked lighting (pre-calculated illumination that is not included in any run-time lighting calculations) and resignation from dynamic lights and using only static ones,
- shadows resolution reduction to minimise processing overhead,
- the use of occlusion culling (disabling rendering of objects not currently seen by the camera because they are obscured by other objects),
- the use of the level of details (LOD) technique (loading meshes in different resolutions depending on the distance from the camera),
- adding a mesh-combine mechanism (to reduce the overall number of draw-calls).

Apart from the changes in the Unity configuration, the application design and the virtual scene, it was necessary to modify the 3D models. The number of triangles for the general simplification of scene construction was limited (specialists working on creating models for the presented application had no previous contact with the limitations required for this type of application (W1)).

Three types of 3D model optimisation procedures were produced, depending on their size, complexity and origin. Objects are obtained using 3D parametric modelling (constructed of surfaces reflecting primitive fragments, e.g. bowl, sphere, cylinder, cone, wedge) were automatically simplified. 3D objects obtained using manual 3D modelling using mesh surfaces (including irregular shapes) were also manually requested – the designer decides on the local density of the mesh to correctly reproduce the shape of the object (however, this method is considerably work-intensive). Objects created from 3D scanners (correct reflection of dimensions and shapes) owe their accuracy to the mesh surface created in the triangulation process, containing a large number of points, and therefore triangles. Automatic reduction of the mesh of objects created from 3D scans can easily lead to excessive simplification (e.g. small rounding). In addition, the available mesh reduction tools do not allow interaction with selected areas of 3D objects. For these objects, a compromise must be made between reflecting the actual shape and the simplicity of the object.

Table 1. The comparison of the 3D model parameters before and after optimisation.

Description	Textured model	Object mesh before optimisation	Object mesh after optimisation
Example		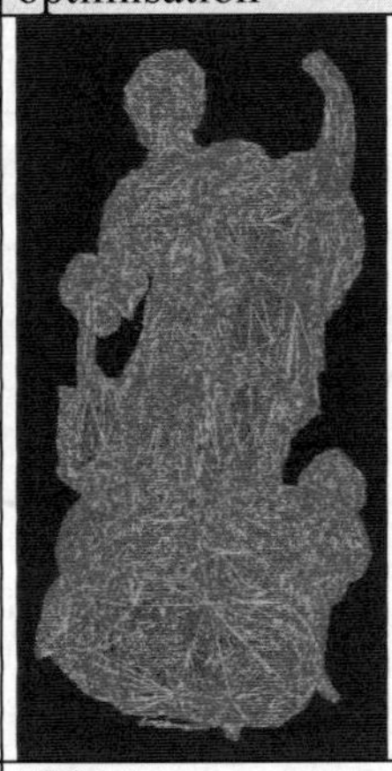	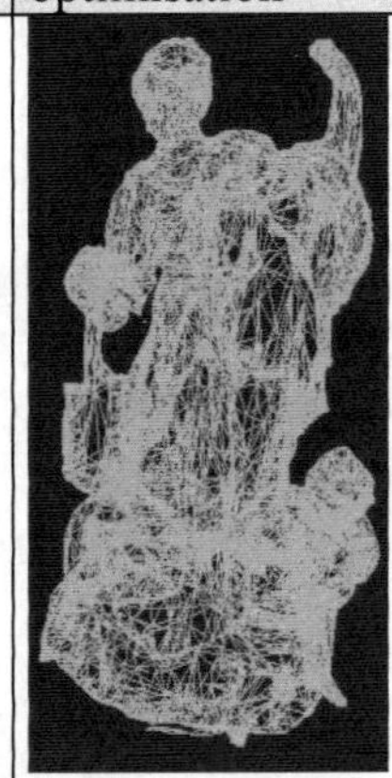
Polygons	Not applicable	73914	30031
Vertices	Not applicable	37041	15100
Size without texture	Not applicable	4.68 MB	1.19 MB

The 3DS Max's Modify> Optimise tool was used, which allows to change the values of vertices and faces in a simplified model relative to the original model (Table 1). the tool was used in several repetitions to test different settings of different values so as to obtain the mesh of low, medium and high (but smaller than the original) number of vertices. In the case of a net with a low value of vertices, the desired value would not significantly affect the deterioration of the quality of the model.

5. Results and observations

Taking into account all the problems that arose during the configuration of the development environment for the purpose of creating a VR mobile application, and subsequent difficulties that occurred during the proper development, the authors state that the creation of such applications may turn out to be a very difficult and even daunting task even for experienced programmers who, however, had not previously created similar interactive visualisations. The correctness of the application depends on many factors (resources coming from many providers), which only a part

IOP Conf. Series: Materials Science and Engineering **710** (2019) 012015 doi:10.1088/1757-899X/710/1/012015

depends on the developer (e.g. optimisation of the application code, quality of the prepared assets). The remaining ones depend on many content providers (e.g. versions of the Unity environment, GVR-SDK versions, JDK versions, etc.).

At every stage of application life (creation, testing, implementation, and exploitation) one can notice the emergence of problems at various levels of abstraction, depending both on the software and hardware layer. A single error made during the configuration of the environment or a single error during the development may require a long backtracking search for sources of errors, effectively hampering the development of the project. In addition, the official documentation of the tools and software used is not always complete and does not provide information on the causes of the problems.

It has been noticed that in the optimisation process the models prepared by classic modelling have a greater advantage because they have a regular grid – this facilitated the use of manual and automatic optimisation methods. In the case of models obtained by 3D scanning, a large number of vertices that did not significantly affect the model's details were observed, which may be a problem during the process of automatic model optimisation [24].

Creating 3D models for use in VR mobile applications requires in-depth knowledge and experience in the field of modelling for projection with limited hardware resources. Otherwise, it will be necessary to modify existing models in such a way that they consist of fewer components and triangles, or in the worst-case scenario – creating models from scratch. The problem is the bigger the later it is identified and the more models it concerns. In both cases, the time of asset creation significantly lengthens and introduces additional work overhead.

As a result of the actions and tests carried out, it can be stated that:

1. The implementation of the optimisation methods described above reduced the average number of triangles in the scene view from 300,000 to 45,000 (over 6 times) and reduced memory usage from 25 MB to 5 MB (fivefold decrease), which resulted in a general acceleration of the application. The average value of fps was obtained at level 40, with an almost two-fold increase in value (21 fps). The value obtained is not fully satisfying because the recommendations of VR device manufacturers indicate the value of 60 fps as the minimum to ensure the proper level of immersion and reduce the risk of virtual disease.
2. As a result of the employed optimisation methods, a decrease in the temperature and an increase in the time of device heating during an uninterrupted VR session were noted. The device temperature elevation is due to the load on the components – processor, internal memory and screen, while the lack of an active cooling system (e.g. a fan), the device is not able to quickly remove the accumulated heat. Before entering the optimisation, the phone could warm up in a way that prevented its operation in a situation where it was necessary to remove it from the frame. Currently, this problem no longer occurs – the device heats up to a slightly higher temperature.
3. The changes introduced in the process of VR scene generation have contributed to the reduction of mobile device battery consumption. At the time when the VR application is running on the device, both the screen and some of the sensors remain in the mode of continuous activity, which directly translates into energy consumption. While these elements can be considered as a fixed part, the necessity of constant conversion of the scene view is variable and can be influenced by manipulating the computational complexity of the interaction and control algorithms used, or the complexity of the virtual scene, which will require the allocation of processor resources and operating memory for each frame, depending on the number and the complexity of the elements necessary to render. Before the introduction of optimisation, the battery discharged after 3 hours of uninterrupted VR sessions. After entering the described settings, this time has increased to 6 hours.
4. The studied group of students did not complain about the disruption of the application's fluency. All other comments of the research participants concerned specific situations and the way the application works. In most cases, they referred to personal aesthetic feelings and their own preferences regarding ergonomics.

The evaluation of the application's overall quality is subjective. It does not depend only on the quality of the software and the device, but also on the individual preferences and perceptions of users. Everyone has different preferences and their own quality thresholds determining the fitness and comfort of the application.

6. Future works

The application with various modifications will be expanded and used in further studies. In order to increase its functionality, it is planned to carry out new optimisation methods, including:

- size of textures mapped to generated models' reduction,
- use of additional maps for materials to increase, compensate for the quality of models with simplified geometry.

The authors are also planning to carry out performance tests on a larger number of devices included in the group of low-end and high-end devices.

7. References

[1] Morrison A, Davies A, Brečević G, Sem I, Boykett T and Brečević R 2010 Designing performativity for mixed reality installations *FORMakademisk* **3**

[2] Kim K H 2016 The Potential Application of Virtual, Augmented, and Mixed Reality in Neurourology *Int. Neurourol. J.* **20** 169–70

[3] Mateu J, Lasala J M and Alamán X 2015 Developing Mixed Reality Educational Applications: The Virtual Touch Toolkit *Sensors* **15**

[4] Szymczyk T 2018 Interactive Educational Application using VR in a Smartphone *12th International Technology, Education and Development Conference* pp 164–9

[5] Minocha S, Tudor A-D and Tilling S 2017 Affordances of Mobile Virtual Reality and Their Role in Learning and Teaching *Proceedings of the 31st British Computer Society Human Computer Interaction Conference* HCI '17 (Swindon, UK: BCS Learning & Development Ltd.) pp 44:1-44:10

[6] Wawrzyk M, Wesołowska K, Plechawska-Wójcik M and Szymczyk T 2019 Analysis of Brain Activity Changes Evoked by Virtual Reality Stimuli Based on EEG Spectral Analysis. A Preliminary Study *Information Systems Architecture and Technology: Proceedings of 39th International Conference on Information Systems Architecture and Technology – ISAT 2018* ed L Borzemski, J Świątek and Z Wilimowska (Cham: Springer International Publishing) pp 222–31

[7] Agnello F, Avella F and Agnello S 2019 Virtual Reality for Historical Architecture *ISPRS - Int. Arch. Photogramm. Remote Sens. Spat. Inf. Sci.* **XLII-2/W9** 9–16

[8] Ball C and Johnsen K 2016 An Accessible Platform for Everyday Educational Virtual Reality *Everyday virtual reality (WEVR), 2016 IEEE 2nd workshop* pp 26--31

[9] Innocenti E, Geronazzo M, Vescovi D, Nordahl R, Seran S, Ludovico L and Avanzini F 2019 Mobile virtual reality for musical genre learning in primary education *Comput. Educ.* **139** 102–17

[10] Barbieri L, Bruno F and Muzzupappa M 2017 User-centered design of a virtual reality exhibit for archaeological museums *Int. J. Interact. Des. Manuf.* **12** 1–11

[11] Carrozzino M and Bergamasco M 2010 Beyond virtual museums: Experiencing immersive virtual reality in real museums *J. Cult. Herit.* **11** 452–8

[12] Moenning A, Turnbull B, Abel D, Meyer C, Hale M, Guerlain S and Brown D 2016 Developing avatars to improve cultural competence in US soldiers *2016 IEEE Systems and Information Engineering Design Symposium (SIEDS)* pp 148–52

[13] Mayr S, Schneider S, Ledit L, Bock S, Zahradnicek D and Prochaka S 2017 Game-based cultural competence training in healthcare *2017 IEEE 5th International Conference on Serious Games and Applications for Health (SeGAH)* pp 1–5

[14] Recupero A, Triberti S, Modesti C and Talamo A 2018 Mixed Reality for Cross-Cultural

Integration: Using Positive Technology to Share Experiences and Promote Communication *Front. Psychol.* **9** 1223

[15] Montusiewicz J, Skulimowski S and Lypovyi T 2017 Simulation of swarm behaviour as a method of examining students' capability to conduct experiments - a case study *11th International Technology, Education and Development Conference* pp 8655–64

[16] Kim J 2016 Modeling and Optimization of a Tree Based on Virtual Reality for Immersive Virtual Landscape Generation *Symmetry (Basel).* **8** 93

[17] Chabal C and Soulabaille Y 2016 Virtual Reality: a way to prepare and optimize operations in decommissioning projects *PREDEC 2016* pp. 2-10

[18] Montusiewicz J, Barszcz M, Skulimowski S, Kayumov R and Buzrukov M 2018 *12th International Technology, Education and Development Conference* pp 353–63

[19] 3D digital silk road 2019 [Internet]. Available from https://silkroad3d.com

[20] Skulimowski S, Montusiewicz J, Badurowicz M and Barszcz M 2019 Virtual reality for cross-cultural education – a case study *Preprint*

[21] Mobile Operating System Market Share Uzbekistan 2019 [Internet]. Available from http://gs.statcounter.com/os-market-share/mobile/uzbekistan

[22] Lai Z, Hu Y C, Cui Y, Sun L, Dai N and Lee H 2019 Furion: Engineering High-Quality Immersive Virtual Reality on Today's Mobile Devices *IEEE Trans. Mob. Comput.* 1

[23] Olly's seven stages of optimizations for mobile VR 2019 [Internet]. Available from https://unity3d.com/how-to/optimize-mobile-VR-games

[24] Smaga M and Plechawska-Wójcik M 2017 Analiza zastosowania okularów 3D w symulacji przestrzennej na przykładzie wizualizacji architektonicznej *J. of Computer Sciences Institute* **10** 187-192

IOP Conf. Series: Materials Science and Engineering **710** (2019) 012016 doi:10.1088/1757-899X/710/1/012016

Neural modelling of cavitation erosion process of 34CrNiMo6 steel

M Szala[1] and M Awtoniuk[2]

[1] Lublin University of Technology, Faculty of Mechanical Engineering, Department of Materials Engineering, 36 Nadbystrzycka Street, 20-618 Lublin, Poland
[2] Warsaw University of Life Sciences, Institute of Mechanical Engineering, Nowoursynowska 164, 02-787 Warsaw, Poland

m.szala@pollub.pl

Abstract. Artificial neural networks (ANN) are commonly used to solve many industrial problems. However, their application for cavitation erosion evaluation is a quite new attempt. Thus, the aim of this work was to elaborate the neural model of the cavitation erosion process of 34CrNiMo6 steel. Cavitation erosion tests were conducted with a usage of the ultrasonic vibratory method with stationary specimen that relies on the ASMT G32 standard. The proceeding damage of marked steel surface area was observed by means of a scanning electron microscope. Wear was evaluated with profiler measurements, image analysis of cavitation worn surface areas and weighing done in stated time intervals. The cavitation erosion results were analysed with Matlab software by Neural Network Toolbox. The developed neural model of cavitation erosion process that combines exposure time, roughness, area fraction of worn surfaces, and mass loss gives promising results.

1. Introduction

The process of cavitation erosion (CE) of structural materials still seems to be a phenomenon, which is not entirely understood. Cavitation consists in formation and subsequent implosion of vapour bubbles due to local pressure fluctuations in liquid that affects the solid surface, resulting in the material surface layer deterioration. Wear due to cavitation is a common problem in fluid machinery, diesel engines, propellers, pipelines, valves and many more machine components [1–8].

The cavitation erosion is considered as a fatigue process which does not present a constant erosion rate [9,10]. Thus, cavitation wear intensity was divided into four stages: incubation, acceleration, deceleration and terminal, schematically plotted in Figure 1. Even though the main goal of CE testing is to calculate the material mass or volume loss, many researchers focus on investigating the incubation process, first initial stage of erosion. Literature on the subject reports that many researchers take an attempt to characterise CE long-time resistance on the basis of the incubation period or discuss behaviour of the material in the initial period of erosion. For instance, Dybowski et al. [11] analysed the incubation period of the magnesium alloy, as well as the incubation period of Al-Si based alloys [12]. Dular et al. [13] developed the original cavitation erosion model on results of study conducted in the incubation period of CE. Franc et al. [14] investigated the incubation period of work hardened materials, while Garcia et al. [15] observed the initial stage of epoxy coatings erosion, and Ning et al. [16] described the initial period of erosion of epoxy resin, ceramic and Polyurethane coatings. It is worth to mention that the incubation stage is easy to observe for metal alloys due to their plastic deformation

[17–20]. Therefore, ceramic materials or thermally sprayed coatings present a limited incubation period of erosion; thus, thermally sprayed ceramic containing coatings exhibits a negligible incubation stage [21–24]. On the other hand, it is acknowledged by the literature that physical vapor deposited ceramic hard films presents superior to steel cavitation erosion [25–27]. It can be concluded that material factors play a crucial role in wear behaviour or materials.

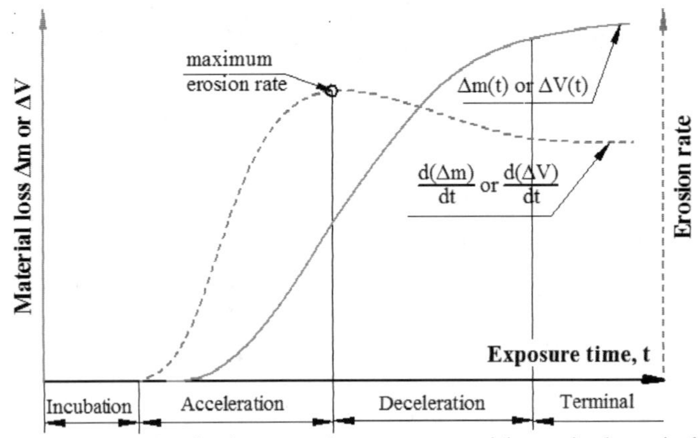

Figure 1. Cavitation erosion curves with typical period names (Δm – mass loss, ΔV – volume loss) [1].

Generally, investigating erosion resistance with a usage of standard test rigs such as vibratory, rotating disc or liquid jet test stand and standardised procedures is time-consuming [28]. Therefore, there still is a demand for elaboration of method that allows to estimate or predict cavitation erosion behaviour of structural metal alloys in a short period of time. Especially, the analysis of wear process done in the incubation period of erosion is considered as a promising solution. The previous paper [1] gave auspicious results for application of computer image analysis software for cavitation erosion description. Thus, in present work the artificial neural network (ANN) was implemented to model CE proses. Especially, that ANN is a promising method for solving problems in various branches and has different applications. ANNs allow to obtain satisfying results for biomedical engineering [29], electrical engineering [30], mechanical engineering [31,32], materials engineering [33,34], pharmacology and pharmacy [35], nutrition and food technology [36]. In addition, Nasiri et al. [37] employed the ANN for cavitation erosion prediction in a centrifugal pump.

Literature on the subject presents application of the ANN for various wear processes prediction [19,38–40]. Even though there are some works presenting its application for modelling the cavitation erosion processes [19] with usage of roughness and residual stress, modelling on the basis of image analysis was not reported in the literature. Therefore, this paper presents the analysis of the cavitation erosion results published in previous paper [1], with the usage of artificial intelligence i.e. ANN.

This paper has two major objectives. First, the main one, is to develop a neural model of the cavitation erosion process of 34CrNiMo6 steel. Second one is to investigate possibilities to use the image analysis as an effective method of creating a useful signal in a process of modelling this phenomenon. The application combination of artificial intelligence with the image analyses processing cavitation damaged worn areas is a new idea, which has not been previously reported in the literature.

2. Material and methods

2.1. Cavitation erosion test of 34CrNiMo6 steel

The description of cavitation erosion experiment is given in the previous work [1]. The 34CrNiMo6 steel was used as a representative material. Total test time lasted for 120 minutes and during time intervals the material mass loss, roughness development and fraction area of damaged surface were

IOP Conf. Series: Materials Science and Engineering **710** (2019) 012016 doi:10.1088/1757-899X/710/1/012016

estimated. The mass loss Δm was estimated with 0.1 mg accuracy and the roughness profile parameter Ra (arithmetic mean of absolute departures form the mean line) was measured with surface profiler. The image analysis was conducted with ImagePro software, which allows to estimate the fraction area of damaged surface vf. Observations were done in the marked sample surface, measurement procedure is described in reference [1]. These three quantitative results were used for the ANN procedure. In addition, the damaged surface was observed with usage of the scanning electron microscope (SEM) equipped with BSE and Topo modes. Observation was conducted in the marked area after stated time intervals of cavitation testing.

2.2. Development of artificial neutrons network model

The ANN model was elaborated with usage of Matlab environment (2017a) and Neural Network Toolbox. The model block diagram is shown in Figure 2. The input signals were cavitation exposure time t [min], roughness parameter Ra [μm] and volume fraction of damaged surface vf [%]. The output signal was the mass loss Δm [mg] measured after each cavitation erosion testing time-interval. Mass loss is a parameter describing the degree of cavitation wear [1].

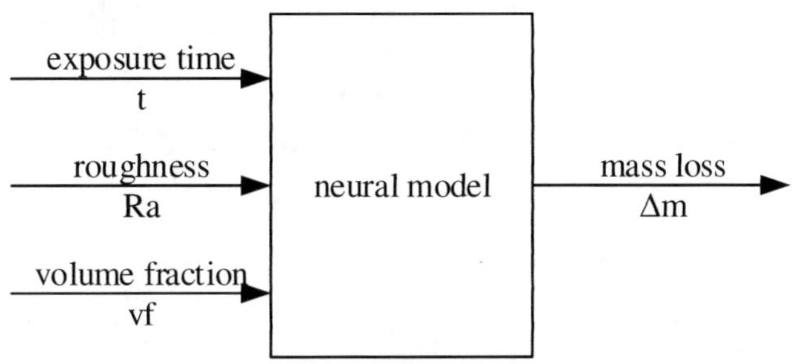

Figure 2. Model block diagram.

The neural model consisted of an input layer, a hidden layer and an output layer. Each of layers was characterised by a certain number of neurons. For example, the model described as 3-2-1 means a network with 3 input neurons, 2 hidden neurons and 1 output neuron. The number of neurons in the input and output layers was related to the number of input and output signals of the model. The number of neurons in the hidden layer was determined experimentally by evaluating the performance of neural networks with different numbers of hidden neurons. Details of this analysis will be described in the Results and discussion section. Learning took place according to Levenberg-Marquardt backpropagation algorithm. The maximum number of epochs to train the network was equal to 15. As a result of the research, $n = 8$ measurement samples were received. Due to the limited data set, k-fold cross validation was decided to be performed. A predicate $k = n$ was assumed, so it was a type of the so-called leave-one-out cross validation [41].

Two model performance evaluation indexes were used: normalised root mean square error NRMSE (informally called fit, which was the name used hereinafter) and root mean square error RMSE. By and large, the model describes the phenomenon in more detail, if the fit value is higher and the RMSE value is lower. The indexes were calculated as follows:

$$\text{fit} = \left(1 - \frac{\|y - \hat{y}\|_2}{\|y - \bar{y}\|_2}\right) \cdot 100\% \qquad \text{RMSE} = \frac{\|y - \hat{y}\|_2^2}{Ns} \tag{1}$$

where: y – output signal (measured), $\hat{y}$ – predicted output signal, $\bar{y}$ – mean of output (measured) signal, Ns – the number of samples.

3. Results and discussion

Cavitation erosion results

Table 1 presents the results of mass loss, roughness measurements and volume fraction of damaged surface areas. The cavitation worn surface of steel is presented in Figures 3 and 4. It is obvious that the incubation period of erosion lasted for 20 minutes of cavitation testing. On the basis of analysis of data from Table 1 and morphology of surface (Figure 3 and 4) versus time, results may be assumed as complementary. Wear parameters were increasing nonlinearly with time. The scanning electron microscope SEM observations allowed qualitative evaluation of damaged surfaces. The results indicate development of cavitation pits, with visible plastic deformation and surfaces roughening that proceeded along with time. The quantitative results form Table 1 were used to build up the ANN model of the cavitation erosion process of 34CrNiMo6 steel.

Table 1. The values of cavitation erosion experiment.

Cavitation exposure time, t [min]	Surface roughness, Ra [μm]	Volume fraction, vf [%]	Mass loss, Δm [mg]
0	0.065	1.55	0
5	0.070	5.77	0
10	0.100	7.34	0
20	0.100	19.64	0
40	0.140	50.42	3
60	0.170	60.18	8
90	0.205	62.50	13
120	0.230	70.20	15

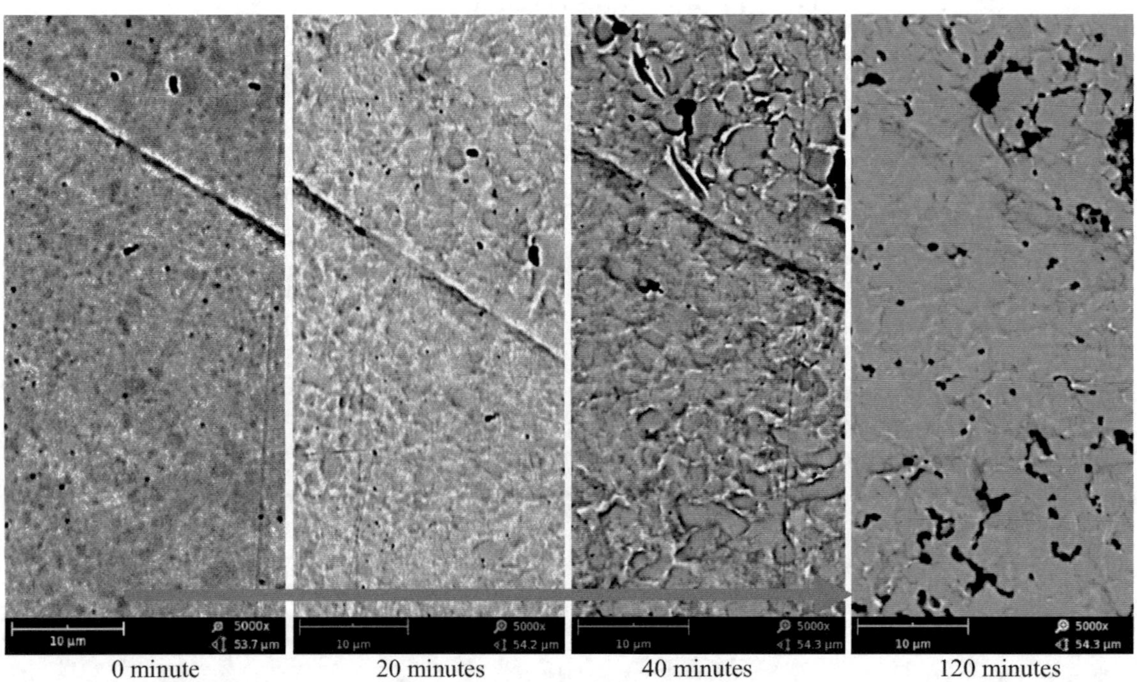

Figure 3. Development of the selected surface area observed after 0, 20, 40, 120 minutes of cavitation testing. Scanning electron microscope (SEM), BSE mode, 5000x.

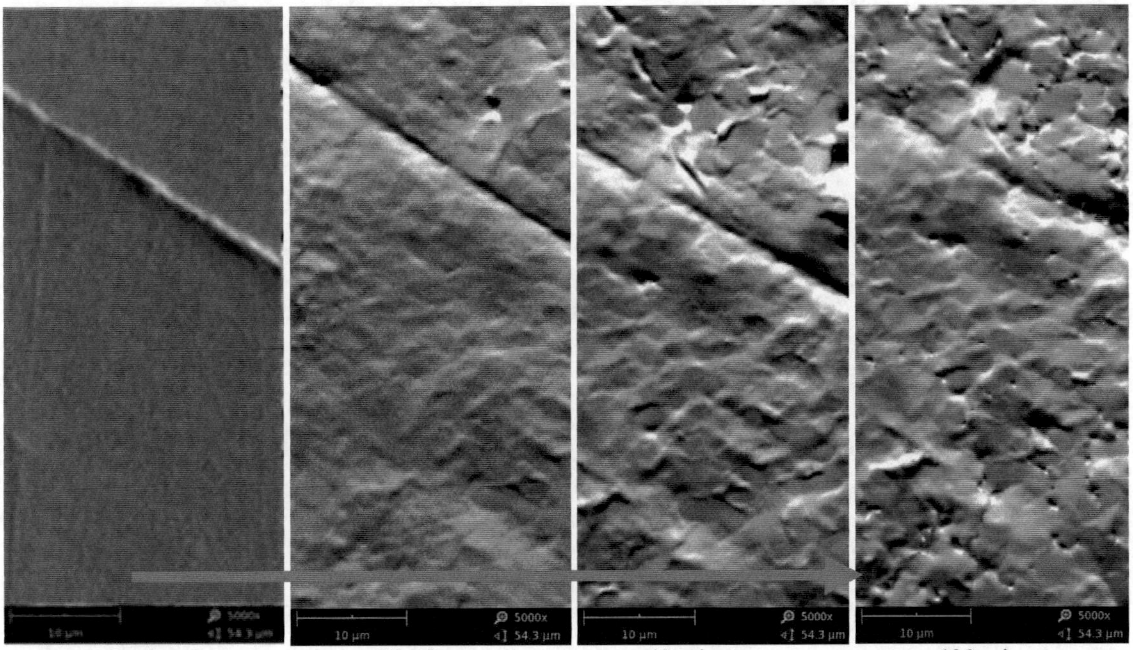

Figure 4 . Development of the selected surface area observed after 0, 20, 40, 120 minutes of cavitation testing. Scanning electron microscope (SEM), SEM-Topo mode (lower row), 5000x.

Modelling the cavitation erosion with ANN

Figures 5-7 show how the model performance has changed depending on the number of neurons in the hidden layer and selected input signals. All networks in the studied range had a high model performance for learning data, i.e. fit index value above 99% and RMSE below 0.013. For this reason, selection of the number of neurons in the hidden layer depended on fitting to the test data. When selecting the size of the hidden network layer, the principle of savings was followed, i.e. network with the highest fit to data with the smallest possible structure was chosen. Finally, neural models M_1, M_2, and M_3 were chosen, having following structures:

- M_1: 3-5-1 for net with time t, roughness Ra, and volume fraction vf as inputs signals (Figure 5),
- M_2: 2-6-1 for net with time t, and roughness Ra as inputs signals (Figure 6),
- M_3: 2-3-1 for net with time t, and volume fraction vf as inputs signals (Figure 7).

Table 2 shows the values of fit and RMSE indexes for selected neural models.

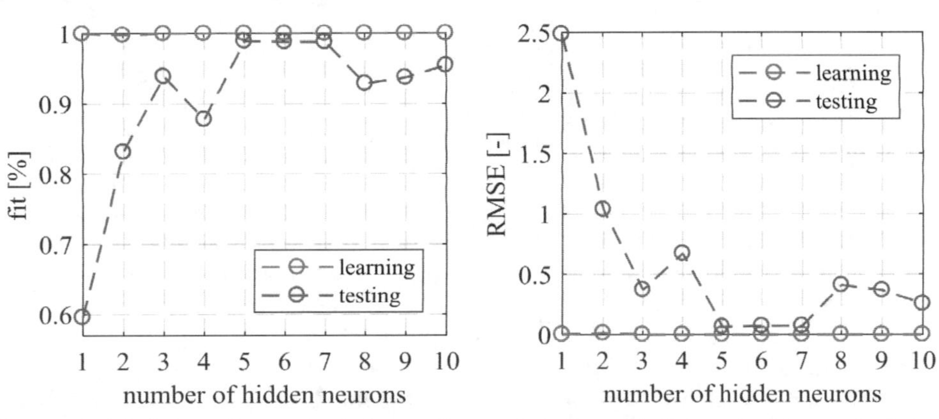

Figure 5. Model performance indexes for model M_1 with 3 inputs (time t, volume fraction vf, and roughness Ra) and different numbers of neurons in hidden layer.

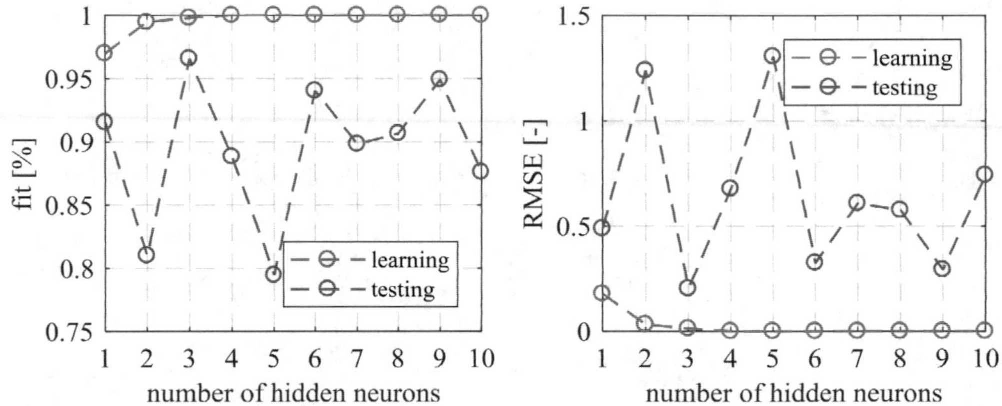

Figure 6. Model performance indexes for model M2 with 2 inputs (time t and roughness Ra) different numbers of neurons in hidden layer.

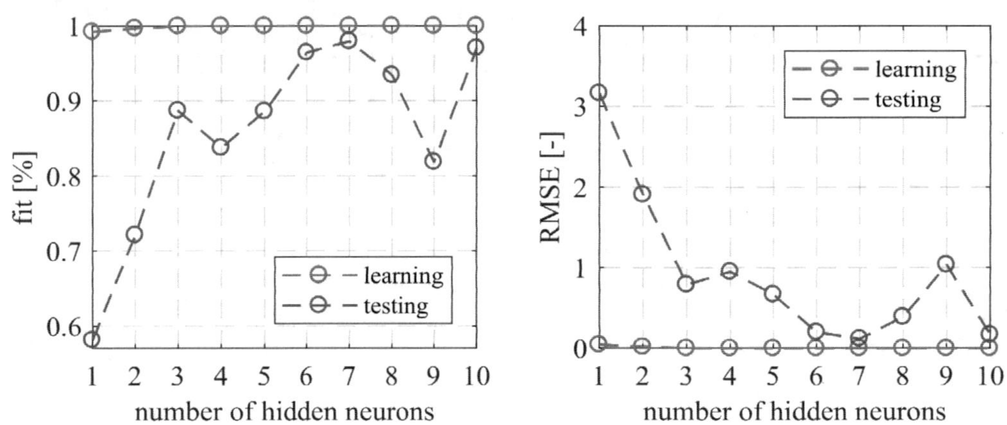

Figure 7. Model performance indexes for model M3 with 2 inputs (time t and volume fraction vf) different numbers of neurons in hidden layer.

Table 2. The values of fit index and RMSE for chosen models.

	fit index [%]		RMSE	
	learning dataset	testing dataset	learning dataset	testing dataset
M1	100	98.9	$6.9\cdot10^{-16}$	0.065
M2	99.8	96.6	0.013	0.204
M3	100	96.5	$8.1\cdot10^{-16}$	0.203

The data shows that the network with three input signals (model M_1) captures cavitation wear most accurately. The model can be reduced to two input signals (M_2 and M_3 models). This will result in a decrease in the fit index by approximately 2.4 percentage points and an increase in the RMSE value (more than 3 times). M_2 and M_3 models differ in one input signal. In the M_2 model, roughness at the input was considered while in the M_3 model it was volume fraction.

Using roughness as a network input is a popular solution. For example, in [19] also a neural model was built with two input signals, i.e. cavitation time and roughness. The network has reached the RMSE value of 0.13. Unfortunately, authors do not specify whether this is a value for learning or testing dataset, so it is difficult to accurately compare the results.

Nevertheless, replacing roughness with a parameter resulting from image analysis, i.e. volume fraction, is a novelty. The results show that it is impossible to clearly judge which version of the network gives more accurate results, M_2 (with roughness as an input) or M_3 (with volume fraction as an input). Therefore, it can be assumed that image analysis is an alternative method of determining cavitation wear.

4. Conclusions

Current paper presents the usage of ANN for describing the cavitation erosion process. The ANN seems to be an auspicious tool for modelling this phenomenon owing to high model performance.

This new attempt that relies on usage of image analysis results i.e. volume fraction (vf) can be successfully used for cavitation erosion phenomena modelling.

Model performance analysis has shown that the most accurate model was the one with three input signals. However, the structure of the model can be successfully reduced to two input signals, i.e. exposure time and roughness or exposure time and volume fraction. Both reduced models have achieved an acceptable performance (fit index was approx. 96.5% and RMSE index 0.204 for testing dataset). It indicates that the models created on the basis of roughness, as well as image analysis results (vf) can be effectively build up and implemented for cavitation erosion process simulation.

ANN model that considers the volume area fraction damaged due to cavitation erosion is an original attempt for description of cavitation erosion process and will be developed in future works.

5. References

[1] Szala M 2017 Application of computer image analysis software for determining incubation period of cavitation erosion – preliminary results *ITM Web of Conferences* **15** 06003

[2] Krella A K and Zakrzewska D E 2018 Cavitation Erosion – Phenomenon and Test Rigs *Advances in Materials Science* **18** 15–26

[3] Francis T & 2014 Cavitation And The Centrifugal Pump: A Guide For Pump Users (Hardback) - Taylor & Francis

[4] Franc J-P and Michel J-M 2004 *Fundamentals of Cavitation* vol 76 (New York, Boston, Dordrecht, London, Moscow: Kluwer Academic Publishers)

[5] Brennen C E 1995 *Cavitation and Bubble Dynamics* (Oxford: Oxford University Press)

[6] Chmiel J, Jasionowski R and Zasada D 2015 Cavitation Erosion and Corrosion of Pearlitic Gray Cast Iron in Non-Standardized Cavitation Conditions *Solid State Phenomena*

[7] Zhang L, Liu Y-H, Luo K-Y, Zhang Y-K, Zhao Y, Huang J-Y, Wu X-D and Zhou C 2018 Tensile Property of ANSI 304 Stainless Steel Weldments Subjected to Cavitation Erosion Based on Treatment of Laser Shock Processing *Materials* **11** 805

[8] Jasionowski R, Przetakiewicz D and Przetakiewicz W 2014 Cavitation Erosion Resistance of Alloys Used in Cathodic Protection of Hulls of Ships *Archives of Metallurgy and Materials* **59** 241–245

[9] Richman R H and McNaughton W P 1990 Correlation of cavitation erosion behavior with mechanical properties of metals *Wear* **140** 63–82

[10] Anon 2010 *ASTM G32-10: Standard Test Method for Cavitation Erosion Using Vibratory Apparatus* (PA, USA: ASTM International: West Conshohocken, Philadelphia)

[11] Dybowski B, Szala M, Kiełbus A and Hejwowski T J 2015 The mechanisms of cavitation erosion of the Elektron21 magnesium alloy *Solid State Phenomena* **229** 99–104

[12] Dybowski B, Szala M, Hejwowski T J and Kiełbus A 2015 Microstructural phenomena occurring during early stages of cavitation erosion of Al-Si aluminium casting alloys *Solid State Phenomena* **227** 255–8

[13] Dular M, Stoffel B and Sirok B 2006 Development of a cavitation erosion model *Wear* **261** 642–655

[14] Franc J-P 2009 Incubation Time and Cavitation Erosion Rate of Work-Hardening Materials *J. Fluids Eng* **131**

[15] García G L, López-Ríos V, Espinosa A, Abenojar J, Velasco F and Toro A 2014 Cavitation resistance of epoxy-based multilayer coatings: Surface damage and crack growth kinetics during the incubation stage *Wear* **316** 124–32

[16] Qiu N, Wang L, Wu S and Likhachev D S 2015 Research on cavitation erosion and wear resistance performance of coatings *Engineering Failure Analysis* **55** 208–23

[17] Romo S A, Santa J F, Giraldo J E and Toro A 2012 Cavitation and high-velocity slurry erosion resistance of welded Stellite 6 alloy *Tribology International* **47** 16–24

[18] Bregliozzi G, Schino A D, Ahmed S I-U, Kenny J M and Haefke H 2005 Cavitation wear behaviour of austenitic stainless steels with different grain sizes *Wear* **258** 503–10

[19] Gao G, Zhang Z, Cai C, Zhang J and Nie B 2019 Cavitation Damage Prediction of Stainless Steels Using an Artificial Neural Network Approach *Metals* **9** 506

[20] Krawczyk J, Jasionowski R, Ura D, Goły M and Frocisz Ł 2018 The effect of cavitation erosion on austenitic-ferritic steel *Zeszyty Naukowe Akademii Morskiej w Szczecinie* **56**

[21] Szala M and Hejwowski T 2018 Cavitation Erosion Resistance and Wear Mechanism Model of Flame-Sprayed Al2O3-40%TiO2/NiMoAl Cermet Coatings *Coatings* **8** 254

[22] Tomlinson W J, Kalitsounakis N and Vekinis G 1999 Cavitation erosion of aluminas *Ceramics International* **25** 331–8

[23] Łatka L, Szala M, Michalak M and Pałka T 2019 Impact of atmospheric plasma spray parameters on cavitation erosion resistance of Al2O3-13%TiO2 coatings *Acta Phys. Pol. A* **136** 342–7

[24] Szala M, Dudek A, Maruszczyk A, Walczak M, Chmiel J and Kowal M 2019 Effect of atmospheric plasma sprayed TiO2-10% NiAl cermet coating thickness on cavitation erosion, sliding and abrasive wear resistance *Acta Phys. Pol. A* **136** 335–41

[25] Krella A and Czyżniewski A 2008 Cavitation erosion resistance of nanocrystalline TiN coating deposited on stainless steel *Wear* **265** 963–70

[26] Krella A K 2011 The new parameter to assess cavitation erosion resistance of hard PVD coatings *Engineering Failure Analysis* **18** 855–67

[27] Szala M, Walczak M, Pasierbiewicz K and Kamiński M 2019 Cavitation Erosion and Sliding Wear Mechanisms of AlTiN and TiAlN Films Deposited on Stainless Steel Substrate *Coatings* **9** 340

[28] Steller J 1999 International Cavitation Erosion Test and quantitative assessment of material resistance to cavitation *Wear* **233–235** 51–64

[29] Sałat R and Sałat K 2013 The application of support vector regression for prediction of the antiallodynic effect of drug combinations in the mouse model of streptozocin-induced diabetic neuropathy *Computer Methods and Programs in Biomedicine* **111** 330–337

[30] Sałat R and Awtoniuk M 2015 Black box modeling of PIDs implemented in PLCs without structural information: a support vector regression approach *Neural Computing and Applications* **26** 723–734

[31] Winiczenko R, Sałat R and Awtoniuk M 2013 Estimation of tensile strength of ductile iron friction welded joints using hybrid intelligent methods *Transactions of Nonferrous Metals Society of China* **23** 385–391

[32] Kosowski K, Tucki K and Kosowski A 2010 Application of Artificial Neural Networks in Investigations of Steam Turbine Cascades *Journal of Turbomachinery* **132**

[33] Zagórski I and Kulisz M 2019 Effect of technological parameters on vibration acceleration in milling and vibration prediction with artificial neural networks ed M Kulisz, M Szala, M Badurowicz, W Cel, M Chmielewska, Z Czyz, K Falkowicz, J Kujawska and T Tulwin *MATEC Web of Conferences* **252** 03015

[34] Zagórski I, Kulisz M, Kłonica M and Matuszak J 2019 Trochoidal Milling and Neural Networks Simulation of Magnesium Alloys *Materials* **12** 2070

[35] Sałat R and Sałat K 2015 Modeling analgesic drug interactions using support vector regression: a new approach to isobolographic analysis. *Journal of pharmacological and toxicological methods* **71** 95–102

[36] Winiczenko R, Górnicki K, Kaleta A and Janaszek-Mańkowska M 2018 Optimisation of ANN topology for predicting the rehydrated apple cubes colour change using RSM and GA *Neural Comput & Applic* **30** 1795–809

[37] Nasiri M R, Mahjoob M J and Vahid-Alizadeh H 2011 Vibration signature analysis for detecting cavitation in centrifugal pumps using neural networks *2011 IEEE International Conference on Mechatronics* 2011 IEEE International Conference on Mechatronics pp 632–5

[38] Kurt H I and Oduncuoglu M 2015 Application of a Neural Network Model for Prediction of Wear Properties of Ultrahigh Molecular Weight Polyethylene Composites *International Journal of Polymer Science*

[39] Humelnicu C, Ciortan S and Amortila V 2019 Artificial Neural Network-Based Analysis of the Tribological Behavior of Vegetable Oil–Diesel Fuel Mixtures *Lubricants* **7** 32

[40] D'Addona D M, Ullah A M M S and Matarazzo D Tool-wear prediction and pattern-recognition using artificial neural network and DNA-based computing *Journal of Intelligent Manufacturing* 1–17

[41] Tangirala A K 2015 *Principles of system identification : theory and practice* (Boca Raton: CRC Press)

IOP Conf. Series: Materials Science and Engineering **710** (2019) 012017 doi:10.1088/1757-899X/710/1/012017

Evaluation of structure stability with the use of remote load

Szymon Molski[1], Paweł Lonkwic[2], Hubert Ruta[1] and Tomasz Krakowski[1]

[1] AGH University of Science and Technology, The Faculty of Mechanical Engineering and Robotics, Department of Machine and Transport Engineering, A. Mickiewicza Av. 30, 30-059 Cracow, Poland
[2] The State School of Higher Education in Chełm The Institute of Technical Sciences and Aviation, 54, Pocztowa Street, 22-100 Chełm, Poland

molski@agh.edu.pl, plonkwic@pwsz.chelm.pl, hubert.ruta@agh.edu.pl, krakowsk@agh.edu.pl

Abstract. It is a quite common practice to place the load at some distance from the machine restraint. In such a case, it is possible to evaluate the values of stresses and displacements by means of numerical analyses. This article presents the numerical analysis results for the carrying structure of the friction lift power unit with the use of the remote load. The impact of the remote load on the strain of the lift power unit carrying structure is also described. Attention is paid to the places of excessive strain of the structure.

1. Introduction

FEM numerical simulations are used to evaluate the structure in terms of its strain, stiffness, safety, correct geometry and prototyping costs reduction. In the publication [1] the authors described the use of numerical methods to evaluate buckling of a thin-walled profile in the shape of a C-beam subject to axial pressing. In the publication [2] the authors used numerical simulations to evaluate the strain of the safety gear structure in the passenger lift. The results of numerical simulations are compared with the results of physical experiment that has given concurring results. In the article [7] authors suggested a 1D/3D finite element model for static analysis of solid frames with thin-walled rods and open cross-section. This model involves standard discretisation using beam thin-walled elements connected with space discretisation of the frame joints. In the publication [3] the authors described the results obtained in the analysis of the safety gear structure by means of the Finite Element Method and Abaqus software. The authors in the paper [8] present a finite element model for analysis of thin-walled structures. Standard discretisation using beam thin-walled elements is connected with space discretisation of some parts of the frame. Transition elements are used for consistent coupling between shell (3D) and beam (1D) elements. The paper [9] presents calculation models used in the analysis of various building structures and soil-structure interaction. Basic schemes, principles and possibilities, which consider the change in time geometric parameters, material and boundary conditions are described. In the paper [13] authors presented the method of strength and dynamic analysis of agricultural rollers. Selected forms of showing the simulation results, that is, stresses, forces displacements etc., are also presented. The practical method of carrying out computer simulations is shown, as well. The descriptions of these methods are extensive enough to show aspects of costs and time reductions, when carrying out these types of analyses. In the publication [4] the authors discussed the method of modelling a composite profile, which was subject to axial pressing. Numerical simulation results are compared with the results received from tensile testing. Authors [10] present a method of strength calculations of truck frames by using integration of MSC Adams software – for dynamics analysis of mechanical systems and NX Nastran/Femap software – for the Finite Element Method analysis. In this study, a method of reduction of freedom degrees has been developed based on the Craig-Bampton approach. The interface has been applied to calculate the strength of the frame of the selected truck running on the test track. The model of the truck can be treated as its virtual prototype useful in the design process. In the publication [5] the authors described the results of numerical simulation conducted to assess some displacements of the machine tool body used for the extruded joints and operated in equal geometric configurations.

This paper [11] presents the study of the vibration characteristics of the truck chassis, that include natural frequencies and mode shapes. The responses of the truck chassis, which include stress distribution and displacement under various load conditions are also observed. In the publication [6] the authors used the Finite Element Method available in Catia program to evaluate strain generated by a carrying frame in a heavy vehicle. Stiffness of the frame elements under load was also evaluated. Applicability of the Finite Element Method is much more extensive in technology, not only for purposes connected with mechanical engineering but also in different fields. Authors in the publication [14] present a problem of the basic concept of using the finite element method in constructing an analytical model for the study of the behaviour of reinforced concrete members. Finite elements chosen to represent concrete, steel reinforcement, and bond links between concrete and steel reinforcement are described. The article [12] presents the calculation model of a structure to drain and disinfect waste containers, which was provided for the fatigue strength analysis. A process of creating the model and discussed techniques, which were used to reflect the kinematics bindings of researched device, are described in detail. In this work [15] the authors focused on solutions for the Euler-Bernoulli and Timoshenko beam theories, in which material behaviour may be elastic or inelastic. Formulation relies on integration of the local constitutive equation over the beam cross section, to develop relations for beam resultants. For this case, axial, bending and shear effects were included. This allows consideration in a direct manner of elastic and inelastic behaviour, with or without shear deformation. The author in the paper [16] reviews basic concepts involved when constructing nonlinear load- deflexion curves of framed structures by the finite element method. The study is limited to the geometrically nonlinear behaviour of elastic structures. Three different procedures are considered: linear incremental method, nonlinear incremental method, and direct method. For each method, governing equations are derived, and the procedure used to plot the load- deflexion curve is outlined. The paper [17] presents a finite element analysis of large deformations in elastoplastic frames. Emphasis is put on the step-by-step analysis of the nonlinear behaviour of such structures. Advantage is taken of the so-called natural formulation in terms of the generalised cross-sectional forces. Limit state functions applied to different loading conditions and frame geometries are implemented into a computer program. This paper [19] presents exploratory stiffness-strength analysis of frame of ARIA sporting car. A conception of innovative car frame and the design assumptions is presented. Moreover, MES models, project modification stages and analysis results are described. The entire paper is summed up by describing reached results and presenting conclusions related with forward progress of construction. The authors presented [18] design of infilled frames to resist lateral loads on buildings in terms of their failure modes, failure loads, and initial stiffness using procedures proposed by previous authors. This verification is made by comparing the results of analytical procedures of previous authors with those of a new finite element model for infilled frames, which are verified using experimental results. To model the interface between the frame and the infill and the mortar joints surrounding the blocks of masonry, a non-associated interface models are formulated using the available test data on masonry joints. This study [20] presents a simple formulation to treat large deflexions by the finite element method. The proposed formulation does not use the displacement concept. It considers position as the main unknown variable of the problem. Strain determination is performed directly from the proposed position concept. Non-dimensional space is created, and relative curvature and fibres length are calculated for both reference and deformed configurations and used to directly compute the strain energy at general points. This paper [21] presents a finite element formulation for the numerical analysis of three-dimensional framed steel, reinforced concrete or composite steel and concrete structures subjected to fire. Within this context, the purpose of this work is to present the steps taken to extend a previously developed static analysis procedure with beam elements, in order to cope with the thermal and structural analysis of structures under fire action.

All publications mentioned above present the use of the FEM method, assuming that the load is axially applied in relation to the structure. This publication demonstrates the FEM method used to evaluate the strain of the power unit carrying structure in the friction lift. The load is located at some distance from the drive attachment that directly applies load to the carrying structure.

2. Numerical model of the power unit carrying frame in the friction lift

The power unit of the friction lift consists of a reducer, where a friction wheel and an electric motor are installed. Depending on the lift structure, the carrying structures can be distinguished considering the engine room location, see figure 1.

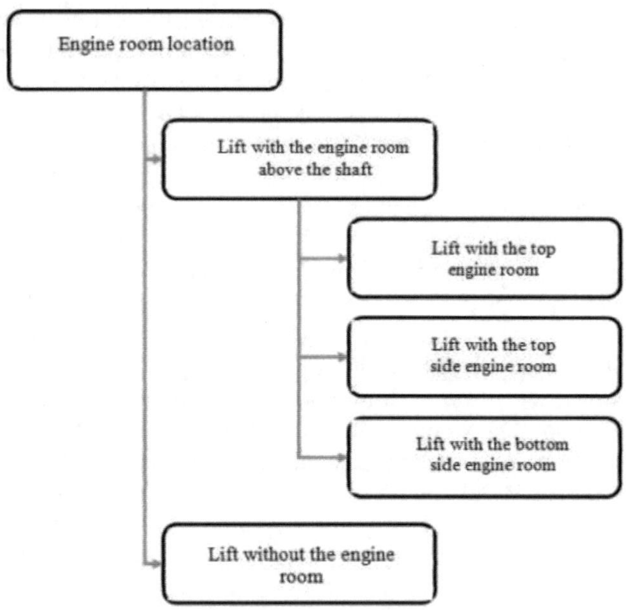

Figure 1. Types of carrying structures of the friction lift power units due to the engine room location.

The friction lift carrying structure that is installed in the bottom side engine room belongs to those which are more complex. Figure 2 presents a three-dimensional model of such a frame with its main subassemblies and main dimensions.

The carrying frame is made of hot rolled C180 type profiles, and the structure is welded. It is mounted to the structural surfaces of the engine room with the interleaved anchors of Hilti type in the places presented in figure 2. These are sheet metal elements with holes on the external side of the structure. A relation between forces acting on the unit and inertial forces is described by means of the below relation [5] in the form of a matrix:

$$f(t) = |M|\{\ddot{u}\} + |C|\{\dot{u}\} + |K|\{u\}$$

where: $\{\ddot{u}\}, \{\dot{u}\}, \{u\}$ – represent vectors of acceleration, speed and displacement respectively, $|M|, |C|, |k|$ – represent the matrices: mass, rigidity and damping, respectively.

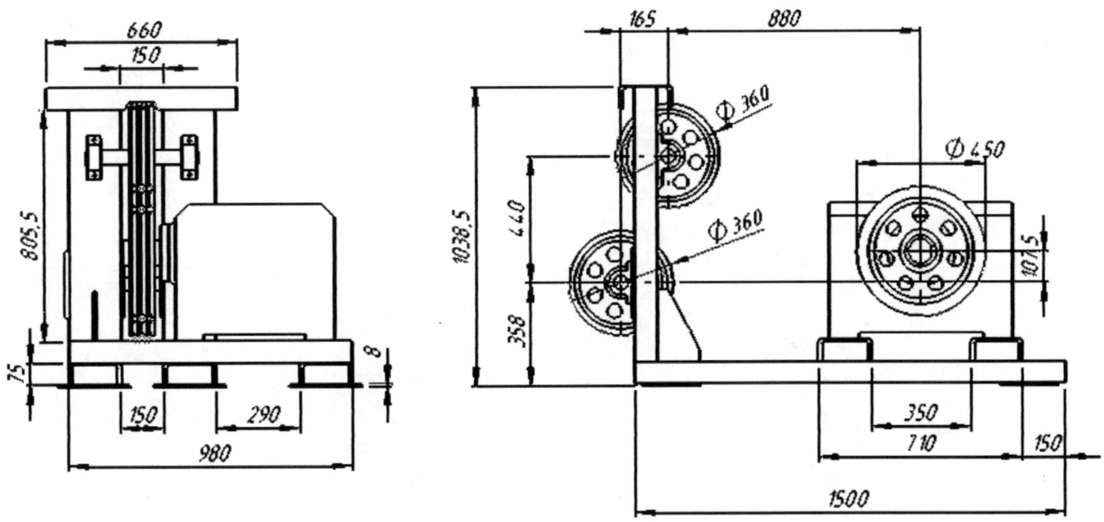

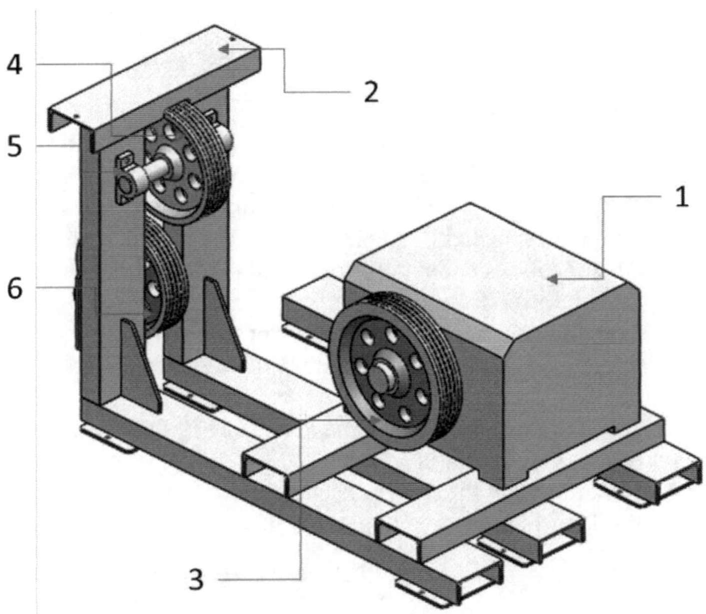

Figure 2. A three-dimensional model of the carrying frame in the friction lift power unit with the bottom side engine room (the authors' private source): 1 – a reducer (without an electric motor), 2 – a carrying frame, 3 – a friction wheel, 4 – a rope pulley axis, 5 – a bearing block, 6 – a directional rope pulley.

All frame components have been designed with the use of steel with the following mechanical properties [22]:

- Young's modulus E = 210 GPa,
- Poisson's ratio $\nu = 0.3$,
- material density $\rho = 7860$ kg/m3,
- yield point $R_e = 230$ MPa

The load system of the modelled unit is presented in figure 3. Loads resulting from the lift operation are generated by the carrying ropes which belt the wheels as presented in the below scheme:

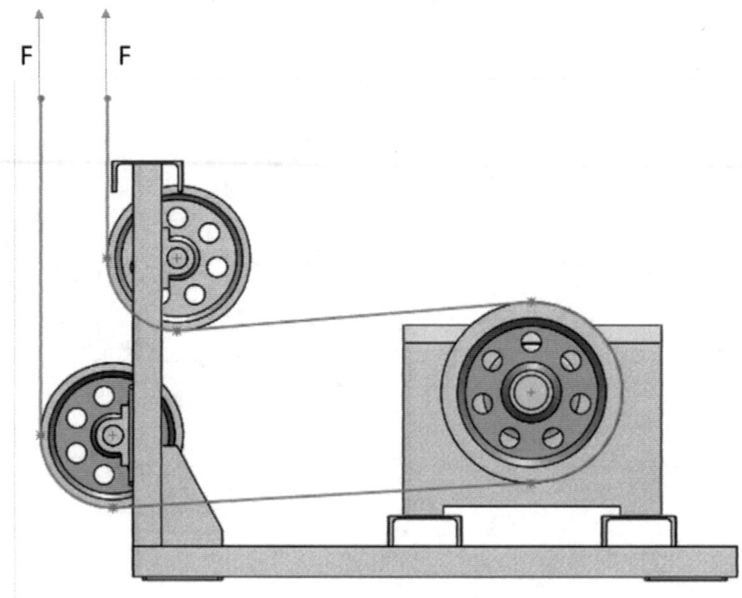

Figure 3. A scheme of the reducer rigging and the carrying ropes load.

The system discretisation process was a primary stage in the FEM analysis, for which the SolidWorks Simulation was used. Contact parameters were based on the mutual interaction of subassemblies in the tangent and regular direction. A total value of 16000 N was taken as the ropes load. The applicable value was a sum of masses: cabin with frame, load capacity and counterweight. A direction of forces generated in the ropes was restrained to the ropes ends (grey rope) and presented in figure 3.

Numerical analysis was conducted for two cases:

Case I – impact of remote load on places where the reducer is mounted;

Case II – impact of remote load on places where the slide bearings are mounted.

In the analysed case, the load impact on the frame is executed as a direct transfer to the place of the reducer and directional rope pulley mounting. Thus, the numerical model of the reducer, the friction wheel and the axis has been omitted in the model discretisation.

As per the 1s[t] case, the load transfer to the place where the reducer is mounted on the frame is presented in figure 4.

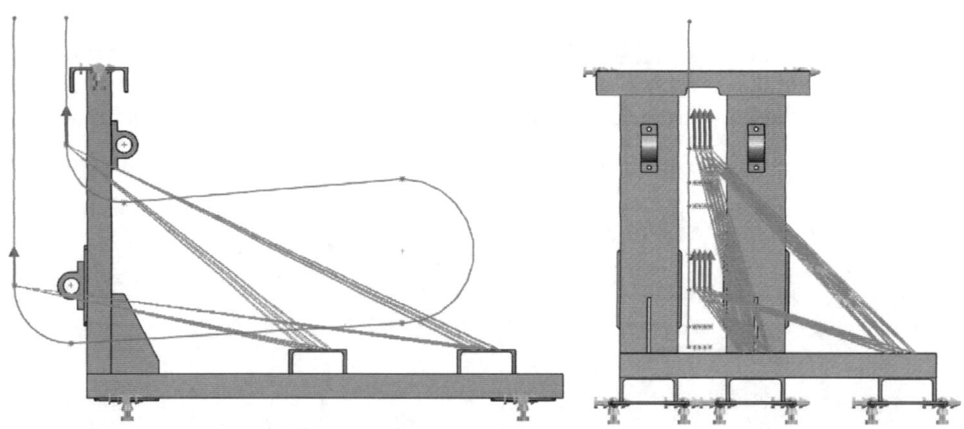

Figure 4. The numerical model boundary conditions of the carrying frame.

IOP Conf. Series: Materials Science and Engineering **710** (2019) 012017 doi:10.1088/1757-899X/710/1/012017

A discrete model of the analysed system is presented in figure 5. A high-quality solid mesh with Jacobian determinants located in nodes was defined for discretisation. The size of the assumed mesh was 15 mm with tolerance of successive iterations of 0.75 mm. Such parameters allowed to create a mesh with 198802 nodes and 100873 elements.

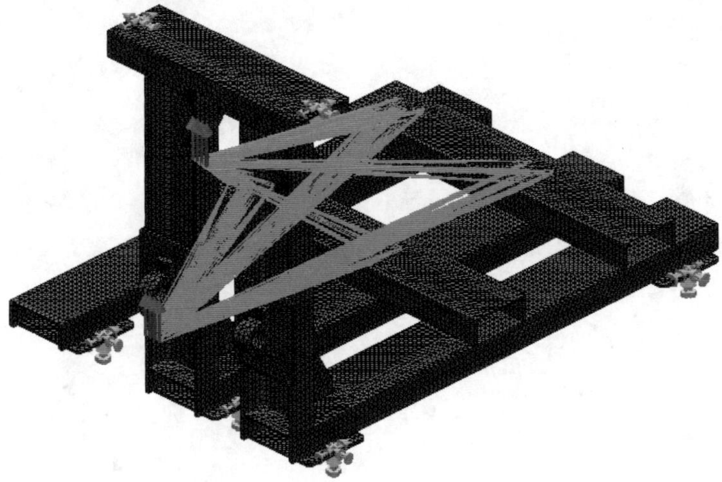

Figure 5. The frame discrete model prepared for the 1st case.

The results of stress numerical simulation for the 1st case are presented in figure 6.

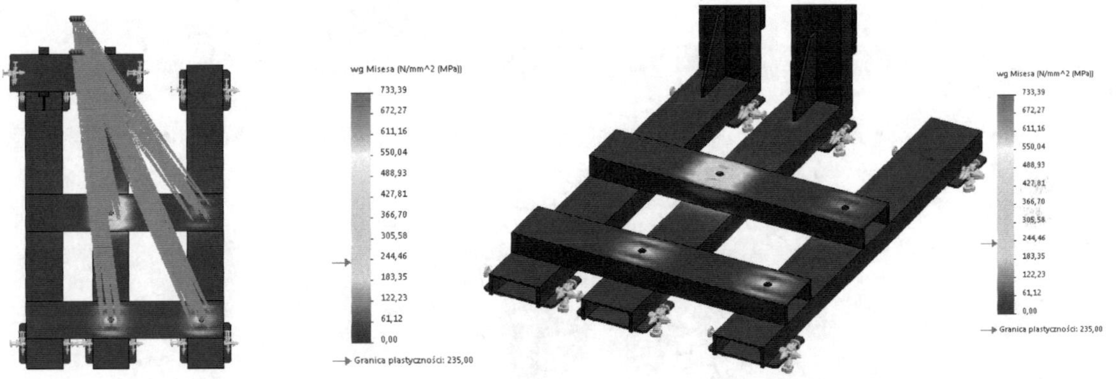

Figure 6. Maximum stresses as per Huber-Mises hypothesis.

Basing on the analysis of the above mentioned figure, it is noticed that the places of the reducer installation are strained excessively. This requires additional analysis of the frame, in terms of its stiffening. The maximum stresses generated under the applied load are 760 MPa, which significantly exceed the yield point of 230 MPa for the assumed material.

Maximum displacements present in the place of the highest strain of the frame are presented in figure 7. The results of calculations regarding a safety factor are presented in figure 8. A red field represents the area with the highest values of stresses; thus, the safety factor is below 1.2. As per the 2nd case, the load transfer to the place where the slide bearings of the directional rope pulleys are installed is presented in figure 9. A discrete model of the analysed system is presented in figure 10. Figure 11 presents the results of numerical simulation for the direct load transfer to the place, where the slide bearings are installed generating load of frame vertical beams.

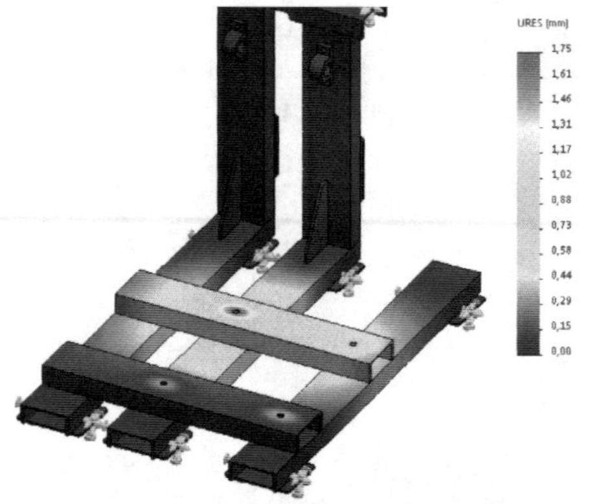

Figure 7. Maximum displacements.

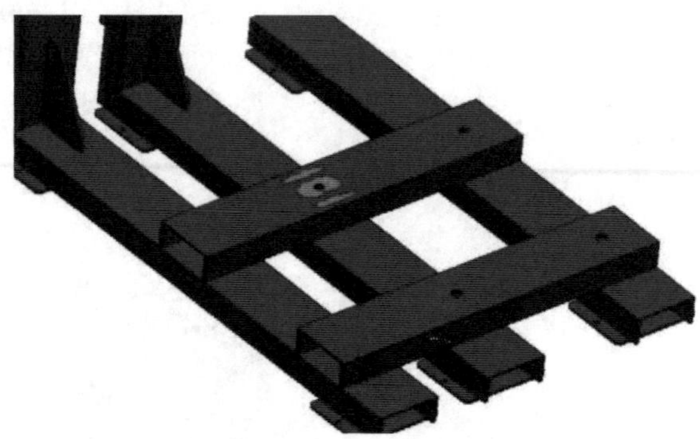

Figure 8. The safety factor of the frame structure.

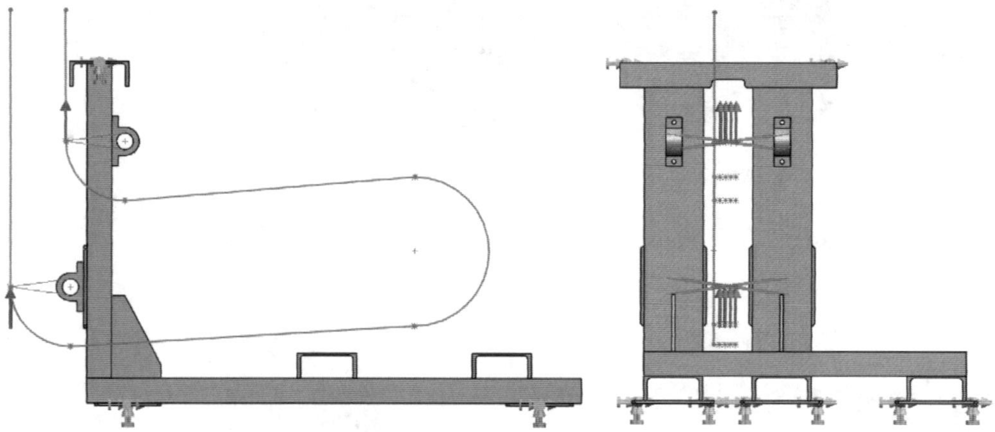

Figure 9. The boundary conditions of the carrying frame numerical model for the 2nd case.

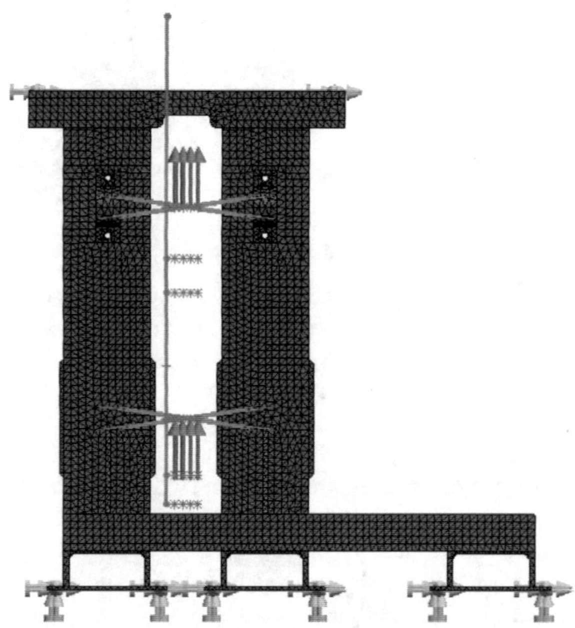

Figure 10. A discrete model of the frame for the 2nd case.

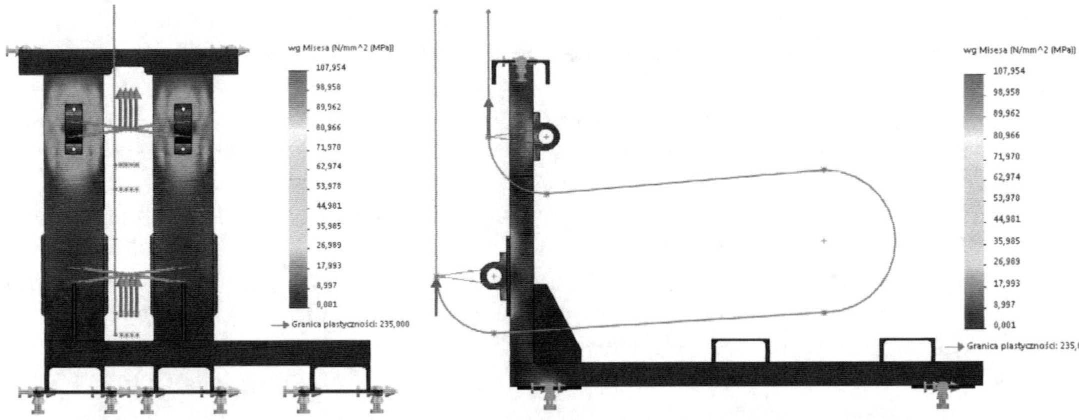

Figure 11. The maximum stresses as per Huber-Mises hypothesis for the 2nd case.

Having analysed the above figure, it can be stated that the places of the reducer installation are strained excessively. Thus, additional analysis of the frame in terms of its stiffening is required. The maximum stresses generated under the applied load are 107 MPa.

Maximum displacements expected in the place of the highest strain of vertical beams are presented in figure 12.

The value of safety factor assumed for this type of structures is 1.2. The calculation results regarding the safety factor are presented in figure 13. In the 2nd case, there are no places where the strain of the frame elements would give the value of safety factor below 1.2, see figure 13.

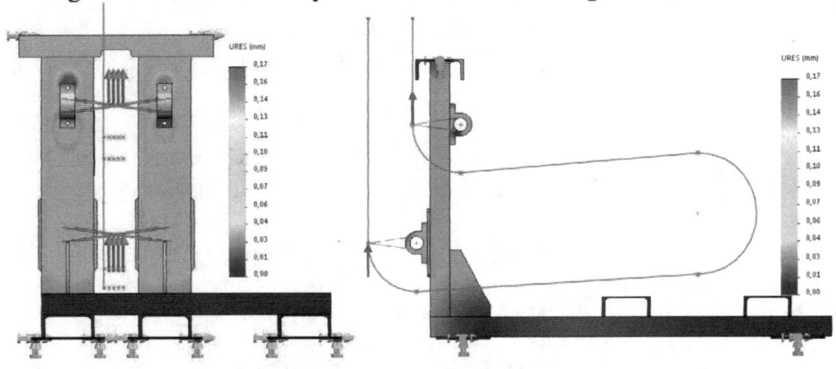

Figure 12. Maximum beams displacements.

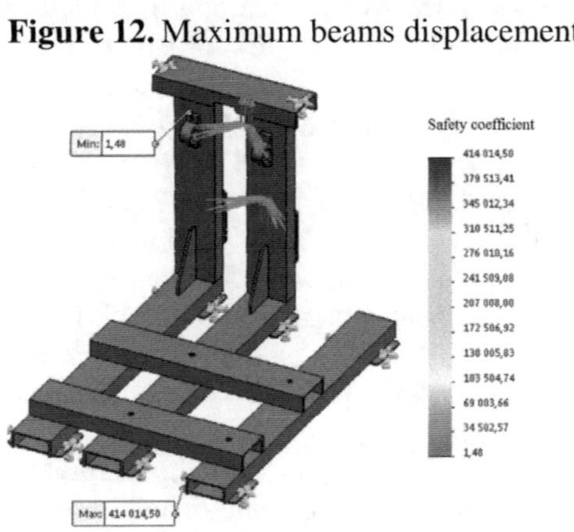

Figure 13. The safety factor of the frame structure.

IOP Conf. Series: Materials Science and Engineering **710** (2019) 012017 doi:10.1088/1757-899X/710/1/012017

3. Conclusions

The numerical analysis of load cases for the friction lift frame structure allows formulation of the following final conclusions:

- The frame structure subject to analysis is used in friction lifts with the bottom side engine room. Due to real way of frame structure loading, the, so called, remote load was assumed for simulation purposes.
- The value of load used in simulation was assumed on the basis of calculation for the friction lift of nominal carrying capacity of 630 kg.
- The assumed frame geometry and the method of its loading were provided in accordance with the design of the friction lift, which is planned to be implemented in the next calendar year.
- Numerical analysis of 1st load case showed that the value of stresses reduced according to the Huber's hypothesis was exceeded in the place of the reducer body restraint. Therefore, appropriate frame modifications involving reduction of stresses should be provided in those places.
- Numerical analysis of 2nd load case showed that maximum values of reduced stresses are 107 MPa and are lower than the yield point. Therefore, the resultant safety factor is 1.48 and is higher than the recommended factor of 1.2.
- Presented numerical analysis results were discussed with the manufacturer of such structures, who indicated that analysis of mesh sensitivity for greater accuracy of results is not necessary at this stage. Such an approach ensues from the necessity to modify the structure of the analysed frame.
- Having analysed the presented case, it can be concluded that the obtained numerical results represent realistic values. However, simulation results should be treated as rough due to the geometric simplification of the entire structure provided during the simulation.
- As a result of the numerical analysis of the structure, it is justified to formulate a conclusion that the discussed frame structure should be modified in terms of reducing the value of stresses, and then simulation calculations should be re-taken for the purposes of verification.

4. References

[1] Różyło, P., Teter, A., Debski, H., et al. (2017) Experimental and Numerical Study of the Buckling of Composite Profiles with Open Cross Section under Axial Compression *Appl Compos Mater* 24 pp 1 – 14 https://doi.org/10.1007/s10443-017-9583-y.

[2] Lonkwic, P., Różyło, P., Dębski, H. (2015) Numerical and experimental analysis of the progressive gear body with the use of finite-element method *Eksploatacja i Niezawodność – Maintenance and Reliability* 17 (4) pp 544–550 http://dx.doi.org/10.17531/ein.2015.4.9.

[3] Lonkwic, P., Różyło, P. (2016) Theoretical and experimental analysys of loading impact from the progressive safety gears on the free fall method *Advances in Science and Technology Research Journal* 10(30) pp 103-109 https://doi.org/10.12913/22998624/62628.

[4] Wysmulski, P., Dębski, H., Różyło, P., Falkowicz, K. (2016) A study of stability and post-critical behaviour of thin-walled composite profiles under compression *Eksploatacja i Niezawodność – Maintenance and Reliability* 18 (4) pp 632–637 http://dx.doi.org/10.17531/ein.2016.4.19.

[5] Markowski, T., Mucha, J., Witkowski, W. (2013) FEM analysis of clinching joint machine's C-frame rigidity *Eksploatacja i Niezawodność – Maintenance and Reliability* 15 (1) pp 21–57.

[6] Supardjo, A., Tri, W. (2018) Finite element analysis of truck frame by using CATIA V5. *AIP Conference Proceedings* 030029.

[7] Cichoń, Cz., Koczubiej, S. (2008) Consistetn FEM model for thin-walled space frames *Czasopismo Techniczne* 1-B, pp 1 – 18.

[8] Koczubiej, S., Cichoń, Cz. (2012) Shell –Beam model in stability analysis of thin-walled structures with open cross section *Modelowanie Inżynierskie* 44, pp 131-138.

[9] Żakowicz, A., Miedziałowski, Cz. (2011) Building structures models taking into account time-dependable geometric and material parameters and statistical schemes *Civil and Environmental Engineering* 2 pp 197-204.

[10] Przemyk, A., Harlecki, A., Tengler, Sz. (2017) Method of strength calculations of truck frames *Autobusy* 1 pp 1252 – 1257.
[11] Fui, T. H., Rahman, R. A. (2007) Statics and dynamics structural analysis of 4.5 ton truck chassis, *Jurnal Mekanikal* vol. 24 pp 56-67.
[12] Spadło, M., Ignasiak, Ł., Szczepaniak, J. (2013) Waste containers for the strength analysis *Technika Rolnicza, Ogradnicza i Leśna* 3 pp 1-7.
[13] Rutkowski, J., Szczepaniak, J. (2005) Simulations and analysis of virtual computer models in application to series of type of agricultural rollers *Journal of Research and Applications in Agricultural Engineering* Vol. 50 (1) pp 1-9.
[14] Ngo, D., Scordelis, A. C. (2012) Finite Element Analysis of Reinforced Concrete Beams *Journal Proceedings* 64(3), pp 152-163.
[15] Taylor, R., Filipou, F., Sarotas, A., Auricchio, F. (2003) A mixed finite element method for beam and frame problems *Computational Mechanics* 31 (1-2) pp 192-203.
[16] Chajes, A., Churchill, J. (2012) Nonlinear Frame Analysis by Finite Element Methods *Journal of Structural Engineering* 113 (6) pp https://doi.org/10.1061/(ASCE)0733-9445(1987)113:6(1221).
[17] Boni, B., Hindenlang, U., Kleiber, M. (1982) Finite element analysis of two- and three-dimensional elasto-plastic frames—the natural approach *Computer Methods in Applied Mechanics and Engineering* 35(2), pp 221-248.
[18] Ghosh, A., Made, A. (2002) Finite Element Analysis of Infilled Frames *Journal of Structural Engineering* 128 (7) https://doi.org/10.1061/(ASCE)0733-9445(2002)128:7(881).
[19] Mańkowsk, J., Błaszczak, B. (2015) Preliminary strength analysis of innovative frame of single seater sporting car *Mechanik* http://dx.doi.org/10.17814/mechanik.2015.2.44.
[20] Greco, M., Coda, H. (2004) A simple FEM formulation for large deflection 2D frame analysis based on position description *Computer Methods in Applied Mechanics and Engineering* 193 (33–35) pp 3541–3557.
[21] Caldas, R., Fakury, R. H, Sousa, J. Jr. (2014) Finite element implementation for the analysis of 3D steel and composite frames subjected to fire *Latin American Journal of Solids and Structures* http://dx.doi.org/10.1590/S1679-78252014000100001
[22] Material tables included in the program SolidWorks 2019.

IOP Conf. Series: Materials Science and Engineering **710** (2019) 012018 doi:10.1088/1757-899X/710/1/012018

Numerical model for thermal calculation analysis of the wheel hub motor for electric car verified by laboratory tests

B Bedkowski, P Dukalski, T Jarek and T Wolnik

Łukasiewicz Research Network - Institute of Electrical Drives and Machines KOMEL, Katowice, Poland

p.dukalski@komel.katowice.pl

Abstract. The article concerns the thermal analysis of an electric motor for mounting in a wheel hub of an electric car. The CFD model was calibrated on the basis of the results of laboratory tests of the prototype electric motor. The analysis was aimed at recognizing the possibility of increasing the load capacity of an electric motor through the use of dielectric materials with higher thermal conductivity than those used in the actual prototype. The authors presented the problem of heat removal from motors in wheel hubs, construction of the electric motor, the calculation model and analysis of calculation results.

1. Introduction

With the growing interest in electric cars, various directions related to electric drives are being developed. Recently, one of the most recognizable trends in the field of traction electric motors are electric motors for mounting in the hub of a vehicle. The solution is not new, it was first used by Ferdinand Porsche at the turn of the 19th and 20th century to create the world's first hybrid car [1]. With the development of electromobility, the idea of placing motors in wheels is being developed again [2, 3, 4, 5, 6, 7, 8, 9]. Motors of this type can be widely used in the electric vehicle industry for application from small city cars, cars and family cars, to delivery vans and buses. Łukasiewicz Research Network – Institute of Electrical Drives and Machines KOMEL has undertaken research and design works that will enable the provision of services related to the design and production of this type of electric motors for various types of vehicles. This article presents analysis of construction of this type of motors in terms of the possibility of effective heat removal, which will allow to obtain the best drive parameters of the electric motor while maintaining the lowest possible mass.

In the case of electric motors for installation in the wheel hubs of cars, the reduction of weight while maintaining high performance parameters is extremely important, because the weight of the electric motor is an additional unsprung mass for the suspension system. Reducing the weight of the electric motor also has a positive effect on the drives of the central structure, because it reduces the mass of the entire drive and increases the space under the hood of the car.

Main advantages of this type of drive are elimination of elements mediating in the transmission of torque between the electric motor and the vehicle wheel, providing space under the hood, introducing new vehicle control capabilities due to application of torque in each motor, and possibility of using hybrid systems. This solution, in order to be able to be used, requires meeting a number of challenges, one of the most important is removal of heat from the motor winding.

One of the main aspects when designing an electric motor with compact design is adequate limitation of power losses in its individual elements, which can affect possible parameters of work.

In the vast majority of design solutions for electric motors, the main source of heat is winding, which maximum operating temperature limits thermal resistance of insulation. In the case of permanent

IOP Conf. Series: Materials Science and Engineering **710** (2019) 012018 doi:10.1088/1757-899X/710/1/012018

magnet motors, in the construction of an electromagnetic circuit, winding occurs only in the stator. Generated power losses in the form of heat are discharged through the air and the magnetic core to the hull of the electric motor or other elements such as support elements on which the stator is mounted, as in the electric motor presented in this article. Thermally, the most loaded parts of winding are its ends, because as a standard, their construction is located outside the stator slot and do not adhere to the core.

Owing to appropriate calculations, similar to other publications [10, 11, 12, 13, 14, 15, 16, 17] the influence of different materials used for construction of the motor on its thermal state can be analysed.

2. Construction of the considered wheel hub motor

The prototype motor is based on bearing and braking system from Fiat Panda III, while the external dimensions were limited by the dimensions of the 17-inch rim (figure 1). The motor consists of two main components: the rotor and the stator. The source of heat in this type of machine are losses in the rotor (magnets, a yoke of the electromagnetic circuit) and in the stator (winding, core). A significant part of losses are those in the stator. To ensure adequate heat recovery from this element in the stator supporting structure, a water jacket was made and the empty space between was filled with a thermoconductive resin. Figure 2 shows the cross section of the 3D model of the electric motor.

Figure 1. A prototype of a wheel hub motor SMZs200S32 for electric car: front view (on the left) and back view (on the right).

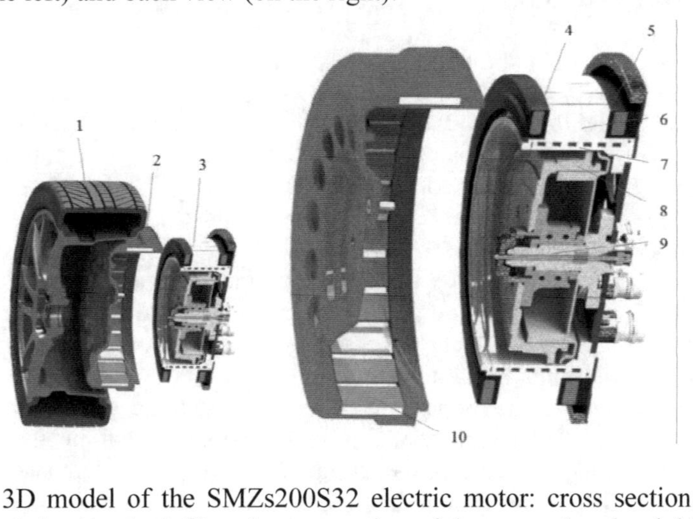

Figure 2. 3D model of the SMZs200S32 electric motor: cross section of motor with rim and tire (on the left) and cross section of the motor (on the right), 1 – rim and tire, 2 – rotor, 3 – stator, 4 – winding ends, 5 – epoxy resin, 6 – stator core, 7 – cooling system jacket with flow channels, 8 – braking drum, 9 – bearing system, 10 – magnet.

IOP Conf. Series: Materials Science and Engineering **710** (2019) 012018 doi:10.1088/1757-899X/710/1/012018

The cooling system of the tested electric motor uses different ways of removing heat from winding. The first is standard for all electric motors and includes a path for heat transfer from winding in the stator slots to the magnetic core of the stator. An important thermal resistance for this path is slot insulation resistance, which protects winding against electric breakdown. Another way to remove heat from winding in the motor under test is the path between the winding end and the cooling system jacket. The path is largely dependent on the thermal resistance of the epoxy resin used. In standard solutions of electric motors, not encapsulated in epoxy resin, it is the temperature of the ends of winding surrounded by air that is the highest.

In the carried out tests, the temperature of stator winding has a significant influence on the work parameters of the electric motor. It is in winding that its parameters, such as rated torque, maximum torque, related to its overload and long-term overload, will depend on it, and more precisely on its insulation class. The torque of the electric motor has a direct effect on the driving characteristics of the car, its acceleration, driving dynamics, the maximum elevation to which it can enter and the maximum speed with which it can move.

As a part of the work carried out, a prototype motor and its laboratory tests were performed. Temperature sensors in key components have been installed in the motor, including stator windings in both magnetic core slots, and in the ends. Based on the results of laboratory tests, the model for thermal calculations has been verified. In the subsequent stages of the work, motor simulations of operation were carried out considering changes in thermal conductivity parameters of the materials: epoxy resin and slot insulation.

Figure 3. Laboratory stand for testing electric motor SMZs200S32.

3. Calculation on the model for CFD analysis

The analysis included in the article was carried out for the motor work point: P_m = 35 kW, T_m = 350 Nm, n = 950 rpm. This is the point corresponding to the Fiat Panda car on a hill with a slope of 12% at a speed of 70 km/h, assuming that the vehicle drive consists of two electric motors mounted in the wheels of the rear axle of the car.

In the CFD analysis, the model was additionally supplemented with a cooling medium in the water jacket channels. It also assumed substitute parameters of thermal resistances: R_S – thermal resistance corresponding to the pressure between the core and the water jacket construction, R_Z – the thermal resistance corresponding to the slot insulation (figure 4).

In the boundary conditions of the model (figure 5), ambient temperature, temperature of the rotor interaction (upper surface of the stator), flow rate of the cooling medium, its type and temperature were considered. The following thermal conductivities of individual elements were assumed: load-bearing element – λ_k = 150 W/mK (aluminium alloy AlSi9Mg), stator core – $\lambda_{bx,y}$ = 25 W/mK (in the plane of electrotechnical sheets), λ_{bz} = 2 W/mK (perpendicular to the plane electrotechnical plates), winding λ_{uz} = 2 W/mK (perpendicular to the wires) and λ_{uxy} = 290 W/mK (along the wires), determined on the basis of the experience of KOMEL Institute and taken from the literature [18, 19, 20, 21, 22, 23].

IOP Conf. Series: Materials Science and Engineering **710** (2019) 012018 doi:10.1088/1757-899X/710/1/012018

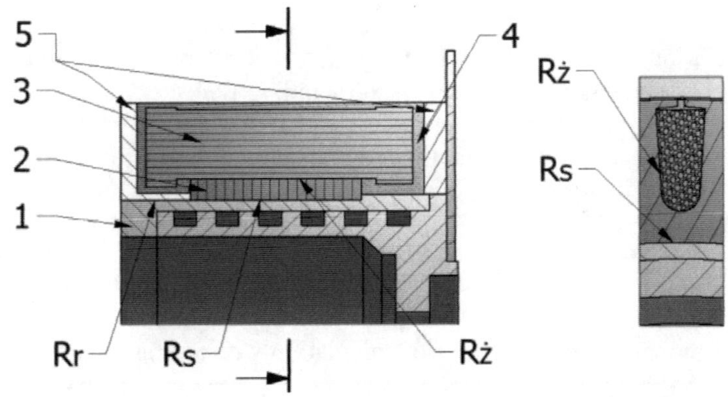

Figure 4. The calculation model of the motor stator: 1 – stator bearing element with a water jacket, 2 – stator core, 3 – winding, 4 – thermo-conductive resin, 5 – radiator, R_z – substitute thermal resistance of slot insulation, R_s – thermal resistance between the jacket and stator core, R_r – thermal resistance between the jacket and the heatsink.

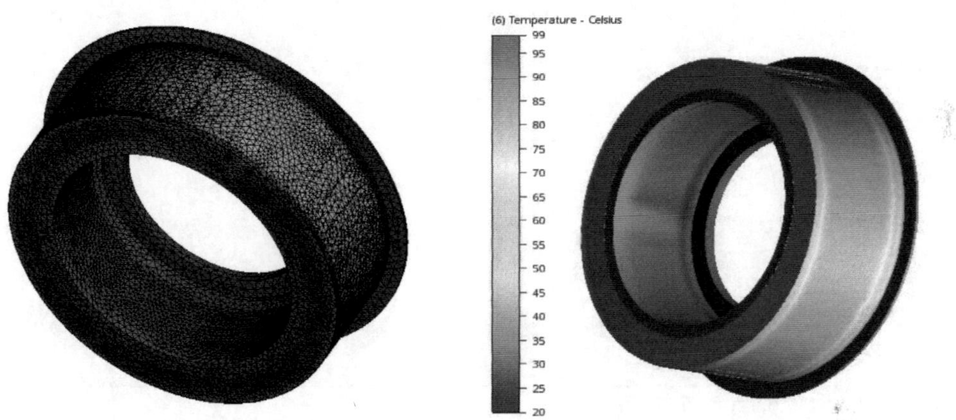

Figure 5. Discrete model for thermal calculations of the considered electric motor.

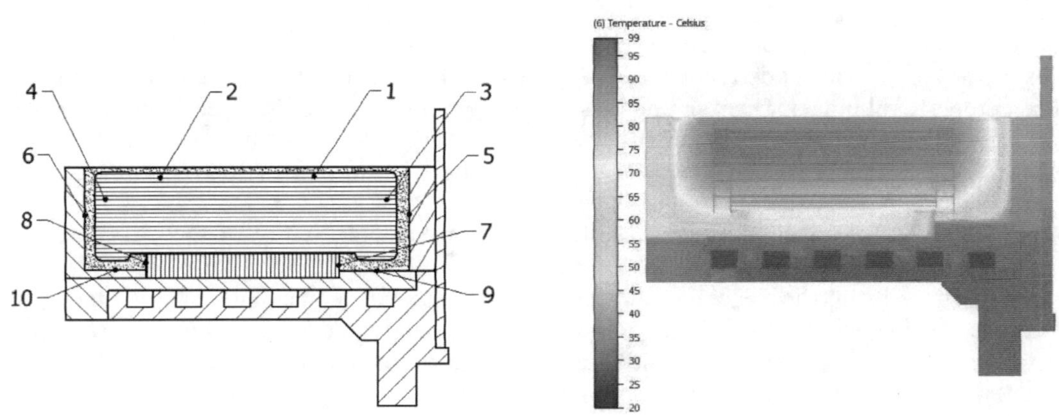

Figure 6. Arrangement of thermocouples in the tested prototype (on the left) and calculated temperature distribution for the working point of the electric motor: P_m = 35 kW (on the right): 1 – winding in the nursery (right side), 2 – winding in the nursery (left side), 3 – winding end (right side), 4 – winding end (left side), 5 – heat sink (right side), 6 – radiator (left side), 7 – stator core (right side), 8 – stator core (left side), 9 – coat (right side), 10 – coat (left side).

During the calibration process, in the stator model, ambient temperature, rotor temperature, coolant temperature, measured and calculated losses in the winding (ΔP_{Cu} = 1560W) and stator core (ΔP_{Fe} = 900W) were taken, and then thermal resistance parameters were corrected in order to obtain the expected convergence of numerical calculation results with an experiment (≤1°C) for a given steady state. As a result of calibration, the following values of thermal resistances were determined: R_z = 0.0018 m²K/W, R_S = 0.00148 m²K/W, R_r = 0.00065 m²K/W. The temperature recorded during the experiment and obtained as a result of calculations using a calibrated model for a power of 35kW are summarized in Table 1 (figure 6).

Table 1. Comparison of the results of calculation with the results of laboratory tests.

	Temperature, ϑ (°C)		
Point (figure 6)	Measurements	Calculations	Difference
1	94.0	93.5	0.5
2	92.3	91.6	0.7
3	95.8	94.8	1.0
4	88.9	88.2	0.7
5	47.5	46.9	0.6
6	57.5	58.5	1.0
7	50.4	49.7	0.7
8	60.8	61.6	0.8
9	36.8	36.5	0.3
10	44.6	45.4	0.8
in[a]	24.1	24.1	-
out[b]	27.7	28.0	0.3
amb[c]	28.0	28.0	-

[a] cooling water temperature at the inlet.
[b] cooling water temperature at the outlet.
[c] ambient temperature.

For the computational model prepared this way, the temperature difference between calculations and results of the experiment does not exceed 1°C, which confirms that the adopted calculation model has been properly calibrated. Then, simulations of the operation of the considered electric motor were performed for the same work point, making changes in the thermal conductivity of selected dielectric materials in the model – an epoxy resin surrounding the winding front and slot insulation in the magnetic core.

Table 2 presents calculation results for different thermal conductivities of the epoxy resin. The results are presented in the function of coolant flow. According to them, for a flow of 0 l/min winding temperature more than twice exceeds the permissible operating temperature of winding (180°C). Application of flow even at 2.5 l/min allows to significantly reduce the temperature. Winding temperature differences for the use of different resins reach about 7 °C, in the case of a difference between a resin with a thermal conductivity of 1.2 W/mK and a resin with a thermal conductivity of 3.5 W/mK. The temperature difference depending on the flow value is up to approx. 12 °C, regardless of the thermal conductivity of the resin.

Figure 7 presents winding temperature characteristics as a function of the coolant flow value for different thermal conductivity of the epoxy resin used. From the course of the characteristics, it can be seen that the coolant flow rate has the greatest impact up to approx. 5 l/min. At a flow rate of 10 l / min, the greater thermal conductivity of the epoxy resin from 1.2 W/mK to 1.5 W/mK allowed for a reduction in temperature of approx. 3°C. The temperature difference between successive resin values is even lower.

The analysis shows that the use of twice the coolant flow from 10 l/min to 5 l/min and the resin with a conductivity of 2.5 W/mK will get the same winding temperature as for the resin used in the prototype of the electric motor (1.2 W/mK). Such a change may allow the use of a smaller pump in the cooling system which may translate into energy savings.

Table 2. Calculated motor winding temperature for different values of thermal conductivity of the epoxy resin as a function of the coolant flow value.

	The temperature of the motor winding ϑ_{Cu} (°C)					
Coolant flow Q (l/min)	$\lambda = 1.2$ (W/mK)	$\lambda = 1.5$ (W/mK)	$\lambda = 2.0$ (W/mK)	$\lambda = 2.5$ (W/mK)	$\lambda = 3.0$ (W/mK)	$\lambda = 3.5$ (W/mK)
10	99.0	96.2	94.5	93.2	92.2	91.3
5	104	102	100	99	98	97
2.5	110	109	107	105	104	103
0	431	429	428	427	426	425

Six slot insulation variants were adopted for calculations. The first variant represents the insulation actually used in the prototype motor. The motor uses insulating material with thickness of 0.3 mm and a thermal conductivity of 0.13 W/mK. On this basis, the insulation resistance R_i has been calculated. Subsequent variants assume that the insulating materials have the same dielectric properties as the original material, for which their thickness in all cases is constant, equal to 0.3 mm. However, the thermal conductivity changes. For materials 2-5, the conductivity varies from 0.3 W/mK to 0.9 W/mK. The values have been selected on the basis of insulating materials available on the market. Materials with a conductivity above 0.5 W/mK are already difficult to access and relatively very expensive. The last material is an innovative solution because it reproduces ceramic insulation made of corundum. Corundum is aluminium oxide (Al_2O_3), it has a high thermal conductivity of up to 35 W/mK. On the basis of the model of substitute thermal resistance of slot insulation and the calculated insulation resistance selected in the framework of the calibration, the resistance of the crevice R_{Z1} was determined. R_{Z1} is the resistance resulting from the technology of making a wound stator that includes, among other things, the contact of insulating material with a magnetic core, which is in fact not ideal. The calculations were made for different values of coolant flow in the cooling system and for the electric motor version with encapsulated winding in epoxy resin and for the version without encapsulated winding.

$$R_i = \frac{d}{\lambda} \tag{1}$$

$$R_Z = R_i + R_{Z1} \tag{2}$$

where:

R_i [m^{2}K/W] – calculated insulation resistance
R_{Z1} [m^{2}K/W] – resistance resulting from the technology, between the slot insulation and the magnetic core
R_Z [m^{2}K/W] – substitute thermal resistance of slot insulation
d [mm] – assumed thickness of slot insulation
λ [W/mK] – thermal conductivity of slot insulation

On the basis of results presented in figure 8, it can be observed that the change in the thermal conductivity of insulation has more impact in the case of the model without resin (resin replaced with air) then in the case with epoxy resin (figure 9). In the case of the model without resin, it can be seen

IOP Conf. Series: Materials Science and Engineering **710** (2019) 012018 doi:10.1088/1757-899X/710/1/012018

that increasing the thermal conductivity of insulation from 0.75 W/mK is no longer effective, only a significant increase, e.g. up to 35 W/mK, allows a noticeable reduction in the winding operating temperature. In the case of a model with winding encapsulated in an epoxy resin, the effect of thermal insulation resistance is much smaller. The use of corundum insulation slightly lowered the temperature of winding in relation to insulation used in the prototype electric motor.

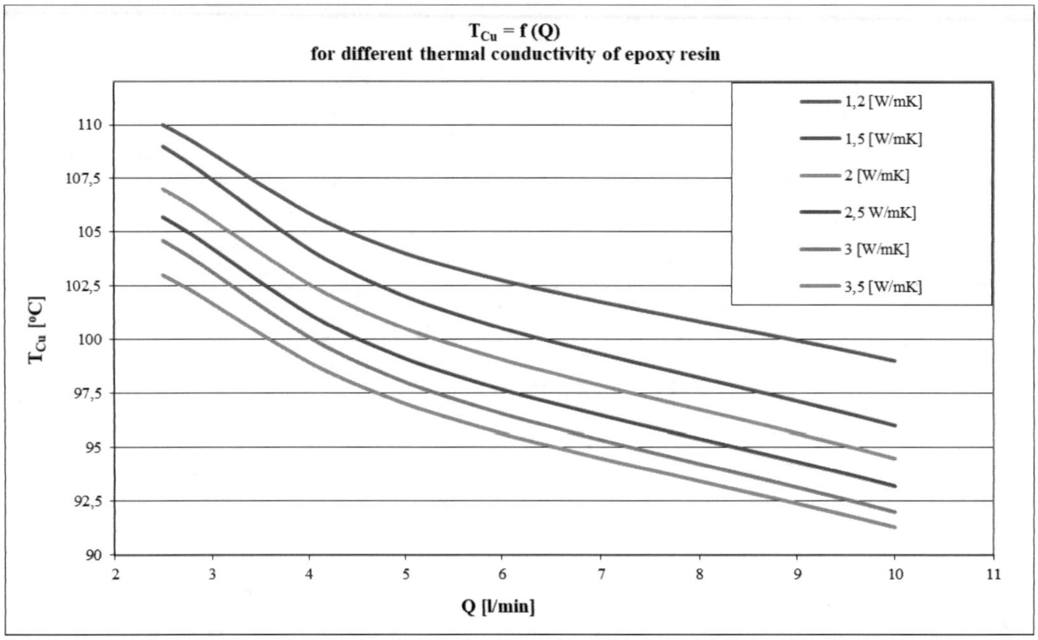

Figure 7. Calculated characteristics of the winding temperature of the electric motor as a function of the coolant flow, for different values of thermal conductivity of the resin.

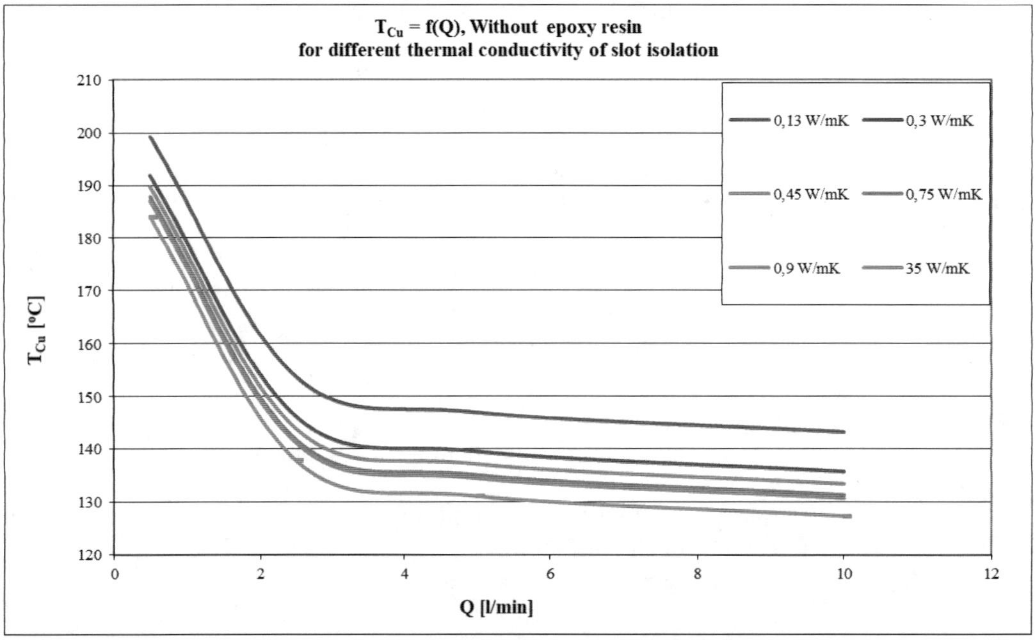

Figure 8. Calculated characteristics of the winding temperature as a function of coolant flow for different thermal conductivities of the slot insulation – model without resin.

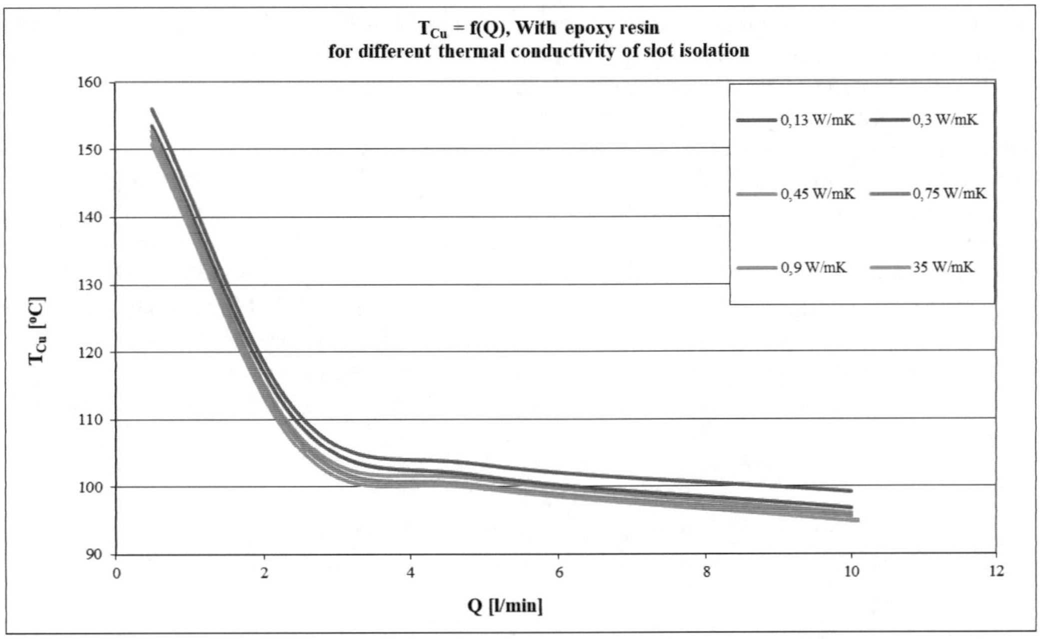

Figure 9. Calculated characteristics of the winding temperature as a function of coolant flow for different thermal conductivities of the slot insulation – model with resin.

Table 3. Calculated resistance used in calculations.

	d (mm)	λ (W/mK)	R_i (m²K/W)	R_{Z1} (m²K/W)	R_Z (m²K/W)
1	0.3	0.13	0.002308	0.000316	0.002624
2	0.3	0.30	0.001000	0.000316	0.001316
3	0.3	0.45	0.000667	0.000316	0.000982
4	0.3	0.75	0.000400	0.000316	0.000716
5	0.3	0.90	0.000333	0.000316	0.000649
6	0.3	35	0.00000857	0.000316	0.000324

4. Conclusions

The numerical calculations of MES and CFD allow to determine the operating parameters of the electric motor. Due to the calculations carried out, the minimum flow values of the cooling medium can be determined considering the maximum permissible operating temperatures of the machine. It is also possible to determine the recommended value of the cooling medium flow, above which there is no significant improvement in the efficiency of the cooling system, but only the hydraulic resistance increases negatively.

Calculations on the calibrated model allow for a wide analysis of the motor structure in terms of thermal aspects. Laboratory tests have made it possible to estimate the thermal resistance RZ1, which is the sum of the technological aspects of stator slot insulation. The extensive analysis allowed to determine the possibility of modifying the motor in the direction of lowering the winding operation temperature by changing the epoxy resin used in the motor to resins with higher thermal conductivities. In the case of commercially available resins, it should be noted that with the increasing thermal conductivity, the mechanical parameters of the resin deteriorate, which directly affects the possibility of its use in an electric motor for electric cars. For resins with higher thermal conductivity, usually the process of potting the winding and stator is more demanding because it is related to the potting regime in the appropriate temperature sequence. Resins with higher thermal conductivity also have less penetration and tendency to crack. It should be remembered that the resin in this type of solutions combines elements with various heat-shrinking properties, i.e. stator magnetic core, copper windings, and aluminium jacket of the cooling system.

Analysis of the change in the slot insulation showed that with such an efficient cooling system, using a labyrinth with cooling liquid flow and epoxy resin, supporting heat from the winding to the cooling system only a significant increase in insulation resistance allows for a noticeable change. Based on the calculations made, the constructor can consider the direction of motor design modification and consider its cost-effectiveness. While the calculations confirmed the applicability of the cooling liquid flow and the use of the resin, the reason for reducing the thermal resistance of the slot and resin insulation is no longer so obvious. A computational model prepared and calibrated on the basis of laboratory tests allows to determine the development directions of a given electric motor structure and save testing costs and costs of making subsequent prototypes, which design changes will have no effect.

Acknowledgments
The project "Innovative Solutions for Direct Drive of Electric Vehicles", financed by National Centre for Research and Development under the LIDER VII program, in accordance with the agreement: LIDER / 24/0082 / L-7/15 / NCBR / 2016 (Poland).

5. References

[1] Gombert B, Fischer R and Heinrich W 2010 Wheel-hub motors: Criteria of construction and vehicle integration *ATZ Elektron Worldw* **5** 4–10

[2] Duvigneau F, Koch S, Daniel C, Woschke E and Gabbert U 2019 Vibration Analysis of an Electric Wheel Hub Motor at Stationary Operating Points *Proceedings of the 10th International Conference on Rotor Dynamics – IFToMM* vol 63, ed K L Cavalca and H I Weber (Cham: Springer International Publishing) 51–64

[3] Freitag G, Klopzig M, Schleicher K, Wilke M and Schramm M 2013 High-performance and highly efficient electric wheel hub drive in automotive design *2013 3rd International Electric Drives Production Conference (EDPC)* 2013 3rd International Electric Drives Production Conference (EDPC) (Nuremberg, Germany: IEEE) 1–7

[4] Hoang N-T and Yan H-S 2018 *On the Design of In-Wheel-Hub Motor Transmission Systems with Six-Link Mechanisms for Electric Vehicles* (ENGINEERING)

[5] Winter O, Ucsnik S, Rudolph M, Kral C and Schmidt E 2012 Ironless in-wheel hub motor design by using multi-domain finite element analyses *International Symposium on Power Electronics Power Electronics, Electrical Drives, Automation and Motion* 2012 International Symposium on Power Electronics, Electrical Drives, Automation and Motion (SPEEDAM 2012) (Sorrento, Italy: IEEE) 1474–8

[6] Kleine A, Rosefort M and Koch H 2014 Lightweight Construction for Electric Mobility Using Aluminium *Light Metals 2014* ed J Grandfield (Cham: Springer International Publishing) 331–5

[7] Dukalski P, Będkowski B, Parczewski K, Wnęk H, Urbaś A and Augustynek K 2018 Analysis of the influence of assembly electric motors in wheels on behaviour of vehicle rear suspension system *IOP Conf. Ser.: Mater. Sci. Eng.* **421** 022004

[8] Dukalski P, Będkowski B, Parczewski K, Wnęk H, Urbaś A and Augustynek K 2018 Dynamics of the vehicle rear suspension system with electric motors mountedin wheels *EiN* **21** 125–36

[9] Tan D and Song F 2017 The advances on the study of heating and cooling issues for in-wheel-motor-driven systems *IJEHV* **9** 121

[10] Będkowski B and Madej J 2018 The efficiency analysis of various structural solutions of the wheel motor cooling system *Zeszyty problemowe - Maszyny Elektryczne* **117** 33–8

[11] Cyganik Ł, Król E, Baranowski J, Drabek T, Dziwiński T and Piątek P 2018 Thermal analysis of the motor housing for the electric torque wrench *Zeszyty problemowe - Maszyny Elektryczne* **117** 39–44

[12] Mejuto C, Mueller M, Shanel M, Mebarki A and Staton D 2008 Thermal modelling investigation of heat paths due to iron losses in synchronous machines *4th IET International Conference on Power Electronics, Machines and Drives (PEMD 2008)* 4th IET International Conference on Power Electronics, Machines and Drives (PEMD 2008) (York, UK: IEE) pp 225–9
[13] Mynarek P and Kowol M 2014 Thermal analysis of a PMSM using fea and lumped parameter modeling *Zeszyty problemowe - Maszyny Elektryczne* **104** 49–54
[14] Nategh S, Wallmark O, Leksell M and Zhao S 2012 Thermal Analysis of a PMaSRM Using Partial FEA and Lumped Parameter Modeling *IEEE Trans. Energy Convers.* **27** 477–88
[15] SanAndres U, Almandoz G, Poza J and Ugalde G 2014 Design of Cooling Systems Using Computational Fluid Dynamics and Analytical Thermal Models *IEEE Trans. Ind. Electron.* **61** 4383–91
[16] Zhang B, Qu R, Xu W, Wang J and Chen Y 2014 Thermal model of totally enclosed water-cooled permanent magnet synchronous machines for electric vehicle applications *2014 International Conference on Electrical Machines (ICEM)* 2014 XXI International Conference on Electrical Machines (ICEM) (Berlin, Germany: IEEE) 2205–11
[17] Huang J, Shoai Naini S, Miller R, Rizzo D, Sebeck K, Shurin S and Wagner J 2019 A Hybrid Electric Vehicle Motor Cooling System—Design, Model, and Control *IEEE Trans. Veh. Technol.* **68** 4467–78
[18] Siesing L, Reinap A and Andersson M 2014 Thermal properties on high fill factor electrical windings: Infiltrated vs non infiltrated *2014 International Conference on Electrical Machines (ICEM)* 2014 XXI International Conference on Electrical Machines (ICEM) (Berlin, Germany: IEEE) 2218–23
[19] Soong W 2008 Thermal analysis of electrical machines: limits and heat transfer principles *Power Engineering Briefing Note Series* 19–21
[20] Miller T J E 2002 *SPEED's Electric Motors* (University of Glasgow: University of Glasgow)
[21] Hendershot J R and Miller T J E 2010 *Design of brushless permanent-magnet machines* (Venice, Florida: Motor Design Books)
[22] Będkowski B and Madej J 2015 Thermal properties of silicon steel core - research and simulations *Zeszyty problemowe - Maszyny Elektryczne* **106** 117–22
[23] Będkowski B and Madej J 2015 Determination of equivalent thermal resistance of slot insulation - research and simulations *Zeszyty problemowe - Maszyny Elektryczne* **106** 123–7.

IOP Conf. Series: Materials Science and Engineering **710** (2019) 012019 doi:10.1088/1757-899X/710/1/012019

Surface quality simulation with neural networks in AZ91D Mg alloy milling

M Kulisz[1], I Zagórski[2] and J Korpysa[3]

[1] Lublin University of Technology, Management Faculty, Department of Organisation of Enterprise, 38 Nadbystrzycka Str., 20-618 Lublin, Poland
[2] Lublin University of Technology, Mechanical Engineering Faculty, Department of Production Engineering, 36 Nadbystrzycka Str., 20-618 Lublin, Poland
[3] PhD student of Mechanical Engineering, Lublin University of Technology, 36 Nadbystrzycka Str., 20-618 Lublin, Poland

m.kulisz@pollub.pl

Abstract. The mathematical model presented in this paper describes the process of milling AZ91D Mg alloy with the TiAlN-coated tool. The experimental data from the neural network training served to determine the effect of various milling parameters on the surface roughness of the workpiece. The 2D roughness measurements were taken on the end-faces of specimens and focused on the set of three parameters – Ra, Rz and RSm. The tests were performed at constant parameters (tool geometry, workpiece strength properties and technological machine properties), variable parameters (cutting speed, feed per tooth, depth of cut) and output variables (Ra, Rz, Rsm). The simulations were executed by means of the artificial neural networks, modelled with Statistica Neural Network software. Two types of neural networks were employed: MLP (Multilayered Perceptron) and RBF (Radial Basis Function). The comparative analysis showed that the discrepancy between the results of simulation and the results obtained from experimental tests is on an acceptable level, not exceeding 15%. The above result confirms that ANNs can be successfully used as a tool for a selection of technological parameters, enabling obtaining the desired values of surface roughness parameters without the need for time-consuming machining tests.

1. Introduction – State of the Art

Magnesium alloys are typically shaped in milling, which may also constitute a finishing operation and, therefore, the high quality of the finish is essential. The primary reason for this are the increasingly strict tolerance requirements for manufactured products, particularly in the case of mating parts. It is essential to bear in mind that the surface roughness structure of parts that are expected to achieve high performance in cooperation with other elements in a system strongly affects their functional properties. What it indicates is that extensive research on surface roughness parameters and technological factors affecting them is very much in place. These studies are conducive to the determination of the combinations of milling data the application of which will result in the improvement of the quality of machined surfaces and the efficiency of the milling process. A more accurate assessment of the impact of milling conditions on the condition of the workpiece surface requires measurements of a wider range of 2D surface roughness and 3D area roughness parameters than in standard tests.

The majority of published studies resolve to focus on one parameter. In our view, this is far from sufficient for a detailed analysis of the surface conditions [1-4]. In addition to dimensional and shape accuracy, surface roughness is among the most frequently analysed indicators of machining [5]. From the literature study, it appears that *Ra* (average roughness) is the standard parameter to consider in tests, as exemplified by the work of Muralidharan *et al.* Their examination set out to determine the surface quality of Mg-SiC/B_4C hybrid composite following milling with nine different coated and uncoated end mill cutters [6, 7]. The results presented in their study [6] show that in the case of milling performed at a range of spindle speed, n = 1500÷2500 r/min, and table feed, v_f = 1200÷2000 mm/min, the surface roughness is affected to a greatest extent by the modification of table feed and burr build-up on the cutting edge. The lowest average roughness, Ra = 0.35 μm, was obtained when milling with a tool with a TiN coating at spindle speed, n = 2000 r/min, and table feed, v_f = 2000 mm/min. In another study [7], carried out at a cutting speed of v_c = 1200 m/min, the high-quality surface finish in milling was shown to result from a suitable choice of technological parameters, among which it was feed per tooth f_z that had a fundamental impact as it has relevance for both 2D surface roughness and 3D area roughness parameters. What is more, axial depth of cut a_p was proven to have a negligible effect on 2D surface roughness parameters, whether considering end or lateral workpiece faces. The study concluded that due to its negligible effect on the surface roughness, axial depth of cut a_p exhibits a decent capacity for increasing milling efficiency while simultaneously not compromising the quality of surface finish.

The search for the optimal combination of milling parameters in dry milling of AM60 Mg alloy with a TiN-coated carbide tool was reported in the work by Sathyamoorthy *et al.* [8]. It was observed that in the analysed range of spindle speed, n = 1000÷2000 r/min, feed per revolution, f_n = 0.1÷0.2 mm/r and cutting depth, a_p = 0.5÷1.5 mm, increasing feed per revolution and depth of cut has a negative effect on the surface roughness; this can be, nevertheless, reduced by increasing the spindle speed. Machining at the top rotational speed, feed per revolution rate f_n = 0.1 mm/r and cutting depth a_p = 1 mm enabled obtaining the smallest roughness *Ra*, of approx. 0.3 μm.

Another relevant work [9] analysed the extent to which the range of machining parameters: spindle speed n = 500÷2000 r/min, table feed v_f = 50÷200 mm/min and cutting depth a_p = 0.5÷1.5 mm, are reflected in the workpiece surface roughness. When milling AZ61 magnesium alloy with a cutting tool with carbide inserts, the *Ra* parameter value was 0.115-0.403 μm in the entire spindle speed range. The roughness remained low regardless of the change in table feed and depth of cut. The results indicate that it is possible to increase the efficiency of the milling process while maintaining high surface quality.

A different investigation into the application of the face milling cutter was reported in the work of Kim and Lee [10]. The study set out to determine the effect of technological parameters and the number of inserts in the cutting tool (z = 1÷6) in milling of AZ31B alloy with air pressure coolant. The surface roughness conditions were shown to deteriorate with the increase in feed per tooth throughout the entire range 0.04÷0.25 mm/tooth and the increase in the number of cutting inserts engaged. On the other hand, the involvement of compressed air was shown to cause a slight decrease in the *Ra* parameter value.

With respect to the workpiece material, the scientific literature tends to show preference to AZ91D/HP Mg alloys, which are widely analysed in a number of studies, including the one by Kuczmaszewski *et al.* [11]. In the said work, AZ91D Mg alloy was machined by means of the milling head in an attempt to test whether face milling could substitute abrasive treatment as a finishing operation. The HSM machining, performed in the range of high cutting speeds, v_c = 900÷1400 m/min, low feed-rate per tooth f_z = 0.03÷0.09 mm/tooth and depth of cut a_p = 0.2÷0.3 mm, confirmed the tested capability as it enabled achieving low values of the *Ra* parameter in the range 0.061-0.133 μm. The lowest surface roughness was obtained at the lowest cutting speed, feed per tooth and depth of cut settings.

The above conclusions are also confirmed by a different study [12], which ran a comparative analysis of 2D surface roughness and 3D area roughness parameters of AZ91HP magnesium alloy measured after it had been subjected to machining with a P320 abrasive cloth, P180 and P240 non-woven abrasive fabrics as well as a carbide milling cutter. The lowest values of the analysed roughness parameters, Ra = 0.124 μm and Rz = 0.952 μm, were produced as a result of milling, while the highest roughness, Ra = 0.81 μm and Rz = 3.82 μm, was observed on surfaces treated with P180 non-woven abrasive fabric. In the case of 3D area roughness parameters, the highest values were characteristic of the surface machined with P320 abrasive cloth, while the lowest values were, as previously, recorded on the surface treated with P180 non-woven abrasive fabric.

The same material was also analysed in the study by Gziut *et al.* [13]. In the study, the tool rake angle γ was shown to significantly affect the roughness of the front and side surfaces of the workpiece machined by means of carbide milling cutters with angles $\gamma = 5^o$ and $\gamma = 30^o$. For both surfaces, lower roughness parameters were recorded when machining with a tool rake angle $\gamma = 5^o$; it was furthermore observed that on the face surfaces the differences were much more significant. In addition, the surface was reported to have a smoother finish when machined at higher cutting speeds (v_c = 400÷1200 m/min), while the increase in feed per tooth (f_z = 0.05÷0.30 mm/tooth) exhibited a counter effect.

Surface roughness is an essential data source of reference for other processes, such as plastic forming as pretreatment of semi-finished materials, which may precede e.g. AWJM [14] or brush treatment (edge deburring) [5]. It is for the given reason that there is a need for a detailed and extensive description of the workpiece surface finish. As previously remarked, current studies in the field are limited to the determination of Ra roughness parameter, which fails to provide an adequate picture of the condition of the magnesium alloy surface considering its utilitarian characteristics.

There is a distinct tendency observed nowadays towards the use of mathematical models to predict the surface roughness after milling. The primary purpose of these simulations is to ensure that the surface produced in an actual process exhibits the necessary quality. The model for the determination of Ra parameter of the surface after milling, proposed by Miko and Nowakowski [15], aid the choice of cutting data according to the desired surface roughness characteristics. Their model describes the stereometry and kinematics of the cutting tool profile on the surface, as it is the case with similar models. The findings from their study indicated that the surface roughness greatly depends on two factors in particular, namely, feed per tooth and the tool corner geometry. The key problem with these models is the inaccuracy of predictions, which results from the simplifying assumptions they employ. Therefore, there is a discrepancy between the modelled and measured surface roughness results, which may be between 1.5-5 times lower in the former case.

Other mathematical models, *e.g.* Zhou *et al.* [16] for 3D surface topography modelling upon helical milling, Urbikain *et al.* [17] for surface topography simulation in flank-milling by means of circle-segment end mills, or Wojciechowski *et al.* [18], accounted for machining parameters, the static tool run-out and dynamic effects of tool deflection. These models, as in the case of the previous paragraph, are not fully comparable with the actual results from machining, which is due to the high complexity and significant randomness factor of the modelled process.

In the face of the inaccuracy challenge, newly produced models employ AI-based techniques that derive from experimental models, which ensure higher precision in the representation of the simulated machining process. The application of neural networks for the prediction of surface roughness is presented in numerous works, *e.g.* by Karabulut [19], using Artificial Neural Network (ANN) modelling and Taguchi method, or by Sangwan *et al.* [20] for the determination of the optimum cutting data with a view to obtaining the minimum surface roughness in a joint effort from ANN and Genetic Algorithm (GA).

Neural networks have been similarly found suitable in the prediction of: dimensional errors [21], surface quality and optimisation of process parameters [22], machining performance [23], or milling chatter [24]. Although the results from simulations may not be capable of providing specific machining parameters, they do however approximate the desired range of cutting data. Therefore, neural networks prove themselves as a solid tool, which can successfully aid technologists in determining optimal machining parameters.

2. Test and simulation methodology

The experimental part of the study was performed with AVIA vertical milling centre VMC800HS, controlled by Heidenhain iTNC 530 system. The milling was performed with a VHM tool –the Ø16 two-edge TiAlN-coated carbide cutter (manufactured by Fenes), which was set in the shrink-fit holder, HSK-A63 Seco – Tools SFD 16x120, for improved stability. The entire tool system was balanced to G2.5, according to balance quality requirements specified in the International Standard ISO 21940-11: 2016. The result of 0.87 g-mm, obtained from the unbalance measurement carried out with CIMAT RT 610 balancing machine, proved the tool-tool-holder system complied with the permissible residual unbalance. The workpiece material subjected to experimental testing and mathematical simulation was AZ91D Mg alloy. The variable machining parameters employed in the study were: cutting speed v_c = 400÷1200 m/min, feed per tooth f_z = 0.05÷0.30 mm/tooth, axial depth of cut a_p = 0.5÷6 mm; whereas the constant milling parameter was radial depth of cut a_e = 14 mm. Figure 1 shows the experimental set-up for the surface roughness test plan.

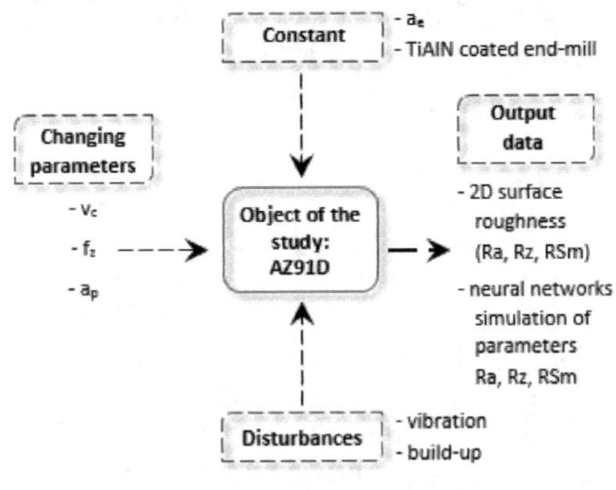

Figure 1. Schematic for surface roughness test plan.

The results from experiments provided the input data fed into the ANN for the simulation of selected surface quality indicators, *i.e.* three surface roughness parameters – average roughness (*Ra*), maximum height of profile (*Rz*) and mean width of profile elements (*RSm*). Given the nonlinear character of the modelled technological process and the resulting problems for its adequate representation by means of mathematical equations, the milling process was approached as the control object and the modelled parameters, *Rz*, *Ra* and *RSm* become the output parameters. As previously stated, the only variable machining parameters were cutting speed v_c, feed per tooth f_z and cutting depth a_p. The schematic model of milling shown in Figure 2 contains the variable input parameters v_c, f_z and a_p, whereas *nn* stands for the output parameter simulated at a given moment; therefore, as three surface roughness parameters were considered in AZ91D alloy milling – three ANNs were produced. For the reason of its performance on the training set, the black box model of ANN was employed in the study. This model is applicable in scenarios when it is difficult to determine mathematical equations describing the analysed process, as in the presented case.

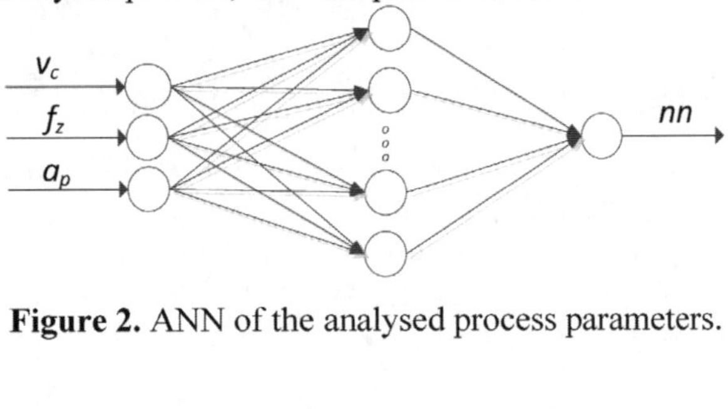

Figure 2. ANN of the analysed process parameters.

The simulations performed as part of this study were executed by means of Statistica Neural Networks software package and employed two ANN types, MLP (multilayered Perceptron) and RBF (Radial Basis Function). The MLP model used linear, exponential, logistic, tanh and sinus activation functions, and was trained with BFGS gradient (Broyden–Fletcher–Goldfarb–Shanno), conjugate gradient and the steepest descent training algorithm. The RBF network model engaged the following activation functions: hidden neurons – Gaussian distribution, output neurons – linear function, and it was trained with RBFT algorithm. The training data set used 75% of data from experimental tests and 25% were used for validation. The test data set was omitted due to insufficient data from milling tests [14].

With respect to ANN structure, for simplicity considerations, one hidden layer was included in both MLP and RBF models, in addition, 3 neurons were in the input layer (cutting speed, feed per tooth and cutting depth) and one neuron in output (*i.e.* the predicted surface roughness parameter). The number of hidden-layer neurons (2÷9) and training epochs (150-250) were determined experimentally. The chosen network types offered optimal training and validation quality and error characteristics. The least-squares algorithm was used for error identification purposes.

3. Results and analysis

3.1. Analysis of results

The results from tests and measurements are presented below as bar charts and include the standard deviation. Figures 3-5 reflect on the impact of cutting speed v_c change on the values of selected 2D surface roughness parameters. The presented values were calculated as an average of five identical measurements for a given set of cutting data. Figure 3a presents the values of *Ra* and *Rz* measured on the face surface of the Mg workpiece for variable cutting speed v_c. For the parameter Ra, similar values of approx. 0.3 μm were obtained in the entire range of cutting speeds; however, they did not exceed 3.12 μm, observed for v_c = 1200 m/min. Considering the *Rz* parameter, major changes in its values were noted in the range of 11.62-13.56 μm. The effect of cutting speed v_c on the values of the *RSm* parameter is shown in Figure 3b. The values ranged from 0.141 to 0.149 mm, and the change was insignificant.

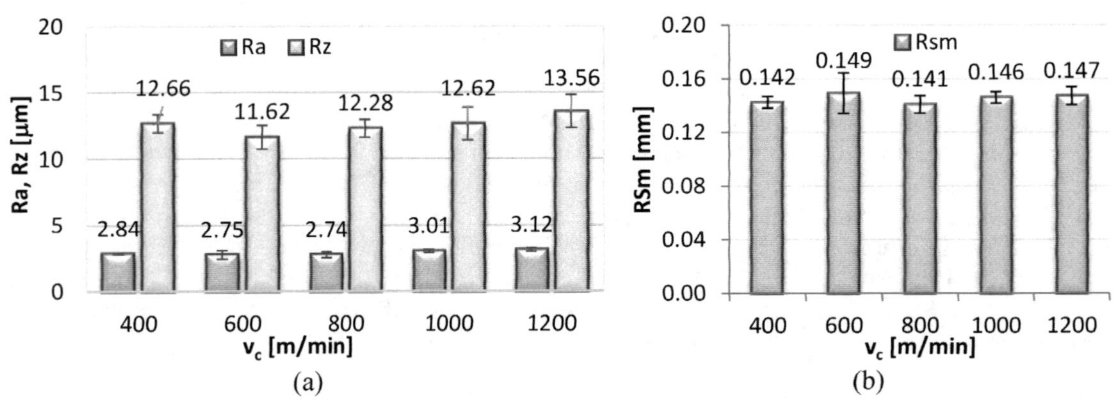

Figure 3. Effect of cutting speed v_c on: a) *Ra* and *Rz*, b) *RSm* (f_z = 0.15 mm/tooth, a_p = 6 mm).

Figure 4a presents the values of *Ra* and *Rz* obtained at variable feed per tooth f_z. For both parameters, the mean values exhibit a linear increase in response to the increase in feed per tooth f_z, to reach the values in the range of 0.53-5.62 μm for the *Ra* parameter and 3.18-27.52 μm for the *Rz* parameter. Figure 4b shows the effect of variable feed per tooth f_z on *RSm*. Increasing the feed-rate per tooth f_z resulted in an increase in the average value from 0.048 mm to 0.264 mm.

IOP Conf. Series: Materials Science and Engineering **710** (2019) 012019 doi:10.1088/1757-899X/710/1/012019

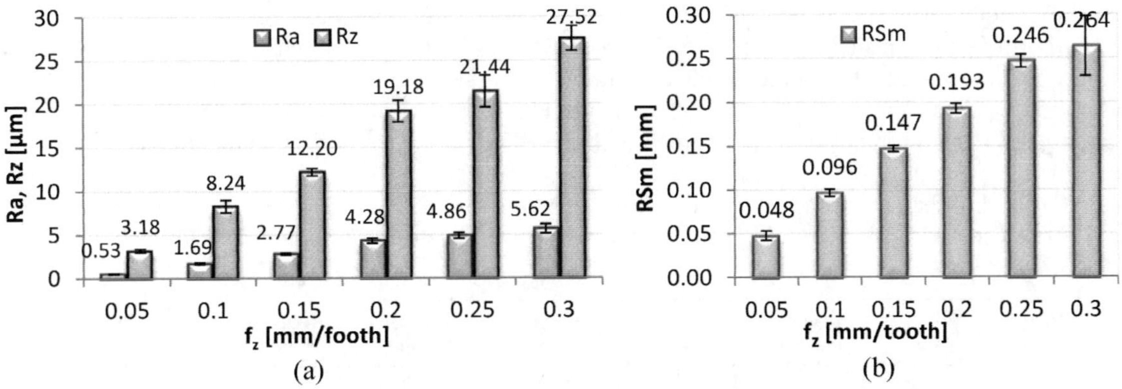

Figure 4. Effect of feed per tooth f_z on: a) *Ra* and *Rz*, b) *RSm* (v_c = 800 m/min, a_p = 6 mm).

Ra and *Rz* parameters measured in the conditions of variable depth of cut a_p are given in Figure 5a. Significantly lower values in the 2.35-2.90 µm range were recorded for the *Ra* parameter, while for the *Rz* parameter the values were within 10.82-12.32 µm. Figure 5b shows the effect of depth of cut a_p on the *RSm* parameter. The values obtained throughout the entire range of depth of cut a_p showed little discrepancy and amounted to 0.137-0.151 mm.

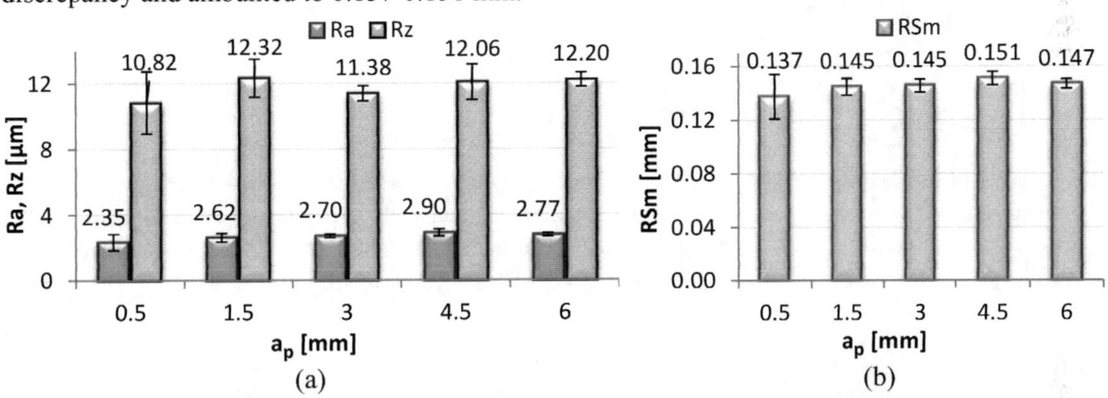

Figure 5. Effect of depth of cut a_p on: a) *Ra* and *Rz*, b) *RSm* (v_c = 800 m/min, f_z = 0.15 mm/tooth).

From the conducted analysis, it can be seen that changes in particular technological parameters investigated in the study have been shown to exert a similar effect on 2D roughness parameters measured on the face surface of the Mg workpiece.

3.2. Simulation results

The experimental research provided data on the AZ91D alloy surface quality indicators (*Rz*, *Ra*, *RSm*). The data were fed into the model as the input for simulation of these quantities using ANNs. Simulations were calculated with the Statistica Neural Networks and employed MLP and RBF networks.

The suitability of a given network in a considered application is assessed by: learning quality, validation quality, learning errors and validation errors identified by means of the least-squares method, all of which are important network quality indicators. Here, 200 networks were produced for each simulation scenario. In order to select the better-suited network type, RBF or MLP, the networks were assessed with the application of the said network quality indicators. Table 1 presents characteristics of the surface roughness parameters analysed in the scope of reported investigations, *i.e. Rz*, *Ra* and *RSm*, measured following milling of AZ91D alloy. The quality assessment indicated that, in general, the RBF network exhibits higher compatibility with *Rz* and *RSm*: in particular RBF 3-

3-1 (three hidden neurons) is optimal for simulation of maximum height of profile (*Rz*), while mean width of profile elements (*RSm*) is best modelled by RBF 3-7-1 (seven hidden neurons) – in both cases the networks were trained with the RBFT algorithm. Considering average roughness parameter (*Ra*) it was the three-neuron network MLP 3-3-1 that provided the best fit with the experimental data; the Multilayered Perceptron network was trained with the BFGS algorithm over 200 iterations.

Table 1. Characteristics of multilayered perceptron (MLP) and radial basis function (RBF) networks for the simulated surface quality indicators (*Rz, Ra, RSm*) for AZ91D alloy.

Network No.	Network Name	Quality (Training,%)	Quality (Validation,%)	Error (Training)	Error (Validation)	Activation (Hidden)	Activation (Output)
			Rz - maximum height of the profile				
1	MLP 3-4-1	99.76	99.48	2.049	6.512	Logistics	Tanh
2	RBF 3-3-1	98.79	99.99	1.256	4.478	Gaussian	Linear
3	MLP 3-8-1	99.75	99.25	2.052	5.960	Logistics	Exponential
			Ra - average roughness				
4	MLP 3-6-1	99.22	97.66	0.099	0.074	Linear	Linear
5	RBF 3-7-1	99.79	97.76	0.082	0.351	Gaussian	Linear
6	MLP 3-3-1	99.96	98.85	0.051	0.112	Exponential	Logistics
			RSm - mean width of profile elements				
7	MLP 3-6-1	99.98	99.85	0.002	0.003	Tanh	Exponential
8	RBF 3-4-1	97.62	99.97	0.002	0.002	Gaussian	Linear
9	RBF 3-7-1	99.46	99.98	0.002	0.002	Gaussian	Linear

The graphical representation of ANN results that showed the best fit with the experimental data is given below. Since three input parameters were considered, each network was shown in two diagrams: v_c/f_z, and v_c/a_p. The results for *Rz* surface roughness parameter produced by the RBF 3-3-1 network are shown in Figure 6, for *Ra* (MLP 3-3-1) – Figure 7, and for *RSm* (RBF 3-7-1) – Figure 8.

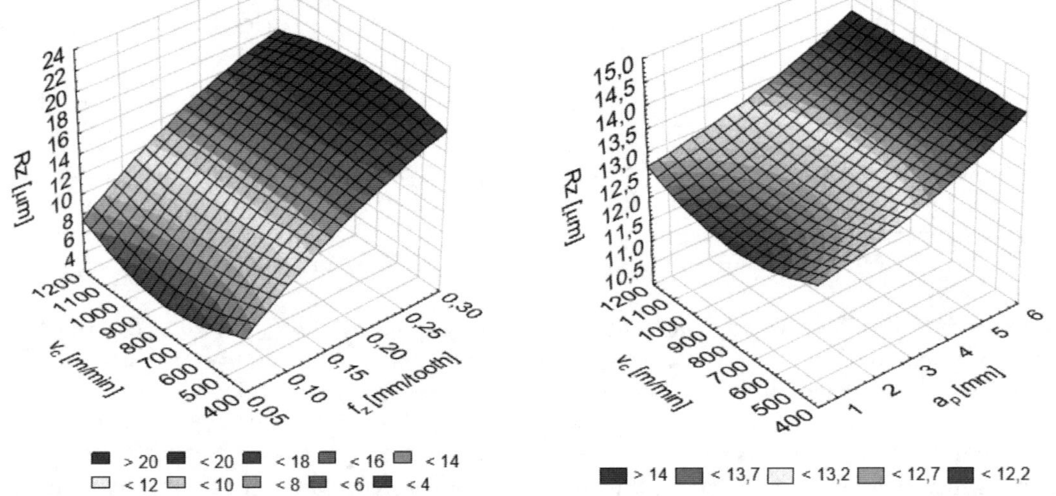

(a) (b)

Figure 6. Numerical results from RBF 3-3-1 network for *Rz* surface roughness parameter depending on the cutting speed v_c and: (**a**) cutting depth a_p=6[mm], (**b**) feed per tooth f_z=0.15[mm/tooth] for AZ91D alloy.

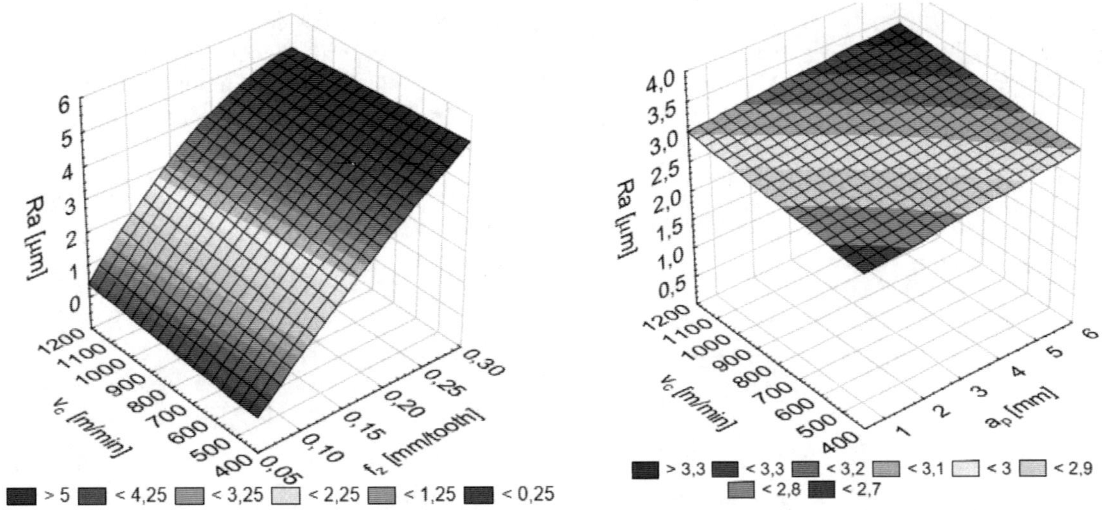

(a) (b)

Figure 7. Numerical results from MLP 3-3-1 network for *Ra* surface roughness parameter depending on the cutting speed v_c and: (**a**) cutting depth a_p=6[mm], (**b**) feed per tooth f_z=0.15[mm/tooth] for AZ91D alloy.

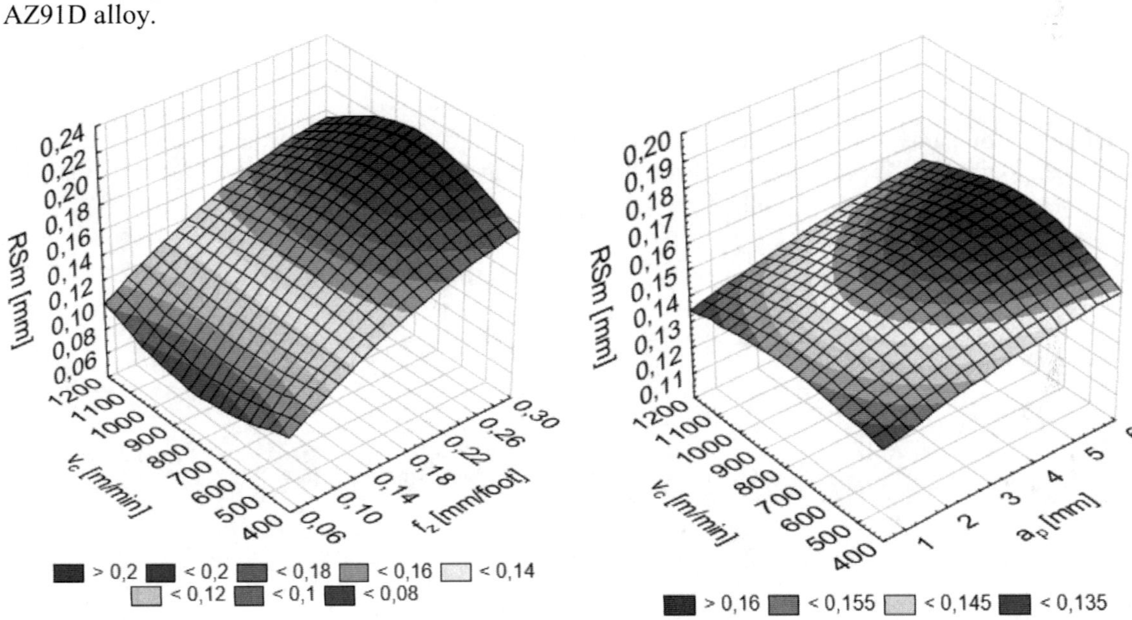

(a) (b)

Figure 8. Numerical results from RBF 3-7-1 network for *RSm* surface roughness parameter depending on the cutting speed v_c and: (**a**) cutting depth a_p=6[mm], (**b**) feed per tooth f_z=0.15[mm/tooth] for AZ91D alloy.

The networks obtained as a result of simulations characterise the behaviour of *Ra, Rz, RSm* surface quality parameters, obtained under variable machining parameters constraint: cutting speed, feed per tooth and cutting depth. The variable parameters (v_c, f_z, a_p) were subsequently fed into Statistica-developed models as network input to produce particular surface roughness parameters (depending on the type of network).

The quality of the networks is shown in Figure 9: it emerges from the comparison of *Rz* and *Ra* parameters for AZ91D alloy material with their simulated counterparts, depending on the cutting speed v_c at f_z = 0.15 m/min and a_p = 6 mm. Figure 10 presents the identical comparison with respect to *RSm*. What is worth noting is that the discrepancy between the actual values of simulated parameters was on an acceptable level: the relative error did not exceed 15%.

Low network error indicator (<15%) indicates that the ANNs implemented in this study have proven their suitability as a modelling tool.

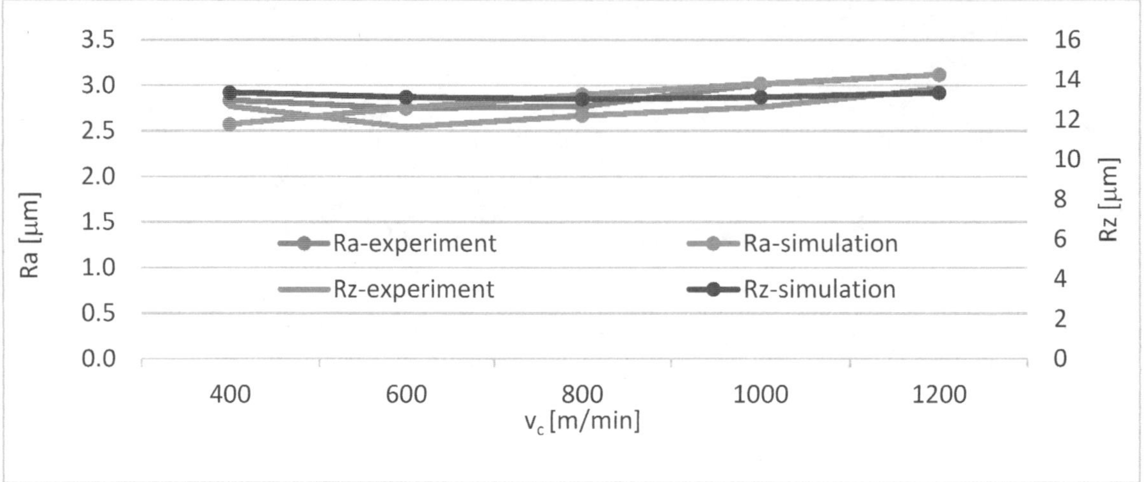

Figure 9. Comparison of experimental and numerical results depending on the cutting speed vc for feed per tooth fz=0.15[mm/tooth] and cutting depth ap=6[mm] for Ra and Rz surface roughness parameters in milling of AZ91D alloy.

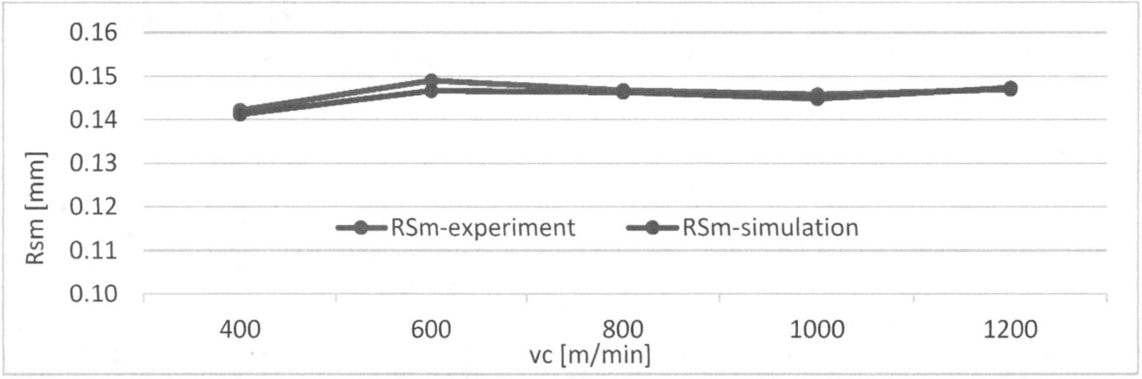

Figure 10. Comparison of experimental and numerical results depending on the cutting speed vc for feed per tooth fz=0.15[mm/tooth] and cutting depth ap=6[mm] for RSm surface roughness parameter in milling of AZ91D alloy.

4. Conclusions

From the conducted experimental and simulation investigation, it is now possible to state that:

- Feed per tooth f_z has the greatest effect on the 2D surface roughness parameters measured on the workpiece face: surface quality dropped when f_z increases.
- The change in cutting speed v_c does not significantly affect the 2D surface roughness parameters on the workpiece face.

- The change in depth of cut a_p has no significant effect on the 2D surface roughness of the workpiece face: therefore, the efficiency of the milling process may be increased without compromising the quality of the machined surfaces by adjusting a_p.
- Considering the results from roughness parameter simulation and how they correlate with the real cutting data, it was shown that the error did not exceed 15%, thus the discrepancy is on an acceptable level. Therefore, ANNs are fully capable of providing accurate data for the preliminary determination of milling data.
- The nonlinear dependence between machining parameters and roughness parameters was successfully represented by means of ANNs. This finding suggests that neural networks are fit for the examination of milling process without the need for time-, labour- and cost-consuming machining tests.
- ANN simulation results may lay the fundaments for designing a modelling tool to describe phenomena occurring in machining. This solution could aid the decision-making process by supplying cutting data that ensure obtaining desired surface roughness. A given set of parameters (v_c, f_z, a_p) might be fed into the system as input for calculations, at the output of which surface roughness parameter values (*Ra, Rz, RSm*) are obtained.

5. References

[1] Grzesik W 2015 Surface topography and utilitarian characteristics of machine parts *Mechanik* **8–9** 587–93

[2] Kłonica M and Kuczmaszewski J 2015 Comparative analysis of the surface energy state of az91hp alloy following abrasive treatment and milling *Mechanik* **8-9** 212-16

[3] Sedlaček M, Vilhena L M, Podgornik B and Vižintin J 2011 Surface topography modelling for reduced friction *Strojniški vestnik – Journal of Mechanical Engineering* **57**(9) 674–80.

[4] Sedlaček M, Gregorčič P and Podgornik B 2017 Use of the roughness parameters Ssk and Sku to control friction – a method for designing surface texturing *Tribol. T.* **60**(2) 260–66

[5] Matuszak J and Zaleski K 2018 Analysis of deburring effectiveness and surface layer properties around edges of workpieces made of 7075 aluminium alloy *Aircr. Eng. Aerosp. Tec.* **90**(3) 515–23

[6] Muralidharan S, Karthikeyan N, Kumar A B and Aatthisugan I 2017 A study on machinability characteristic in end milling of magnesium composite *International Journal of Mechanical Engineering and Technology* **8**(6) 455–62

[7] Zagórski I and Korpysa J 2019 Surface quality in milling of AZ91D magnesium alloy *Advances in Science and Technology Research Journal* **13**(2) 119-29 doi:10.12913/22998624/108547

[8] Sathyamoorthy V, Deepan S, Sathya Prasanth S P and Prabhu L 2017 Optimization of machining parameters for surface roughness in end milling of magnesium AM60 alloy *Indian Journal of Science and Technology* **10**(32) 1–7 doi:10.17485/ijst/2017/v10i32/104651

[9] Alharti N H, Bingol S, Abbas A T, Ragab A E, El-Danaf E A and Alharbi H F 2017 Optimizing cutting conditions and prediction of surface roughness in face milling of AZ61 using regression analysis and artificial neural network *Adv. Mater. Sci. Eng.* **2017** 7560468

[10] Kim J D and Lee K B 2010 Surface roughness evaluation in dry-cutting of magnesium alloy by air pressure coolant *Engineering* **2** 788–92

[11] Ruslan M S, Othman K, Ghani J A, Kassim M S and Haron C H C 2016 Surface roughness of magnesium alloy AZ91D in high speed milling *Jurnal Teknologi* **78**(6–9) 115–19

[12] Kuczmaszewski J, Pieśko P and Zawada-Michałowska M 2016 Surface roughness of thin-walled components made of aluminium alloy EN AW-2024 following different milling strategies *Advances in Science and Technology Research Journal* **10**(30) 150–58 doi:10.12913/22998624/62515

[13] Gziut O, Kuczmaszewski J and Zagórski I 2015 Surface quality assessment following high performance cutting of AZ91HP magnesium alloy *Management and Production Engineering*

Review **6**(1) 4-9

[14] Zagórski I, Kłonica M, Kulisz M and Łoza K 2018 Effect of the AWJM method on the machined surface layer of AZ91D magnesium alloy and simulation of roughness parameters using neural networks *Materials* **11**(11) E2111 doi:10.3390/ma11112111

[15] Miko E and Nowakowski E 2015 Models for prediction of Ra roughness parameters of milled surfaces *Mechanik* **8-9** 82-90

[16] Zhou Y, Tian Y, Jing X, Wang F and Liu Y 2018 *Int. J. Adv. Manuf. Technol.* **95** 4561 doi:10.1007/s00170-017-1516-2

[17] Urbikain G and de Lacalle L N L 2018 Modelling of surface roughness in inclined milling operations with circle-segment end mills *Simul. Model. Pract. Theory* **84** 161–76

[18] Wojciechowski S, Twardowski P, Pelic M, Maruda R W, Barrans S and Krolczyk G M 2016 Precision surface characterization for finish cylindrical milling with dynamic tool displacements model *Precision Engineering* **46** 158–65 doi:10.1016/J.PRECISIONENG. 2016.04.010

[19] Karabulut S 2015 Optimization of surface roughness and cutting force during AA7039/Al2O3 metal matrix composites milling using neural networks and Taguchi method *Measurement* **66** 139-49 doi:10.1016/j.measurement.2015.01.027

[20] Sangwan K S, Saxena S and Kant G 2015 Optimization of machining parameters to minimize surface roughness using integrated ANN-GA approach *Proc. CIRP* **29** 305-310 doi:10.1016/j.procir.2015 .02.002

[21] Arnaiz-González Á, Fernández-Valdivielso A, and Bustillo A. et al. 2016 *Int. J. Adv. Manuf. Technol.* **83** 847 doi:10.1007/s00170-015-7543-y

[22] Kaviarasan V, Venkatesan R and Natarajan E 2019 Prediction of surface quality and optimization of process parameters in drilling of Delrin using neural network *Prog. Rubber. Plast. Re.* **35**(3) 149-69 doi:10.1177/1477760619855078

[23] Zerti A, Yallese M A, Zerti O, Nouioua M and Khettabi R 2019 Prediction of machining performance using RSM and ANN models in hard turning of martensitic stainless steel AISI 420 *P. I. Mech. Eng., Part C* **233**(13) 4439–62 doi:10.1177/0954406218820557

[24] Zagórski I, Kulisz M, Semeniuk A and Malec A 2017 Artificial neural network modelling of vibration in the milling of AZ91D alloy *Advances in Science and Technology Research Journal* **11**(3) 261-69 doi:10.12913/22998624/76546

IOP Conf. Series: Materials Science and Engineering **710** (2019) 012020 doi:10.1088/1757-899X/710/1/012020

Study of operating parameters of a plate conveyor used in the food industry

J Caban[1], A Nieoczym[2], W Misztal[1] and D Barta[3]

[1] University of Life Sciences in Lublin, Głęboka 28, 20-612 Lublin, Poland
[2] Lublin University of Technology, Nadbystrzycka 36, 20-618 Lublin, Poland
[3] University of Zilina, Univerzitná 8215/1, 010 26 Žilina, Slovakia

jacek.caban@up.lublin.pl

Abstract. Internal transport (means of continuous transport) is an inseparable element of technological processes in the food industry. The selection of appropriate parameters of the transport equipment depends on the product being manufactured, its design features, packaging and transport conditions. The paper presents the characteristics of a plate conveyor, selection of operating parameters in the juice transport process in a glass bottle. The paper presents selected test results depending on the linear speed of the plate conveyor, the angle of inclination of the load mass transport module and the position of the gravity centre. The work defined the linear speed ranges of the conveyor and the angle of inclination of the conveyor transporting module for transported glass bottles filled with juice.

1. Introduction

A proper choice of technological and organisational parameters of production processes enables more efficient and effective production of goods [1]. Traditionally, research on manufacturing processes was mainly conducted to improve efficiency and accuracy, as well as to lower costs [2]. Production operations within an automated production line would be impossible if they did not cooperate with means of continuous transport and robots. Robots replace human employees in harmful working environments, where highly qualified personnel is scarce [3]. Robots have become a common solution in a wide range of industrial processes and have therefore contributed to boosting efficiency and flexibility of production, increasing work safety, allowing to achieve high quality of production at higher reliability and lower cost [4]. Whether or not a production process to be executed is capable of achieving the assumed performance parameters depends, among others, on reliability of machines and technological devices that make up the system under design [5].

There are various methods of transport systems used in the food industry. Usually, fluid products are transported by different pumping systems [6], solid and powder by different type of conveyors. Operation time of machine components is strongly affected by resistance to different types of wear likewise sliding, abrasive, corrosive, particle erosion or even cavitation wear [7]. Thus, to prevent damage of transportation system, it is crucial to use proper material, or to adjust accurate operating parameters that prevents from wear.

Transport works play an important role in the warehouse systems and logistics of manufacturing systems, as well as distribution [8, 9, 10, 11]. Conveyors are counted among the means of continuous transport; whose task is to regularly feed and receive the transported cargo. Depending on the

IOP Conf. Series: Materials Science and Engineering **710** (2019) 012020 doi:10.1088/1757-899X/710/1/012020

construction, they are used to transport unit loads, as well as loose materials carried in bulk. Commonly used are belt conveyors mounted in a transport system, consisting of conveyor modules with and without drive, turntables or arches. In [12] a construction of a belt conveyor was proposed, enabling the drive of straight sections and curves with one drive. Support and drive elements are subjected to variable loads resulting from the possibility of transporting loads of various shapes and weights on one conveyor [13]. In order to increase reliability of conveyors, models describing the dynamic impact load processes [14] have been developed, the aim of which is to create a mathematical description of the conveyor operation under varying loading conditions.

To assess the value of operating conditions, a number of approaches is used, due to the fact that in practice there are different circumstances allowing (or not) the use of certain methods [15]. The additional operating conditions of conveyors are normative requirements related to the reduction of noise and vibrations [16, 17]. In [18], algorithms and design procedures are described, as well as ways to control the assembled transport system in terms of eliminating bottlenecks causing queues at junction points. The problem of selection of the kinematic structure of sorting devices cooperating with the conveyor is described in [19]. The goal was to determine efficiency of the overload conveyor work on the conveyor from the weight of the loaded elements. In addition, the technical conditions of the sorting process on the conveyor and at buffer points were described on the basis of characteristic features recognised by scanning and detection system. In order to increase efficiency of transport, as well as correct localisation and tracking of carried goods, UHF-RFID technology is being used more and more often [20]. Identification results enable on-line efficiency analysis and are used to quickly identify a specific item in a transport system.

Plate conveyors are included in the group of wire conveyors. The load-bearing elements of plate conveyors are train chains, usually two-legged, connected to each other by plates, on which the transported material is moved. These conveyors are often used for transporting glass packaging. They can be used not only as transport means, but also as buffering conveyors or cooperating with additional technological devices for filling, labelling, closing, capping, sorting, etc. Plate conveyors are often used where high temperature or chemical resistance is required.

The paper presents the characteristics of a plate conveyor PP-4/2014, selection of operating parameters in the juice transport process in a glass bottle. The goal was to identify variables that affect work efficiency when transporting glass packaging. The article describes regression models used to identify basic technical parameters of transfer points in operational conditions. The highlighted parameters are conveyor efficiency, unit load of the conveyor belt and the possibility of queuing. The test results are to be the basis for formulating the requirements for the use of this type of conveyor on the production line. The requirements should relate to reliability and efficiency, correctness of construction and the possibility of incorporating the conveyor into the production system.

2. Methodology, test stand and material

The conveyor as a mechatronic device consists of a physical mechanical model, a motion device (motor + gearbox) and a controller that controls the movement of the motor and the conveyor itself [21]. The research on determining transport efficiency of glass packaging has been carried out on the PP-4/2014 plate conveyor (Figure 1).

The metrological values were measured using:

- ArcoMaster electronic protractor with a spirit level,
- electronic calliper,
- AXIS BA30C platform scale.

The conveyor is driven by an electric motor type SK1SI40-IEC80-80S/4 TF from "NordDrivesystems" with a capacity of 0.55kW and a rated voltage of 380V. Using a control and measurement system consisting of an iG5A inverter, the engine speed and controller with display have been adjusted. Liquid crystal, which is equipped with time measurement (Figure 2.).

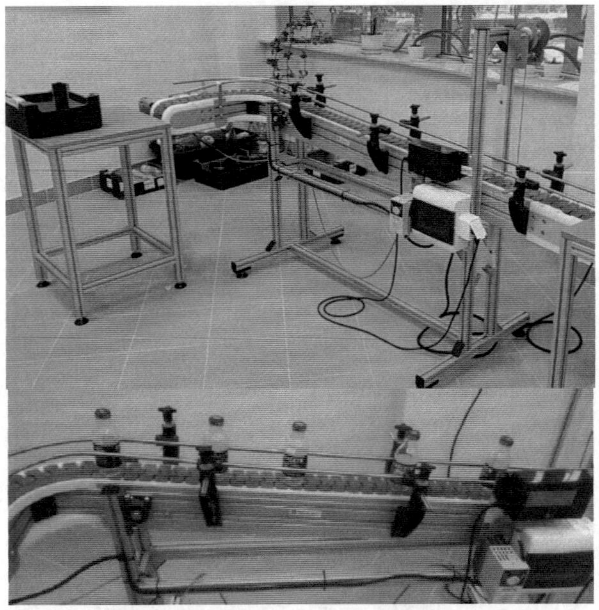

Figure 1. Plate conveyor – general view and bottle transport.

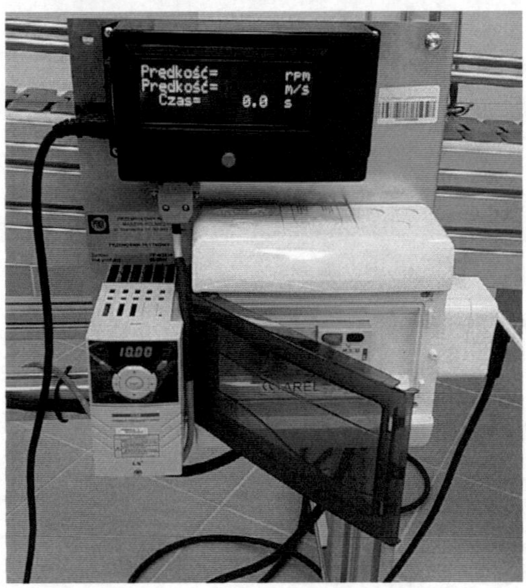

Figure 2. Measurement controller and inverter.

Transport efficiency tests were conducted for piece cargo (a glass bottle with juice of 0.33 litre capacity) - Table 1, with three angles of inclination of the transport module 0°, 3° and 6°. During measurements, the following values were recorded: frequency of the inverter, linear speed of the conveyor belt, rotational speed of the drive motor of the conveyor belt, loading time, general transport time, distance between loads.

Table 1. Characteristics of bottles used during tests.

Weight (kg)	Height (mm)	The diameter of the package		Contact area (mm)
		Min. (mm)	Max. (mm)	
0.16	165	30	63	22.76

3. Results and discussion

Tables 2-4 present the results of transport conditions tests on a plate conveyor for glass bottles.

Table 2. The results of measurements of transporting glass bottles at the angle of inclination of the transporting module is equal to 0°.

No.	Frequency of the inverter (Hz)	Loading time (s)	General transport time (s)	Distance between loads (m)	Conveyor speed	
					(m/s)	(rpm)
1	10	1.2	16.5	0.25	0.281	39
2	11	1.2	16.5	0.27	0.31	43
3	12	1.1	15.3	0.28	0.34	47
4	13	1.0	14.7	0.3	0.37	51
5	14	1.0	14.4	0.32	0.4	55
6	15	0.9	13.0	0.32	0.43	60
7	16	0.9	13.3	0.4	0.46	64

Table 3. The results of measurements of transporting glass bottles at the angle of inclination of the transporting module is equal to 3°.

No.	Frequency of the inverter (Hz)	Loading time (s)	General transport time (s)	Distance between loads (m)	Conveyor speed	
					(m/s)	(rpm)
1	10	1.1	16.8	0.25	0.281	39
2	11	1.1	17.0	0.27	0.31	43
3	12	1.1	16.1	0.30	0.34	47
4	13	1.1	15.8	0.30	0.37	51
5	14	1.0	15.2	0.33	0.40	55
6	15	1.0	15.1	0.40	0.43	60
7	16	1.0	14.2	0.42	0.46	64

Table 4. The results of measurements of transporting glass bottles at the angle of inclination of the transporting module is equal to 6°.

No.	Frequency of the inverter (Hz)	Loading time (s)	General transport time (s)	Distance between loads (m)	Conveyor speed	
					(m/s)	(rpm)
1	10	1.1	18.4	0.23	0.281	39
2	11	1.1	17.8	0.25	0.31	43
3	12	1.1	17.0	0.30	0.34	47
4	13	1.1	17.1	0.34	0.37	51
5	14	1.1	16.5	0.37	0.40	55
6	15	1.1	16.7	0.42	0.43	60
7	16	1.1	16.5	0.50	0.46	64

Based on the results of the tests, the conveyor efficiency Q was calculated (1):

$$Q=G\ v/a \tag{1}$$

where:

G – unit weight of the load (kg),

v – linear speed of the conveyor belt (m/s),

a – distance between loads (m).

IOP Conf. Series: Materials Science and Engineering **710** (2019) 012020 doi:10.1088/1757-899X/710/1/012020

Figures 3 - 6 show the results of calculations of conveyor efficiency at three values of the conveyor transport module inclination angle and the variable frequency settings of the inverter.

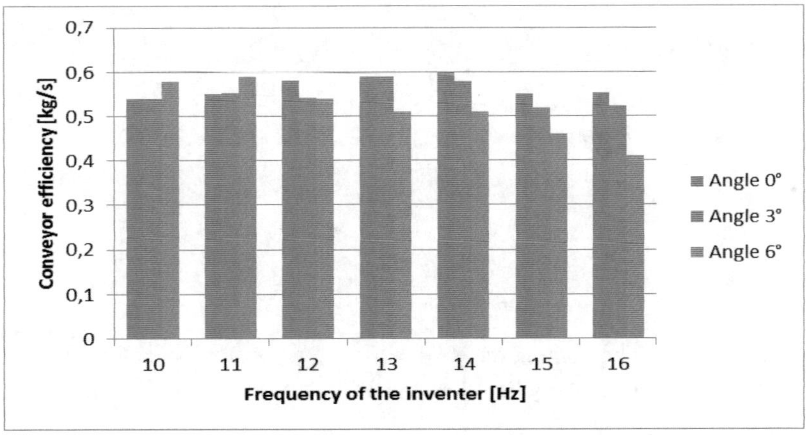

Figure 3. Graph of conveyor operation efficiency depending on the frequency of the inverter and the angle of inclination of the conveyor transport module.

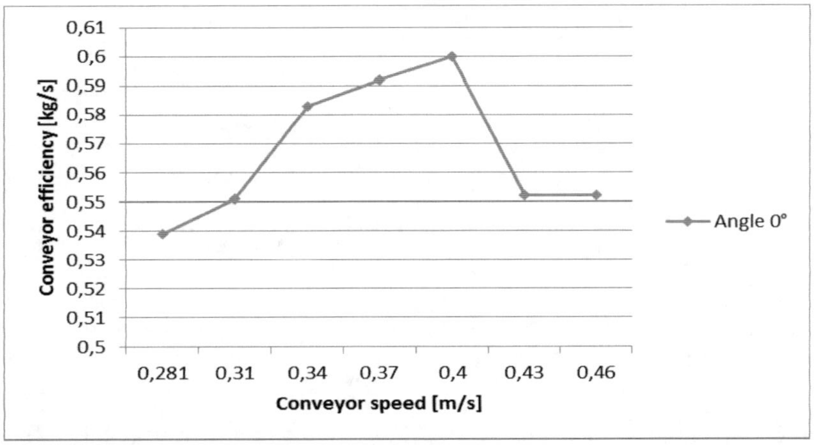

Figure 4. The dependence of efficiency on the speed of the conveyor transporting the glass bottles at an angle of 0°.

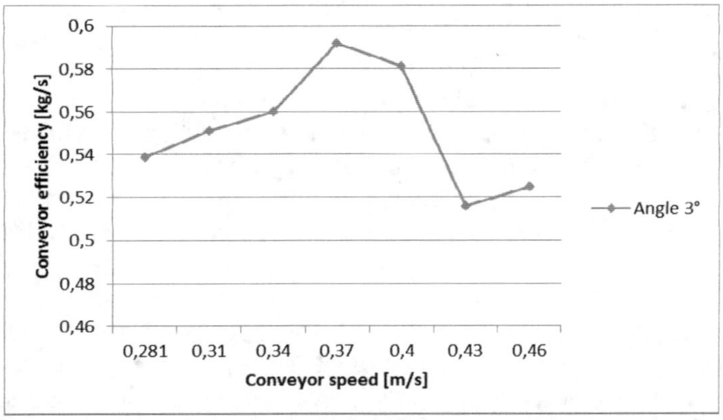

Figure 5. The dependence of efficiency on the speed of the conveyor transporting the glass bottles at an angle of 3°.

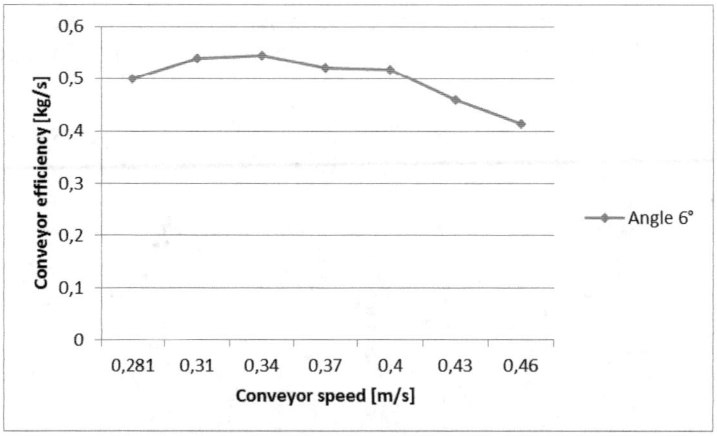

Figure 6. The dependence of efficiency on the speed of the conveyor transporting the glass bottles at an angle of 6°.

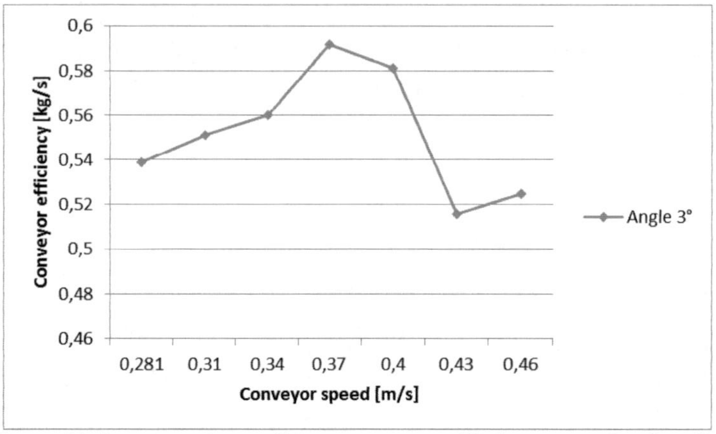

Figure 7. A comparison of the conveyor efficiency results for all tested transport module angles.

Analysing the above graphs, their course for glass bottles retains repeatability for the inclination angle of the 0° and 3° transport module. In both graphs, the extreme is located in the speed range from 0.37 m/s to 0.43 m/s. The curve at an angle of 6° has a stable drop without sudden fluctuations, except for 0.31 m/s where a slight increase in performance is observed.

4. Conclusions

Conclusions regarding correctness of the transport process:

- During transportation of glass bottles, a slip was observed between the conveyor plate and the load contact surface. This phenomenon intensified with the increase of the speed of the transport device. Slip occurred mainly when placing bottles on the conveyor belt. This phenomenon is caused by the difference in velocity vectors and the direction of the bottle being put on the tape and the speed of the belt movement. It is also important to locate the centre of gravity of the packaging and shape of its bottom.
- Based on the conducted tests, it was noticed that with the increase of the inclination angle of the transport module, efficiency of the device decreases. The most noticeable is the angle of inclination of the transport module of 6°.The area between 3 and 6 degrees is interesting, which would have to be investigated, because in this range there was the greatest reduction in the efficiency of the plate conveyor.

Conclusions regarding the construction of the conveyor:

- Efficiency measurements were carried out in laboratory conditions, but the results obtained can be approximated and used in industrial applications [22, 23].
- In order to increase work efficiency while increasing the conveyor speed and elevation angle:
- use overlays on the conveyor plates, whose task is to increase the coefficient of friction between the bottle base and the plate surface [24, 25].
- do not use rubber aprons on the side surface of the conveyor side. As previous studies have shown, rubber aprons pressing the bottle from both sides cause an increase in resistance when moving the bottle and as a result the bottles fall over [25, 26].
- One way to increase efficiency when moving bottles at an angle greater than 3o and when increasing speed is to build a grip conveyor into the production line. Details of this solution are presented in the paper [26].

Conclusions regarding the use of the conveyor in the technological line:

- The feeder is not an isolated element but is a part of the transport system. The work efficiency of the conveyor is to be correlated with efficiency of loading, receiving and additional devices, e.g. control devices, labelling machines. For this reason, subsequent research should focus on reliability of the conveyor, including temporary booking. Mathematical analysis is performed using Markov processes considering the states of work, renewal and standstill [27, 28, 29].
- Another analysis that should be carried out in relation to the considered conveyor as a part of the production line is the analysis of operational reliability by calculating the value of the OEE (Overall Equipment Effectiveness) indicator. The OEE indicator is the basis for setting the direction of activities to improve production processes. It enables to identify bottlenecks, as well as technical and organisational problems of the enterprise. OEE is also a measure of implemented improvements and allows you to calculate the benefits of improving processes and eliminating individual problems [30, 31 32, 33].

5. References

[1] Rudawska A, Flechsig A and Miturska I 2019. Technical and Organisational Changes in a Selected Production Process. *MATEC Web of Conferences* **252**, 06014.

[2] Linke BS, Corman GJ, Dornfeld DA and Tonissen S 2013. Sustainability indicators for discrete manufacturing processes applied to grinding technology. *Journal of Manufacturing Systems*, **32**, 556– 563.

[3] Czarnecka A, Sobaszek Ł & Świć A 2018. 2D image-based industrial robot end effector trajectory control algorithm. *Applied Computer Science*, **14**, 1, 73–83. doi:10.23743/acs-2018-07

[4] Hajduk M and Koukolová L 2015. Trends in Industrial and Service Robot Application. *Applied Mechanics and Materials*, **791**, 161–165. doi:10.4028/www.scientific.net/AMM.791.161

[5] Gola A 2019. Reliability analysis of reconfigurable manufacturing system structures using computer simulation 321 methods. *Eksploatacja i Niezawodnosc – Maintenance and Reliability*, **21**, 1, 90–102.

[6] Szala M and Łukasik D 2016. Cavitation wear of pump impellers. *Journal of Technology and Exploitation in Mechanical Engineering*, **2**, 1, 40-44.

[7] Szala M, Walczak M, Pasierbiewicz K and Kamiński M 2019. Cavitation Erosion and Sliding Wear Mechanisms of AlTiN and TiAlN Films Deposited on Stainless Steel Substrate. *Coatings*, **9**(5), 340.

[8] Blatnicky M, Dizo J, Blatnicka M and Svoboda M 2017. Design of a robot manipulator working screw revolutions. 23rd International Conference on Engineering Mechanics Svratka, Czech Republic, May 15-18, 2017. *Engineering Mechanics*, 166-169.

[9] Burduk A and Musiał K 2016. Optimization of chosen transport task by using generic algorithms. *Lecture Notes in Computer Science*, **9842**, 197-205.

[10] Drożdziel P, Winska M, Madlenak R and Szumski P 2017. Optimization of the position of the local distribution centre of the regional post logistics network. *Transport Problems*, **12**, 3, 43–

50.
[11] Stopka O and Luptak V 2018. Optimization of Warehouse Management in the Specific Assembly and Distribution Company: a Case Study. *Nase More*, **65**, 4, 266-269.
[12] Cempírek V, Gašparík J, Zitrický V amd Blaho P 2018. Control of modular conveyor and automated handling devices interconnection. *Advanced in Science and Technology Researched Journal*, **12**, 3, 210–221. doi: 10.12913/22998624/94963
[13] Kobayashi Y, and Toya K 2007. Effect of belt transport speed and other factors on belt mistracking. *Microsystem Technologies*, **13**, 8-10, 1325–1330. doi: https://doi.org/10.1007/s00542-006-0357-y
[14] Grincova A, Andrejiova M and Marasova D 2015. Measuring and comparative analysis of the interaction between the dynamic impact loading of the conveyor belt and the supporting system. *Measurement*, **59**, 184–191.
[15] Marczuk A, Caban J, Savinykh P, Turubanov N and Zyryanov D 2017. Maintenance research of a horizontal ribbon mixer. *Eksploatacja i Niezawodnosc – Maintenance and Reliability*, **19**, 1, 121–125.
[16] Liptai P, Lumnitzer E, Moravec M and Piňosová M 2018. Analysis and Classification of Noise Sources of Conveyor Systems by Sound Visualizing on the Postal Package Sorting Line. *Advances in Science and Technology Research Journal*, **12**, 4, 172–176. doi: https://doi.org/10.12913/22998624/100348
[17] Piňosová M, Andrejiová M, Liptai P and Lumnitzer E 2018. Objective and subjective evaluation of the physical risk factors near a conveyor system. *Advanced in Science and Technology Researched Journal*, **12**, 3, 188–196. doi: 10.12913/22998624/94964
[18] Johnstone M, Creighton D and Nahavandi S 2015. Simulation-based baggage handling system merge analysis. *Simulation Modelling Practice and Theory*, **53**, 45–59.
[19] Piątkowski T, Sempruch J and Tomaszewski T 2013. Zagadnienie bezpieczeństwa ładunków jednostkowych w zautomatyzowanych liniach sortujących. *Autobusy*, **3**, 883–891.
[20] Buffi A, Nepa P and Lombardini F 2015. A Phase-Based Technique for Localization of UHF-RFID Tags Moving on a Conveyor Belt: Performance Analysis and Test-Case Measurements. *IEEE Sensors Journal*, **15**,1, 387–396. doi: 10.1109/JSEN.2014.2344713
[21] Ružarovský R, Holubek R, Rolando Delgado Sobrino R and Janíček M 2018. The Simulation of Conveyor Control System Using the Virtual Commissioning and Virtual Reality. *Advanced in Science and Technology Researched Journal*, **12**, 4, 164–171. doi: https://doi.org/10.12913/22998624/100349
[22] Kowalski A and Marut T 2012. Hybrid methods aiding organisational and technological production preparation using simulation models of nonlinear production systems. Proceedings of the 7th International Conference on Hybrid Artificial Intelligent Systems, part II.
[23] Jasiulewicz-Kaczmarek M and Bartkowiak T 2016. Improving the performance of a filling line based on simulation. *Materials Science and Engineering*, **145**, 024-042.
[24] Ault H 2006. Design of a bottling line mechanism. Worcester Polytechnic Institute 2006, 113.
[25] Dallinger N and Hübler J 2017. Simulation of bottle conveyors – opportunities of the discrete element method (DEM). Conference: 3. International Symposium Plastic-Slide-Chain an Tribology in Conveyor Systems, Chemnitz, 91-100.
[26] Gannavarapu K, Chand G and Gopal R 2014. Modeling and analysis of side grip conveyor for bottle inspection machine. *IPASJ International Journal of Mechanical Engineering (IIJME)*, **2**, 12, 5-11.
[27] Beaurepaire P, Valdebenitob M, Schuëller G and Jensen H 2012. Reliability-based optimization of maintenance scheduling of mechanical components under fatigue. *Computer Methods Applied Mechanical Engineering*, **221**, 24–40.
[28] Nieoczym A and Gola A 2017. Mathematical modelling the assembly cell operation in the context of increasing its reliability. Conference: International Conference on Intelligent Computing and Information Engineering (ICIE), **1**, 124-132.

[29] Shin YW and Moon DH 2014. Approximation of throughput in tandem queues with multiple servers and blocking. *Applied Mathematical Modelling*, **38** (24), 6122–6132.

[30] Gola A and Nieoczym A 2017. Application of OEE Coefficient for Manufacturing Lines Reliability Improvement. Proceedings of the 2017 International Conference On Management Science And Management Innovation (MSMI 2017). Edited by: Wee HM and Yuan C. AEBMR-Advances in Economics Business and Management Research, **31**, 189–194.

[31] Mansur A, Rayendra R and Mastur M 2016. Performance Acceleration on Production Machines Using the Overall Equipment Effectiveness (OEE) Approach, IOP Conf. Series: Materials Science and Engineering, **105**, 012019.

[32] Ahire CP and Relkar AS 2012. Correlating Failure Mode Effect Analysis (FMEA) and Overall Equipment Effectiveness (OEE). *Procedia Engineering*, **n38**, 3482-3486.

[33] Relkar AS and Nandurkar KN 2012. Optimizing and analyzing Overall Equipment Effectiveness (OEE) through Design of Experiments (DOE). *Procedia Engineering*, **n38**, 2973-2980.

IOP Conf. Series: Materials Science and Engineering **710** (2019) 012021 doi:10.1088/1757-899X/710/1/012021

Analytical method for projecting the buckling form of composite palates with a cut-out

K Falkowicz and K Szklarek

Lublin University of Technology, Faculty of Mechanical Engineering, Department of Machine Design and Mechatronics, Nadbystrzycka 36, 20-618 Lublin, Poland

k.falkowicz@pollub.pl

Abstract. The work focused on the original concept of a thin-walled plate element with a cut-out, for use as a spring or a load-bearing element. The subject of the study were rectangular plates with a cut-out with variable geometrical parameters and with a variable angle of fibre arrangement, made of a carbon-epoxy composite with high strength properties in an asymmetrical lay-up, subjected to uniform compression. The paper presents a method of predicting the buckling form of composite plates with a cut-out ensuring the stable nature of the structure's work. The influence of geometrical parameters of the cut-out and the angle of fibre arrangement on the value of the critical load of the structure and buckling form was investigated. The commercial ABAQUS programme using the finite element method was used to develop the discrete model and perform numerical calculations.

1. Introduction

Thin-walled structural components are load-carrying structures that can work after the loss of stability, provided that they remain in the elastic state [1-9]. This is a significant limitation regarding components made of conventional structural materials such as metals, as their work in the post-critical state is determined by the yield strength of the material. The group of materials which maintain their elastic characteristics in the post-critical state, virtually until failure, are composite materials, particularly glass fibre reinforced plastic (GFRP) and carbon fibre reinforced plastic (CFRP). Due to their favourable strength properties in relation to density, these materials are now widely used in modern thin-walled structures. Another characteristic of structural elements made of fibrous composites is their wide margin of structural load capacity, i.e. the ability to work after the loss of stability, sometimes exceeding the critical load values two or three times [10-13]. There are numerous studies that investigate the behaviour of thin-walled laminated elements in the pre-critical and post-critical states [14-16]. The wide margin of load capacity of composite elements increases the safety of work of the structure, as even after the first symptoms of failure, the composite structure maintains its high stiffness until achieving the value of failure load [14]. These characteristics of fibrous composites combined with a low density of material offer numerous applications for these materials in many fields, such as aviation, construction industry [17], automotive or aerospace industries [18], where design solutions based on the use of plate and thin shell elements are dominant.

Uniform thin plates are relatively cheap to manufacture, however, due to their low bending stiffness, they can carry relatively small loads [19]. The slightest form of plate stability loss, which is the bending form, leads to its rapid destruction due to the increase in high deflection, with a slight increase in compressive load – low rigidity of the structure. However, the stiffness characteristics of the structure change significantly when the work of the plate is forcing according to the higher flexural-torsional form of buckling. The plate is able to carry a much higher compressive load while maintaining much stiffer work characteristics in the post-critical range (figure 1).

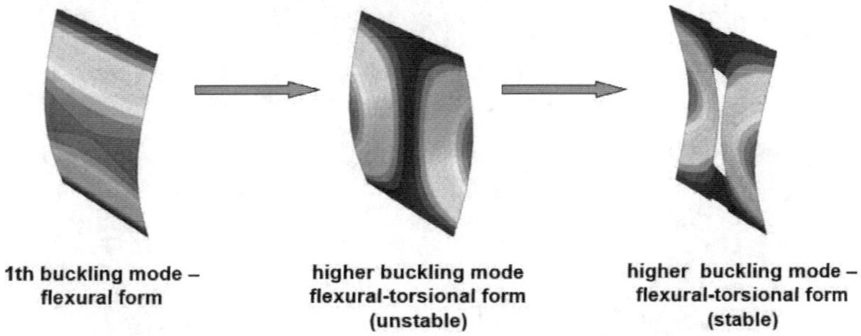

Figure 1. The possibility of using thin-walled plate elements as the elastic elements (own study).

The calculations carried out in former studies [20,21] show that plates with a forced higher form of buckling are characterised by stable, progressive paths of post-critical equilibrium, enabling their use as elastic elements. The characteristics of such elements can be designed in a wide range by changing the geometrical parameters of the cut-out, i.e. height and width [22], as well as by changing the angle of fibre arrangement [21]. The paper presents an uncomplicated method for the estimation of buckling forms of tested plates for different variables. The calculations presented in the work, were performed by means of Abaqus software, which is at present widely applicable in many fields of science [12,23]. The obtained results are of significant practical importance in the design of structures with elastic elements, allowing engineers and designers to achieve the required maintenance characteristics of the device.

2. Research subject and methodology

2.1. Object of the study

The subject of the study was a rectangular plate of fixed dimensions of 160x80 [mm] with a central cut-out of variable geometrical parameters and the corner rounding radius R=5 [mm] (figure 2).
The plate was made of carbon-epoxy composite with the following properties: Young's modulus, E1 = 130.71 GPa and E2 = 6.36 GPa, Poisson's ratio, ν12 = 0.32, shear modulus, G12 = 4.18 GPa and ply thickness, t= 0.131 [mm] in an asymmetrical lay-up of layers $[0_3/\Theta/-\Theta/0/-\Theta/\Theta/-\Theta/\Theta/-\Theta/\Theta/90/\Theta/-\Theta/90_3]T$, where Θ – means angle (figure 3).

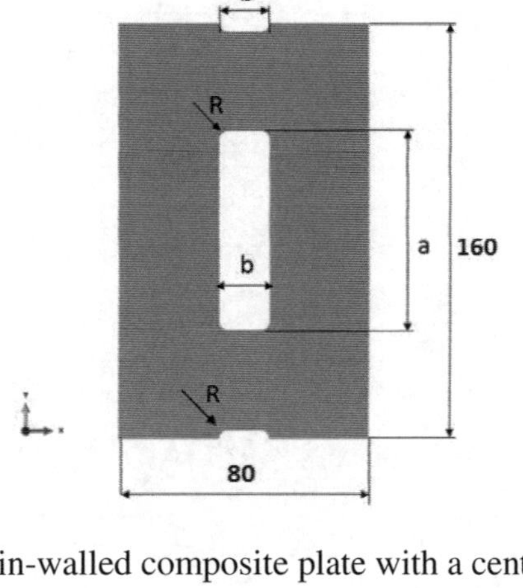

Figure. 2. The tested thin-walled composite plate with a central cut-out (own study).

IOP Conf. Series: Materials Science and Engineering **710** (2019) 012021 doi:10.1088/1757-899X/710/1/012021

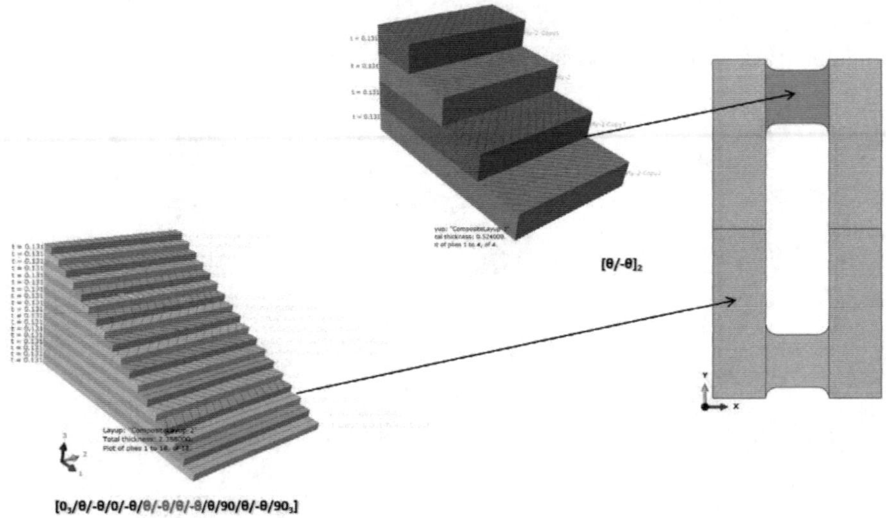

Figure. 3. Composite layers lay-up after the plate (own study).

2.2. Research methodology

In order to determine the form of plate buckling, linear calculations were carried out. A more detailed description of the numerical analysis along with boundary conditions is provided in other works [21,24], whereas a similar modelling method was used in the study by Rozylo [25].

Table 1 summarises the results of the critical state numerical analysis for various geometrical parameters of the cut-out and for various fibre angles. The height of the cut-out was changed in the range of 80÷120 [mm], the width of the cut-out in the range of 20 ÷ 40 [mm], while the angle of fibre arrangement was changed at every 150. The penultimate column contains the critical force values, while the last column contains the buckling forms for given cases.

Table 1. The results of the numerical analysis for the critical state (own study).

a [mm]	b [mm]	θ °	Critical Force [N]	1st buckling mode
100	20	15	1909.4	Flexural-torsional
100	20	30	1660.5	Flexural-torsional
100	20	45	1378.1	Flexural
100	20	60	1129.3	Flexural
100	20	75	1021	Flexural
100	20	90	999.6	Flexural
100	30	15	1477.2	Flexural-torsional
100	30	30	1244.4	Flexural-torsional
100	30	45	949.87	Flexural-torsional
100	30	60	770.6	Flexural-torsional
100	30	75	721.98	Flexural-torsional
100	30	90	711	Flexural-torsional
100	40	15	1151.1	Flexural-torsional
100	40	30	961.78	Flexural-torsional
100	40	45	711.85	Flexural-torsional
100	40	60	543.07	Flexural-torsional
100	40	75	488.87	Flexural-torsional
100	40	90	479	Flexural-torsional

80	30	15	1524.4	Flexural-torsional
80	30	30	1412.6	Flexural-torsional
80	30	45	1107.1	Flexural-torsional
80	30	60	913.31	Flexural
80	30	75	827	Flexural
80	30	90	812	Flexural
120	30	30	1185.91	Flexural-torsional
120	30	45	841.42	Flexural-torsional
120	30	60	598.56	Flexural-torsional
120	30	75	526.46	Flexural-torsional
120	30	90	523.86	Flexural-torsional

The diagram below (Figure 4) shows the distribution of the buckling form which the test object assumes in a three-dimensional spatial system, depending on parameters such as the angle of the fibres θ, the height of the cut-out a and the width of the cut-out b.

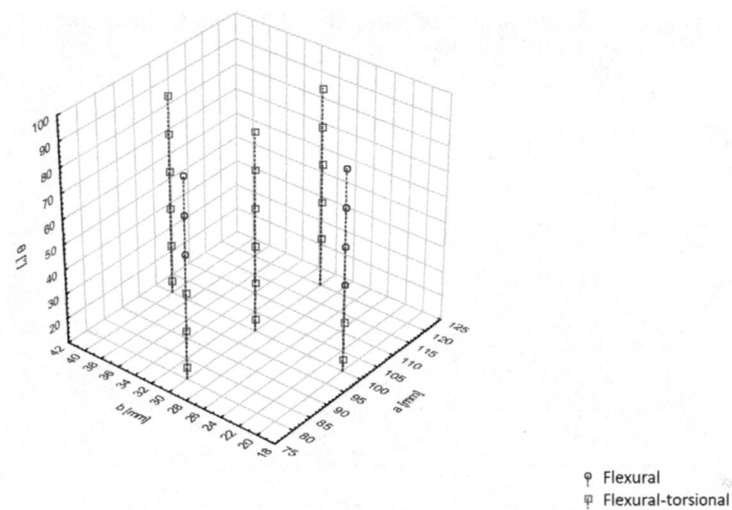

Figure 4. Dispersion of buckling form depending on the angle of fibre lay-up *Θ* and the height *a* and width *b* of the cut-out (own study).

From figure 4, it can be seen that bending forms occur in plates with a 100x20 [mm] and 80x30 [mm] cut-out. In the case of 100x20 [mm] cut-out, the bending form occurs for four variants of the fibre angle, i.e. 45°, 60°, 75° and 90°, while in the case of a 80x30 [mm] cut-out for three variants of the angle: 60°, 75°, 90°. In all cases, the bending form occurred for a fibre angle equal to or greater than 45°. Based on the values given in Table 1 and the dispersion obtained, it can be stated that the lowest buckling form depends on the above-mentioned parameters. Presenting the obtained results in two-dimensional space, depending on the height of the cut-out a/width of the cut-out b to the angle Θ (figure 5 and figure 6), it emerges that in some cases the points corresponding to different buckling forms are projected at the same points, which makes them difficult to use for further predictions. Based on the charts and the limited number of data, it is difficult to clearly determine the form that the element will take for a different set of parameter values.

As already mentioned, the buckling form depends on the height and width of the cut-out and the angle of the fibre. By creating a new variable, whose value depends on two selected parameters, we can present the distribution of buckling form in a more intelligible way, e.g. by calculating the ratio of the height to the width of the cut-out (figure 7).

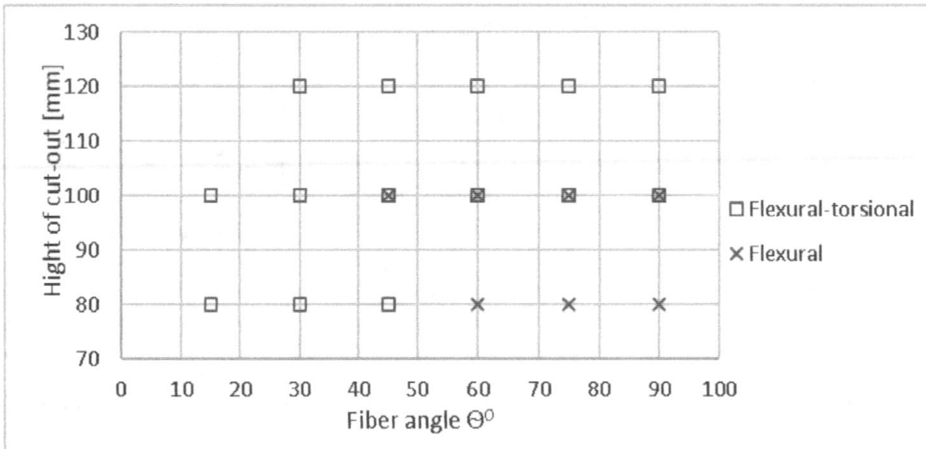

Figure 5. Graph of buckling form depending on the height of the cut-out *a* and the angle of the fibres *Θ* (own study).

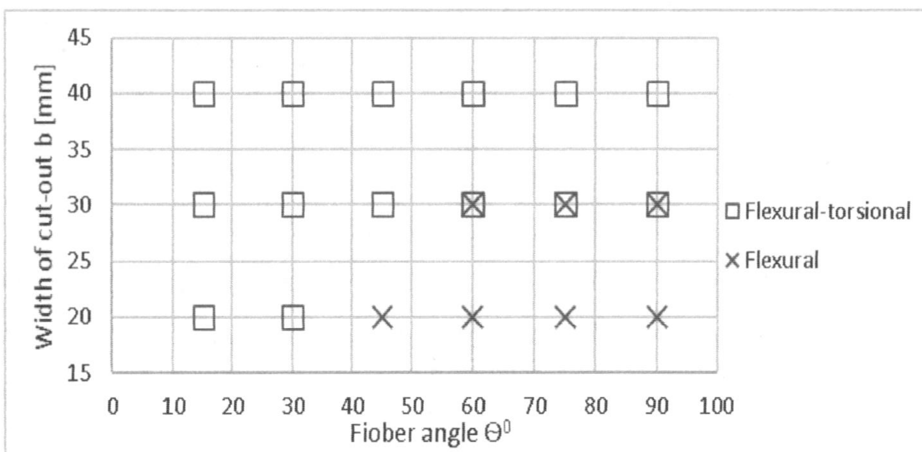

Figure 6. Graph of buckling form depending on the width of the cut-out *b* and the angle of the fibres *Θ* (own study).

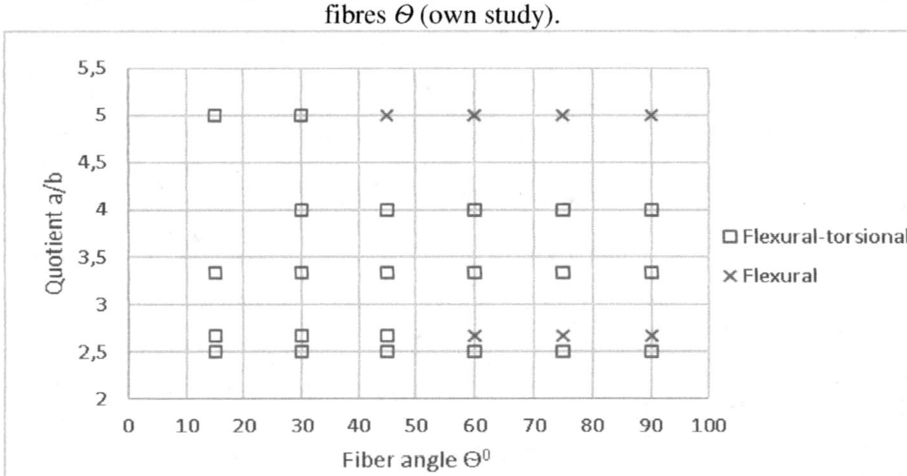

Figure 7. Graph of buckling form depending on the ratio *a* / *b* and the angle of fibre arrangement *Θ* (own study).

IOP Conf. Series: Materials Science and Engineering **710** (2019) 012021 doi:10.1088/1757-899X/710/1/012021

In this way, the points representing the different buckling forms do not overlap. However, despite the fact that the obtained graph is clearer, it is somewhat difficult to see the dependencies and describe them. It is difficult to predict with certainty when the appropriate form of buckling will occur. A much better result was obtained by multiplying the height and width parameters of the cut-out, calculating the area of the cut-out (figure 8).

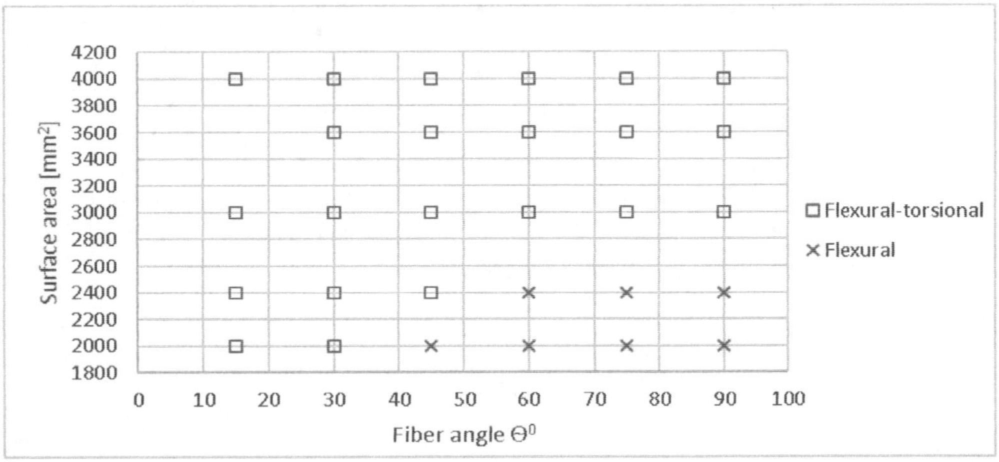

Figure 8. Graph of buckling form depending on the area of the cut-out and the angle of fibre arrangement Θ (own study).

From the graph, it is possible to read what form can occur at a given value of the cut-out surface area and the angle of the laminate fibres. Bending forms occur for larger values of fibre arrangement angles and for smaller surface areas. As the height and width of the cut-out increases, the likelihood of flexural-torsional forms increases. Identical conclusions can be formulated thoroughly by analysing the first graph (figure 4); however, these relationships are better-visible in the two-dimensional graph. In comparison with the previous graph (figure 7), in the graph in figure 8 the points corresponding to individual buckling forms are ordered and separable, which can be done using any line or function. This graph form allows us to make a prediction regarding the form that occurs at a given value of the cut-out surface area and the angle of the laminate fibres. Nevertheless, it is a graphical method.

On the basis of the obtained results and the analysis, it was found that the form which the tested element assumes can be described by a function depending on the angle of the fibre and the height and width of the cut-out:

$$P = f(\theta\ , a, b) \tag{1}$$

where:
P – form of buckling
θ – laying angle of laminate layers
a - height of cut-out
b – width of cut-out

By dividing the obtained surface area values by the fibre angle, a new parameter p (equation 2) was obtained, which takes into account all factors affecting the buckling form of the element. On the buckling dispersion chart, depending on the area of the cut-out and the new variable *p* (figure 9), the points representing the respective buckling forms are grouped and form sets. These sets can be separated by a straight line drawn between the extreme points on the borders of the set. The separation of sets can be made using any line or function and this could be done already on the dispersion plot of the cut-out surface area and the layer arrangement, however, the obtained relationship would only take into account the surface area of the cut-out and forms of buckling would be constant for all lay-ups. In addition, the use of functions other than rectilinear function would require the use of additional transformations and mathematical operations. The advantage of this method is to prevent the random determination of the coefficients' value of straight, thus preventing the discrepancy in later results. Nonetheless, the straight line created by this method does not separate the sets throughout the entire range. The value of parameter *p* corresponding to the cut-out 100x30 [mm] and Θ equal to 90° is below the straight line and equals 33.3 (3), while the value of the parameter corresponding to the point on the dividing line equals 33.375.

$$p = \frac{a+b}{A} \tag{2}$$

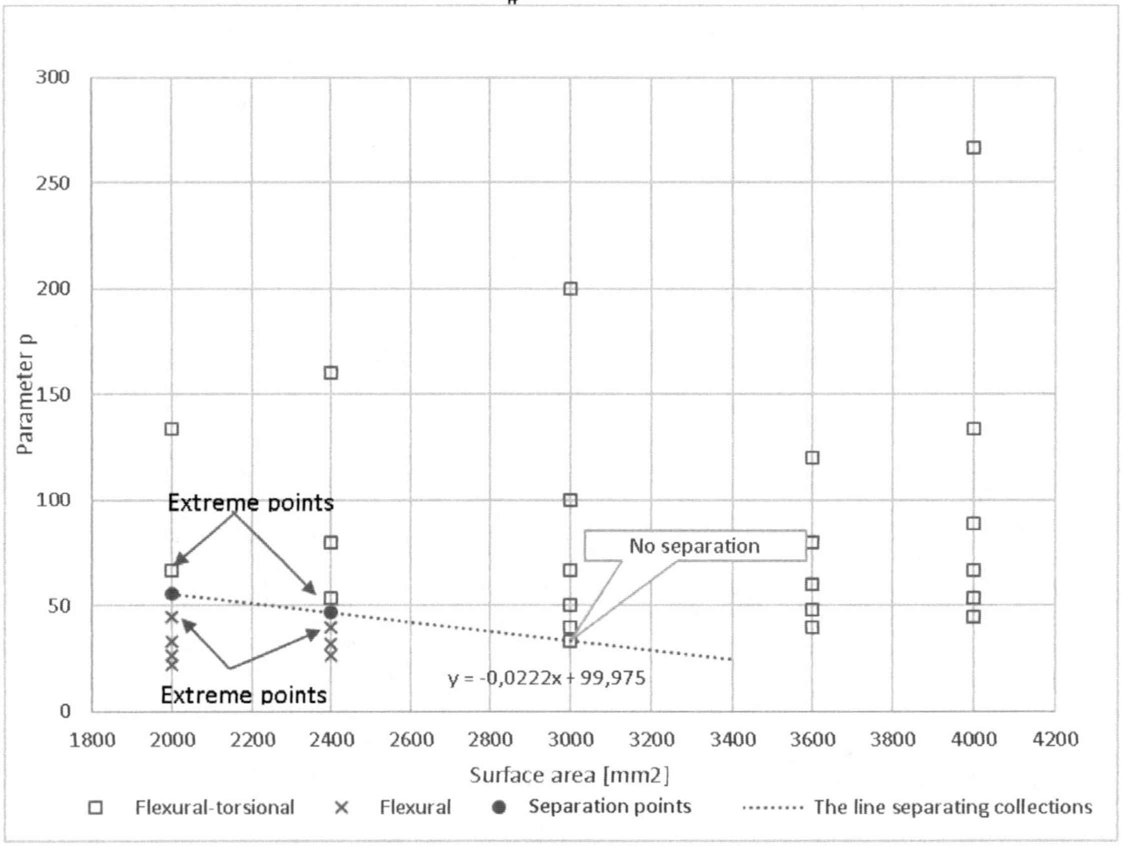

Figure 9. Graph of buckling form depending on the area of the cut-out and parameter *p* (own study).

In order to compensate for this situation, the parameter p was subjected to a mathematical procedure consisting in the numerator rooting in the formula. In this way, a new parameter s was obtained, which is calculated from:

$$s = \frac{\sqrt{a+b}}{A} \tag{3}$$

Proceeding analogously as in the previous case, a separating line was drawn, which separates the sets in the whole range (figure 10).

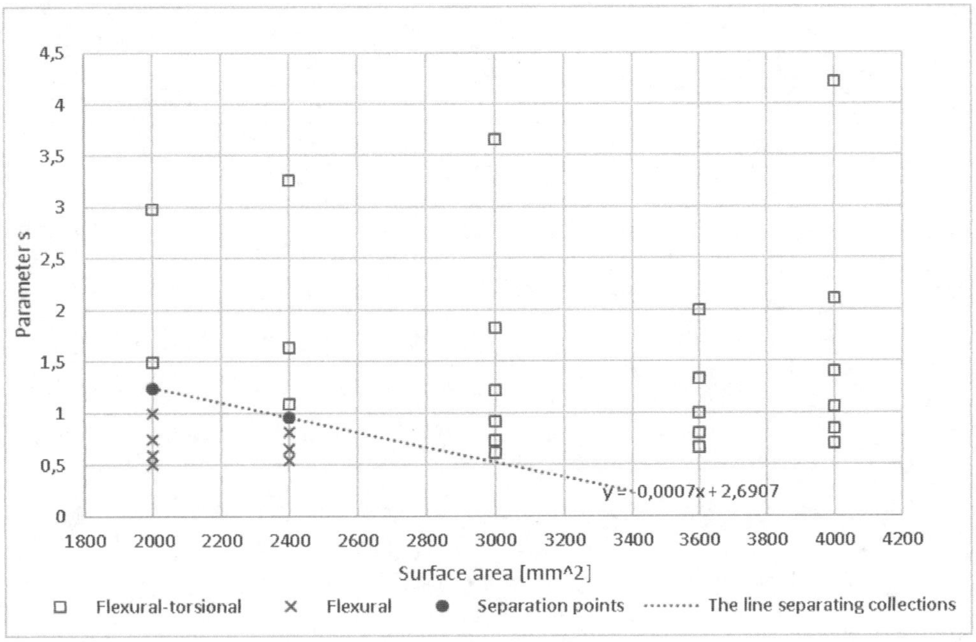

Figure 10. Graph of buckling form depending on the area of the cut-out and parameter *s* (own study).

The resulting straight line is described by the equation:

$$s = -0.0007a * b + 2,69 \tag{4}$$

By modifying the equation of a straight line, we obtain the formula for the next parameter *f*, on the basis of we can determine (estimate) what buckling form the composite element will have:

$$f = \frac{s-2.69}{a*b} = \frac{\frac{\sqrt{a*b}}{g}-2.69}{a*b} \tag{5}$$

The value of the parameter *f* for flexural-torsional forms is greater than the value of the directional coefficient of the straight line used to separate sets is equal to 0.000724202 (figure 11), while for bending forms, it is smaller. The value of the directional coefficient of the straight line, with using the approximation method of separating sets, is the limit (reference) value, on the basis of which is possible to estimate the buckling form. The advantage of this division is that we include all variable parameters in the predictions.

Taking into account the results obtained and the methodology adopted, we arrive at the function *P* for the buckling form:

$$P = \begin{cases} flexural - torsional\ form\ f > -0.0007 \\ flexural\ form\ f < -0.0007 \end{cases} \tag{6}$$

The above function does not account for the scenario in which the value of the parameter f is equal to the directional coefficient of the straight line. This was done on purpose because the presented method is only an approximate method, the accuracy of which may be boosted by increasing the amount of available data. When we us this method, it should be borne in mind that with values close to the limit value, the results may be questionable.

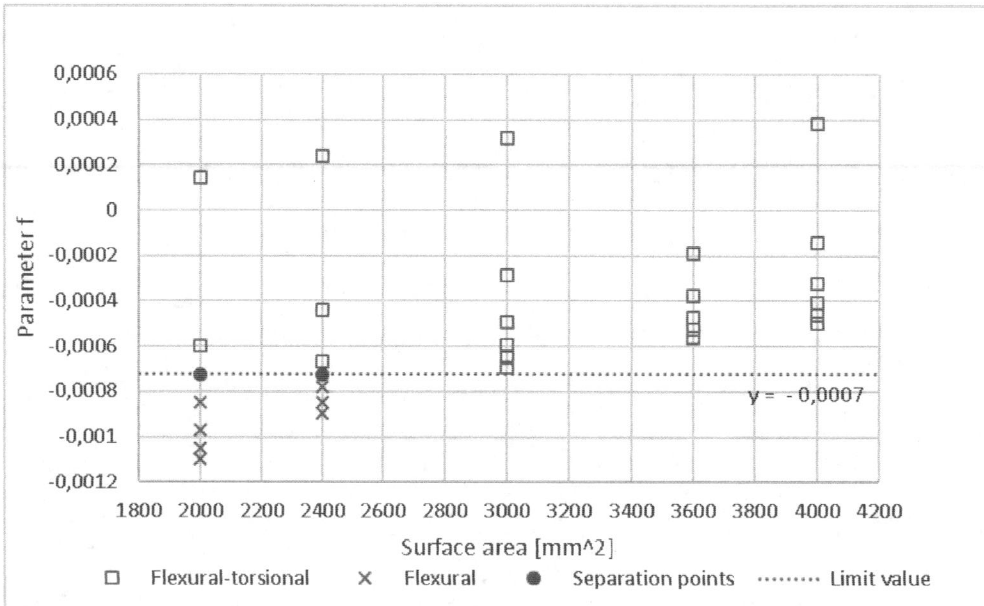

Figure 11. Graph of buckling form depending on the area of the cut-out and parameter *f* (own study).

3. Conclusions

The paper presents a method for simple estimation of buckling forms of tested plates for different variables. The obtained results are of significant practical importance in the design of structures with elastic elements, as they lead to obtaining the required maintenance characteristics of the device.

Acknowledgment

The paper was produced under the project UMO-2017/25/N/ST8/01066 financed by the National Science Centre Poland.

4. References

[1] Koiter W T 1963 *Elastic stability and post-buckling behavior* (In Proceedings of the Symposium on Non-linear Problems Wisconsin: Univ. of Wisconsin Press) p 257-75

[2] Singer J, Arbocz J and Weller T 1998 *Buckling Experiments. Experimental methods in buckling of thin-walled structure. Basic concepts, columns, beams, and plates* (New York: John Wiley & Sons Inc)

[3] Falkowicz K, Debski H, Wysmulski P and Rozylo P 2019 *Composite Structures* **214** 406-13 DOI: 10.1016/j.compstruct.2019.02.001

[4] Falkowicz K 2017 *Archives of Civil Engineering* **63**(4) 161-172. DOI: 10.1515/ace-2017-0047

[5] Rozylo P, Wysmulski P, and Falkowicz K 2017 *International Journal of Applied Mechanics and Engineering* **22**(2) 393-402 DOI: 10.1515/ijame-2017-0023

[6] Chróścielewski J, Miśkiewicz M, Pyrzowski Ł, Rucka M, Sobczyk B and Wilde K 2018 *Composites Part B* **151** 245-255 DOI: 10.1016/j.compositesb.2018.06.016

[7] Campbell FC 2006 *Manufacturing Technology for Aerospace Structural Materials* (Elsevier)

[8] Ferdynus M Kotełko M and Urbaniak M 2019 *Thin-Walled Structures* **144** 106239. https://doi.org/10.1016/j.tws.2019.106239

[9] Darvizeha M, Darvizeha A, Ansaria R and Sharma CB 2004 *Composites Structures* **63**(1) 69-74 https://doi.org/10.1016/S0263-8223(03)00133-8

[10] Bazant ZP andCedolin L 1991 *Stability of structures. Elastic, inelastic, fracture and damage*

theories (Oxford University Press)
[11] Coan JM 1951 *Journal of Applied Mechanics* **18** 143-51
[12] Wysmulski P, Teter A and Debski H 2018 *Effect of load eccentricity on the buckling of thin-walled laminated C-columns* (AIP Conference Proceedings)1922 p 080008 DOI: 10.1063/1.5019079/
[13] Falkowicz K and Debski H 2019 *The work of a compressed, composite plate in asymmetrical arrangement of layers* (AIP Conference Proceedings) 2078 p 020005 DOI: 10.1063/1.5092008
[14] Tereszkowski Z 1970 *Archive of Mechanical Engineering* **3** 485-493
[15] Falkowicz K, Debski H and Teter A 2018 *Design solutions for improving the lowest buckling loads of a thin laminate plate with notch* (AIP Conference Proceedings) 1922 p 080004 DOI: 10.1063/1.5019075
[16] Karpiński R, Jaworski Ł, Jonak J and Krakowski P 2019 *Stress distribution in the knee joint in relation to tibiofemoral angle using the finite element method* (MATEC Web of Conferences) 252 p 07007 https://doi.org/10.1051/matecconf/201925207007
[17] Simitses GJ and Hodges DH 2006 *Fundamentals of structural stability* (Amsterdam: Elsevier/Butterworth-Heinemann)
[18] Królak M and Mania R (Eds.) 2011 *Statics, Dynamics and Stability of Structures Vol. 1, Stability of Thin-Walled Plate Structures* (Lodz: Technical University of Lodz - A series of Monographs)
[19] Singer J, Arbocz J and Weller T 2002 *Buckling Experiments. Experimental methods in buckling of thin-walled structure. Shells built-up structures, composites and additional topics* (New York: John Wiley & Sons Inc)
[20] Van der Heijden AMA, editor 2009 *W.T. Koiter's elastic stability of solids and structures* (Cambridge University Press)
[21] Karpiński R, Jaworski Ł, Jonak J and Krakowski P 2019 *The influence of the nucleus pulposus on the stress distribution in the natural and prosthetic intervertebral disc* (MATEC Web of Conferences) 252 p 07006 https://doi.org/10.1051/matecconf/201925207006
[22] Banat D and Mania R 2019 *Failure analysis of thin-walled GLARE members during buckling and post-buckling response* (AIP Conference Proceedings) 2060 p 020001 https://doi.org/10.1063/1.5086132
[23] Dębski H, Rozylo P and Gliszczynski A 2018 *Composite Structures* **184** 883-93
[24] Banat D and Mania R 2019 *Composite Structures* **212** 338-45
[25] Rozylo P 2017 *Archives of Materials Science and Engineering* **84**(2) 58-64.

IOP Conf. Series: Materials Science and Engineering **710** (2019) 012022 doi:10.1088/1757-899X/710/1/012022

Numerical evaluation of friction effects on the energy absorption of square profiles under bending load

Q Estrada[1], D Szwedowicz[2], J Vergara-Vazquez[2], A Rodriguez-Mendez[3], J Silva-Aceves[1], LC Wiebe[1] and L Escarcega[1]

[1] Universidad Autónoma De Ciudad Juárez (UACJ), Instituto de Ingeniería y Tecnología, Ciudad Juárez, Chihuahua, México
[2] Centro Nacional de Investigación y Desarrollo Tecnológico TecNM/Cenidet, Departamento de Ingeniería Mecánica, Cuernavaca, Morelos, Mexico
[3] University of California, Department of Mechanical Engineering, Berkeley, CA., USA

quirino.estrada@uacj.mx

Abstract. Friction is often associated with undesirable effects on mechanical systems. However, friction could be useful to increase the energy absorption capacity of thin-walled structures subjected to bending loads. The current article proposes the use of friction as an additional energy dissipating mechanism to plastic deformation. For this purpose, three different friction fixtures are presented and numerically evaluated using Abaqus/Explicit software. The friction fixtures were placed at the ends of square profiles generating internal, external and combined friction. In all cases, the square profiles were made of aluminium 6063-T5. The reliability of the numerical results was experimentally validated by a modified three-point bending test with friction. The numerical results for the three arrangements were compared with the results obtained for a profile without friction. According to our results, in all cases friction improves the energy absorption performance of the square profiles in a range from 48.71% to 179.19%. Finally, the best crashworthiness performance was obtained when external friction is generated at the ends of the tube. An increase in the crush force efficiency (CFE) parameter up to 71.47% respect to a typical profile was computed. Thus, the effectiveness of the friction mechanism was demonstrated.

1. Introduction

Among the passive dissipating systems, the use of plastic deformation of thin walled-structures in car collisions is increasing [1]. For the specific case of lateral car collisions, automobile structural elements such as door-impact beams and pillars should be designed to experience bending loads [2, 3]. In order to understand the bending and energy absorption behaviour of thin-walled structures, several studies have been carried out [4, 5]. Many parameters affect the crashworthiness performance of these structures. However, the cross-section [6] material [7, 8,] and span [9] are among the most relevant. In all cases, the authors have focused their efforts on optimising the energy absorption of structures by plastic deformation [10]. Moreover, the incorporation of additional passive energy systems to plastic deformation also begins to be studied. In this way, viscoelastic and friction mechanisms are used [11, 12, 13]. The energy absorption by the viscoelastic mechanism is obtained when foams are used to fill the thin-walled structures. As a result, an increase in bending resistance is obtained by a change of the structural stiffness [14, 15]. In this context, the bending crashworthiness performance of empty and foam-filled circular profiles was analysed [16]. The study was conducted by quasi-static and dynamic three-point bending tests. Three different arrangements were evaluated including single and double tubes. It was concluded that the best crashworthiness performance and bending resistance is obtained when a double tube is filled with aluminium foam. On the other hand, optimisation procedures have been proposed, in this way, Jiang [17] carried out a multi-objective optimisation for a foam-filled double cylindrical tube design. The analysed structure was formed by two concentric cylindrical tubes with structural epoxy foam acting as filler between the tubes. Parameters such as diameter and thickness for the inner and outer tubes and density of the foams were explored. Relative to the original design, the optimised structure allowed an increase in specific energy absorption up to 57.17% and a decrease in peak load of 25.75%. Li [11] conducted an experimental investigation of the bending resistance and the energy absorption effectiveness of empty and foam-filled square and circular tubes. According to their experimental observations, the use of aluminium foam allowed and improvement in energy absorption capacity with a high resistance to bending deformation for double tubes. However, the best energy absorption capacity of 295.10 J was obtained by the foam-filled double square arrangement. Regarding the friction mechanism, this has been reported only for thin-walled structures subjected to axial compression. In this sense, the authors of [12] proposed an arrangement of two concentric tubes. When the compression load occurs, a rigid steel ring is press-fitted on top of a circular aluminium tube. Then, an expansion of the ductile structure is carried out. During this process, energy absorption by friction is obtained. Once the stroke of the steel tube has been reached, both tubes are plastically deformed. Lastly, it was reported that additional energy absorption mechanisms are plausible to improve the crashworthiness capacity of thin-walled structures. However, for the specific case of friction, this is mostly associated with axial compression and barely to structures subjected to bending load. Thus, the current article studies the effect of friction mechanism on the bending crashworthiness performance of square profiles. In this way, three friction fixtures are proposed and numerically evaluated using Abaqus/Explicit software. The friction fixtures were placed at the end of the square tubes. Finally, the assessment of the arrangements is carried out by a modified three-point bending test.

1.1. Dimensional and dimensionless crashworthiness parameters

The assessment of the profiles was carried out by calculating the energy absorption (E_a), mean force (P_m), peak load (P_{max}), specific energy absorption (SEA) and crush force efficiency (CFE) [18]. The mathematical expressions for each case are described in equations (1-4), where F is force, δ displacement and m mass.

$$E_a = \frac{1}{2}\sum_{i}^{n-1}(F(\delta)_{i+1} + F(\delta)_i)\cdot(\delta_{i+1} - \delta_i) \tag{1}$$

$$P_m = \frac{E_a}{\delta} \tag{2}$$

$$SEA = \frac{E_a}{m} \tag{3}$$

$$CFE = \frac{P_m}{P_{max}} \tag{4}$$

2. Modified three-point bending test with friction

Despite the efficiency of Abaqus/explicit to model the plastic deformation of thin-walled structures, the addition of friction effects needs to be validated. Thus, a first discrete model was developed and compared to experimental results. The model consists of a square tube with friction pads evaluated by a modified three-point bending test. The friction pads were placed at the bottom and top faces of both ends of the profile with a length of 50 mm. In order to avoid slipping effects and plastic deformation before the test, a pressure of 1 Nm was applied to the pads. The square profile was modelled with S4R elements and conferred with the elastoplastic properties for Al 6063-T5 described in table 1. Meanwhile, the friction pads were model with mechanical properties for steel using C3D8R elements. A Young modulus equal to 200 GPa, Poisson coefficient of 0.3 and density 7850 kg/m^3 were used. Lastly, from a mesh sensitivity analysis, a general element size of 3 mm was applied. Details of the discrete model are shown in figure 1.

Table. 1. Input material parameters for Al 6063-T5, [19].

Elasticity	Young modulus [MPa]	Poisson coefficient	Density [kg/m^3]
	66940	0.33	2700
Plasticity	Yield stress S_y [MPa]		
	158.79		

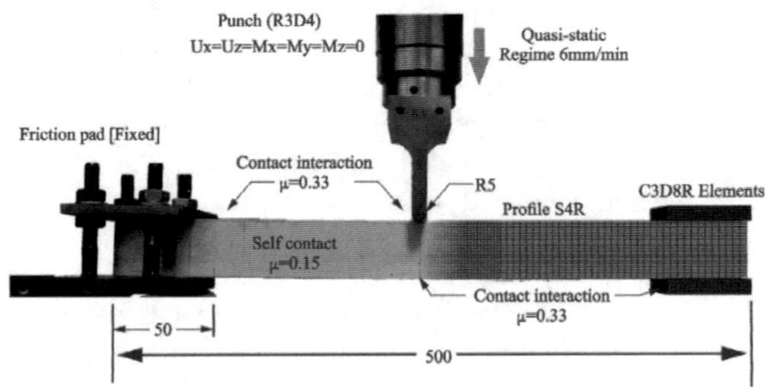

Figure 1. Modified three-point bending test for a square profile with friction pads, units in mm.

The bending force versus displacement curves is presented in figure 2a. Both curves are characterised by an initial increase in bending force (P_{max}) up to ~6 kN. As the ends of the profile are forced to slip, friction forces and bending deformation are observed. Friction forces contributed to keeping the P_{max} almost constant until a 15 mm of displacement is reached. Subsequently, a smooth decrease inin the force is observed. The area under both curves, which indicates the energy absorption, is similar for both models with a value close to 250 J. Respect to the theoretical bending deformation mode for a square structure described by Estrada et al. [6], the evaluated arrangement presents the same pattern at the mid-section. However, an additional deformation is present at the ends of the profile as illustrated in Figure 2b. Besides minor differences in E_a and P_{max} of less than 5% the numerical results correctly predict the plastic deformation behaviour of the structure. Thus, we continue with our numerical study of friction fixtures to improve the energy absorption capacity of square profiles.

a) b)

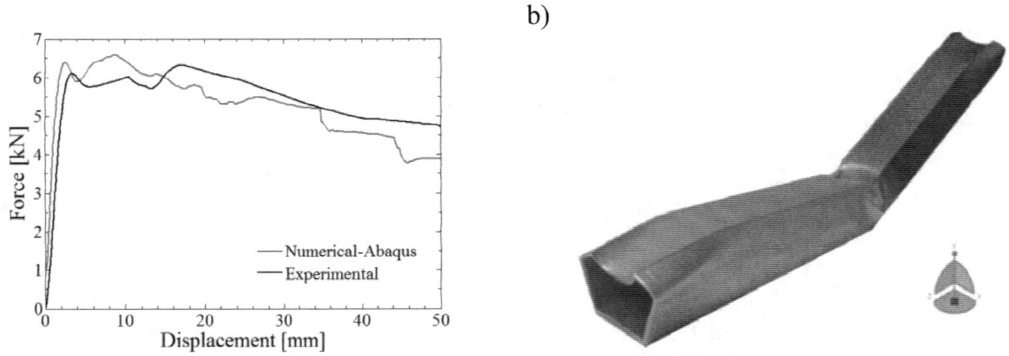

Figure 2. a) Force-displacement curve and b) final deformation state of a square profile under a modified three-point bending test.

3. Numerical simulations

The main objective of this article is to analyse friction effects on the bending and crashworthiness capacity of square tubes. For this purpose, three different friction fixtures inspired in [20] were designed and evaluated by a modified three-point bending test. The friction fixtures were placed at the ends of the square profile. The dimensions of the tube's cross-section are 40.9 x 40.9 mm, length equal to 500 mm and thickness of 1.40 mm. In order to evaluate just the friction and avoid a pre-stress state at the ends, a perfect assembly was considered. The friction mechanisms are designed to generate internal, external and combined friction. For comparison, a typical square profile without friction was also evaluated. Details of the studied arrangements are presented in table 2.

Table 2. Numerical setup, units in mm.

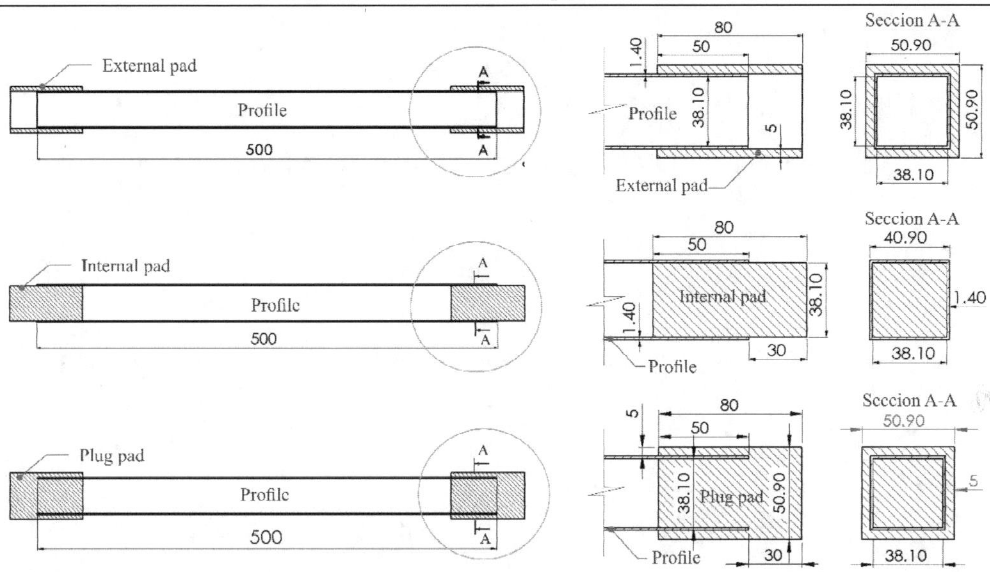

Arrangement code	Cross-section	Thickness of the tube [mm]	Mass of the Tube [kg]	Generated friction	Total contact area [mm²]
S-00		-	-	-	-
S-IF		1.437	43.815	Internal	7620
S-EF		1.442	41.910	External	7620
S-CF		1.445	40.005	Combined	15240

4. Results

The assessment of the arrangements with and without friction was carried out by the study of the bending force v displacement curves. As illustrated in Figure 3, at the beginning of the bending process all structures presented an increase in force close to 7 kN. Once this value is reached two behaviours were observed. The first group is formed by the typical square profile (S-00) and the arrangement with external friction (S-EF). In both cases, after reaching the maximum peak load (P_{max}), a stabilisation process of the force was observed. The opposite case was observed for the second group (S-IF and S-CF) where, contrary to what was expected, a continuous increase in the bending force is achieved. The explanation for this behaviour is associated with the pads acting as a restraining force on both ends. Thus, as the bending test is carried out, friction forces and plastic deformations at the ends are generated. Meanwhile, the S-00 profile presented only bending and crushing deformation at its mid-section.

The mechanical response of the structures is complemented by calculating the energy absorption which is presented in figure 4. Although all the square profiles present the same mass (0.290 kg), the use of a frictional pad increased the energy absorption capacity of the arrangements in all cases. However, the best E_a capacity is obtained when internal and external frictional forces are generated on arrangement S-CF.

IOP Conf. Series: Materials Science and Engineering **710** (2019) 012022 doi:10.1088/1757-899X/710/1/012022

The amount of energy absorbed by the structure is determined by its plastic deformation and the frictional forces. Friction of the fixture at the ends of the profile modifies the bending deformation of a typical square profile without friction. As seen in figures 5-6 the frictional pads induce additional plastic deformation at the ends. Hence, as the frictional contact area increases, higher resistance to bending deformation beneath the punch is observed. Thus, minor plastic deformation is observed at the ends.

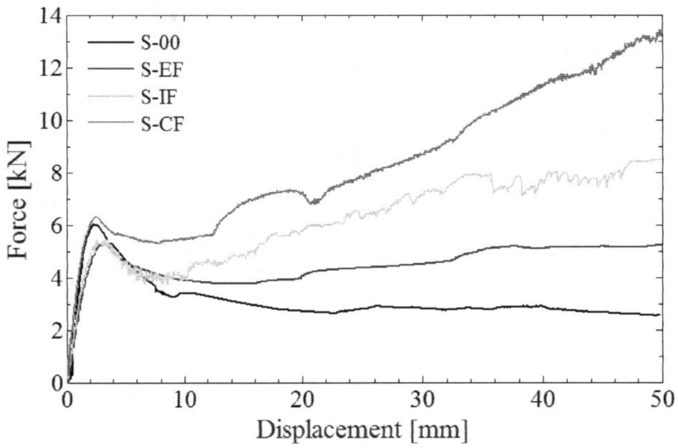

Figure 3. Comparison of force vs. displacement curves for the evaluated structures.

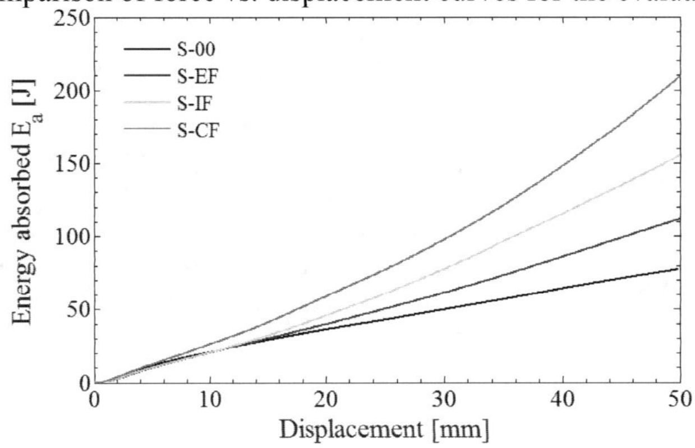

Figure 4. Energy absorption for evaluated structures.

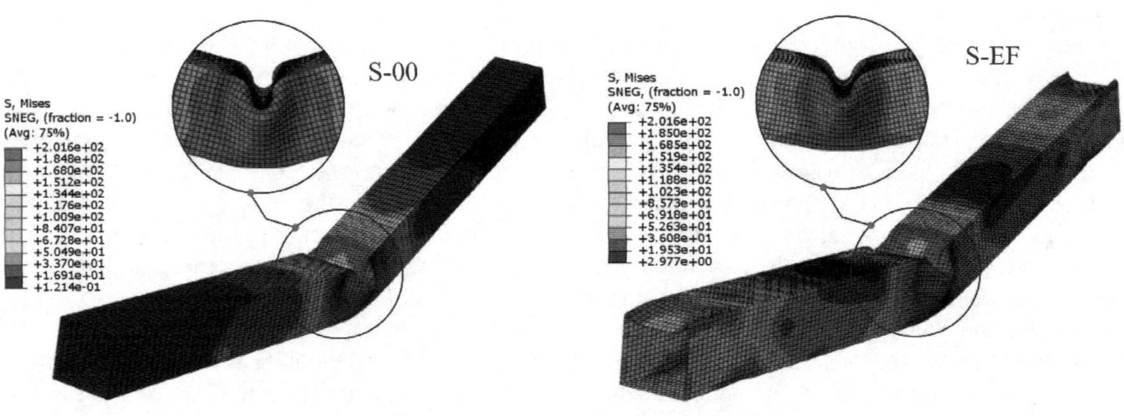

Figure 5. Final deformation mode at 50 mm of displacement I.

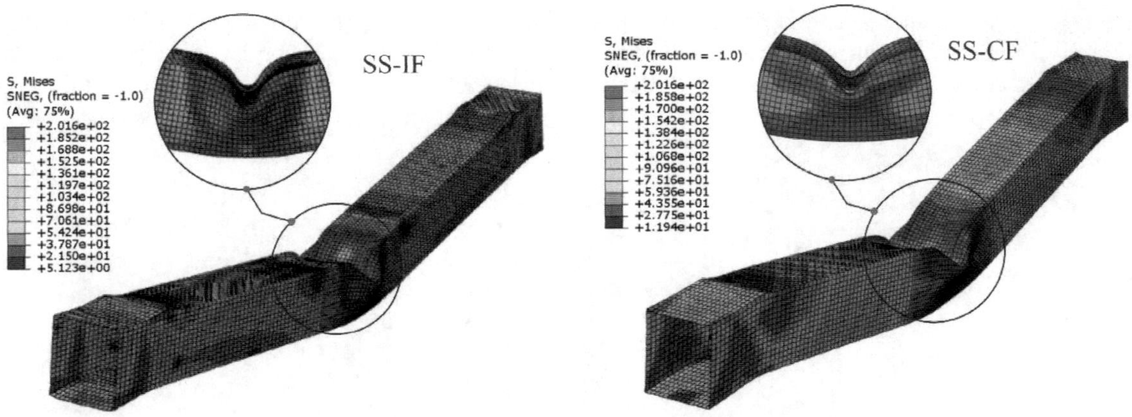

Figure. 6. Final deformation mode at 50 mm of displacement II.

The obtained results are shown in table 3. Accordingly, as the ends of the profile are fully constrained, an increase in P_{max} is obtained. The structure with internal and external friction (S-CF) presents the highest P_{max} value of 8.40 kN. This value represents an increase in 120% relative to the arrangement without friction (S-00). Meanwhile, the lowest value was computed by the structure with external frictional pads (S-IF) with 5.31 kN. In all cases, frictional fixtures improved the energy absorption capacity (E_a) in a range from 48.71% to 179.19%. The largest energy absorption capacity (419.96 J) is obtained when internal and external friction (S-CF) is generated at the ends of the profile. This improved behaviour was also reflected by calculating the highest mean force (P_m) and specific energy absorption (SEA) for S-CF with 8.40 kN and 1.448 J/g, respectively. This suggests higher resistance to the formation of travelling and stationary hinge lines during the bending process. Additionally, energy dissipated by friction effects and secondary hinge lines close to the ends were also observed.

Table 3. Summary of numerical results for the evaluated arrangements.

Code	P_{max} [kN]	P_m [kN]	E_a [J]	SEA [J/g]	CFE
SQ	6.09	3.01	150.42	0.518	0.49
S-EF	5.31	4.47	223.69	0.771	0.84
S-IF	8.51	6.22	310.96	1.072	0.73
S-CF	13.41	8.40	419.96	1.448	0.62

The effectiveness of the three frictional pads to improve the energy absorption capacity of a typical square profile has been confirmed. However, a large energy absorption capacity does not mean the best crashworthiness performance. To find the optimal behaviour of the arrangements, the crush force efficiency (CFE) is introduced in figure 7. Optimal performance is obtained when CFE is equal to 1. Considering the CFE value obtained for a typical profile (SQ), an increase in CFE from 26.53% to 71.47% was achieved by the arrangement with frictional pads.
According to Figure 7, and contrary to expectations, an increase in contact area of the frictional fixtures does not guarantee a better CFE performance. The combination of internal and external friction resulted in the poorest CFE (0.62). Despite the presence of the typical deformation at the middle section which is characterised by a central hinge line and two outward folds, a larger work was required to form the plastic hinges. Then, the maximum crush force was the highest, thus obtaining a lower CFE performance. On the other hand, external friction pads (S-EF) added resistance during the initial bending force, this resistance also contributed to generate frictional forces at the ends. Likewise, the ends not being fully constrained allowed the plastic deformation of the bottom edge of the beam's ends, dissipating a large amount of energy with a relative low P_{max}. This contributed to achieving the best CFE of 0.84. The opposite case was observed for structures with internal and combined friction since an

increase in E_a also increased the peak load. Thus, lower CFE values were obtained. Lastly, from our numerical findings presented in table 3 and figure 7, the use of friction as a secondary energy absorption mechanism was confirmed, especially when external friction is generated. Thus, this kind of arrangement (S-EF) should be considered for future mechanical designs of thin-walled absorbers.

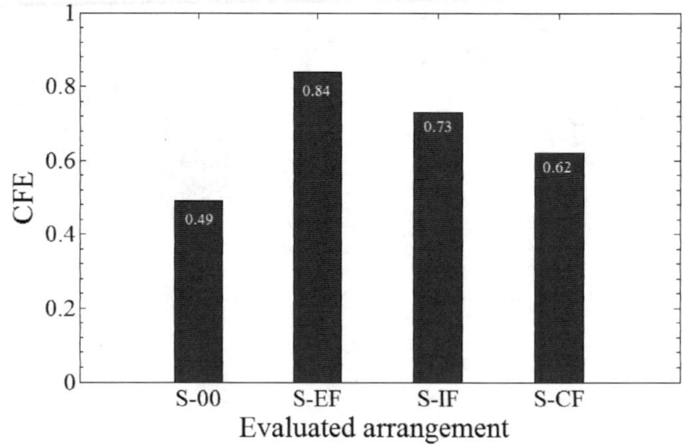

Figure 7. CFE of the four evaluated arrangements.

5. Conclusions

A numerical study of friction effect on crashworthiness performance of square profiles under bending load was carried out. The main objective of this article was to investigate the effectiveness of frictional pads to improve the crashworthiness and bending capacity of mechanical structures. From our numerical study we conclude:

1. Relative to a profile without friction (S-00), the use of frictional pads in any configuration increases the energy absorption capacity (E_a) in a range from 48.71% to 179.19%.
2. From figure 5-6 and table 3, the ability to dissipate energy is more a function of the constraint at the ends of the profile than of the contact area. This can be observed by comparing E_a for S-EF (223 J) and S-IF (310 J). In both cases, the arrangement presented the same friction contact area.
3. The use of frictional pads improves the total energy absorption capacity. However, it also induces an increase in peak load (P_{max}). Hence, when internal and external friction are generated at the same time, an increase in P_{max} up to 120% is computed with respect to a profile without friction (S-00).
4. Regard the CFE parameter, the best crashworthiness performance was obtained when external friction (S-EF) is generated at the ends of the profile. This condition obtained the highest CFE value of 0.84.
5. External friction forces add stiffness to the system and at the same time allows a secondary deformation at the ends. Then, a balance of P_{max} and E_a is obtained. From this mechanical behaviour configuration, S-EF could be used to counter the harmful effects of lateral crashes.

6. References

[1] Tran T 2017 *Compos. Strcut.* **160** 734-47
[2] Lilehkoohi A H, Faieza A A, Sahari B B, Nuraini A A and Halali M. 2014 *Appl. Mech. & Mat.* **663** 552–56
[3] Sun G, Pang T, Zheng G, Song J and Li Q. 2016 *Int. J. Mech. Sci.* **115** 465-80
[4] Wang Z, Li Z and Zhang X. 2016 *Thin Walled Struct* **107** 287-99.
[5] Tang T, Zhang W, Yin H and Wang H. 2016 *Thin-Walled Struct.* **102** 43–57
[6] Estrada Q, Szwedowicz D, Vergara-Vazquez J, Solis José, Paredes A M, Silva J and Wiebe L. 2019 *Appl. Comput. Sci.* **15(2)** 31-41
[7] Yin H, Chen C, Hu T and Wen G. 2018 *Int. J. Crashworthiness* **23(4)** 446–60
[8] Ghadianlou A and Abdullah S B. 2013 *Thin Walled Struct* **67**, 25-33

[9] Beng Y K, Dalimin M N, Wahab M A and Lai H. 2018 *J. Mech. Sci. Technol.* **32(8)** 3611–16
[10] Kotełko M, Ferdynus M and Jankowski J. 2018. *Acta Mech. Autom.* **12(1)** 54-9
[11] Li Z and Lu F. 2015 *J Reinf. Plast. Comp.* **34(9)** 761–68
[12] Salehghaffari S, Tajdari M, Panahi M and Mokhtarnezhad F. 2010 *Thin-Walled Struct.* **48(6)** 379–390
[13] Zhang X, Zhang H and Leng K. 2018 *Thin-Walled Struct* .**127** 728-40
[14] Sampath V, Rao C L, and Reddy S. 2017. *Procedia Manufacturing*, **7** 225-33
[15] Li Z and Lu F. 2015 *J. Reinf. Plast. Compos*. **34(9)** 761-68
[16] Li Z, Zheng Z, Yu J and Guo L. 2013 *Materials & Design (1980-2015)* **52** 1058–64
[17] Jiang P, Wang Q, Yin A, Hu J and Gu X. 2018 *Adv. Mech. Eng.* **10(12)** 168781401881123
[18] Ferdynus M, Kotełko M, and Urbaniak M. 2019 *Thin-Walled Struct.* **144**, 106239
[19] Estrada Q, Szwedowicz D, Gutierrez-Wing E, Silva-Aceves J, Rodriguez-Mendez A, Elias-Espinosa M and Bedolla-Hernandez J. 2019 *P. I. Mech. Eng. D-J Aut.* **233(8)** 2120–38
[20] Mamalis, A G, Manolakos D E, Baldoukas A K and Viegelahn G L. 1989 *P. I. Mech. Eng. C-Mech. Eng. Sci*. **203(6)** 411–17.

IOP Conf. Series: Materials Science and Engineering **710** (2019) 012023 doi:10.1088/1757-899X/710/1/012023

Calculations of exhaust emissions produced by vehicle with petrol engine in urban area

B Šarkan[1], A Kuranc[2] and Ľ Kučera[1]

[1] University of Zilina, Faculty of Operation and Economics of Transport and Communications Department of Road and Urban Transport, Univerzitná 1, 010 26 Žilina, Slovak Republic
[2] University of Life Sciences, Faculty of Production Engineering, Department of Energy and Vehicles, Akademicka 13, 20-618 Lublin, Poland

branislav.sarkan@fpedas.uniza.sk

Abstract. The paper deals with the method of calculating exhaust emissions of a passenger vehicle during drive in urban area. For this calculation, a running test was conducted in which exhaust gas volume data and engine control data were recorded. The evaluated exhaust gas components are nitrogen oxide, hydrocarbon, carbon monoxide, carbon dioxide. The engine speed, intake air quantity, vehicle speed and intake air temperature were recorded from the engine control unit. The exhaust components are calculated in gram per kilometre of vehicle drive.

1. Introduction

Air pollution contributes significantly to global environmental problems such as climate change. The transport sector presents the major factors in energy issues and environmental problems, because it is one of the biggest consumers of fossil energy sources [1-3].

Greenhouse gases are classified as gases that cause global warming. It is one of the most serious environmental problems of modern times. These greenhouse gases include carbon dioxide CO_2, methane CH_4, nitrous oxide N_2O and fluorine compounds [4-6].

Over the past few decades, the European Union has introduced regulations and standards with clearly defined limits for individual exhaust gas components to eliminate emissions from road vehicles (CO, CO_2, HC, NO_x and PM) [7]. This was due to the reduction in the volume of greenhouse gases produced by the transport sector. The European Union is committed to reducing the fuel consumption of road vehicles in order to reduce greenhouse gas emissions. It is in line with long-term requirements to prevent temperatures increase caused by climate change. The European Union's aim is to reduce emissions by 80% by 2050 comparing to 1990 values. CO_2 emissions which are related to transport should be reduced by 60% by 2050 comparing to 1990 [8].

Some components of emissions from transport in Slovakia are being gradually eliminated. It is mainly CO, NOx, VOC and SO_2. On the other hand, components of emissions such as CO_2, PM10 and PM2.5 are constantly increasing year by year. This applies to urban areas with increased traffic due to missing or inconvenient road infrastructure [9-10].

The European Commission set up the Real Driving Emissions legislation (RDE) to ensure that cars produce low pollutant emissions values in real driving. This legislation came into force in March 2016. RDE test is carried out by vehicle manufacturers as part of type-approval tests. Under RDE, a vehicle is driven on public roads and exposed to a wide range of different conditions. Specific equipment installed on the vehicle collects data to verify that legislative caps for pollutants such as NO_x are not exceeded [11-12].

During measurement, the vehicle is equipped with a Portable Emission Measuring System (PEMS). It is a combination of complex measuring devices that integrate advanced exhaust gas analysers, exhaust mass flowmeters, weather stations, GPS (Global Positioning System) and vehicle network connections. These measuring devices are very expensive, and they are mostly operated by type-approval authorities, vehicle manufacturers, or major research and development centres.

2. Measurement methodology

The Department of Road and Urban Transport has its own measurement technology, with which it is possible to determine the amount of exhaust gas emissions produced while driving a vehicle in grams per kilometre using known calculation procedures [13].

The main measuring device is the Maha MGT 5 exhaust gas analyser. This gas analyser is capable to measure HC, CO, CO_2, O_2 and NO_x emissions. The analyser operates on the principle of selective absorption which means that each component of the exhaust is assessed in the infrared range. The tested exhaust gases are conducted from a vehicle exhaust pipe to an exhaust probe that is connected to the analyser by a hose. The whole scheme of data acquisition and processing is presented in Figure 1 [14] [15].

At first, H_2O water vapor is separated from exhaust gases, which then are led to the measuring chamber. The infrared light beam in the direction of the measuring element is weakened by the gas. The amount of attenuation of this light beam is manifested by a different wavelength depending on the type of gas. Such a method is the measured amount of HC, CO, CO_2. On the other hand, O_2 and NO_x are measured by electrochemical detection. The measured data is from the emission analyser evaluated on a portable computer with Maha Emission Viewer software, which allows emissions to be recorded during whole driving of the vehicle [16].

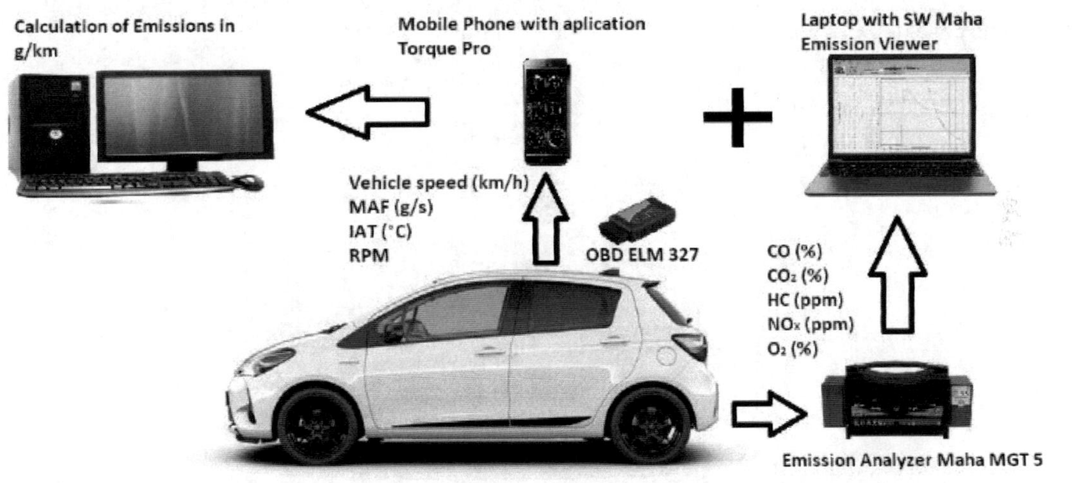

Figure 1. Scheme of data acquisition and processing (Source: authors).

A simple Bluetooth OBD ELM327 paired with a mobile phone was used to record data from the engine control unit. The mobile phone was processing and storing data through application Torque Pro. During the measurements data about the intake air mass MAF (g/s), vehicle speed (km/h), and RPM engine speed and intake air temperature IAT (°C) were recorded and stored [17-18].

The used application allows also recording the GPS position of the vehicle from which the vehicle route data can be evaluated. All measured and stored data are transferred to an evaluation computer, where vehicle emissions in g/km are evaluated according to known methodologies [19].

The data from emission analyser is delayed by 8 seconds because of transmission of emissions from the exhaust pipe to the analyser. The data recording frequency is 1 Hz [20-21]

IOP Conf. Series: Materials Science and Engineering **710** (2019) 012023 doi:10.1088/1757-899X/710/1/012023

Emissions measurements while driving were carried out on a Toyota RAV 4 which was made in 2015. This vehicle is powered by a 1987 cc petrol engine with a maximum power of 112 kW and torque of 195 Nm. It is subject to the EURO 6 emission standard with the following emission values: CO = 0.384 g/km, NO_x = 0.012 g/km, HC = 0.032 g/km, CO_2 = 159 g/km. The vehicle was operated at the urban ring road in Žilina as it is shown in figure 2. The route was 9.3 km long and the measurement was repeated three times.

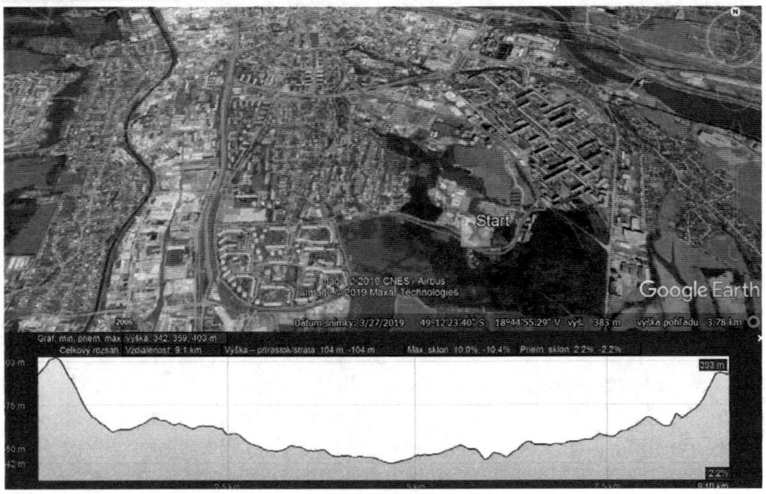

Figure 2. The route of urban ring road in Žilina (Source: authors).

3. Results

The measured exhaust emission values in volume expression had to be converted into mass expression. For this purpose, the methodology published by the author was used [19].

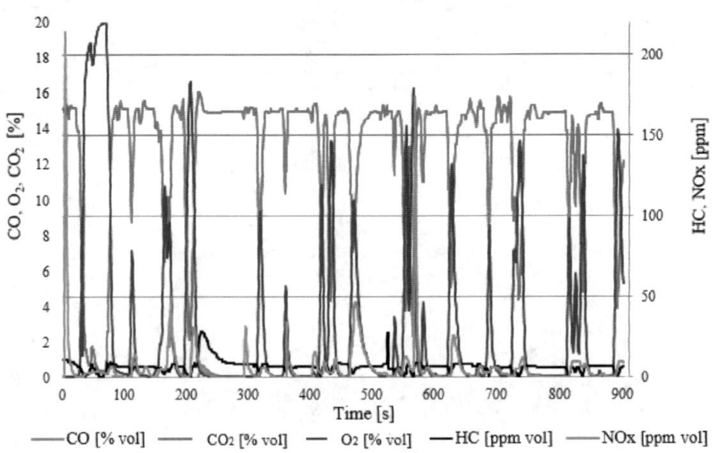

Figure 3. The course of the individual components of the emissions while driving the vehicle (Source: authors).

The cumulative values of the individual emission components are processed by this methodology for all three measurements. The course of the individual components of the emissions while driving the vehicle is shown in figure 3.

The parameters of each driving were affected by the traffic situation. During three measurements were achieved maximum speeds of 66, 70 and 69 km/h and average vehicle speeds of 36.11 km/h, 33.96 km/h and 37.89 km/h. The whole speed course of the vehicle during measurement No.1 is shown in figure 4.

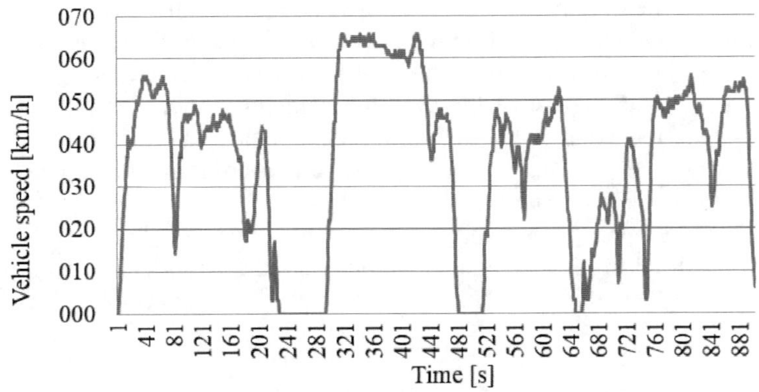

Figure 4. The speed course of the vehicle during measurement No.1. (Source: authors).

In the following Table 1, the exhaust gas components are calculated for all three measurements in grams per kilometre and these values are compared to the vehicle type approval values. As can be seen, especially the HC values are significantly different from those measured on dynamometer in a laboratory.

Table 1. The comparison components of emissions values (Source: authors).

Emissions components	Values from the measurements			Type approval values
	No. 1	No. 2	No. 3	
CO_2 [g/km]	169.26	156.40	170.35	159.00
O_2 [g/km]	0.0164	0.012	0.0150	-
CO [g/km]	0.165	0.1253	0.1039	0.384
HC [g/km]	0.0167	0.0178	0.0240	0.032
NO_x [g/km]	0.0066	0.0073	0.0101	0.012
Average speed [km/h]	36.11	33.96	37.89	

The next part of the results is aimed at comparing different driving modes of the vehicle within the first emission measurement. The measurement No. 1 took a total of 902 seconds and this time period was divided into four different driving modes. Then, exhaust gas components were calculated for each driving mode separately. Four modes were selected: standing of the vehicle (idle mode), steady drive mode, acceleration mode and finally – vehicle deceleration mode.

Table 2. Emission values during different vehicle travel modes during measurement No.1 (Source: authors)

Emissions components	Idle	Steady drive	Steady drive	Acceleration	Deceleration
Whole Drive 902 s	114 sec	396 sec	396 sec	210 sec	182 sec
CO_2 [g]	69.491	671.27	671.27	704.115	129.278
O_2 [g]	0.957	68.25	68.25	13.303	70.083
CO [g]	0.1505	0.3401	0.3401	0.8993	0.1471
HC [g]	0.0087	0.0612	0.0612	0.0722	0.0139
NO_x [g]	0.00199	0.00998	0.00998	0.03402	0.01601

Up to 13% of the total measurement time, the vehicle was in idle mode, which was mainly due to the traffic situation on a given measurement route. This was also due to the presence of light-signalling devices on the measuring route. During the first measurement, the vehicle was stopped three times with a total duration of 114 seconds. Almost half of the total measurement time, the vehicle was moving at steady speed (steady drive) and the rest of the time the vehicle was accelerated and decelerated. To compare the components of the exhaust, the individual components had to be calculated in grams (not in grams per kilometre). This is in order to make it possible to compare the production of emissions while the vehicle is stationary (in idle mode).

In order to be able to compare the driving modes with each other, not only the total emissions production has been quantified in the driving modes, but also the individual components of emissions were recalculated per one second. These values are shown for each components of emissions at following figures 5, 6, 7 and 8.

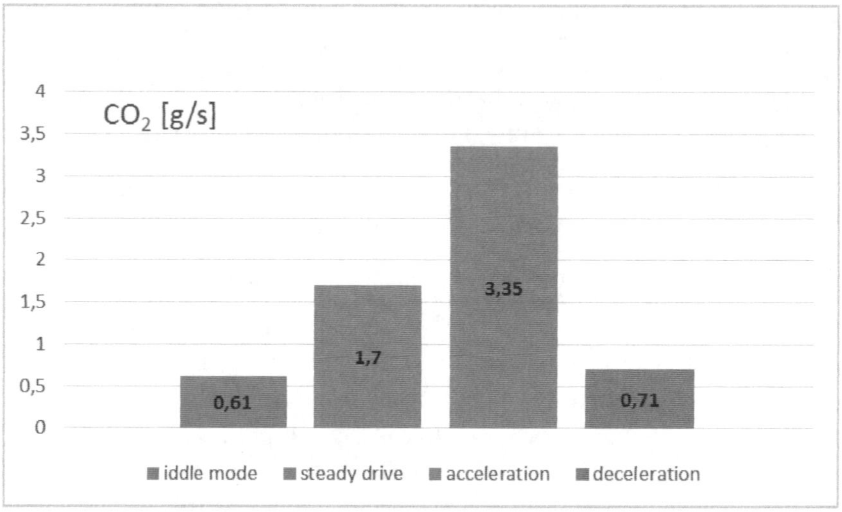

Figure 5. The amount of CO_2 (g/s) (Source: authors).

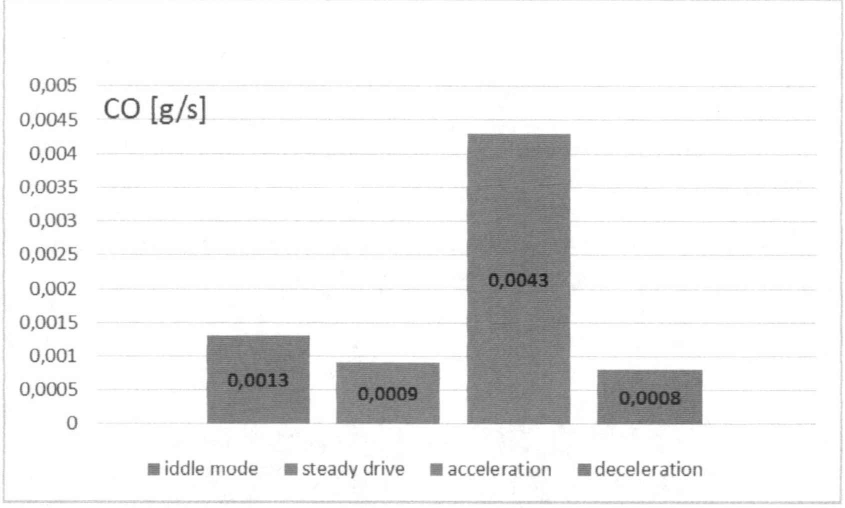

Figure 6. The amount of CO (g/s) (Source: authors).

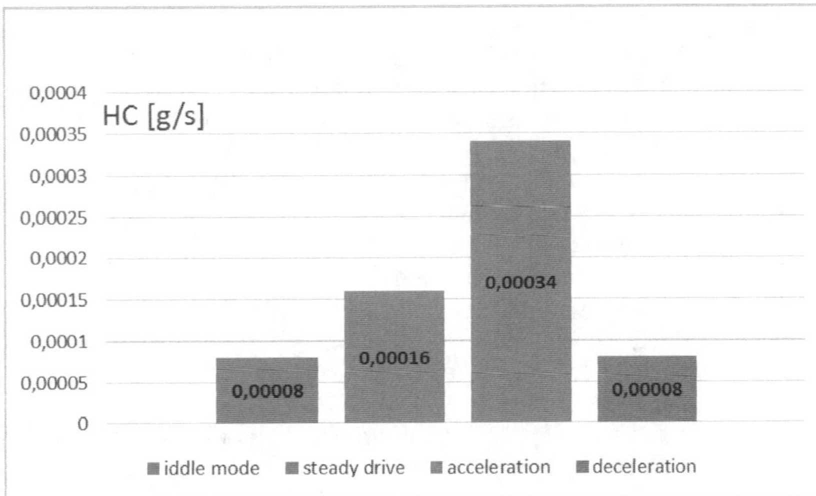

Figure 7. The amount of HC (g/s) (Source: authors).

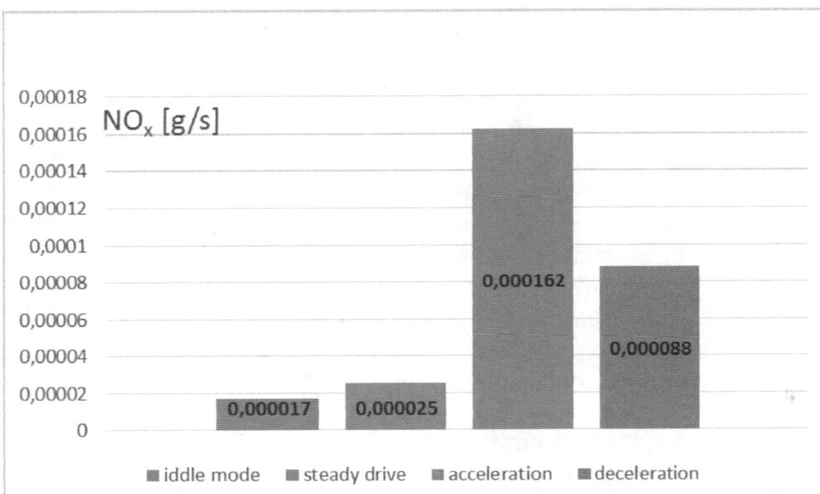

Figure 8. The amount of NO_x (g/s) Source: authors).

4. Conclusion

The method of measuring the emissions of passenger cars during their urban and extra-urban operation is nowadays an obligation in the vehicle type-approval process. The method of calculating emissions through a simple exhaust gas analyser is an alternative to the PEMS method, especially in scientific researches. It propose a relatively fast and cheap measurement method. Results from this kind of measurement can be used to determine the impact of road transport on selected areas of cities.

The results show that the way the vehicle is driven has a significant impact on the amount of produced emissions. If a vehicle in one area of the city has to constantly accelerate, this will affect the amount of emissions produced in that area. For the purpose of quantifying the impact of road transport on the environment, it is necessary to have the most accurate data about traffic flow structure. There are differences in emissions in the case of trucks and in the case of passenger cars. However, differences can be observed in vehicles with petrol engine and diesel engines, too.

Data from such types of emission measurements should in future serve, for example, to decision of cities where to introduce emission zones and the prohibition of entry of vehicles of particular category or emission system.

IOP Conf. Series: Materials Science and Engineering **710** (2019) 012023 doi:10.1088/1757-899X/710/1/012023

5. References

[1] Slovak Environment Agency – Sector Indicator Report - Transport, *Transport and Environment in Slovakia* 2017 available at: http://www.enviroportal.sk /uploads/report/ 8063.pdf

[2] Caban J and Ignaciuk P 2018 Technical-economic aspects of CNG gas usage in buses of urban communication. *Engin. for Rural Development, 17th Int. Scient. Conf.* 2129-33

[3] Vong C M, Wong P K and Li Y P 2006 Prediction of automotive engine power and torque using least squares support vector machines and Bayesian inference *Eng Appl Artif Intel* **19**(3) 277–87 10.1016/j.engappai.2005.09.001.

[4] Koreňová Ľ 2016 Greenhouse gas emissions from transport Slovak Environment Agency available at: http://www.enviroportal.sk/indicator/detail?id=1081

[5] Bontoft C A, Carbonr R, Almena M D C, Farenback - Brateman J H , Gueit J, Hovius H and Zemroch P J 2016 Phase 1: Effect of fuel octane on the performance of two euro 4 gasoline passenger cars *Concawe Reports* **13**

[6] Lizbetin J, Stopka O and Kurenkov 2019 Declarations regarding the energy consumption and emissions of the greenhouse gases in the road freight transport sector *The Archives of Automotive Engineering – Archiwum Motoryzacji* **83**(1) 59-72

[7] Rimkus A, Matijosius J, Bogdevicius M, Bereczky A and Torok A 2018 An investigation of the efficiency of using O_2 and H_2 (hydrooxile gas-HHO) gas additives in a ci engine operating on diesel fuel and biodiesel *Energy* **152** 640-51

[8] European Environment Agency *Explaining road transport Emissions* 2016 (Public Office of the EU: Luxembourg)

[9] Osipowicz T, Abramek K F, Matuszak Z, Jaśkiewicz M, Ludwinek K A and Lagowski P 2018 The Concept of Annular Channels Application on the Spraying Nozzle Needle of Modern Fuel Injector in the Aspect of Combustion Process Improvement *11th Inter. Scien. and Technical Conf. on Automotive Safety* (Casta Papiernicka, Slovakia)

[10] Machalikova J, Sejkorova M, Livorova M, Kocourek L and Corny S 2010 Possibilities of reducing the consumption of lubricants in traffic *The 4th Czech-Slovak Scien. Conf.: Transport, Health and Environment* p 185-94

[11] ACEA, 2019: https://www.acea.be/publications/article/access-to-euro-6-rde-monitoring-data.

[12] Kubíková S, Kalašová A and Černický Ľ 2014 Microscopic simulation of optimal use of communication network. Telematics - Support for Transport. TST 2014 *Communications in Computer and Information Science* p 414-23 ISBN 978-3-662-45316-2. DOI: https://doi.org/10.1007/978-3-662-45317-9_44.

[13] Skrucany T, Kendra M, Kalina T, Jurkovic M, Vojtek M and Synak F 2018 Environmental comparison of different transport modes *Nase More* **65**(4) 192-6.

[14] Hudak M and Madlenak R 2016 The research of driver distraction by visual smog on selected road stretch *Slovakia Proc. of the 16th Inter. Scien. Conf. Reliability and Statistics in Transportation and Communication (RELSTAT) Proc. Engineering* **178** p 472-9

[15] Lizbetin J, Stopka O and Nemec F 2016 Methodological Assessment of Environmental Indicators in Combined Transport in Comparison with Direct Road Freight Transport *20th International Scien. Conf. on Transport Means 2016* (Juodkrante, Lithuania) p 151-5

[16] Hu B, Turner J W G, Akehurst S, Brace C and Copeland C 2017 Observations on and potential trends for mechanically supercharging a downsized passenger car engine *Proc. of the Institution of Mechanical Engin. Part D: Jour. of Automobile Engin.* **231** p 435-6. DOI: 10.1177/0954407016636971

[17] Sebok M, Gutten M, Kucera M and Korenciak D 2019 Thermal and Electrical Analysis of Automotive Ignition System *The Archives of Automotive Engineering – Archiwum Motoryzacji 2019* **83** (1) 113–21 DOI: https://doi.org/10.14669/AM.VOL83.ART8

[18] Simkova I, Konecny V, Liscak S and Stopka O 2015 Measuring the quality impacts on the performance in transport company *Transport Problems* **10** (3) 113-124 DOI: 10.21307/tp-2015-039

[19] Kuranc A 2015 Exhaust emission test performance with the use of the signal from air flow meter *Eksploat Niezawodn* **17(**1)

IOP Conf. Series: Materials Science and Engineering **710** (2019) 012023 doi:10.1088/1757-899X/710/1/012023

[20] Vojtisek – Lom M, James T and Cobb Jr 1997 Report: Vehicle mass emissions measurement using a portable gas exhaust analyzer and engine computer data, Poor man's PEMS - low cost, simple portable on-board exhaust emissions monitoring systems
[21] Vojtíšek M and Kotek M 2014 Estimation of engine intake air mass flow using a generic speed-density Method *MECCA 01* 2014 p 15 available at: https://www.degruyter.com/downloadpdf /j/mecdc.2014.12.issue-1/mecdc-2014-0002/mecdc- 2014-0002.pdf.

IOP Conf. Series: Materials Science and Engineering **710** (2019) 012024 doi:10.1088/1757-899X/710/1/012024

AI-aided e-nose calibration

A Malec[1], Z Suchorab[1], Ł Guz[1] and C Caserta[2]

[1] Lublin University of Technology, Faculty of Environmental Engineering, Nadbystrzycka 40B, 20-618 Lublin, Poland

[2] Mediterranean University of Reggio Calabria, Via dell'Università, 25, 89124 Reggio Calabria, Italy

aga_malec@interia.pl

Abstract. The development of the semiconductor technology and computer methods contributes to the improvements in measurement techniques of air pollution. Besides traditional techniques of air pollution measurement, such as chromatography or dynamic olfactometry, the electronic nose (EN) is attracting an increasing amount of attention in this area. EN is a device which imitates the human sense of smell. The main component of the device is an array of many non-specific gas sensors. The aim of this article is to present the result of calibration of EN composed of MOS (metal-oxide semiconductor) sensor with the use of the artificial neural networks. For data analysis, linear and artificial intelligence algorithms were employed, whose accuracy was compared.

1. Introduction

An artificial neural network is a general name for mathematical structures and their software or hardware models that perform calculations or signal processing through rows of processing elements. These elements are known as artificial neurons and they perform certain basic operations at their input. The inspiration for such a structure was the construction of natural neurons, the synapses linking them, and the nervous system, particularly the brain. The common feature of all neural networks is that their structure consists of neurons connected by synapses. There are weights connected with synapses, which are numerical values whose interpretation depends on the model [1, 2].

The mathematical model describing the principle of information processing is based on the determination of the block that weights the input signals [3, 22], which resembles the biological synaptic connection. Low-capacity dendrites are represented in the model by low-capacity scales. The summing element sums up the weighted signals. If the sum of weighed signals is greater than the activation threshold set at the input, the signal is transferred by activation functions [4, 15]. As a result of this operation, we obtain an output signal which corresponds to the signal flowing through the cellular axon.

Artificial Neural Networks (ANNs) are relatively new computational tools which have been widely used to solve many complex problems in the real world. The attractiveness of ANNs lies in their information processing characteristics, mainly related to nonlinearity, high parallelism, resistance to errors and noise, and learning and generalisation capabilities [5, 20].

In order for the network to function properly, it must go through a learning process [6, 24]. Teaching a neural network consists in selecting weights so that the output signal is within the limits of the accepted error with respect to the expected signal. The object to be taught in such a network is a matrix, the values of which should be selected so that the network realises the assumed form of mapping the input signals to the output signals. Network training can be carried out with or without supervision by an algorithm. One of the teaching algorithms is the backpropagation (BP) algorithm of error, which consists in the fact that having determined the error occurring during the implementation of this step of the learning process in the neuron with the number of m - we can "project" this error back to all those neurons whose signals were inputs for the m-th neuron. This algorithm is used in multi-layer networks with the so-called hidden layers [7].

Considering the cases of such networks, it can be concluded that in many of them, with simple input-output relationships, one layer of hidden neurons is sufficient.

In other cases, the use of additional hidden layers might be necessary. There is a general suggestion from Nielsen [8, 25] that the number of hidden neurons should be less than $2n + 1$, where n is the number of inputs. A network with a small number of hidden neurons does not allow the modelling of complex input-output data, but on the other hand networks with many neurons in hidden layers have weak generalising capabilities [9, 23].

The neural networks have been used in many areas of life, due to the constant development of the field of knowledge they build. So far, they have found application in: electronic system diagnostics, psychiatric research, stock market forecasts, sales forecasts, oil exploration, interpretation of biological and medical research, price forecasts, spectral analysis, waste disposal optimisation, selection of raw materials, selection of investigation targets in forensics and industrial process control [10, 21]. Currently, much attention is paid in the scientific literature to the identification of samples emitting volatile air pollutants by means of different sensor matrices, for example for the identification of gaseous pollutants, poor quality food products or cereals infected with fungi [11, 12, 13]. Despite its usefulness, ANN also has some disadvantages. Network learning usually requires large amounts of data. The output information is not very accurate, since they are only approximations, so it is not suitable for precise engineering calculations. From the practical point of view, networks are treated as "black boxes" because it is difficult to trace the transformation of the input information into a useful output result. For example, they were used to identify pollutants from the group: four [2, 5, 14], five [7, 27] or eight compounds [14, 19]. Due to the limited number of components, the effectiveness of identification was high.

2. Materials and methods

Sensors

The device used in tests was a gas sensor array consisting of 17 MOS (metal oxides semiconductor) sensors manufactured by TGS Figaro producer:

- TGS-2600-B00 – general air pollution sensor 1-30 ppm (for H_2)
- TGS-2602-B00 – toxic air pollutants (NH_3, H_2S, C_2H_5OH, C_6H_5-CH_3) 1-30 ppm (for C_2H_5OH)
- TGS-2610-D00 – propane and butane sensor with carbon filter (C_3H_8, C_4H_{10}) 500-10000ppm
- TGS-2611-E00 – methane sensor with carbon filter (CH_4) 500-10000ppm
- TGS-2620-C00 – ethanol sensor, 50-5000 ppm
- TGS-4161 – carbon monoxide 350-10000 ppm
- TGS-2444 – ammonia 10-100 ppm
- TGS-2442-B02 – carbon monoxide sensor CO 30-1000 ppm
- TGS-800 – general air pollution 1-1000 ppm (for C_4H_{10})
- TGS-825-A00 – hydrogen sulphide 5-100 ppm
- TGS-813-A00 – flammable gas sensor 500-10000 ppm
- TGS-821 – hydrogen 10-5000 ppm
- TGS-823-A00 – solvent vapours 50-5000 ppm
- TGS-812 – sensor for explosive and toxic gases 100-5000 ppm (for CH_4, C_4H_{10}, H_2)
- TGS-830 – chlorofluorocarbons 100-3000 ppm
- TGS-832-A00 – chlorofluorocarbons 100-3000ppm
- TGS-2106 – exhaust gas sensor, diesel exhaust gas sensor 0.1-10 ppm (for NO_2).

The general scheme of the stand is presented in figure 1. Switching between synthetic air and the connected sample took place automatically according to the set protocol.

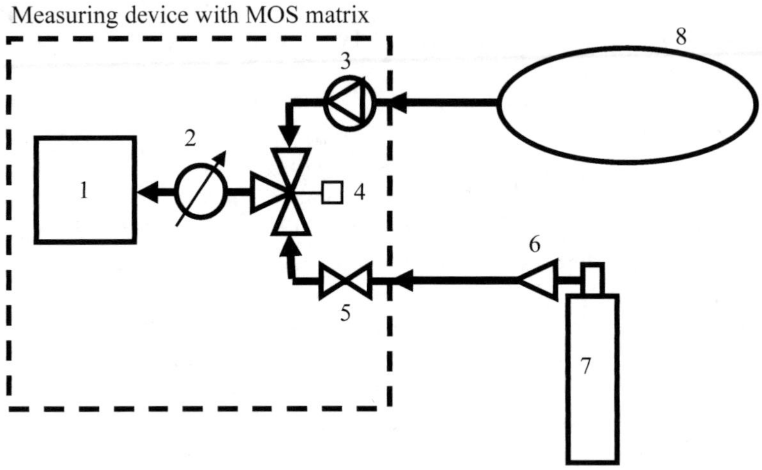

Figure 1. General diagram of the measuring station: 1) sensor matrix, 2) flow meter, 3) micro membrane pump, 4) switchover valve, 5) control valve, 6) reducer, 8) air cylinder, 7) bag.

Three measurements were taken for each sample of a specific concentration. A single measurement lasted 90 seconds and consisted of a 420-second phase of clean air rinsing and 480-second measurement. The flush of airflow and sample collection were set to 0.5 l/min. An example of voltage change for the TGS-800 sensor during butanol measurements is shown in figure 2.

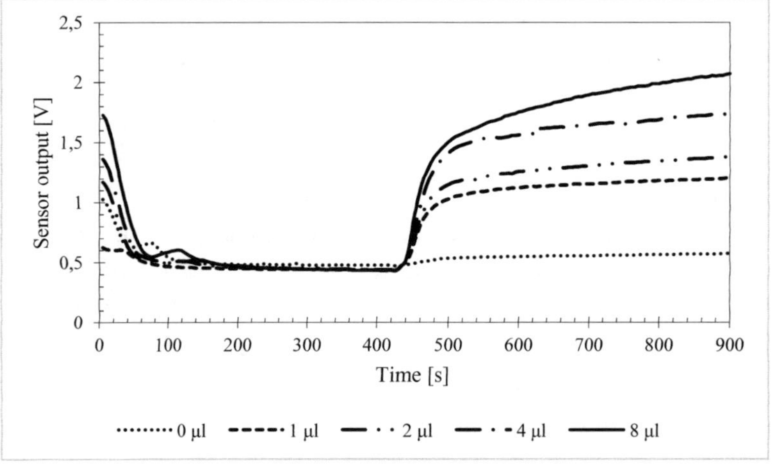

Figure 2. Voltage change for TGS-800 sensor during 1-butanol measurements.

Samples

Specific doses of chemical compounds: n-1-butanol, acetone, benzaldehyde, dimethylamine, toluene were dosed with a chromatographic syringe (10 µL) into a 16-litre Tedlar bag with synthetic air, with a hydrocarbon content of less than 0.1 ppm. The doses were respectively: 0 (blank samples), 1, 2, 4 and 8 µL. After evaporation of the dosed substance, the bag was connected to the sampling port of the e-nose. The detected concentrations are given in the table below.

Table 1. Concentrations of chemical compounds.

Compound	Dispensed volume (μl)	Concentration (ppm)
benzaldehyde	1	0.0666
	2	0.1341
	4	0.2682
	8	0.5366
acetone	1	0.0608
	2	0.1341
	4	0.2430
	8	0.4860
toluene	1	0.0671
	2	0.1341
	4	0.2683
	8	0.5366
dimethylamine	1	0.0671
	2	0.1342
	4	0.2684
	8	0.5367
1-butanol	1	0.0671
	2	0.1341
	4	0.2683
	8	0.5365

Analysis

Strict results of measurements obtained with the use of electronic nose contain signals from individual sensors. Finding important relationships in such multidimensional sets is not easy because of their multidimensionality and often redundancy of information. This requires the use of appropriate statistical tools which, for instance, reduce the multidimensionality of the data set, such as the analysis of the main components, the results of which must be further analysed. Artificial neural networks, on the other hand, immediately provide useful information about the tested sample. For this reason, artificial neural networks were used to analyse the data presented in this paper.

The statistical package Statistica 13.1 was used for the analysis. Individual variables (results from sensors) were standardised according to their dependence: $s = (S - \bar{\bar{S}})/\delta$ where: s - standard value, S - raw value, S - mean value of the sensor, δ - standard deviation.

PCA analysis based on covariance matrix was used for initial visualisation. The number of main components was limited to two.

For further analysis, (ANN) artificial neural networks were used. First of all, supervised classification networks were used to identify chemical compounds. Five different MLP (Multilayer Perceptron) networks were determined (table 2), each with five outputs, corresponding to particular compounds.

Afterwards, upon the chemical compound detection, subsequent neural networks were used in the tested sample to estimate the concentration of the gas sample. For each chemical compound, five different regression neural networks were also determined. The use of several neural networks in each application was intended to exclude the possibility of obtaining random results of the analysis.

In the Statistica software, the Automated Neural Networks module was used. For neural network analysis, all measurements from the range of 800-900 sec. were selected. The data was randomly divided into a training subset (75%), a test subset (15%) and a validation subset (15%). These samples are used to 1) train the network, 2) verify (test) the performance of the networks while under training, and 3) perform *n* final validation test to determine how well the network predicts "new" data that was neither used to train the model or to test its performance when trained. The networks used had one layer of hidden neurons with the minimum and the maximum number of neurons, respectively 17 and 34. With fewer neurons, network error increases, and with a much larger number of neurons, networks lose their ability to generalise. All the neural networks used had 17 inputs. As the training algorithm BFGS

(Broyden–Fletcher–Goldfarb–Shanno) was used. For error evaluation Entropy and SOS (Sum of Squares) functions were applied.

3. Results

PCA analysis

The cumulative value of the first two main components is 83.19%. The use of subsequent main components makes a much smaller contribution to the description of the variance of the data set, which is visible on the landfill slope diagram (figure 3a).

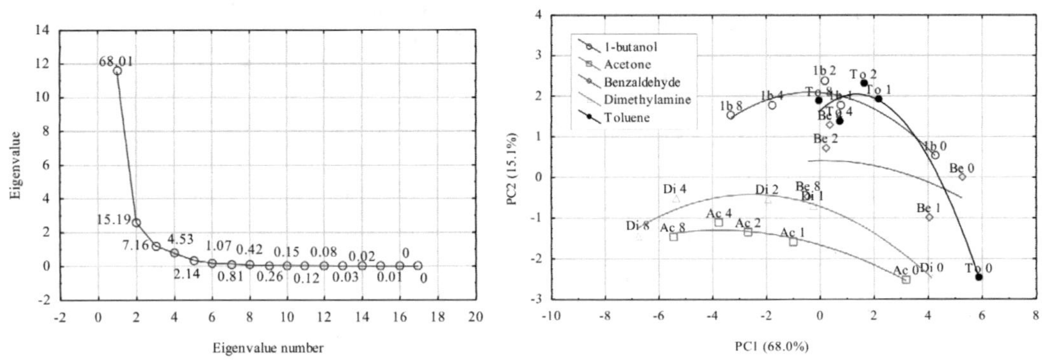

Figure 3. Analysis of the obtained data using PCA method. a) change in percentage contribution to the description of the total variance of the dataset depending on the number of main components; b) PCA diagram of the analysed compounds. On the left: n – 1-butanol, Ac – acetone Be – benzaldehyde, Di – dimethylamine, To – toluene. The number on the right denotes dose.

The two-dimensional PCA analysis graph is shown in figure 3b. The graph shows the average values from the standardised data. The separation of individual chemicals is visible. The points representing individual concentrations tend to be arranged along the line, and ordered cording to increasing concentrations. Measuring points for clean air are grouped in the right part of the graph. The lowest regression coefficient was observed for benzaldehyde.

Identification of chemical compounds using neural networks

The data computed by means of the artificial neural networks for detection of the particular chemicals, regardless of sample concentration are presented in table 2.

Table 2. Neural networks applied for compound identification.

Net. name	Training accuracy (%)	Test accuracy (%)	Validation accuracy (%)	Training algorithm	Error function	Hidden activation	Output activation
MLP 17-29-5	97.8716	95.9142	88.2410	BFGS 21	Entropy	Logistic	Softmax
MLP 17-31-5	94.2857	87.6857	81.5477	BFGS 83	SOS	Exponential	Exponential
MLP 17-34-5	99.2764	92.3271	86.7874	BFGS 22	Entropy	Linear	Softmax
MLP 17-32-5	92.2044	90.3603	89.4567	BFGS 20	Entropy	Linear	Softmax
MLP 17-21-5	95.9478	89.2315	82.9853	BFGS 28	Entropy	Logistic	Softmax

The graph below shows the result of the identification of chemical compounds. On the cut-off axis, there are chemical compounds that have been measured with e-nose and then classified with ANN.

Neural network responses are indicated on the ordinate axis. 1-butanol was correctly identified as 1-butanol, but also as acetone and dimethylamine. All the networks correctly identified benzaldehyde, despite the fact that the PCA analysis graph showed the highest data scattering for a given chemical compound.

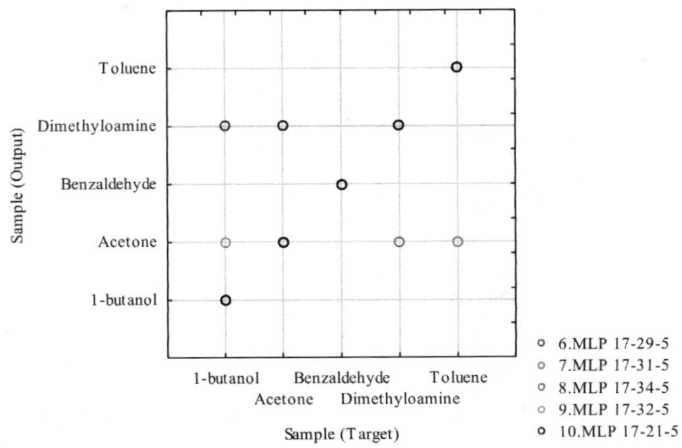

Figure 4. Identification of chemical compounds using particular neural networks.

Evaluation of particular compounds concentration using neural networks

The potential of multisensory arrays to evaluate particular compounds concentrations (1-butanol, acetone, benzaldehyde, dimethylamine, toluene) is presented in the tables and diagrams below. The tested chemical compounds were selected because of their characteristic, perceptible odour. In addition, 1-butanol is the reference substance in dynamic olfactometry (EN-13725).

Table 3. Identification of tested substances concentration using particular networks.

Net. name	Training accuracy	Test accuracy	Validation accuracy	Training error	Test error	Validation error	Training algorithm	Error function	Hidden activation	Output activation
					1-butanol					
MLP 17-19-1	0.999372	0.999121	0.999755	0.006487	0.005904	0.004877	BFGS 24	SOS	Linear	Tanh
MLP 17-24-1	0.999533	0.999042	0.999745	0.005605	0.006511	0.005779	BFGS 30	SOS	Linear	Tanh
MLP 17-17-1	0.999979	0.999911	0.999984	0.000175	0.000539	0.000149	BFGS 97	SOS	Tanh	Logistic
MLP 17-18-1	0.999997	0.999982	0.999997	0.000023	0.000100	0.000032	BFGS 220	SOS	Tanh	Tanh
MLP 17-24-1	0.999288	0.998739	0.999678	0.006829	0.007918	0.005385	BFGS 25	SOS	Logistic	Tanh
					acetone					
MLP 17-24-1	0.999965	0.996825	0.996412	0.000277	0.033676	0.034449	BFGS 39	SOS	Linear	Exponential
MLP 17-22-1	0.998054	0.995334	0.993337	0.015533	0.043683	0.059666	BFGS 6	SOS	Exponential	Linear
MLP 17-23-1	0.965803	0.969746	0.966735	1.409294	1.867394	1.519765	BFGS 4	SOS	Logistic	Linear
MLP 17-17-1	0.999911	0.948048	0.942155	0.000712	0.507327	0.508122	BFGS 27	SOS	Exponential	Logistic
MLP 17-26-1	0.999996	0.974881	0.972070	0.000033	0.245970	0.245975	BFGS 179	SOS	Tanh	Tanh
					benzaldehyde					
MLP 17-24-1	0.999989	0.999986	0.999967	0.000098	0.000158	0.000164	BFGS 167	SOS	Tanh	Linear
MLP 17-22-1	0.999999	0.999949	0.999981	0.000009	0.000377	0.000114	BFGS 267	SOS	Tanh	Tanh
MLP 17-27-1	0.999994	0.999980	0.999977	0.000055	0.000133	0.000107	BFGS 105	SOS	Exponential	Logistic
MLP 17-20-1	1.000000	0.999995	0.999959	0.000002	0.000034	0.000176	BFGS 369	SOS	Exponential	Tanh
MLP 17-25-1	0.999989	0.999962	0.999956	0.000100	0.000250	0.000263	BFGS 122	SOS	Logistic	Logistic
					dimethylamine					
MLP 17-26-1	0.999974	0.999996	0.990211	0.000200	0.000046	0.093301	BFGS 86	SOS	Logistic	Logistic
MLP 17-30-1	0.999995	0.999999	0.986592	0.000036	0.000010	0.129585	BFGS 122	SOS	Logistic	Logistic
MLP 17-28-1	0.999986	0.999972	0.986598	0.000106	0.000257	0.129745	BFGS 76	SOS	Tanh	Exponential
MLP 17-31-1	0.999981	0.999975	0.992708	0.000158	0.000268	0.067533	BFGS 52	SOS	Exponential	Tanh
MLP 17-23-1	0.999991	0.999997	0.990497	0.000067	0.000030	0.083596	BFGS 88	SOS	Exponential	Logistic
					toluene					
MLP 17-20-1	0.999974	0.999968	0.997244	0.000197	0.000256	0.029879	BFGS 105	SOS	Logistic	Exponential
MLP 17-25-1	0.999993	0.999995	0.999473	0.000050	0.000047	0.005711	BFGS 268	SOS	Tanh	Linear
MLP 17-21-1	0.999991	0.999991	0.997046	0.000066	0.000078	0.033534	BFGS 141	SOS	Logistic	Logistic
MLP 17-17-1	0.999992	0.999989	0.996964	0.000062	0.000084	0.033414	BFGS 84	SOS	Tanh	Logistic
MLP 17-18-1	0.999978	0.999988	0.996962	0.000164	0.000169	0.032799	BFGS 92	SOS	Tanh	Exponential

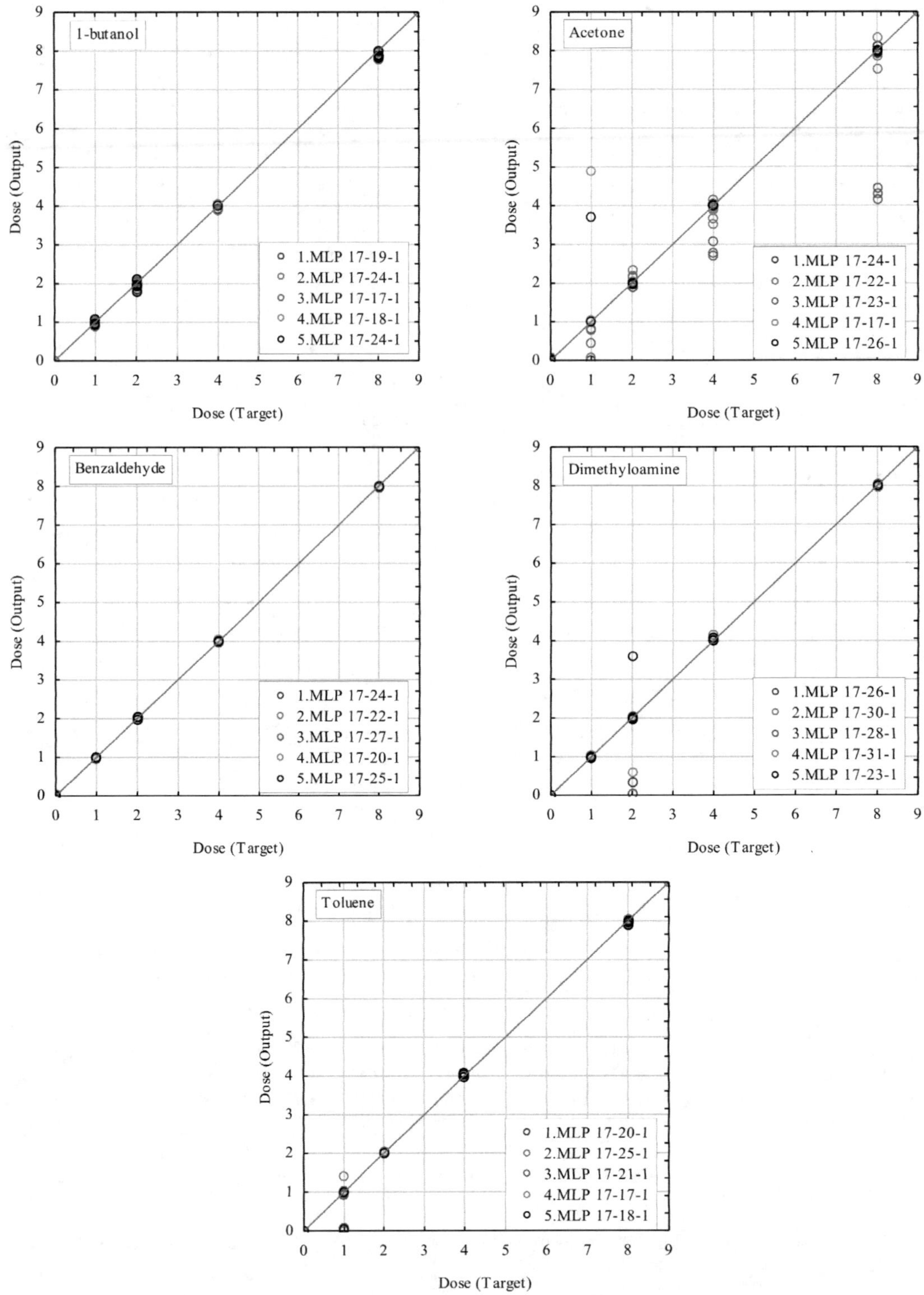

Figure 5. Efficiency of the particular networks in particular substance concentration detection.

IOP Conf. Series: Materials Science and Engineering **710** (2019) 012024 doi:10.1088/1757-899X/710/1/012024

4. Discussion

The PCA diagrams show that it is possible to distinguish particular compounds. The position in the PCA space is different for each tested substance. It is also possible to distinguish substance concentration. In the applied reference system, it was noticed that the increase in the concentration level was inversely proportional to the PCA component.

The use of ANN also enabled to achieve high values of accuracy in the detection of particular substances and their concentrations detection. Five output neurons (one output for each compound) were applied there for the compound identification. The training accuracy was between 92.2% and 99.3%, test accuracy: 87.7% and 95.9%, and finally validation accuracy was between 81.5% and 89.5%. For 1-butanol the best match (0.999997) and the lowest error (0.000032) for validation data were obtained for MLP 17-18-1 network, which has tanh as a function of activation of hidden and output neurons. For acetone, MLP 17-24-1 with linear and exponential functions, matching and error of 0.996412 and 0.034449 respectively. Benzaldehyde: MLP 17-27-1 with exponential and logistic functions, matching and error of 0.999977 and 0.000107 respectively. Dimethylamine: MLP 17-31-1 with exponential and tanh functions, matching and error of 0.992708 and 0.067533 respectively. For toluene, MLP 17-25-1 network with tanh and linear activation functions with 0.999473 matching and 0.005711 error. The obtained results did not indicate any activation function that would be optimal for hidden and output neurons.

In the case of the concentration detection, these coefficients were higher for the particular compounds and achieved values between 0.94 and 0.99.

The results from the study prove that the network has good generalisation capabilities, i.e. it correctly interprets new data, not used previously in ANN training. It is a characteristic feature of a network with an optimal number of hidden neurons. Networks with too many hidden neurons tend to remember the learners' datasets, so they are unable to interpret new data correctly. Too few hidden neurons make the network incapable of recognising complex relationships between input and output data. According to Basheera and Hajmeerb, ANNs are more effective at classifying and determining complex relationships than other methods of analysis [9]. The efficiency of ANN was also described by Fu et al., Hetch-Nielsen and Jiang et al. in the articles concerning architectural design with the use of artificial neural networks and their capacity for learning [3, 10, 16]. Comparable effectiveness of chemical compound identification was achieved by Srivastava [14].

The detection effectiveness measured at this stage of research showed very good results in comparison with the results of other studies published in the scientific literature. The reported high identification efficiency was achieved only in the case of studies with the number of compounds below five [8, 17, 18]. With 6 chemical compounds, the classification accuracy reported by J.F.et al. [3] was between 92.9% and 100%, [3], while for 10 chemical compounds in the experiment conducted by Santos et al. [11], the average efficiency of classification was only 71.9%.

5. Conclusions

This paper shows the potential of the Artificial Neural Networks in detection of certain chemical compounds present in the tested air and, in addition, their concentration. According to the conducted measurements and the ANN analysis, the following conclusions may be formulated:

- Artificial Neural Networks provide better results compared to the PCA analysis.
- The multi-sensor device in use enables the detection of individual substances with validation accuracy in the range of 81.5 and 89.5%.
- The applied device evaluates particular substance concentration using the ANN with validation accuracy in the range of 0.94 and 0.99%.

6. References

[1] Gardner J W and Bartlett P 1994 *Sensor Actuat B-Chem* **18**(1–3) 210–11
[2] Villarubia G, De Paz J F, Chamoso P and De la Prieta F 2018 *Neurocomputing* **272** 10-6
[3] Fu J, Li G, Qin Y and Freeman W J 2007 *Sensor Actuat B-Chem* **125**(2) 489–97

[4] Haugen J E and Kvaal K 1998 *Meat Sci* **49**(98) 273-86
[5] Barisci J N, Wallace G G, Andrews M K, Partridge A C and Harris P D 2002 *Sensor Actuat B-Chem* **84**(2) 252–57
[6] Guz L, Łagód G, Jaromin-Gleń K, Suchorab Z, Sobczuk H, Bieganowski A 2015 *Sensors* **15**(1), 1-21
[7] Hamilton S, Hepher M J and Sommerville J 2005 *Sensor Actuat B-Chem* **107**(1) 424–32
[8] Ding H, Ge H and Liu J 2005 *Sensor Actuat B-Chem* **107**(2) 749–55
[9] Basheera A and Hajmeerb M 2000 *J Microbiol Meth* **43**(1) 3-31
[10] Hecht-Nielsen R 1987, *Proc. of the IEEE* (San Diego, USA) p 11–3
[11] Negri R M and Reich S 2009 *Sensor Actuat B-Chem* **75**(3) 172–78
[12] Peris M and Escuder-Gilabert L 2009 *Anal Chim Acta* **638**(1) 1–15
[13] Sardi S, Vardi R, Sheinin A, Goldental A and Kanter I 2017 *Sci Rep-UK* **7**(1) 18036
[14] Lawson D and Marion G 2008, *Bioinf Stat* **53**(4) 543-56
[15] Tadeusiewicz R 1993 *Sieci neuronowe* (Kraków: Akademicka Oficyna Wydawnicza)
[16] Jiang J, Trundle P and Ren J 2010 *Comp Med Imag Grap* **34**(8) 617-631 10.1016/2007.07.003,
[17] Llobet E, Brezmes J, Vilanova X, Sueiras J E and Correig X 1997 *Sensor Actuat B-Chem* **41**(1–3) 13–21
[18] Szczurek A and Maciejewska M 2004 *Talanta* **64**(3) 609–17
[19] Santos J P, Aleixandre M and Cruz C 2012 *Chem Engineer Trans* **30** 181-6
[20] Schaller E, Bosset J O and Escher F 1998 *LWT-Food Sci Technol* **31**(4) 305–16
[21] Stetter J R, Findlay M W, Schroeder K M, Yue C and Penrose W R 1993 *Anal Chim Acta* **284**(1) 1–11
[22] Srivastava A K 2003 *Sensor Actuat B-Chem* **96**(1–2) 24–37
[23] Zhang Z 2018, *Multivariate Time Series Analysis in Climate and Environmental Research* (Cham: Springer)
[24] Yang Y M, Yang P Y and Wang X R 2000 *Sensor Actuat B-Chem* **66**(1–3) 167–170
[25] Lee D S, Jung J K, Lim J W, Huh J S and Lee D D 2001 *Sensor Actuat B-Chem* **77**(1) 228–236
[26] Kadmon J and Sompolinski H 2016, *Adv Neur In*, 29 1-5
[27] Johnson J L, Schamschula M P, Inguva R and Caulfield H J 1998 Pulse-Coupled Neural Networks *Neural Networks and Pattern Recognition*, ed Omidvar O., Dayhoff J (San Diego: Akademic) 1 p 1-56.

IOP Conf. Series: Materials Science and Engineering **710** (2019) 012025 doi:10.1088/1757-899X/710/1/012025

Movement analysis of the semitrailer with the tank-container at hard braking - the case study

J Vrabel[1], T Skrucany[1], L Bartuska[2] and J Koprna[1]

[1] University of Zilina, Univerzitna 1, 010 26 Zilina, Slovakia
[2] Institute of Technology and Business in Ceske Budejovice, Czech Republic

jan.vrabel@fpedas.uniza.sk, tomas.skrucany@fpedas.uniza.sk,
bartuska@mail.vstecb.cz

Abstract. Transport of liquid, gaseous, powdered or granulated substances is an integral part of the industry in each country. This type of cargo is most often transported by companies that are a part of petrochemical, food or also other types of industry. Increased attention should be paid to transport of tank-vehicles and tank-containers from the point of view of road safety and environmental protection. The volume of such types of transport is growing every year. The extensive trend relates mainly to the developed world economies, which are experiencing economic growth, which includes also Slovakia according to the Statistical Office of the Slovak Republic. This article deals with evaluating the movement of semi-trailer trucks with a tank container during and immediately after hard braking. The movement of the semi-trailer truck with a tank container will be analysed and evaluated immediately after the vehicle is stopped. The movement of the vehicle will be analysed by video analysis. The focus is on analysing behaviour of the semi-trailer truck with a tank container during and immediately after braking. The article compares semi-trailer trucks with tank container with breakwater plates and tank containers without breakwater plates. An important feature of this comparison was processing of videos capturing these vehicles during hard braking in video analysis program.

1. Introduction

Transport of goods in tank vehicles and tank containers is important for several types of industry in the Slovak Republic and whole Europe [1–3]. This article discusses in detail issues that directly affect safety of goods transport in tank containers, especially the situation that occurs during hard braking of such vehicles transporting liquid substances [4].

It is important to compare behaviour during braking of a semi-trailer truck with a tank container with breakwater plates inside, a tank truck without breakwater plates and a semi-trailer truck with tarpaulin transporting bulk goods [5]. We assume that tank trucks without breakwater plates will, due to their behaviour when the truck is braking sharply, represent a significantly higher level of road traffic hazard than the other two mentioned types of semi-trailer trucks.

Theoretical foundations from which we have started include mainly the knowledge of the issue of transport of dangerous goods by the European Agreement concerning the International Carriage of Dangerous Goods by Road (ADR), the knowledge of fluid movement in a closed container while this container is moving, the basic knowledge of constructions of tank containers, as well as the knowledge of the dynamics of semi-trailer trucks movement under hard braking.

Safety during transport of dangerous goods in tanks and tank containers is, according to the European Agreement on the Transport of Dangerous Goods by Road (ADR), achieved by a number of measures and trainings of employees [6-10]. However, these trainings are more theoretical than practical, and do not pay sufficient attention to the driver's practical skills and their training considering abnormal behaviour of the load during hard braking of such vehicles [7, 11 – 13]. So far, no one has specified and dealt with this issue in Slovakia, so it is not possible to build on existing research in this area. Given the theoretical background, we assume that the effect of fluid movement in a tank container on vehicle movement during sharp braking will be less pronounced for a tank container with breakwater plates than without them [6]. We also assume that driving behaviour of a tank container truck with breakwater plates inside will be similar to that of an ordinary tarpaulin semi-trailer truck transporting non-liquid goods (under hard braking) [14].

2. Methodology

The case study carried out for the purposes of this article may be described as an applied form of research, since its results and resulting information can be practically applied during transport of liquid goods [15].

The planning and implementation part of the research was focused on collection of data on the movement of tank semi-trailer trucks with breakwater plates, tank semi-trailer trucks without breakwater plates and tarpaulin semi-trailer trucks. The phase of data collection using cameras represented the most difficult part in the framework of the research task, as the video quality has a major impact on the possibility of further relevant image processing.

Recording of braking process (video) and data collection on the movement of semi-trailer trucks with tank containers were carried out with every change in the quantity of goods in the tank (with or without breakwater plates). It should be noted that, between the individual braking events of the trucks, a part of the transported fluid was gradually removed from both the tank with breakwater plates and the tank without them. Subsequently, the video was analysed and evaluated using a program that serves to track the movement of objects, to create kinematic and dynamic models of mass points, and to create data.

After that, further braking was performed with a conventional tarpaulin semi-trailer truck on which non-liquid goods were transported in order to compare behaviour of this type of semi-trailer truck with a tank semi-trailer truck. The videos were re-analysed and evaluated using the above-mentioned computer program.

3. General information (the procedure)

A total of 31 videos of a semi-trailer truck with tank container with or without breakwater plates were recorded. On four of the videos, the movement of the tarpaulin semi-trailer truck transporting non-liquid goods was recorded [14]. After analysing all the videos for evaluation, two videos of the tarpaulin semi-trailer truck carrying non-liquid goods, thirteen videos of the semi-trailer truck with tank container with breakwater plates, and nine videos of the semi-trailer with tank container without breakwater plates were selected. Other videos were discarded for several reasons, for example, due to an insufficient length of recording to draw relevant conclusions, or due to camera movement during video recording of the semi-trailer truck to such an extent that it could not be used for further analysis of vehicle movement. Another reason for discarding a video for detailed analysis and data acquisition was that the camera could not focus on a moving semi-trailer truck in a quality that would give a meaningful analysis result.

All the selected videos were inserted into the program, where the analysis of the movement of the semi-trailer truck was carried out. After its completion, obtained outputs could have been inserted into tables. This procedure was repeated for each video being evaluated. Based on the tables, graphs with the course of the acceleration change over time were developed [16].

4. Measurement results

The analysis in the program yielded output values that were arranged in tables. From these data, graphs of the courses of speed and acceleration achieved at the end of braking and immediately after stopping of the semi-trailer trucks were made. Each measurement is marked so that it can be identified easily and so that it is clear from its name what type of graphical representation it refers to. Designation consists of the measurement number, designation of the variable being measured (speed/acceleration), designation of direction of the vehicle (forward/reverse) and initial speed.

The following part of the paper will provide the analysis of the movement of all three types of semi-trailer trucks during braking and their subsequent movement immediately after stopping. With all the trucks, the vehicle was stopped, and the brake pedal was released. The analysis in the program shows that the tarpaulin semi-trailer truck behaved in such a way that the change in the acceleration value of the tarpaulin semi-trailer truck after stopping was low. This means that after the driver stopped and released the brake pedal, the truck no longer tended to move forward or backward. The course of speed of the semi-trailer truck from two selected measurements is shown in Figure 1 and the course of acceleration of the same measurements is shown in Figure 2.

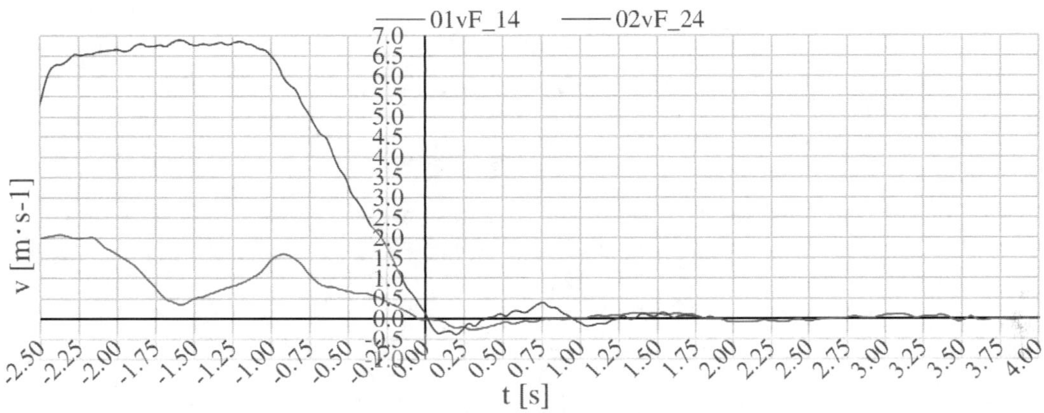

Figure 1. The course of speed of tarpaulin semi-trailer truck over time within measurements 01F_14 and 02F_24.

Braking of the semi-trailer truck with a tank container is specific because fluid in the container moves during braking and, after stopping, generates a pressure wave on the container wall. In order to reduce the undesirable effect of the pressure wave, various means have been introduced for reducing them, such as partitions, breakwater plates, outer or inner rings. In terms of video evaluation, the movement of the semi-trailer truck with a tank container with breakwater plates (transverse breakwater plates in all cases) was analysed first and then the analysis of movement of the semi-trailer truck without breakwater plates followed. Because two identical semi-trailer trucks were compared differing only in equipping/not equipping the container with breakwater plates, only speed changes and the change in acceleration over time are evaluated only after the trucks stopped.

Following graphs show the course of speed (Figure 3 and Figure 4) over time for nine different measurements of the semi-trailer trucks with tank containers with breakwater plates just from the moment after trucks stopped. A clear subsequent forward or reverse movement of the truck can be observed. However, it does not exceed the value of 0.3 m· s-1 in the positive direction and the value of -0.2 m· s-1 in the negative direction. Although the truck is moving, the movement can be considered slight.

For better illustration of the findings, we also provide a graphical representation of the course of speed before stopping and after stopping the truck with a tank container with breakwater plates. The change in speed after stopping the vehicle does not show large deviations in relation to the course of speed at which the semi-trailer trucks moved before stopping.

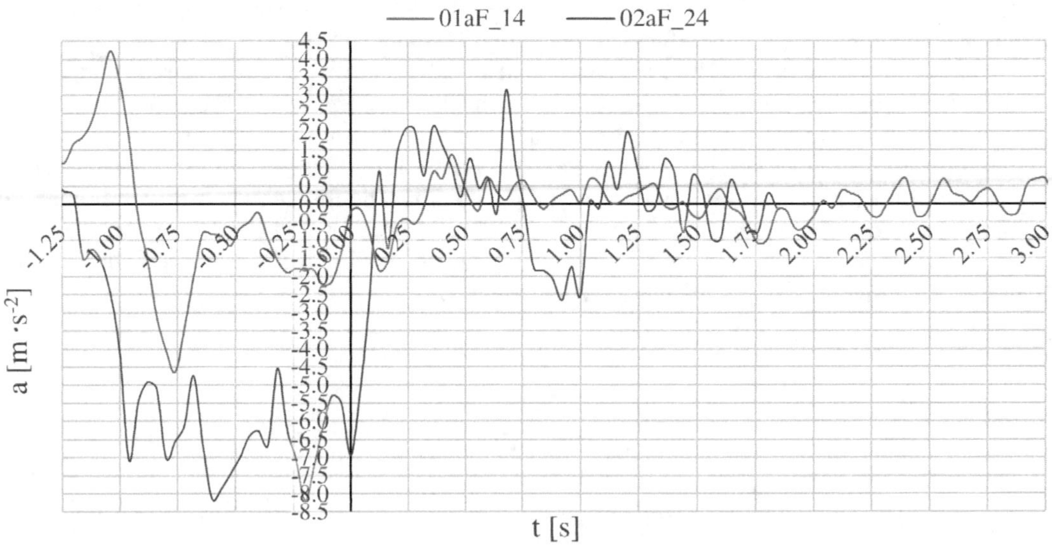

Figure 2. The course of acceleration of tarpaulin semi-trailer truck over time within measurements 01aF_14 and 02aF_24.

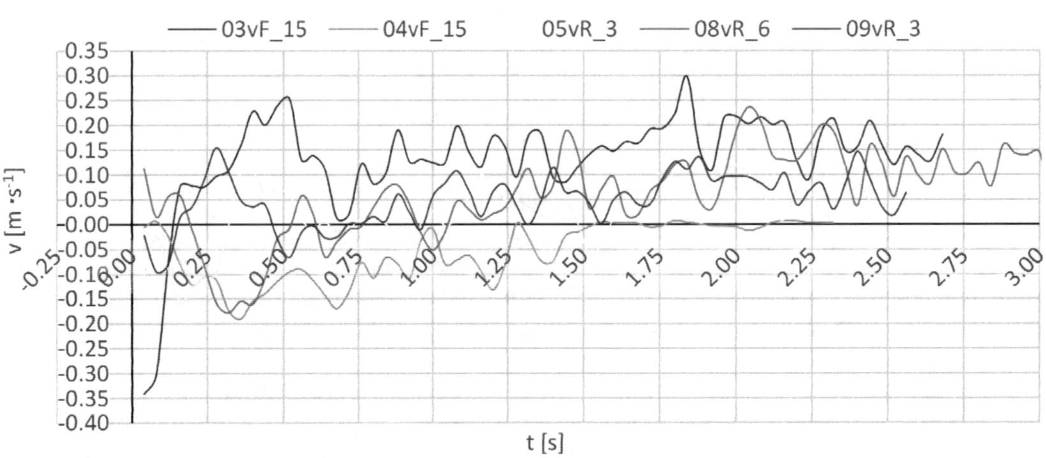

Figure 3. The course of speed of semi-trailer truck with tank container with breakwater plates over time for measurements 03vF_15, 04vF_15, 05vR_3, 08vR_6, and 09vR_3.

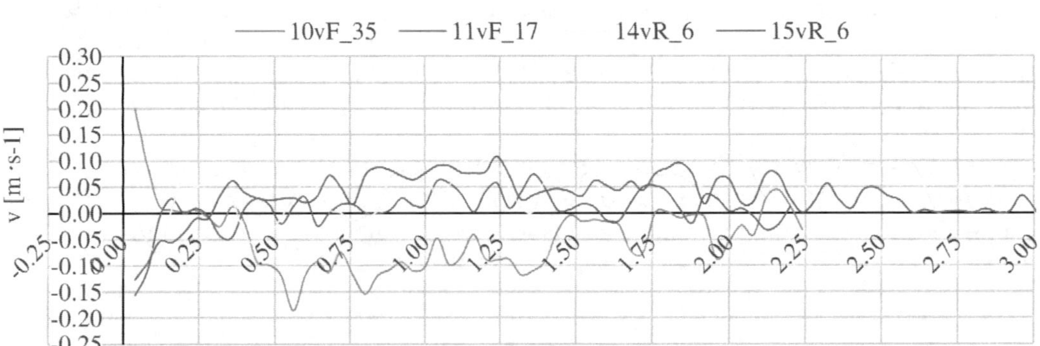

Figure 4. The course of speed of semi-trailer truck with tank container with breakwater plates over time for measurements 10vF_35, 11vF_17, 14vR_6, and 15vR_6.

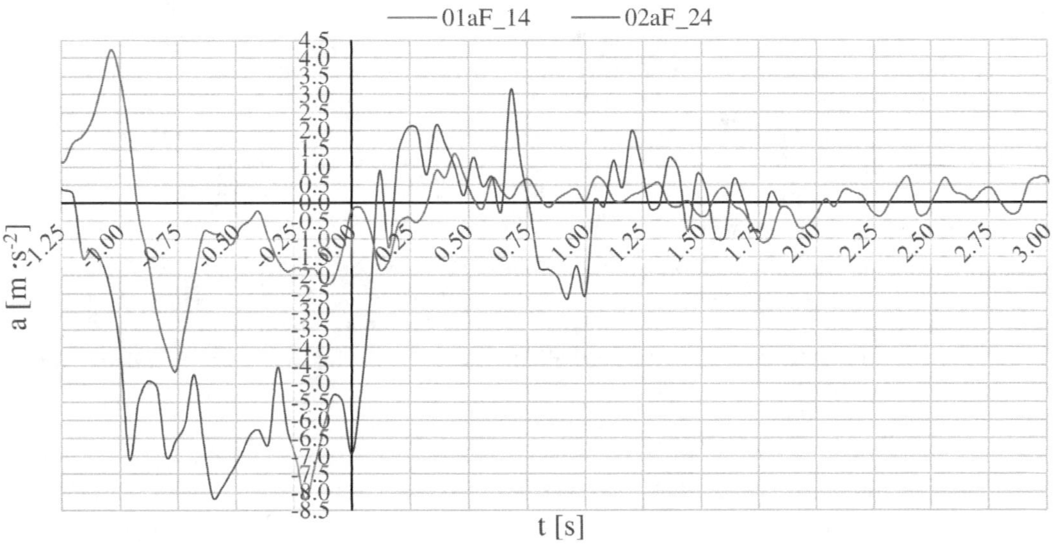

Figure 5. The course of speed of semi-trailer truck with tank container with breakwater plates over time for measurements 06vF_21, 07vF_20, 12vR_8, and 13vR_6.

In the following graph (Figure 6), we can see the course of speed over time for four measurements of a semi-trailer truck with a tank container without breakwater plates from the moment the vehicle stopped. The graph clearly shows the significant movement of the vehicle just after stopping and releasing the brake pedal of the semi-trailer truck. The movement of the whole vehicle was caused by the forward and backward movement of liquid inside the tank container.

It should be noted that in case of stopping the vehicle in forward direction and subsequent releasing of the brake pedal, the movement of the liquid inside the tank container caused that the whole vehicle moved in the backward direction. Quite the opposite, that is the subsequent movement of the vehicle forward, happened when the vehicle was stopped when moving in backward direction (i.e. when reversing).

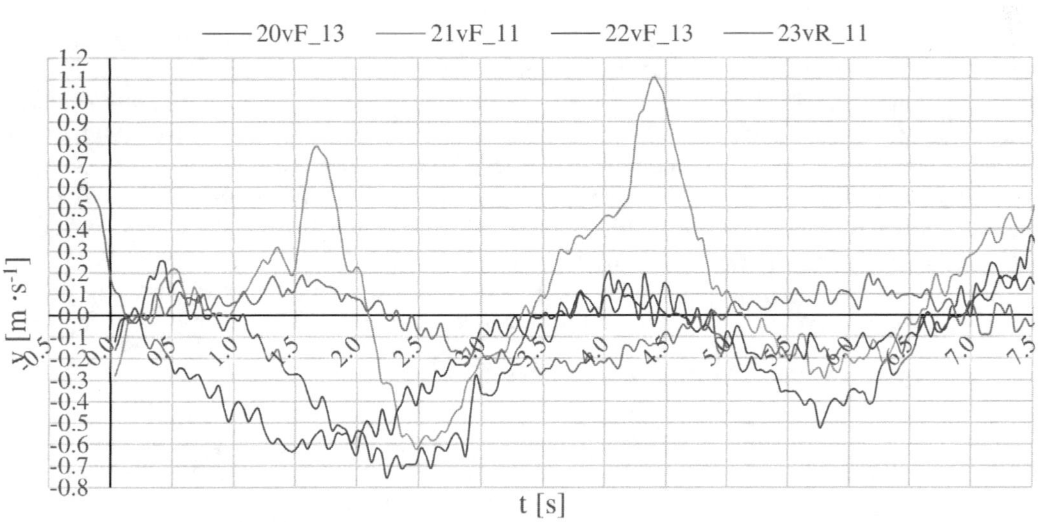

Figure 6. The speed of semi-trailer truck with tank container without breakwater plates over time for measurements 20vF_13, 21vF_11, 22vF_13, and 23vR_11.

The following graph (Figure 7) also shows the course of speed over time for five measurements of a semi-trailer truck with a tank container without breakwater plates. In these measurements, the subsequent movement after stopping is not as considerable as in the previous graph. This is because, during the measurements, the tank container was filled either to the maximum possible total volume or, on the contrary, to the minimum allowable volume of the container. In both of these cases, the free surface area of transported fluid in the tank container was smaller in comparison to the free surface area when the tank container is filled to about 50%. Thus, during the measurements, the wave motion of liquid did not manifest in full force and fluctuations were lower than -0.3 m· s-1 to 0.2 m· s-1. Measurement no. 18vF_22 was carried out with a fully filled tank container of approximately 26,000 litres. Approximately 13,000 litres of liquid were drained from the tank container prior to measurement 20vF13. Thus, this measurement was performed with an approximately half-filled tank container.

Approximately 5,000 litres of liquid were drained from the tank container prior to measurement 22vF-13. Thus, measurements 22vF_13 and 23vR_11 were performed with a tank container filled with approximately 8,000 litres of liquid.

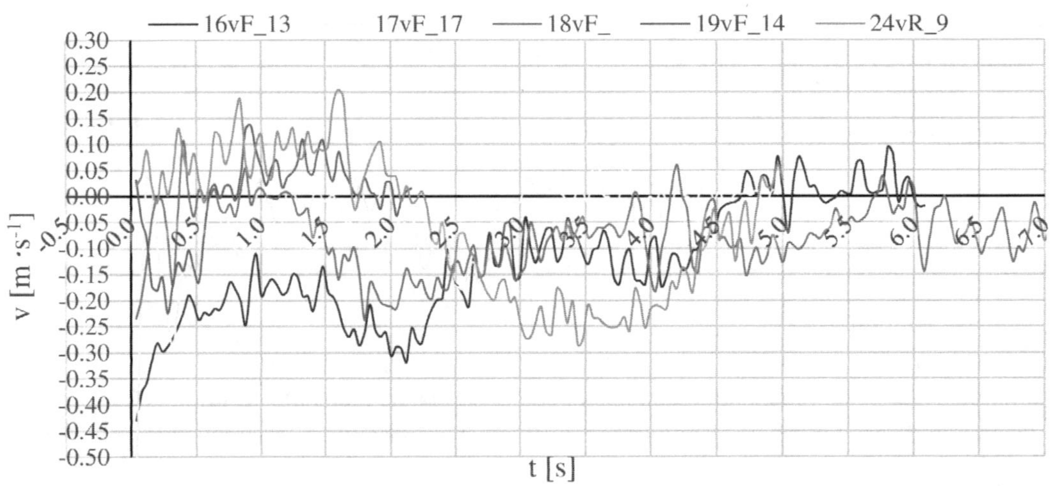

Figure 7. The course of speed of semi-trailer truck with tank container without breakwater plates over time during measurements 16vF_13, 17vF_17, 18vF_22, 19vF_14, and 24vR_9.

During measurements 20vF_13 and 21vF_11, the tank container was approximately half-full, and the free area of liquid was the largest when compared to other measurements. Therefore, the highest amplitudes, within the range of -0.63 m· s-1 and 1.11 m· s-1, occur during these measurements.

In the last graph (Figure 8) we compare three most significant courses of speed for all three types of semi-trailer trucks. The graph shows the course of speed after stopping the vehicle for the semi-trailer truck with tarpaulin indicated by blue colour (02vF_24). The semi-trailer truck with a tank container with breakwater plates is shown in green colour (03vF_15). The semi-trailer truck with a tank container without breakwater plates is shown in red colour with (21vF_11).

Although the lowest speed (only 11 km· h-1) was reached during braking in 21vF_11, speed variations achieved after stopping the vehicle show the greatest dispersion. From this it can be concluded that this type of semi-trailer truck is most risky with regard to road safety.

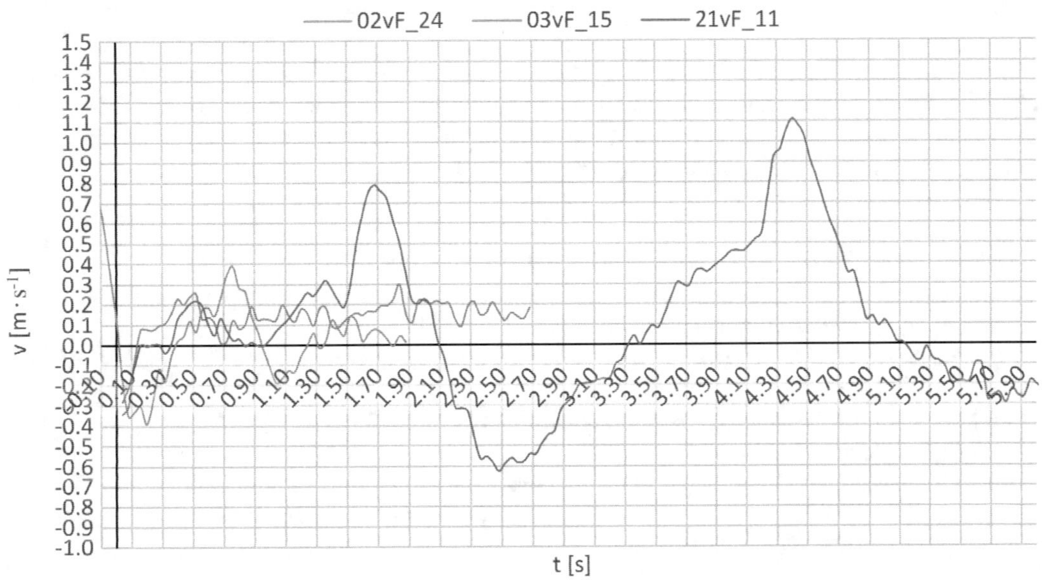

Figure 8. The course of speed over time after stopping the vehicle and releasing the brake pedal. Semi-trailer truck with tarpaulin – in blue; semi-trailer truck with tank container with breakwater plates – in green; semi-trailer truck with tank container without breakwater plates – in red.

5. Conclusions

After comparing the results of graphs from Figure 6 and Figure 7, which show the subsequent movement after stopping the vehicle with the tank container without breakwater plates, it can be stated that when the tank container is filled to the maximum possible extent, there is smaller subsequent movement of the vehicle immediately after stopping in comparison to a tank container filled approximately to 50 % of its total volume [17]. From the safety point of view, as well as economical use of tank containers, it is necessary to fill tank containers to the maximum possible volume, i.e. to about 95 percent. It is not allowed to fill the tank container to 100% of its volume, as the transported fluid needs some space to expand [18].

The comparison of graphs in Figure 1 and Figure 6 reveals that after stopping the semi-trailer truck with tarpaulin, a speed change range of -0.39 m· s-1 to 0.28 m· s-1 was achieved. In case of the vehicle with a tank container without breakwater plates, these values ranged from -0.76 m· s-1 to 1.1 m· s-1 in each measurement. In case of the semi-trailer truck with tank container with breakwater plates, values in the range -0.19 m· s-1 to 0.14 m· s-1 were achieved. After the analysis of the measured values, it was found that during braking with a semi-trailer truck with a tank container without breakwater plates, the subsequent movement of the vehicle immediately after stopping it is more extreme than the movement of a semi-trailer truck with tarpaulin carrying non-liquid goods [19], The movement of the tank container without breakwater plates after stopping has a significant repetitive forward and reverse movement over time. The movement of the tarpaulin semi-trailer truck is shorter in time and less rapid than the tank truck without breakwater plates [20]. The semi-trailer truck with a tank container with breakwater plates behaves similarly to the conventional semi-trailer truck with tarpaulin. This can be clearly seen in Figure 8. At approximately 2.5 seconds after stopping, the tarpaulin semi-trailer truck and the tank-container semi-trailer truck with breakwater plates were completely stationary. On the other hand, the semi-trailer with a tank container without breakwater plates moved forwards and backwards for a long time [21].

The evaluated results of the semi-trailer truck with a tank container with breakwater plates are very close to the measured results of the tarpaulin semi-trailer truck. This is because the breakwater plates mounted inside the tank container, by their design, almost completely eliminate the undesirable waves caused by the movement of the truck. Elimination of waves occurs because the breakwater plates inside the tank divide the interior into smaller sections and thus, reduce the wave motion force caused by the truck movement.

The results of this research can be put directly into practice. However, further research is needed in the field of tank-containers and tank-vehicles transport, for example, in relation to the shape of individual intersections and possibilities of these vehicles at a given intersection to bring the vehicle to a stationary condition, i.e. to a complete stop.

6. References

[1] Triebel, H. G.: (2007). *Aufbaukurs Tank: Erstschulung für den Gefahrgut-Fahrzeugführer. Zusatzkurs – Tankwagen. Zuzsatzgrundkurs Tanktransport und Aufbaukurse Klasse 2,3,5.1,6.1, und 8.* München: Vogel. ISBN: 9783574232206.

[2] Jagelčák, J.: (2015). *Nakladanie a upevňovanie v cestnej doprave*. Žilina: EDIS. ISBN: 978-80-554-1083-8.

[3] Varchola, M.: (1993). *Mechanika tekutín*. Bratislava: Slovenská technická univerzita. ISBN: 80-227-0553-5.

[4] MDVRRSR: (2011). *Vestník Ministerstva dopravy, výstavby a regionálneho rozvoja Slovenskej republiky*. Bratislava: Ministerstvo dopravy, výstavby a regionálneho rozvoja Slovenskej republiky. ISSN 1335-9789.

[5] Marczuk, A., Caban, J., Savinykh, P., Turubanov, N. & Zyryanov, D. (2017). Maintenance research of a horizontal ribbon mixer. *Eksploatacja i Niezawodnosc. Vol.: 19, Iss.:1, p.: 121-125*, ISSN: 15072711, DOI: 10.17531/ein.2017.1.17

[6] Jagelcak, J., Zamecnik, J., Kiktova, M. (2017). Potential for Intermodal Transport of Chemical Goods in Slovakia. *MATEC Web of Conferences Vol.:134 Art. Nr.: 00021*. DOI: 10.1051/matecconf/201713400021

[7] Skrúcaný T., Vrábel J., Kendra M. & Kažimír P., (2017). Impact of cargo distribution on the vehicle flatback on braking distance in road freight transport. *MATEC web of conferences, LOGI 2017 - 18th international scientific conference: České Budějovice, Czech Republic, Vol. 134, art. no. 00054,* October 19, ISSN 2261-236X., DOI: 10.1051/matecconf/201713400054

[8] Klockiewicz, Z., Ślaski, G., Spadło, M. (2019). The Influence of the Conditions of Use and the Type of Model Used on the Vertical Dynamic Responses of a Car Suspension. *The Archives of Automotive Engineering – Archiwum Motoryzacji, 85*(3), 57-82. https://doi.org/10.14669/AM.VOL85.ART5

[9] Stopka, O. (2019). Design of placing the container reloading station by applying the Operations Research methods: a research study. *The Archives of Automotive Engineering – Archiwum Motoryzacji, 85*(3), 123-136. https://doi.org/10.14669/AM.VOL85.ART9

[10] Stopka, O., Kampf, R.(2018). Determining the most suitable layout of space for the loading units' handling in the maritime port. *Transport 33(1), 2018, pp. 280-290.* DOI:10.3846/16484142.2016.1174882.

[11] Figlus T., Kuczynski, L. (2018). Selection of a semi-trailer for the haulage of long oversize loads, considering an analysis of operational damage. *11th International Scientific and Technical Conference on Automotive Safety*. Casta Papiernicka, Slovakia, APR 18-20, 2018

[12] Gnap, J.; Kubanova, J. (2017). Selected Options for Modelling of Transport Processes Particularly in Relation to Intermodal Transport. *MATEC Web of Conferences Vol.:134, 00015*. DOI: 10.1051/matecconf/201713400015

[13] Szala, M., Kot, E. A. (2017). Influence of repainting on the mechanical properties, surface topography and microstructure of polyester powder coatings. *Advances in Science and Technology Research Journal, 11 (2) 159-165*, DOI: 10.12913/22998624/69680

[14] Ondruš, J., Hockicko P. (2015) Braking deceleration measurement using the video analysis of motions by Sw tracker. In: Transport and Telecommunication, Vol: 16, pp: 127 – 137.

[15] Jurecki, R. S., Jaskiewicz, M et al. (2012). Driver's reaction time under emergency braking a car - research in a driving simulator. *Eksploatacja i niezawodnosc – Maintenance and reliability. Vol: 14 Iss: 4 p: 295-301.*

[16] Matuszak, Z., Jaskiewicz, M. et al. Remarks to the Reliability Assessment and to Human Actions - Especially Car Driver *MATEC Web of Conferences. Vol: 134 Art Nr: 00036* DOI: 10.1051/matecconf/201713400036

[17] Morales, J., Martinez, J. L. and Mandow, A. et al., (2013) Static Tip-Over Stability Analysis for a Robotic Vehicle With a Single-Axle Trailer on Slopes Based on Altered Supporting Polygons, IEEE – Asme transactions on mechatronics. vol: 18, pp: 697 – 705.

[18] Hac, A., Fulk, D. and Chen, H., (2009) Stability and control considerations of vehicle-trailer combination, SAE International Journal of Passenger Cars - Electronic and Electrical Systems Vol: 1, pp: 925 – 937.

[19] Moreno, G., Nicolazzi, L. and Vieira R., et al., (2015) Three-dimensional analysis of the rollover risk of heavy vehicles using Davies method, in proc. 14th International Federation for the Promotion of Mechanism and Machine Science World Congress, 10.6567/IFToMM.14TH.WC.PS4.006.

[20] Haugland, O. (2013) Car stopping distance on a tabletop. In: Physics teacher, 51, 268 - 268.

[21] Marczuk, A., Caban, J., Savinykh, P., Turubanov, N., Zyryanov, D., (2017) Maintenance research of a horizontal ribbon mixer. *Eksploatacja i niezawodnosc – Maintenance and reliability; 19 (1): 121–125*, http://dx.doi.org/10.17531/ein.2017.1.17.

IOP Conf. Series: Materials Science and Engineering **710** (2019) 012026 doi:10.1088/1757-899X/710/1/012026

Modeling sequencing batch reactor operational conditions depending on oxygen concentration

G Łagód[1], A Piotrowicz[1], W Cel[1], J Zaburko[1] and J Drewnowski[2]

[1] Lublin University of Technology, Faculty of Environmental Engineering, Nadbystrzycka 40B, 20-618 Lublin, Poland
[2] Gdansk University of Technology, Faculty of Civil and Environmental Engineering, Narutowicza 11/12, 80-233 Gdansk, Poland

g.lagod@pollub.pl

Abstract. Sequencing batch reactors (SBR) can be used as a fill-and draw activated sludge system for wastewater treatment with considerable operating flexibility and the possibility to conduct experiments under standard conditions and extreme case scenarios. Mathematical modeling and computer simulations provide an opportunity to implement existing wastewater processes in modeling software and evaluate different modifications at low costs and no disturbances for on-going processes of full scale WWTP. Additionally, the used model can be calibrated and validated against experimental data from laboratory scale devices. The aim of this study was to simulate the processes occurring in laboratory scale SBR under different aeration strategies. The results include the analysis of the adaptation period of the activated sludge biomass in the SBR, as well as the case of breakdown of treatment process due to stoppage of raw wastewater inflow and the interruption of the aeration and/or mixing. As a result, it can be stated that the oxygen transfer rate should be incorporated in the calibration of biological nutrient removal model in order to effectively visualize the individual contributions of each process.

1. Introduction

Sequencing batch reactors (SBRs) are a type of activated sludge process for wastewater treatment [1,2]. These are usually characterized by the five sequential stages of filling, reacting, settling, decanting and idling, allowing considerable flexibility for the design and operation of different biological wastewater treatments [3-8]. This versatility allows to carry out experiments under standard and/or variable conditions at full and laboratory scale, with the possibility for implementing new methods of measurement, technological solutions or exploitation strategies as well as to test extreme conditions during different treatment processes (e.g. failure of a reactor component) [9-18]. Additionally, SBR reactors at large scale, usually require a higher level of sophistication than conventional systems, with complex timing units and controls to achieve optimized management of performance over time [19-22]. Thus, properly set experiments allow to address questions on how to respond to different factors affecting wastewater treatment process in the short time.

The design of SBR bioreactors for nutrient removal, including nitrogen and phosphorus, is based on the selection of parameters such as sludge age, volume exchange ratio, cycle time and sludge volume index, as described for instance, in the ATV guidelines [23]. In reality, the resulting wastewater treatment plants (WWTPs) are mostly oversized to sustain efficient carbon and nutrient removal under varying environmental and operating conditions. Nowadays, the achievement of this goal can be aided by computer softwares using mathematical models, e.g. IWA's ASM, which reflect the kinetics and stoichiometry of wastewater treatment processes [24-26]. Mathematical modelling and computer simulations of the projected or existing WWTP provide an opportunity to study current treatment and evaluate potential modifications at little financial costs and no disturbances of the on-going processes [27,28]. However, although the system might be designed with safety margins, attention should be paid to the operating conditions, which may positively or negatively affect the overall system performance. The optimization of operating parameters, e.g. aeration control, can be an asset to benefit from simultaneous nitrogen removal coupled with enhanced biological phosphorus removal (EBPR) [29]. Moreover, computer modelling can help establish optimal procedures in case of system failure in the treatment or can help predict problems with the quantity or quality of inflowing wastewater. The models used for simulations can also be calibrated and validated using data obtained from experiments carried out at the laboratory scale [30,31]. This approach has the advantage of testing for hypotheses and collecting data that otherwise would not be implementable or achievable at the full-scale WWTP.

The aim of this study was to simulate processes occurring in an laboratory scale SBR used for nutrients removal under different aeration strategies. Here we present the model used for the SBR operation along with the results of the simulations. The results were used to evaluate the adaptation period of the biomass and biological wastewater treatment processes in SBRs, as well as the period of breakdown of treatment process caused by the stoppage of raw wastewater inflow and interruption of the aeration and/or mixing system.

2. Material and methods

2.1. Laboratory experiments

The experimental setup consisted of three SBRs with an active and total volume of 8 and 10 liters, respectively. In the first part of the study, in order to simulate a conventional biological wastewater treatment process, a standard schedule was applied. The reactor operating cycle consisted of six consecutive phases: filling (30 min), mixing (120 min), aeration (420 min), sedimentation (90 min), decantation (50 min) and idle (10 min). The simulation was based on an analysis of the system assuming similar parameter concentrations for the wastewater flow during each cycle. Aeration (7 hours) was carried out at constant dissolved oxygen and controlled by an on/off aeration valve. Oxygen was kept at around 2 mg/L with an on/off controller and a 0.5 mg/L dead band. Electrodes for pH, DO (dissolved oxygen) and temperature measurements, were installed and connected to each individual SBR system. Experimental data were collected and the status and fluctuations of the reactor were constantly monitored. Time series of the electrode signals were stored in a data log-file. The temperature during the whole experiment was kept at 20-21°C. The Hach test-in-tube Hach DR2800 spectrophotometer was used to determine the concentrations of phosphorus, total nitrogen, ammonia as well as nitrites, nitrates, COD, BOD_5 and TSS. These were systematically measured at the beginning and at the end of each SBR cycle (each cycle lasted 12 hours). The WTW CellOx® 325 DO sensor was used to measure DO concentration and to establish set point for each SBR.

In order to simulate operational challenges of limiting substrates inflow, during the second stage of the experiment, the SBR was challenged with different operational strategies:

- SBR1, Aeration: SBR1 was constantly aerated and maintained a high DO;
- SBR2, Mixing: SBR2 used only mechanical mixing with a final DO level around 0.5 mg/L;
- SBR3, Control: SBR3 was used as a control reactor and neither aeration nor mixing was used.

Described laboratory SBRs and conducted experiments gives the data for build computer model and introduce parameters characterizing realized processes, activated sludge and treated wastewater. The rest of the paper presents the results of computer simulations.

2.2. Model layout

A computer model was developed and simulations were carried out using the modelling platform GPS-X v.7.0.1 (Hydromantis, Canada). A schematic of the laboratory SBR based on the GPS-X modeling platform can be found in Figure 1. In order to describe the processes involving activated sludge, the ASM2d model was applied. The general model calibration of laboratory scale SBR with nutrient removing processes under standard conditions as well as limiting substrates inflow, limiting aeration and mixing were applied. Based on process measurements in SBR and influent wastewater characterization, the ASM2d model input data were verified by Influent Advisor module linked to the GPS-x program. Other parameters used to build the model for simulations are presented in detail in Table 1.

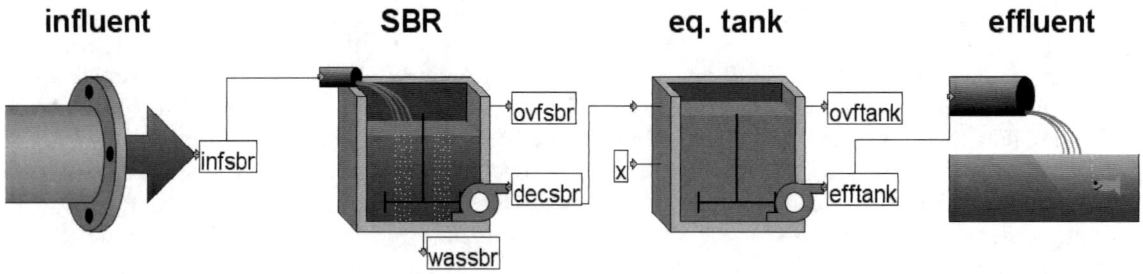

Figure 1. Schema of the laboratory SBR based on the GPS-X modelling platform.

Table 1. List of data used in GPS-X.

Type of data	Description
Library	CNPlib (Carbon-Nitrogen-Phosphorus library)
Model components	Influent, SBR (Advanced), Equalization tank, Discharge (Figure 1)
Sub-models used	Influent – COD fractions, Advanced SBR – ASM2d, Equalization tank – No react, Discharge– Default
Influent quality parameters	COD, TKN (Total Kjeldahl Nitrogen), Phosphorus (Total Phosphorus, Orthophosphates), Nitrogen (Total Nitrogen, Nitrate & Nitrite, Ammonia), TSS
Number of cycles per day	2 cycles, 12 hours each
Duration of phases	Variable and based on schedule
Influent characteristics	Influent Advisor – GPS-X accompanying application

3. Results

Sample results of the simulations in GPS-X during the first stage of experiment are presented in Figure 2. The concentration profiles of nitrite, ammonia nitrogen and total nitrogen were analysed (Figure 2a). Additionally, profiles of biological and chemical oxygen demand (BOD_5, COD) as well as TSS are presented in Figure 2b. A general stability of the treated wastewater leaving the equalization tank can be remarked.

According to simulations during the first part of the experiment, the adaptation of activated sludge and stabilization of process parameters were achieved after 20 cycles (10 days) of SBR work. This adaptation period is generally coherent with literature data [32,33] and experiments carried out in our laboratory [3].

The second stage of the scenario started after the initial adaptation period. In order to determine the process efficiency, key parameters such as total COD, total carbonaceous BOD_5 and TSS were analysed (Figure 2b).

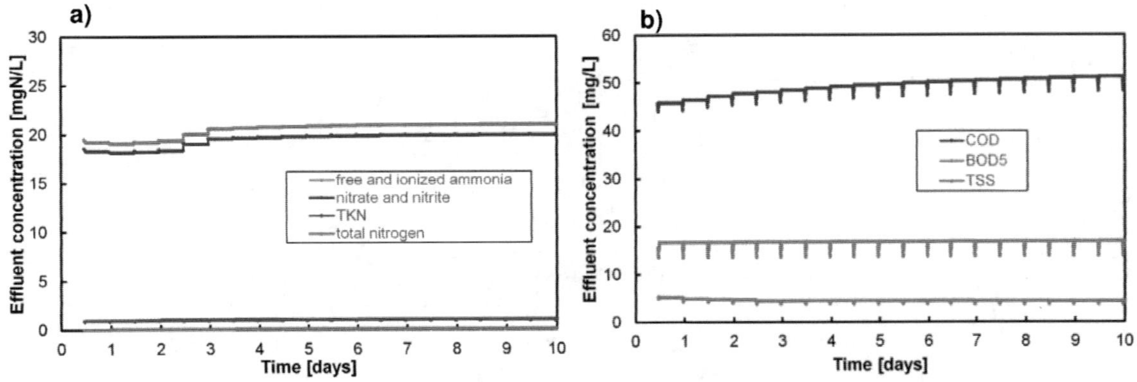

Figure 2. Simulated effluent concentrations of (a) nitrogen species and (b) effluent COD, BOD_5 and TSS.

Figure 3 shows the results of simulation concerning situation in SBR1 where after stopping of substrates inflow continuous aeration was conducted. Under aerobic conditions and the lack of an external carbon source, during the first three days, the accumulation of nitrates and nitrites can be observed. Moreover, due to loss of alkalinity, a lower rate of nitrification with an increase of free and ionized ammonia was observed (Figure 3a). At the same time a high depletion of heterotrophic biomass was observed along with a lower decrease of autotrophic and P-accumulating organisms. This situation appear in aerobic condition regardless level of DO, but it was more pronounced at higher oxygen concentration (6 – 7 mg/L) rather than at standard oxygen concentration (2 mg/L).

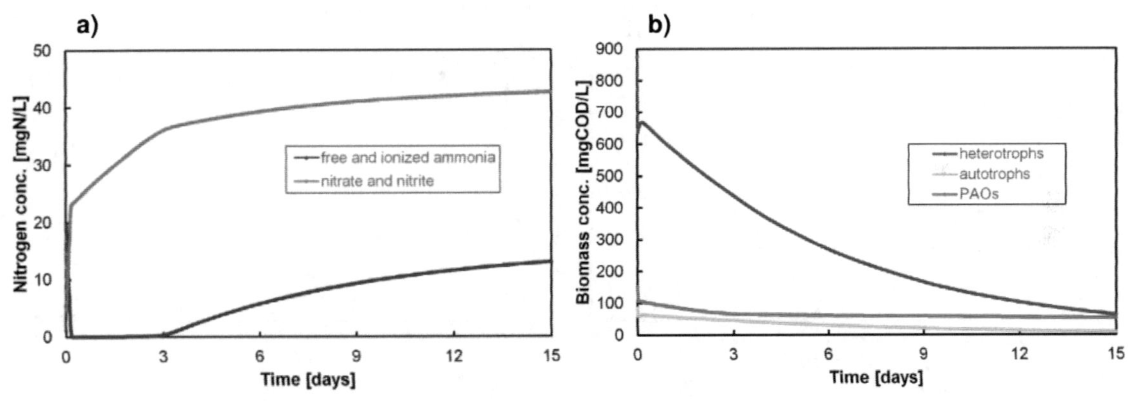

Figure 3. Concentrations of (a) nitrogen species and (b) biomass in SBR1.

Figure 4 shows the case scenario where SBR2 did not receive any sewage inflow and the only parameter used was mixing. Since mixing allowed some oxygen transfer, this was also considered as a simulation scenario. When the oxygen concentration reached 0.4 – 0.6 mg/L, the nitrate and nitrite levels stabilized after around seven days, and a small increase in ammonia was also observed (Figure 4a). The change in biomass concentration (Figure 4b) showed similar trends to the case scenario with constant aeration (Figure 3b). However, when mixing and a "zero" oxygen concentration was imposed, the results showed only an increase in ammonia without accumulation of nitrite and nitrate within reactor (Figure 4c). This scenario showed a much faster decrease of heterotrophic biomass with a slower decrease of P-accumulating (Figure 4d).

IOP Conf. Series: Materials Science and Engineering **710** (2019) 012026 doi:10.1088/1757-899X/710/1/012026

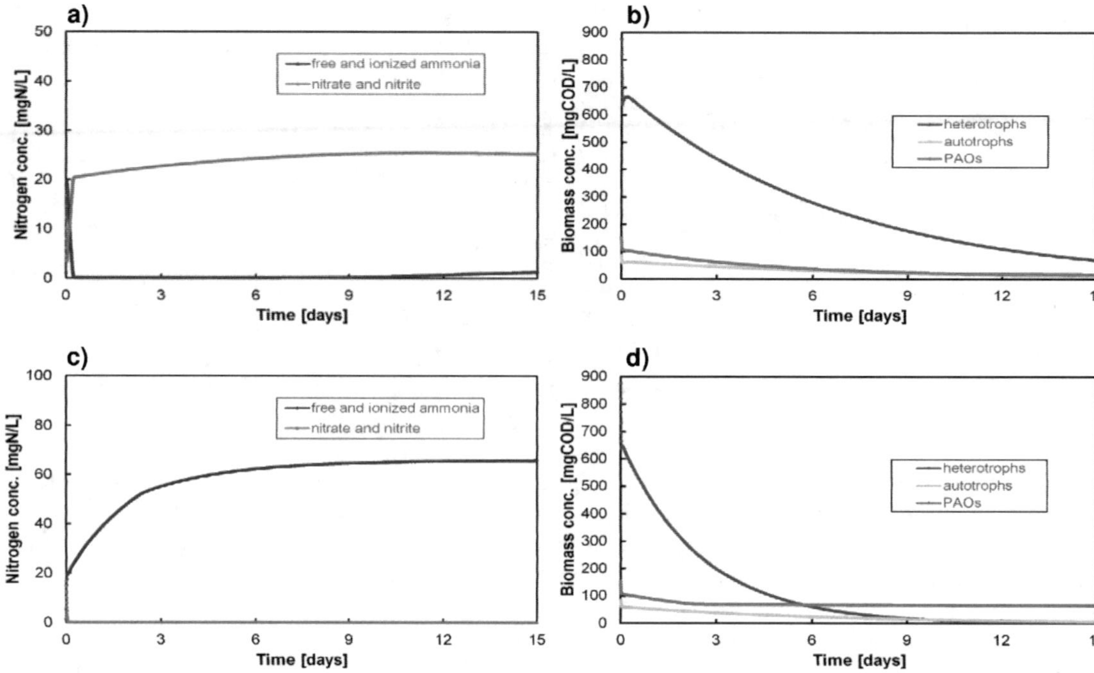

Figure 4. Nitrogen concentrations in SBR2 in presence of (a) low DO and (c) zero DO. Biomass concentrations in SBR2 in presence of (b) low DO and (d) zero DO.

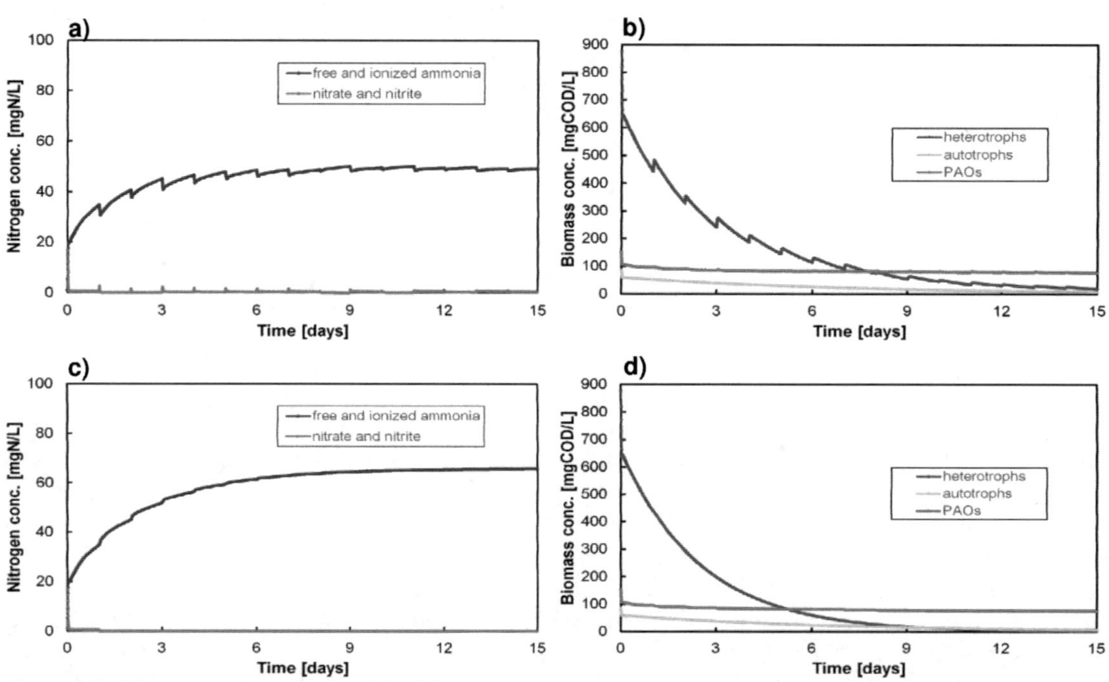

Figure 5. Nitrogen concentrations in SBR3 (a) low DO, (c) zero DO. Biomass concentrations in SBR3 (b) low DO and (d) zero DO.

The most interesting results are based on the simulation of the reactor with no continuous aeration or mixing. The first scenario shows more realistic conditions where complete mixing of the reactor occurs once a day before activated sludge and sewage were sampled for analysis (Figure 5a, b). In this case, a transient (15 min) low DO concentration (0.4 – 0.6 mg/L) characterized this phase with subsequent short drops of the ammonia concentration, increase of nitrite and nitrate concentration as well as an overall decrease of biomass amount. The second scenario was based on applying only mixing with no oxygen in the reactor. In this case, a greater increase of ammonia concentration occurred along with no formation of nitrite and nitrate and a faster loss of biomass from the reactor (Figure 5c, d).

The processes that occurred in SBR3 (with predominant sedimentation) as well as in reactors with aeration and mixing were presented with the data averaged for 10 modelled layers of SBR. In this circumstances it is worth to mention in case of SBR3. The results for the 4th layer (near the middle of the cross section of the reactor), where in mixing scenario whole volume of reactor is homogenised once for day (Figure 6a and b) – then in 4th layer appear also biomass, before sediment at the bottom (Figure 6b). The same layer without mixing only showed dissolved compounds – e.g. free and ionised ammonia nitrogen (Figure 6c and d). This confirms that in computer simulation time and place of sampling is very important for proper data analysis.

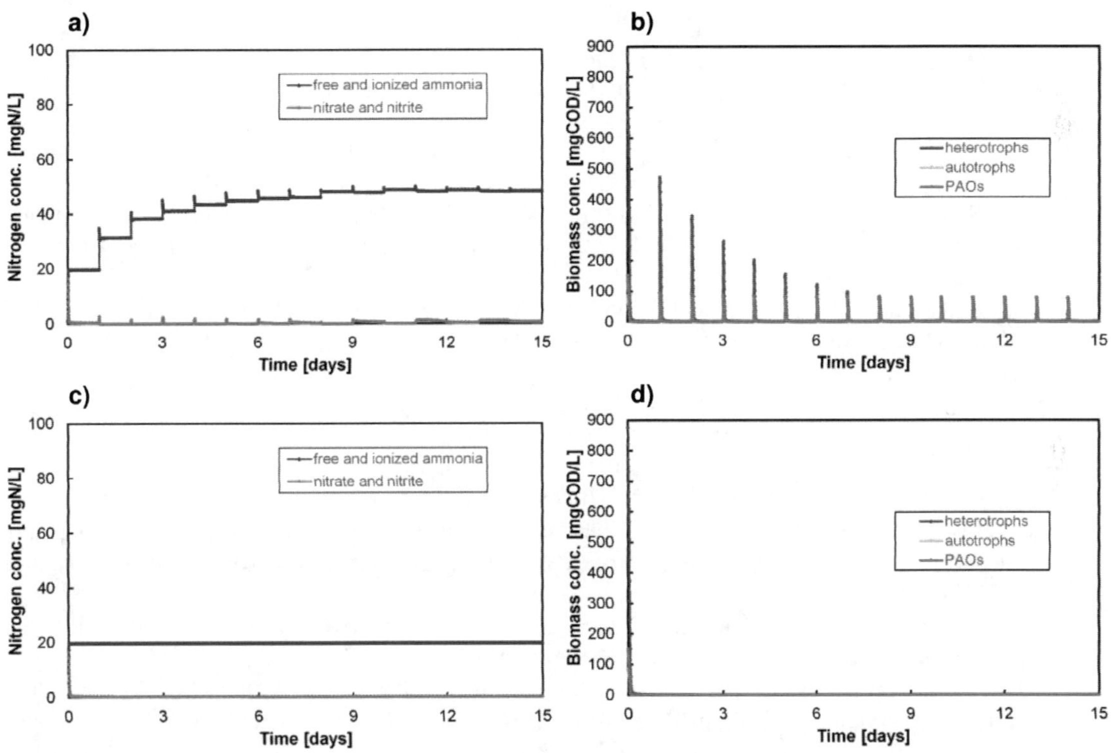

Figure 6. Nitrogen concentrations in 4th layer of the SBR3 - with mixing (a), without mixing (c) and biomass concentration in the SBR3 with mixing (b), without mixing (d).

4. Conclusions

From this study, the following conclusions can be derived:

- GPS-X allowed to carry out different simulations and evaluate the adaptation period of the SBR with the cycle of scheduled sequential phases and also within the scenario where different aeration/mixing conditions were applied.
- Modeling SBRs, where it is difficult to achieve steady state operation, requires accurate setting of the initial conditions. This will significantly affect the simulation results.

- A general model calibration methodology was applied using data from a laboratory scale SBR. Simulations showed a simultaneous nitrification-denitrification process together with biological phosphorus removal. However, here we mainly focused on the nitrogen compounds.
- The biological processes described in the model were found to occur simultaneously under limiting aeration conditions. A low oxygen transfer inherently differentiates the SBR behaviour from systems under traditional design calculations.
- After a general calibration of the model the best compatibility between simulation results and measurements in laboratory was achieved for concentration of free and ionized ammonia. This was particularly accurate for SBR3, in scenario where mixing was applied once per day.
- Computer simulations conducted with model calibrated using data from laboratory experiment show, that in case of inflow interruption, the biomass is at the highest concentration during aeration. A fast biomass depletion occurs when activated sludge settled at the bottom of reactor.

Acknowledgements
The authors would like to thank Fabrizio Sabba for scientific and language consultation.

5. References

[1] Irvine R, Wilderer P and Flemming H 1997 Controlled unsteady state processes and technologies - An overview *Water Sci. Tech.* **35**(1) 1-10
[2] Ketchum L 1997 Design and physical features of sequencing batch reactors *Water Sci. Tech.* **35**(1) 11-18
[3] Babko R, Kuzmina T, Jaromin-Gleń K and Bieganowski A 2014 Bioindication assessment of activated sludge adaptation in a lab-scale experiment *Ecol. Chem. Eng. S.* **21**(4) 605-16
[4] Babko R, Kuzmina T, Łagód G and Jaromin-Gleń K 2014 Changes in the structure of activated sludge protozoa community at the different oxygen condition *Chem. Didact. Ecol. Metrol.* **19** (1-2) 87-95
[5] Babko R, Jaromin-Gleń K, Łagód G, Pawłowska M and Pawłowski A 2016 Effect of drilling mud addition on activated sludge and processes in sequencing batch reactors *Des. Water Treat.* **57**(3) 1490-98
[6] Chan Y, Chong M, Law C and Hassell D 2009 A Review on anaerobic-aerobic treatment of industrial and municipal wastewater *Chem. Eng. J* **155** 1-18
[7] Dutta A and Sarkar S 2015 Sequencing batch reactor for wastewater treatment: Recent advances *Curr. Pollution Rep.* **1** 177-90 DOI 10.1007/s40726-015-0016-y
[8] Lackner S and Horn H 2012 Evaluating operation strategies and process stability of a single stage nitritation-anammox SBR by use of the oxidation-reduction potential (ORP) *Bioresour. Technol.* **107** 70-7
[9] Carvalho G, Lemos P, Oehmen A and Reis M 2007 Denitrifying phosphorus removal: Linking the process performance with the microbial community structure *Water Res.* **41** 4383-96
[10] Cydzik-Kwiatkowska A, Bernat K, Zielinska M, Bulkowska K and Wojnowska-Baryla I 2017 Aerobic granular sludge for bisphenol A (BPA) removal from wastewater *Int. Biodeterior. Biodegrad.* **122** 1-11 DOI: 10.1016/j.ibiod.2017.04.008
[11] Czarnota J, Maslon A and Zdeb M 2018 Powdered keramsite as unconventional method of AGS technology support in GSBR reactor with minimum-optimum OLR *E3S Web of Conferences.* **44** 00024 DOI: 10.1051/e3sconf/20184400024
[12] Guz Ł, Łagód G, Jaromin-Gleń K, Suchorab Z, Sobczuk H and Bieganowski A 2015 Application of gas sensor arrays in assessment of wastewater purificationeffects *Sensors* **15** 1-21
[13] Maslon A, Tomaszek J, Zamorska J, Zdeb M, Piech A, Opalinski I and Jurczyk L 2019 The impact of powdered keramsite on activated sludge and wastewater treatment in a sequencing batch reactor *J Environ. Manage.* **237** 305-12 DOI: 10.1016/j.jenvman.2019.02.035
[14] Nasr M, Moustafa M, Seif H and El Kobrosy G 2011 Modelling and simulation of German BIOGEST/EL-AGAMY wastewater treatment plants - Egypt using GPS-X simulator

Alexandria Eng. J. **50** 351-57
[15] Oehmen A, Lemos P, Carvalho G, Yuan Z, Keller J, Blackall L and Reis M 2007 Advances in enhanced biological phosphorus removal: From micro to macro scale *Water Res.* **41** 2271-300
[16] Sytek-Szmeichel K, Podedworna J and Zubrowska-Sudol M 2016 Efficiency of wastewater treatment in SBR and IFAS-MBSBBR systems in specified technological conditions *Water Sci. Technol.* **73**(6) 1349-56
[17] Waclawek S, Grubel K, Chlad Z, Dudziak M and Cernik M 2016 The Impact of Oxone on Disintegration and Dewaterability of Waste Activated Sludge *Water Environ. Res.* **88**(2) 152-57
[18] Werle S, Dudziak M and Grubel K 2016 Indirect methods of dried sewage sludge contamination assessments *J Environ. Sci. Health Part A-Toxic Hazard. Subst. Environ. Eng.* **51**(9) 754-8 DOI: 10.1080/10934529.2016.1170449
[19] Cohen A, Hegg D, de Michele M, Song Q and Kasabov N An intelligent controller for automated operation of sequencing batch reactors *Water Sci. Tech.* **47**(12) 57-63
[20] Man Y, Shen W, Chen X, Longa Z and Corriou J 2018 Dissolved oxygen control strategies for the industrial sequencing batch reactor of the wastewater treatment process in the papermaking industry *Environ. Sci.: Water Res. Technol.* **4** 654
[21] Piotrowski R, Paul A and Lewandowski M 2019 Improving SBR performance alongside with cost reduction through optimizing biological processes and dissolved oxygen concentration trajectory *Appl. Sci.* **9** 2268 DOI:10.3390/app9112268
[22] Traore A, Grieu S, Puig S, Corominas L, Thiéry F, Polit M and Colprim J 2005 Fuzzy control of dissolved oxygen in a sequencing batch reactor pilot plant *Chem. Eng. J* **111**(1) 13-19
[23] Teichgräber B, Schreff D, Ekkerlein C, Wilderer P 2001 SBR technology in Germany - an overview *Water Sci. Tech.* **43**(3) 323-30
[24] Andrews J 1993 Modeling and simulation of wastewater treatment processes *Water Sci. Tech.* **28** (11/12) 141-50
[25] Drewnowski J and Mąkinia J 2013 Modeling hydrolysis of slowly biodegradable organic compounds in biological nutrient removal activated sludge systems *Water Sci. Tech.* **67**(9) 2067-74
[26] Swinarski M, Makinia J, Czerwionka K, Chrzanowska M and Drewnowski J 2012 Modeling external carbon addition in combined n-p actvated sludge systems with an extension of the IWA activated sludge models *Water Environ. Res.* **84**(8) 646-55
[27] Billing A and Dold P 1988 Modelling techniques for biological reaction systems. 1. Mathematic description and model representation *Wat. SA* **14**(4) 185-92
[28] Gernaey K, van Loosdrecht M, Henze M, Lind M and Jorgensen S 2004 Activated sludge wastewater treatment plant modeling and simulation: state of the art *Environ. Model. Soft.* **19**(9) 763-83
[29] Daigger G and Littleton H 2000 Characterization of simultaneous nutrient removal in staged, closed-loop Bioreactors *Wat. Env. Res.,* **72**(3) 330-39
[30] Tykesson E, Blackall L, Kong Y, Nielsen P and la Cour Jansen J 2006 Applicability of experience from laboratory reactors with biological phosphorus removal in full-scale plants *Water Sci. Tech.* **54**(1) 267-75
[31] Yagci N, Insel G, Tasli R, Artan N, Randall C and Orhon D 2006 A new interpretation of ASM2d for modeling of SBR performance for enhanced biological phosphorus removal under different P/HAc ratios *Biotechnol. Bioeng.* **93**(2) 258-70
[32] Barbusiński K 1991 Adaptation of activated sludge to laboratory research conditions *Environ. Protect. Eng.* **17**(3-4) 57-65
[33] Thiem L and Alkhatib E 1988 In situ adaptation of activated sludge by shock loading to enhance treatment of high ammonia content petrochemical wastewater *J Wat. Pollut. Control Fed.* **60**(7) pp 1245-52

IOP Conf. Series: Materials Science and Engineering **710** (2019) 012027 doi:10.1088/1757-899X/710/1/012027

The effect of the 3D printout filling parameter on the impact strength of elements made with the FDM method

S Korga[1], M Barszcz[1] and Ł Zgryza[2]

[1] Lublin University of Technology, Department of Computer Science, Nadbystrzycka 38, 20-618 Lublin, Poland
[2] Lublin University of Technology, Fundamentals of Technology Faculty, Nadbystrzycka 38, 20-618 Lublin, Poland

s.korga@pollub.pl

Abstract. The paper discusses the aspects of potential use of 3D printing technology in the process of manufacturing elements with different layer fillings. In 3D printing technology, the denser the percentage of filling a material, the higher the mechanical strength of a given element. However, in the literature on the subject, the dependence on these parameters for additive technology has not been unequivocally determined. Therefore, the aim of the study is to find a correlation between the internal structure of the elements understood as a percentage of filling the material produced by 3D printing technology and the impact strength achieved by these elements. To carry out the research process, sets of polymeric samples were prepared using 3D FDM printing. The shape and dimensions of the samples were determined in accordance with the PN-EN 10045-1 standard. The samples were made by means of the Printo H3 device with a closed working chamber, using ABS polymer filament. The samples were filled with a percentage interval of 10%. Impact strength tests of polymeric samples were carried out on a Charpy hammer stand adapted for this purpose. In this way, the results concerning the breaking energy were collected for particular groups of samples. The collected data allowed us to determine the relationship between the percentage of filling the sample and the energy absorbed by the sample. The results provided the basis for conclusions and recommendations regarding one of the most important technological parameters – the percentage of printed element filling.

1. Introduction

Plastics are a relatively new group of materials which have become highly popular in recent years. They are important for the manufacture of lightweight structures and components. Their special properties, such as corrosion resistance, low density, low cost, flexibility of machining, good mechanical properties in relation to weight, good tribological properties make them alternative materials to traditional metal and ceramic materials [1]. The properties of plastics are modified by the use of various additives [2, 3] to obtain various composite polymeric materials. This broadens the scope of their engineering applications in various branches of industry, opening up a whole range of possibilities for individual adaptation of materials to specific purposes [4, 5, 6, 7, 8, 9].

Plastic products (components) are manufactured using various methods, e.g. extrusion, injection moulding, casting or machining. An additional method that has appeared in recent years and is being dynamically developed is 3D printing technology. It has gained acceptance and popularity in production [10], educational and home applications. FDM printers are the most common type of equipment for manufacturing components. This is a method in which the thermoplastic polymer film is plasticised and applied in layers until an object is created. The basic advantages of FDM printing include: relatively low cost of manufacturing an element, lightness and speed of production without the need to use a mould, wide range of printing parameters, the possibility of obtaining complex shapes and structures, elimination of machining or additional finish processing, etc. Currently, the number of polymers used for 3D printing is limited due to the special properties needed for successful printing. These are: relatively low glass transition temperature, melting point and low tendency to shrink after solidification. The most commonly used materials for 3D printing are acrylonitrile butadiene styrene (acrylonitrile butadiene styrene), acrylonitrile-butadiene-styrene (ABS) and polylactic acid (PLA). It is caused by their technological properties – dimensional stability and low glass transition temperature. Other printable polymers include polycarbonate (PC), polyvinyl alcohol (PVA) and polyetherimide (Ultem). Year by year the development of new polymeric composites with a wider range of physical properties is visible, which increases their applicability in 3D printing. These are polymeric composites in which the matrix material is a printable polymer. This is dictated by the fact that it is compatible with 3D printers. There are many examples of successful implementation of such material modifications for the use in FDM [11]. Mechanical properties of 3D printing technology products depend mainly on the properties of the substrate material [12]. In addition, these properties are also affected by the parameters of the printing process itself, i.e. density and pattern of material filling, angle of printing elements, the thickness of layers, etc. Due to the fact that the use of this technology increases in the most diverse areas, it requires continuous improvement of knowledge and technology. Therefore, these issues attract increased attention from scientists and researchers [13]. In their work, Mitrović et al. [14] attempted to establish a functional relationship between the angle of printing 3D ABS samples and the tensile strength under static axial load conditions. In turn, simulation and experimental studies of deformation and cracking behaviour of the crystal lattice of printed 3D samples under uniaxial tensile load were carried out by Xu et al. [15]. Two types of printed elements were considered: elements printed horizontally and vertically. Whereas, mechanical properties of 3D printed composite samples reinforced with continuous carbon fibres were characterised in the paper by Werkena et al. [16].

The aim of this study was to find a correlation between the percentage filling of the material of the 3D print samples and one of the mechanical properties parameters – impact strength.

2. Object and research methodology

The impact tests were carried out with the use of 3D printing technology. The samples were printed on a Printo H3 device (figure 1) with a closed working chamber using the FDM method. For printing, a thermoplastic filter made of acrylonitrile-butadiene-styrene (ABS) with a relative density of 1.06 and melting point >80 °C was used. Varying percentage of internal filling of the sample material was applied. The minimum filling density was 10% and the maximum of 100%. The density was increased by 10% in relation to the previous one.

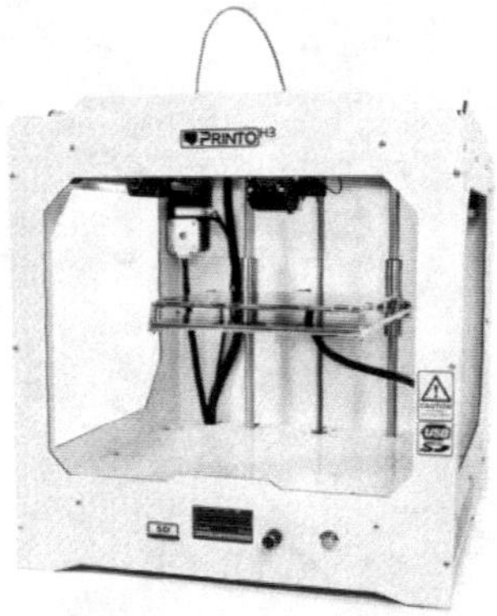

Figure 1. View of a 3D Printo H3 printer.

Digital 3D models of samples were made in Autodesk Inventor according to PN-EN 10045-1. It contains detailed data concerning the preparation of samples for the testing process. 3D models of

IOP Conf. Series: Materials Science and Engineering **710** (2019) 012027 doi:10.1088/1757-899X/710/1/012027

55x10x10 mm samples with two types of notches (V and U-shaped) with a depth of 2 mm were prepared (figure 2).

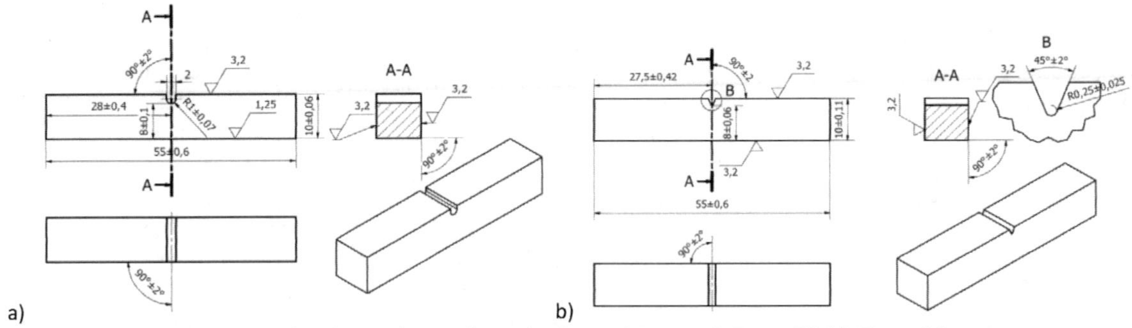

Figure 2. Dimensions of notched specimens: a) Type U, b) Type V.

The geometry models developed are saved in *.STL files so that they can be exported to a slicer program. Cura version 2.3.1 was used, and configured with the 3D printer used. After loading, the model was placed in the working space of the printer table with the notch facing up and the dimensions interpreted by the program were checked for correctness. The grid pattern of sample filling was selected due to the fact that it is most frequently used in the printing of elements. With these settings, a sequence of instructions was generated for the printer in the form of separate g-code files for all sample options (different fill levels and different notches). The data estimated by the program for each printout, such as object weight and print time (without taking into account the warm-up time of the printer elements) were saved. Figure 3 shows an overview of the grid distribution for 3D models with different fillings, which was generated by the program. The blue lines on the view indicate the idle movements of the printer head. The distribution of the filled grid is the same for samples with V and U notches.

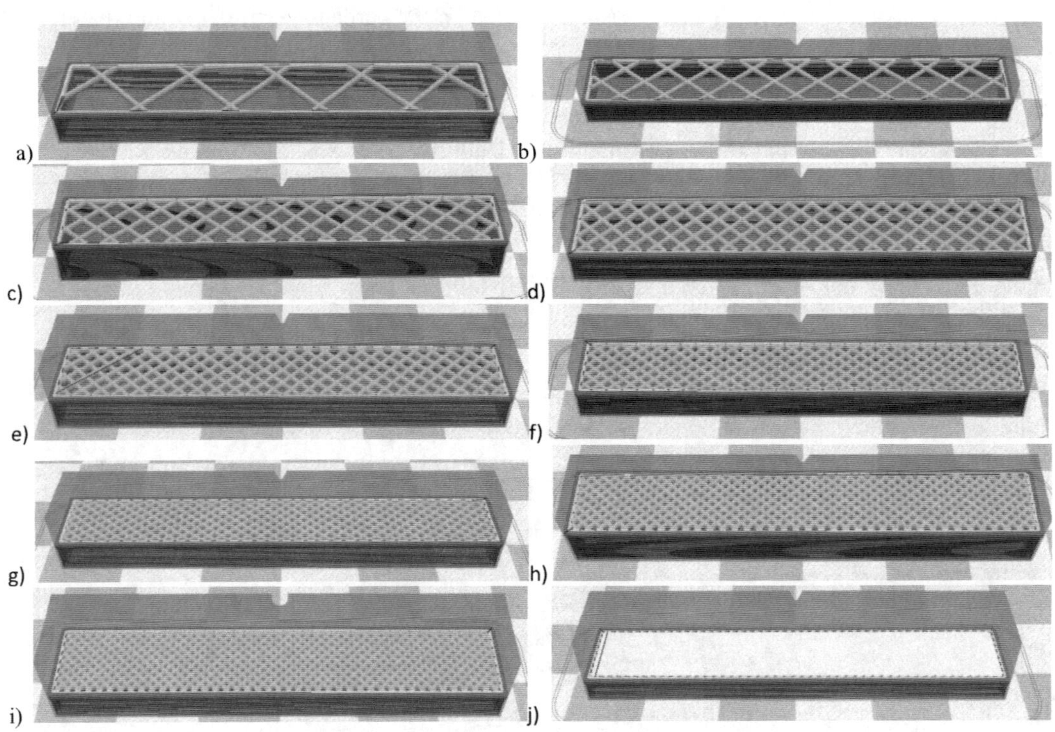

Figure 3. 3D models with the following filling values: a) 10%, b) 20%, c) 30%, d) 40%, e) 50%, f) 60%, g) 70%, h) 80%, i) 90%, j) 100%.

The 3D printing process was carried out using the parameters presented in Table 1.

Table 1. Summary of the most important parameters of the sample printing process.

Parameter name	Value
Layer height	0.2 mm
First layer height	0.2 mm
Number of wall outlines	2
Pattern of filling	GRID
Material filling density	from 10% to 100% (every 10%)
Head temperature	235 °C
Table temperature	95 °C
Retraction distance	4 mm
Retraction speed	60 mm/s
Total print speed	50 mm/s
Filling print speed	60 mm/s,
Outside wall print speed	25 mm/s
Cooling of the printout	off
Supports	none
Raft	none
Scale of the object	100%

The impact tests were carried out using a Charpy pendulum hammer. A hammer with a rated initial energy of 1.5 kpm was used. Using the following conversion factor: 1 kpm = 9,806 J, the rated initial energy of the hammer was determined as 14,709 J. The laboratory stand consisted of a pendulum with a nominal arm length of 380 mm (from the beginning of the arm to the middle point of the hammer), at the end of which there was a hammer with a nominal mass of 2,035 kg. The stand was equipped with an analogue scale that enabled the reading of the energy absorbed by the breaking of the sample and the angle of the arm's inclination after breaking. The maximum swing angle of the arm when released was 160° and corresponded to the free fall of the hammer without the specimen. The scale was attached to the fixture body in such a way that it was possible to change its position if the scale zero point needed to be moved. Figure 4 shows the Charpy hammer stand with a visible scale.

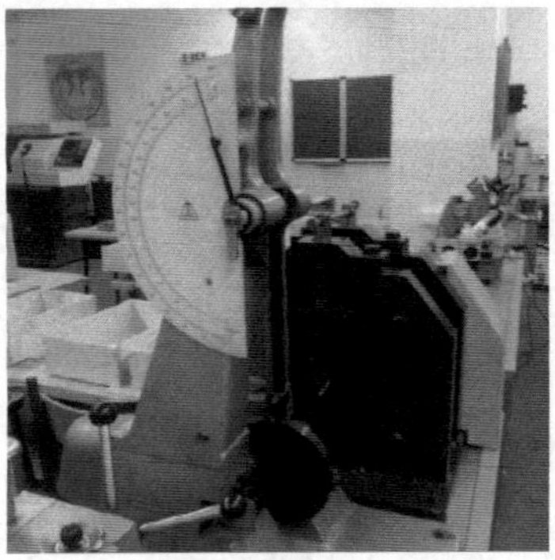

Figure 4. Charpy's hammer used to break the samples.

The base of the device was a cast-iron block which allowed it to maintain the stability of the position of the device during the tests. The equipment was used to break samples made of various materials, including metallic materials. The specimens were placed in such a way that the hammer struck from the opposite side of the notch. The device was activated by manually unlocking the hammerlock with a lever.

3. Results

The results of the research allowed to determine the correlation between the internal structure of the elements (understood as the percentage of material filling) made with 3D printing technology using the FDM method and ABS polymer filament, and the mechanical properties parameter – impact strength.

The results from the tested processes conducted for selected samples are presented in Table 2.

Table 2. Summary of fracture energy and angle of deflection for a group of samples tested.

Sample determination	Sample filling [%]	Fracture energy [kpm]	Fracture energy [J]	Hammer angle after fracture [°]
1pV10_02	10	0.07	0.68642	148
2pV10_02	10	0.075	0.73545	147.5
3pV10_02	10	0.07	0.68642	148
4pU10_02	10	0.08	0.78448	147
5pU10_02	10	0.08	0.78448	147
6pU10_02	10	0.07	0.68642	148
7pV20_02	20	0.06	0.58836	149
8pV20_02	20	0.07	0.68642	148
9pV20_02	20	0.07	0.68642	148
10pU20_02	20	0.07	0.68642	148
11pU20_02	20	0.07	0.68642	148
12pU20_02	20	0.08	0.78448	147

Samples from a single test series show similar results when broken. Small differences in impact strength occur between specimens with the same percentage filling but with a different notch shape. Figure 5 shows averaged data for the groups of samples tested.

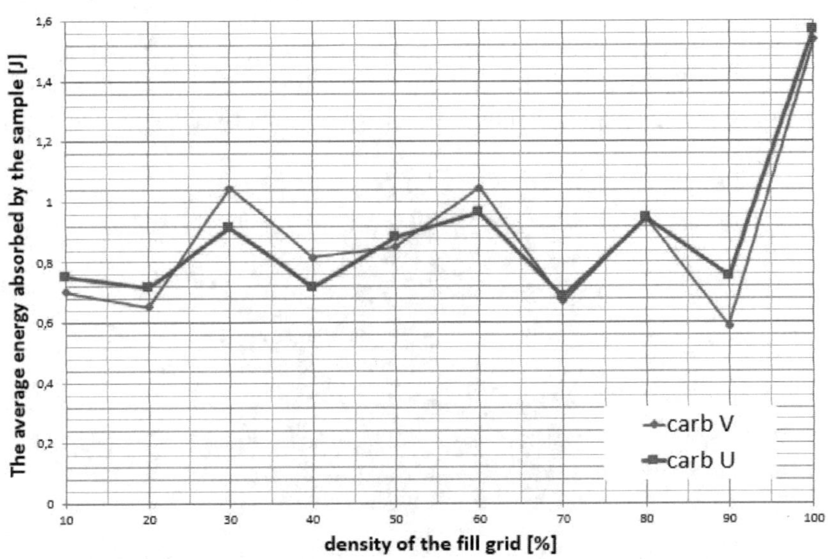

Figure 5. The relationship between grid density and average energy absorbed by the specimens in a fracture test.

The scraps of samples after the tests differed from each other. Samples with lower grid density had irregular edges at the breaking point. Denser filled samples had a regular fracture edge. Some of the samples did not break completely in two. At the point of impact, a joint consisting of not completely torn first layers of material on the side of impact remained. On this basis, such scrap was characterised as stratified scrap. White discolourations were visible on the specimen material, probably in the places with the highest material stresses. Scrap from the test specimens is shown in figure 6.

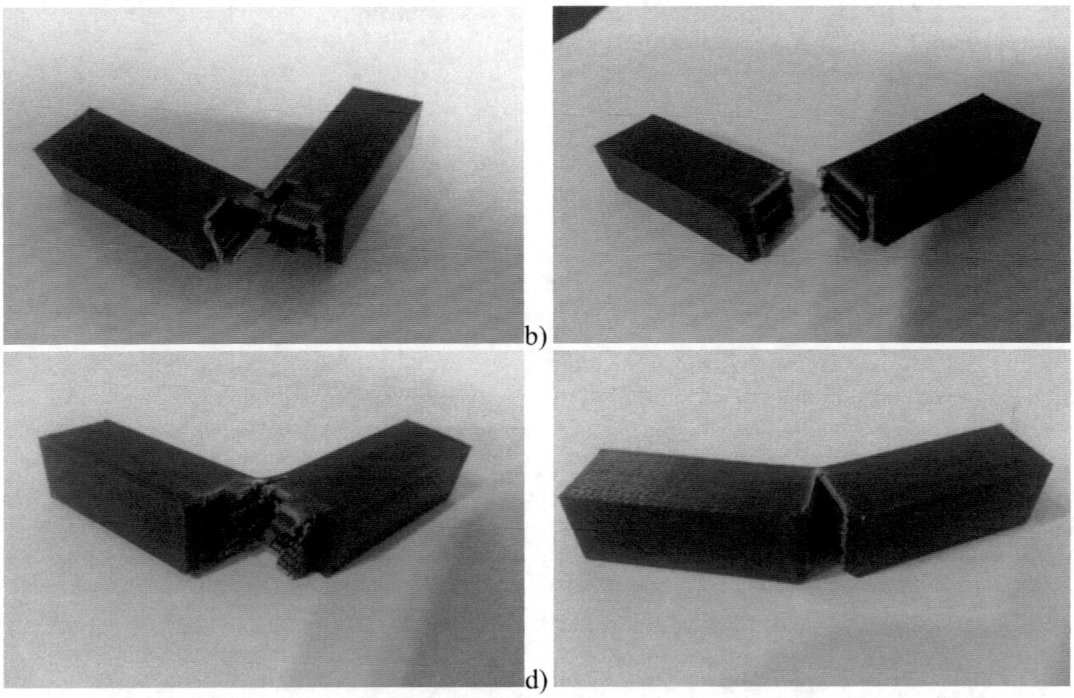
a) b) c) d)

Figure 6. Sample scraps a) 12pU20_02, b) 21pV40_02, c) 47pU80_02, d) 49pV90_02.

Figure 6a shows a scrap specimen with visible jagged edges at the breaking point. Figure 6b shows a completely broken scrap. Figure 6c shows a specimen with an irregular rupture of the inner mesh of the filling. Figure 6d shows scrap metal with a visible regular edge of the broken specimen. This condition of scrap indicates that in many cases the specimen was first bent before breaking. The fracture occurred only after crossing the yield point. In addition, the delamination, which results in non-separation of the specimen into two separate elements, confirms the high anisotropy of the specimens. The collected data were used to determine the impact strength of the samples. For its calculation, formula No. 1 was used.

$$KC = \frac{K}{S} \qquad (1)$$

where: KC – impact strength, K – impact work, S – starting surface at the notch location.

The impact strength of all the other samples was calculated in the same way, the selected calculation results are presented in Table 3.

Table 3. Summary of impact strength values for selected groups of samples.

Sample determination	Sample fill [%]	Impact rate $[\frac{J}{cm^2}]$	Average impact rate for the series $[\frac{J}{cm^2}]$
1pV10_02	10	0.858025	
2pV10_02	10	0.9193125	0.878454167
3pV10_02	10	0.858025	
4pU10_02	10	0.9806	
5pU10_02	10	0.9806	0.939741667
6pU10_02	10	0.858025	
7pV20_02	20	0.73545	
8pV20_02	20	0.858025	0.817166667
9pV20_02	20	0.858025	
10pU20_02	20	0.858025	
11pU20_02	20	0.858025	0.898883333
12pU20_02	20	0.9806	

The results for the analysed series of samples in the form of the functional dependence of average impact strength and grid density are presented in figure 7.

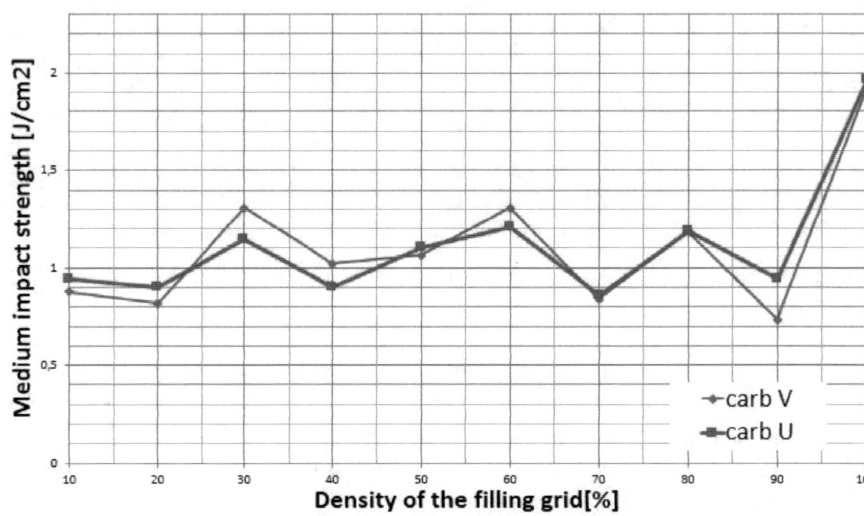

Figure 7. The functional dependence of average impact strength and grid density.

The impact energy absorbed by GRID filled specimens varies disproportionately to the filling density, with no clear increase in energy absorbed as the filling grid density increases, with the exception of 100% filled specimens, which exhibit superior strength to any other specimen.

4. Conclusion

Within the framework of the work, the production parameters were determined for two basic groups of samples – with the U notch and V notch. For each variant samples were prepared, which were produced using the 3D FDM printing method. In total, 60 samples were produced and tested for impact strength. On the basis of the collected data the following conclusions were drawn:

- Most of the impact energy was absorbed by 100% filled specimens, the average values for this series are respectively: 1.92 J/cm^2 (V notch), and 1.96 J/cm2 (U notch).
- The curves of the relationship between the filling density and the impact strength obtained for the V notch and U notch are similar in shape but differ in values at certain points.

- It is likely that parameters other than the filling density of the grid such as the local arrangement of the grid at the impact point, the pattern of the grid, the distribution of the external elements of the object (base and wall) are decisive for the impact strength of the specimen.
- The weight of the specimens increases proportionally with the density of the filling.
- The linear relationship between the fill density and the weight of the specimens allows for an estimation of the grid density at which the specimen was filled.
- FDM printed parts are highly anisotropic.
- The best ratio of mass to the energy absorbed during the impact was achieved with samples filled with 10% (U notch) and 30% (V notch).
- The best weight/impact ratio of fully fractured specimens was achieved with 20% (U notch) and 60% (V notch).
- No proportionality was found between the weight of the specimen and the impact strength of the specimen.
- Samples with the same notch were not visually different in any way.
- The breakage of V-notched samples was much more frequent than that of U-notched samples.
- All samples with a filling density between 40% and 70% V notched were completely broken.

The data collected relate to one of the many materials available in the ABS category. This group of materials from other manufacturers can show different results. In addition, samples were produced with a specific 3D printer, and printouts from other printers may show different results. Further differences may be due to different software used to generate a sequence of instructions for the device and different print parameters. All of the limitations should be kept in mind when interpreting the data collected.

5. Future works

On the basis of the assembled data and the conclusions, possible fields of research expansion in the future were also identified. These are presented below:

- Giving up U-notch samples in favour of more V-notch ones. V-notch specimens were more reliable (more fully fractured specimens).
- Employing a hammer with lower impact energy, because only 12% of the initial energy of the test stand in question was used.
- Examining more points on the graph, increasing the filling density with a smaller stroke, every 5% or even every 1%.
- Examining other filling patterns.
- Examination of samples with different orientation of the layers in relation to the direction of impact.
- Examination of other types of materials.
- Investigating the influence of other parameters, such as layer height, or the number of contours.
- Carrying out tests at different temperatures of the tested elements.
- Fractographic analysis of selected samples.

6. References

[1] Avalle M and Romanello E 2018 Tribological characterization of modified polymeric blends *Procedia Structural Integrity* **8** 239-255

[2] Czerniec M, Shil'ko S, Paszeczko M and Barszcz M 2018 Wear resistance of glass- and carbon-filled polyamide composites for metal-polymer gears *Journal of Friction and Wear* **39(5)** pp. 361-364

[3] Kiran M D, Govindaraju H K, Jayaraju T and Kumar N 2018 Review-effect of fillers on mechanical properties of polymer matrix composites *Materials Today: Proceedings* **5** pp. 22421-22424

[4] Akhtar S, Saad M and Pandey P 2018 Overview of current advances in the development of polymer composite in biomedical applications *Materials Today*: *Proceedings* **5(9)** pp. 20217-20223.

[5] Friedrich K 2018 Polymer composites for tribological applications *Advanced Industrial and Engineering Polymer Research* **1(1)** pp. 3-39

[6] Foard J H D, Rollason D, Thite A N and Bell C 2019 Polymer composite Belleville springs for an automotive application *Composite Structures* **221** pp. 1-9

[7] Yadav R, Tirumali M, Wang X, Naebe M and Kandasubramanian B 2019 Polymer composite for antistatic application in aerospace *Defence Technology* in press

[8] Yuan S, Shen F, Chua C K and Zhou K 2019 Polymeric composites for powder-based additive manufacturing: Materials and applications *Progress in Polymer Science* **91** pp. 141-168

[9] Zhao W, Liu L, Zhang F, Leng J and Liu Y 2019 Shape memory polymers and their composites in biomedical applications *Materials Science and Engineering*: C **97** pp. 864-883

[10] Ruszniak P, Józwik J B, Dziedzic K and Ostrowski D 2017 Shearing strength test of orthopedic titanium alloy screw produced in the process of 3D printing technology *Advances In Science And Technology Research Journal* **11(1)** pp. 128–137.

[11] Torrado Perez A R, Roberson D A and Wicker R B 2014 Fracture surface analysis of 3D-printed tensile specimens of novel ABS-based materials *Journal of Failure Analysis and Prevention* **14(3)** pp. 343-353

[12] Szczepanik S and Bednarczyk P 2017 Bending and Compression Properties of ABS and PET Structural Materials Printed Using FDM Technology *Journal of Casting & Materials Engineering* **1(2)** pp. 39-42

[13] Korga S, Barszcz M and Dziedzic K 2019 Development of software for identification of filaments used in 3d printing technology *Applied Computer Science* **15(1)** pp. 74-83.

[14] Mitrović R, Mišković Ž, Ristivojević M, Dimić A, Danko J, Bucha J and Milesich T 2018 Statistical correlation between the printing angle and stress and strain of 3D printed models *Procedia Structural Integrity* **13** pp. 475-482

[15] Xu Y, Zhang H, Šavija B, Figueiredo S C and Schlangen E 2019 Deformation and fracture of 3D printed disordered lattice materials: Experiments and modeling *Materials & Design* 1**62** pp. 143-153

[16] Werkena N, Joel Hurley J, Khanbolouki P, Sarvestani A N, Tamijani A Y and Tehrani M 2019 Design considerations and modeling of fiber reinforced 3D printed parts *Composites Part B: Engineering* **160** pp 684-692

[17] PN-EN 10045-1:1994 Metale - Próba udarności sposobem Charpy'ego - Metoda badania.

IOP Conf. Series: Materials Science and Engineering **710** (2019) 012028 doi:10.1088/1757-899X/710/1/012028

Numerical analysis of the thin-walled structure with different trigger locations under axial load

M Rogala, J Gajewski and M Ferdynus

Lublin University of Technology, Department of Machine Design & Mechatronics, 36 Nadbystrzycka Street, 20-618 Lublin, Poland

michal.rogala@pollub.edu.pl

Abstract. This paper presents the results of the numerical analysis of thin-walled structures with a square cross-section. Aluminium models contain different types of triggers and distance to the base. Postbuckling analyses were compared with the form of the buckle to find the most beneficial location of the trigger. The study problem was approachedwith FEM Abaqus.Numerical analysis results are shown in theLoad–Shortening diagram. The data obtained during numerical analysis were used to determine the crushing efficiency indicators such as SE, CLE, TE. The aim of the research was to obtain the best efficiency indicators for the tested positions of triggers.

1. Introduction

One of the most important factors in the transport industry is passenger safety. For years, the construction of thin-walled profiles has been evolving in order to improve their efficiency or, as it was originally thought, to absorb the greatest amount of energy. As early as in the 1980s, Wierzbicki and Abramowicz [1] dealt with the mechanism of thin-walled structures destruction, where they described the behaviour of plastic joints for thin-walled structures with a square section. The mathematical model of energy-absorbing structure destruction has been described in detail [2,3,4]. The studies have shown that the length of a plastic hinge is relative to the wall thickness and the w0idth of the profile. Along with evolving knowledge of thin-walled model analysis, not only the energy absorbed by the specimen, but also the maximum forces generated by the model during the test proved to be an important aspect of the impact tests. Peak crushing force (PCF) [5,6], was an important factor in the impact analysis. Generating excessive overload could kill the passengers, despite the model absorbing a very large amount of energy. The initiation of crushing was managed by the holes incorporated in the profile in order to reduce its stiffness locally and induce to crush at a given point.

Nowadays, examination often concern thin-walled elements made of composite materials [15, 16]. In this type of profile, the most important factor influencing the form of buckling and the mechanism of destruction is the layout of laminate layers. Due to the multiplicity of composite layer layouts, artificial neural networks are used [14]. Prediction has become a popular tool with the development of software.

Another way of initiating crushing at a selected location and decreasing the peak crushing force generated, most frequently, in the initial stage of crushing, was to crush the side walls [7, 10]. Trigger made in such a way allowed for easy formation of the first fold. With the development of technology, models with fillings began to appear in order to increase the efficiency of the absorber. At the turn of the 20th century two Hanssen's papers [8, 9] were published in which aluminium profiles with foamed aluminium filling were described. There, the dispersed energy increased, however, it did not affect the maximum destructive force. Similar conclusions can be drawn from the study by Mohammadiha et al. [11], where the aluminium profile was filled with a honeycomb structure. This material increased the amount of absorbed energy, but due to it is structure it could slightly increase the PCF due to increased stiffness in the direction of impact.

The effectiveness of crushing is influenced by the position of the triggers. Such tests were conducted by Ferdynus [12, 21], where the initiator in the form of dents was placed on the top and bottom of the sample. In the studies, there was no arrangement of the triggers from intermediate distances, which would give a full picture of the influence of the position of the triggers on the coefficients of effectiveness.

The scientific team of the Department of Machine Design and Mechatronics has extensive experience in performing analyses of finite element methods [17-20]. Researchers of the departmentare engaged in numerical analyses of various areas of mechanical engineering.

2. Energy absorption indicator

In order to correctly interpret the results of the dynamic analysis, it is necessary to describe the course of crushing with several parameters. The coefficients describing the impact analysis can be obtained by analysing the force - shortening diagram. An exemplary characteristic describing the crushing is shown in figure 1.

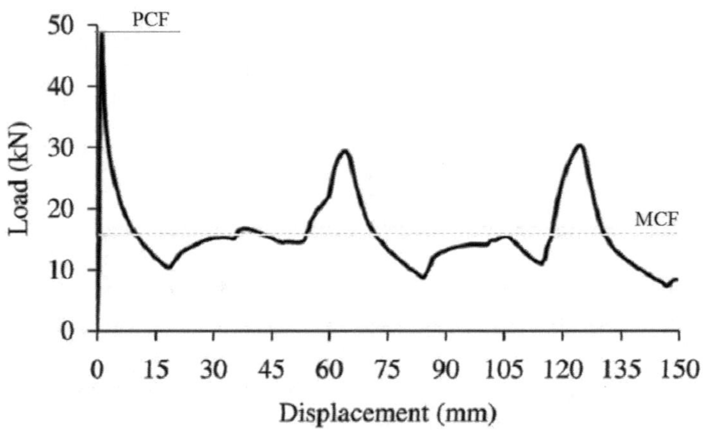

Figure 1. Exemplary force – shortening diagram of a thin-walled structure [13].

The main quantity determining the crush is the absorbed energy, which is defined as the area of the surface under the function, and is given by the equation(1).

$$EA(d_x) = \int_0^{d_x} F(x)dx \quad (1)$$

where the d_x, is crushing distance, and the F(x) is crushing force given as a function.
Another factor is Specific Absorbed Energy (SEA), which is energy per unit mass (2).

$$SEA = \frac{EA}{m} \quad (2)$$

These two values are not direct measures of crashworthiness. There are several other coefficients that help to understand the process of crushing thin-walled profiles.

To determine the first factor it is necessary to determine the Peak Crushing Force (PCF) which is the maximum force and the Mean Crushing Force (MCF) which is given by formula 3.

$$MCF = \frac{EA(d_x)}{d_x} \tag{3}$$

The crash load efficiency is determined by comparing the mean crushing force with the peak crushing force and is given by the formula below.

$$CLE = \frac{MCF}{PCF} \cdot 100\% \tag{4}$$

The value describing the crush length is SE, which compares the length of the specimen after crushing with its initial length.

$$SE = \frac{U}{L_o} \tag{5}$$

The last factor is TE (6), which is the product of CLE and SE. The coefficient was first described by Hanssen [9]. Nowadays, it is often used in literature to describe the crushing process.

$$TE\ [\%] = CLE \times SE \tag{6}$$

3. Numerical analysis

The numerical analysis was carried out using Abaqus CAE. The subject of research was the aluminium column with a different type of trigger. The models were generated as a *shell element*. The square section of the specimen is 40x40 mm, andthe length of the column is 200mm. The shell thickness is 1.2mm. The model has a notch in various configurations. The length of notch is changing from 10 to 40mm, as well asthe location of the trigger from 20 to 180mm. Every described dimension is presented in figure 2.

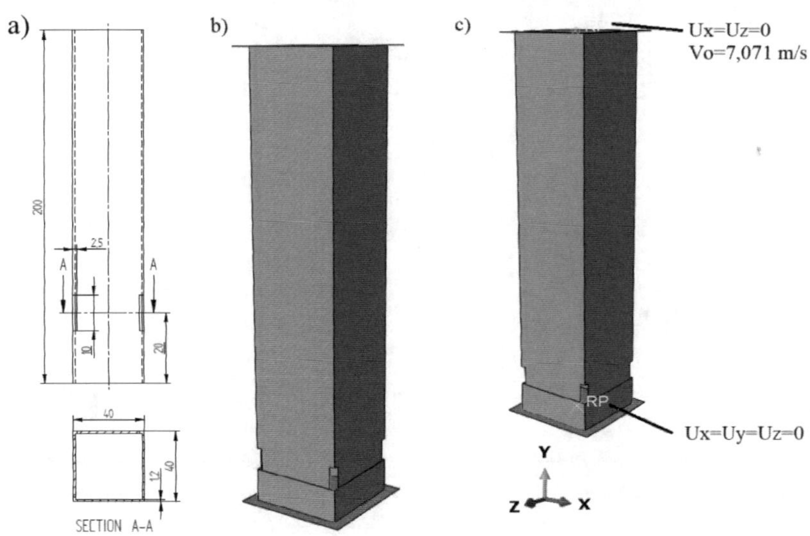

Figure 2. a) detailed drawing b) general view c) boundary condition.

The boundary conditions were defined using the top and the bottom plates. The energy was simulated as a mass assigned to the reference point as well as the speed of the top plate. Both plates were connected to the column with *tie* contact option. The numerical discretisation was related to the use of the appropriate mesh type. In this analysis, two types of elements were used:S4R is a 4-node, quadrilateral, stress–displacement element with reduced integration, and the other element – R3D4, has 4nodes within a single element, and 3 degrees of freedom in each node. The column was subjected to impact load of kinetic energy E=1500J, which corresponds to the mass of m=60 kg dropping with initial velocity V=7.071 m/s.

The material model of aluminium is based on elastic-plastic characteristics. The elastic properties are described by two coefficients –Young's modulus, and Poisson's Ratio. Plastic ones are defined by bilinear characteristics. Material data are presented below in table1.

Table 1. Aluminium column material properties (own study).

Al-6061 aluminium material properties	
Density	2700
Young Modulus [MPa]	70000
Poisson ratio v [-]	0.33
Yield point R_e [MPa]	200
Tensile Strength R_m [MPa]	279.98
Elongation A% [%]	5.98

The analysis was modelled in two steps. The first one is a buckling analysis, the second one is a dynamic analysis based on the model obtained in the first step.

4. Results of numerical analysis

The group of tested models was a column with a trigger in the form of a 10-mm-long cut-out at the edges. A variable parameter was the distance between the centre of the trigger and the base of the element. The distance changed from 20 mm to 180 mm. The behaviour of the model in the different stages presented below will allow a better understanding of the energy absorption process.

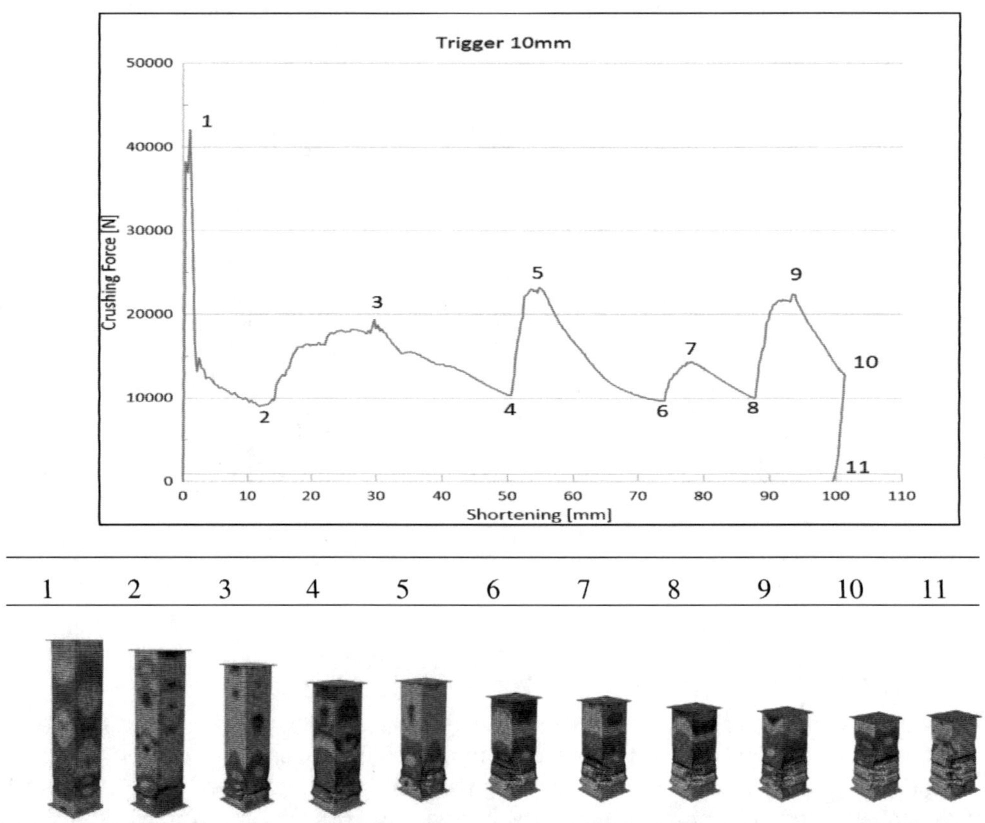

Figure 3. Force – Shortening diagram and stages of impact crushing of the specimen C10-30.

The force shortening diagram (figure 4) below allows indicates which models absorbed the energy most quickly, which translates into crushing efficiency indicators. The mechanism of damage to the structure is the same for all models. The first fold occurs in the place of the trigger. Differences in the shortening result from the fact that the defect is implemented into the model by the mode of buckling (figure 5).

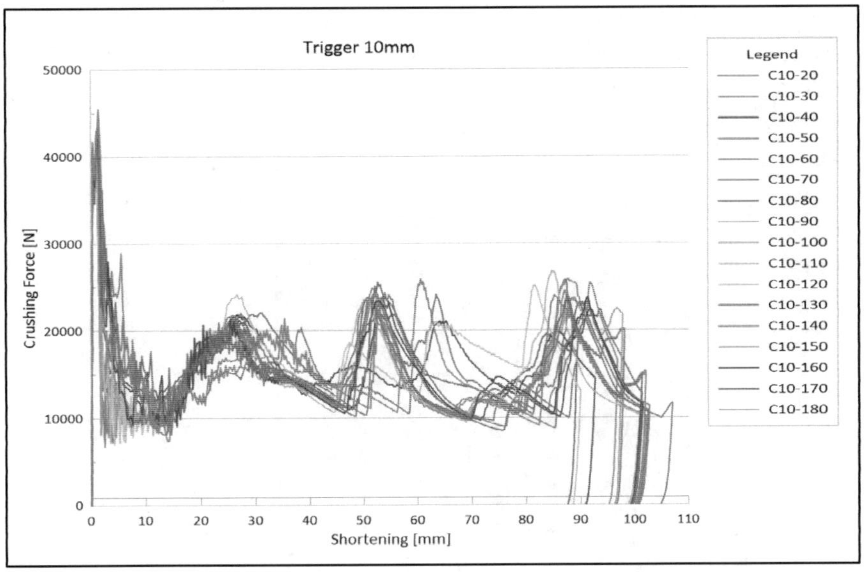

Figure 4. Characteristic curve force shortening for models with 10-mm-long trigger.

The models that absorbed energy the fastest are those that have the initiator of the crushing at a distance of 50 mm and 110 mm from the base. The model that has been shortened the most during the dynamic analysis is the one with the trigger located at 170 mm from the base. The biggest shortening results in the worst ST and MCF ratio. These results allow us to state that the position of the trigger affects the form of buckling and thus changes the course of the impact crush.

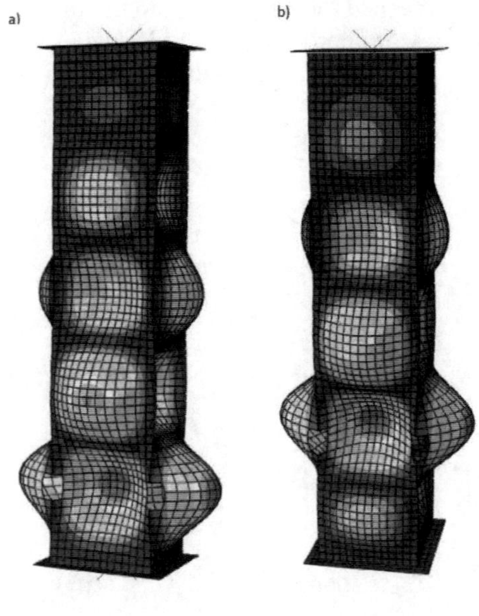

Figure 5. Different mode of the buckle a) 20 mm to the base b) 50 mm to the base.

The presented buckling forms show that locating the trigger too close to the bottom plate causes the first fold's potential to be underutilised. Positioning the initiator 30 mm higher causes the plastic hinge to form in the place where the trigger occurs and free folds to appear above and below the crush initiator.

5. Studies of crashworthiness indicators

To obtain a complete image of the analysis, use the indicator described in the second chapter. Coefficients based on the characteristics of the crush behaviour together with data from the previous chapter will determine the most advantageous position of the trigger.

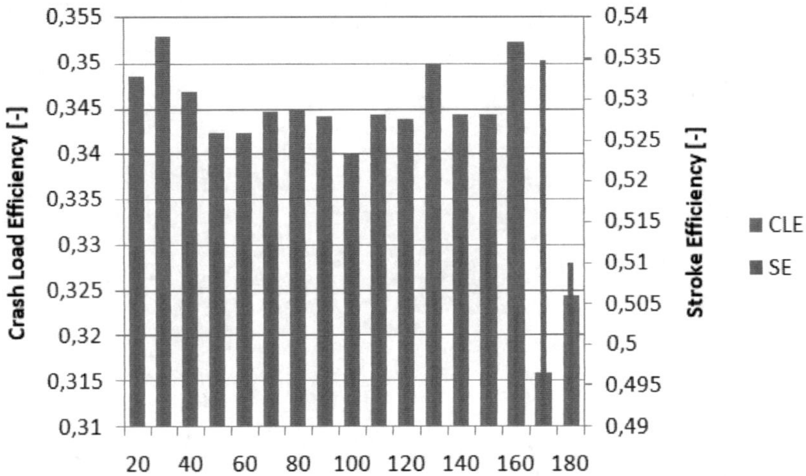

Figure 6. The CLE and SE factors for a sample with a 10 mm trigger.

The graph above shows that the models with the best CLE factor are the columns with a 30 mm, 130 mm, and 160 mm distance triggers. This indicator is based on both the height of the force peak as well as the average force during the analysis. The lowest value was obtained for models with a trigger located near the top plate (170 mm, 180 mm).

The highest SE factor was reached by the model with the initiator located at 170 mm from the base. Other models achieved an index of approximately 0.51. The model achieves the best results when the CLE is the largest and the SE the smallest. The models that received the best combination of indicators are 30 mm, 110 mm and 160 mm.

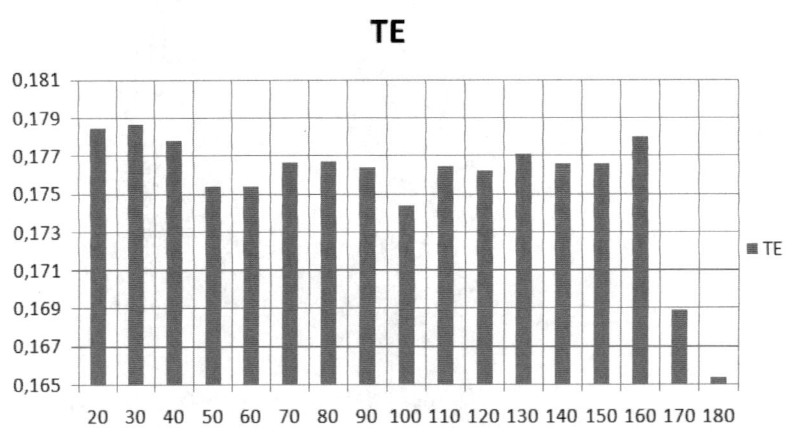

Figure 7. TE coefficient for the specimen with a 10–mm trigger.

The last coefficient is the value resulting from the multiplication of CLE and SE. The course of TE (figure 7) values changes depending on the position of the trigger. The highest values of the coefficient are around 30 mm, 80 mm, 130 mm and 160 mm.

6. Conclusions

This paper presents the numerical analysis of a thin-walled profile. The sample had a trigger in the form of a 10-mm-long cut-out. The distance between the trigger and the base changed from 20 mm to 180 mm every 10 mm. The change of initiator's position did not affect the PCF force, the value fluctuated around 42.5 kN for all models. The analysis showed that placing the cut-out too close to the top plate has a negative effect on its performance. The CLE is about 11% higher for the 160–mm model than for the 170–mm model. The maximum difference in shortening the models was 17 mm, i.e. 8.5%. The TE coefficient reaches its maximum for models with distances of 30 mm, 80 mm, 130 mm, 160 mm. Placing the trigger between the optimal values reduces the efficiency of the applied trigger.

7. References

[1] Wierzbicki T and Abramowicz W 1983 *J. Appl. Mech.* **50** 727–34
[2] Bai Y, Bao Y and Wierzbicki T 2006 *International Journal of Impact Engineering* **32** 671–701
[3] Sun G, Pang T, Xu C, Zheng G and Song J 2017 *Thin-Walled Structures* **118** 214–28
[4] Kopczyński A and Rusinek E 2010 *Bezpieczeństwo bierne. Pochłanianie energii przez profile cienkościenne* (Wrocław: Oficyna Wydawnicza Politechniki Wrocławskiej)
[5] Jones N 2003 *Structural Impact* (Cambridge University Press)
[6] Abramowicz W 2003 *Thin- Walled Struct.* **41**
[7] Ming S, Zhou C, Li T, Song Z and Wang B 2019 *International Journal of Mechanical Sciences* **158** 150–64
[8] Hanssen AG, Langseth M and Hopperstad OS. 1999 *Int J Mech Sci* **41** 967-93
[9] Hanssen AG, Langseth M and Hopperstad OS. 2000 *Int J Mech Sci* **24** 347-83
[10] Ferdynus M 2019 *Numerical Analysis of energy absorbers in the form of aluminium square section columns with round indentations* AIP Conference Proceedings **2078** 020020
[11] Mohammadiha O, Beheshti Haji H and Aboutalebi F 2015 *International Journal of Crashworthiness* **20:1** 44-59 DOI: 10.1080/13588265.2014.970398
[12] Ferdynus M, Kotełko M and Urbaniak M 2019 *Thin-Walled Structure* **144**
[13] Dimas D, Pereira P, Soares D, Vilarinho C 2012 *Experimental study of impact energy absorption in aluminium square tubes with thermal triggers* (Portugalia)
[14] Szklarek K, Gajewski J and Valis D 2019 *Estimation of critical force of the buckling composite structures using modelling methods* MATEC Web of Conferences **252** 03004
[15] Rozylo P, Debski H, Wysmulski P and Falkowicz K 2018 *Composite Structures* **204** 207-16
[16] Wysmulski P, Debski H, Falkowicz K and Rozylo P 2019 *Composite Structures* **213** 98-107
[17] Gajewski J, Podgorski J, Jonak J and Szkudlarek Z 2008 *Materials Science* **43** 115-18
[18] Falkowicz K, Ferdynus M and Debski H 2015 *Eksploatacja i niezawodność* **17(2)** 222-27
[19] Gajewski J, Golewski P and Sadowski T 2017 *Composite Structures* **159** 589-99
[20] Karpiński R, Jaworski Ł, Jonak J and Krakowski P 2019 *The influence of the nucleus pulposus on the stress distribution in the natural and prosthetic intervertebral disc* MATEC Web of Conferences **252**, 07006 DOI: 10.1051/matecconf/201925207006
[21] Ferdynus M and Rogala M 2019 *Advances in Science and Technology Research Journal* **13(3)** 144-51 doi:10.12913/22998624/110611.

IOP Conf. Series: Materials Science and Engineering **710** (2019) 012029 doi:10.1088/1757-899X/710/1/012029

Thermal bridge occurrence in straw-bale timber frame walls

P Brzyski[1], P Kosiński[2] and M Nadratowska[2]

[1] Lublin University of Technology, Faculty of Civil Engineering and Architecture, Department of Construction, Nadbystrzycka 40, 20-618 Lublin, Poland
[2] University of Warmia and Mazury in Olsztyn, Institute of Building Engineering, Faculty of Geodesy, Geospatial and Civil Engineering, Jana Heweliusza St. 10, 10-724 Olsztyn, Poland

p.brzyski@pollub.pl

Abstract. Timber frame walls usually require additional filling that improves their thermal insulation. Straw bales, which constitute a waste product from cereals, are a sustainable solution employed for that purpose. The temperature distribution in a wall is influenced by the heterogeneity of the partition that comprises the elements characterised by higher thermal conductivity, i.e. timber frame. Thermal bridges (i.e. thermally weaker areas) in walls are conducive to the mould growth as well as condensation; thus, it is important to take these phenomena into account at the wall designing stage. The work describes a 2D heat-transfer analysis carried out with the finite element method by means of THERM software. The analysis involved several different variants of external walls and corners. The thermal parameters of the investigated straw were derived from our former research. The obtained results were shown as the values of thermal transmittance coefficient and linear thermal transmittance equivalent to timber construction. The temperature distribution for the investigated wall was presented in a graphical form. The study also involved examining the possibility of surface condensation.

1. Introduction

Straw is a waste product from cereal crops, e.g. rye or wheat, therefore, its use is consistent with the principles of sustainable development. In addition, the crop is an annual plant and, in Poland, it is grown in every province, thus it can be locally sourced. Straw is primarily composed of cellulose, hemicellulose and lignin [1].

Straw is used in construction mainly as a wall filling in a timber frame construction [2]. It is used in the form of compressed bales of various sizes, tied by steel wire or polypropylene twine. Bales with a density of 80-100 kg/m^3 are preferred due to their optimal insulation parameters [3]. Straw bales of larger sizes are used as a load-bearing material, transferring loads from the roof and ceilings. Straw is also used as a thermal insulation filler in wall composites based on lime binder [4], in addition to hemp shives [5, 6] and flax shives [7].

Thermal properties of straw bales depend on the moisture content, the density, the porosity, the fibre and bale orientation [2]. Munch [8] showed the values of thermal conductivity of straw bales at different straw bale orientations and densities (62-150 kg/m^3) in the range of 0.038-0.082 W/(m·K). Similar observations are reported by Schiavoni et al. [9]. In turn, the heat capacity of straw is given in the literature [10] in the range of 1338-2000 J/(kg·K). The porosity of straw ranges between 46 and 84 % [11]. Straw cubes for use in construction must meet numerous requirements. Mass moisture should not exceed 14% [12] due to the risk of biological corrosion development. Under the conditions of high humidity, the optimal temperature at which biological corrosion may occur is 20-70°C [13]. Long-term exposure to high levels of humidity (above 80% RH) may present an increased risk of straw degradation [14]. In addition to the moisture content in the straw bale, its pH value is also important, and 7.29 is typical for straw [15].

In some areas of the frame, increased cross-sections of timber elements are used in the construction of walls. Despite the satisfactory thermal insulation properties of wood, they can form thermal bridges. The effect of increased heat flow may be the occurrence of condensation of water vapour, mould growth, which results in deterioration of the indoor air quality [16, 17]. It is particularly dangerous in the case of organic materials because they are particularly exposed to biological corrosion [18, 19].

The paper presents the two-dimensional heat-transfer analysis based on the finite element method, using THERM software. The subject of the analysis involves external walls and corners based on a timber frame construction filled with straw bales. The temperature distribution was analysed for several variants of the external wall and corners construction solutions, which enabled to select the most favourable solution due to the heat transfer.

2. Materials and methods

2.1. Straw

The study used rye straw from the crops of the Lublin province (Poland), mowed in August, during the harvest period. The diameter of the straw was approx. 1.0-2.5 mm. A sample of straw used in tests is shown in Figure1.

Figure 1. Straw used in tests.

2.2. Thermal conductivity of straw

The thermal conductivity of rye straw was investigated in accordance with EN 12667 standard, using a Laser Comp Fox 602 plate apparatus and the heat flow meter method. WinTherm32 software was used for measurements of the thermal conductivity coefficient of cereal straw.

Throughout the test, the average reference temperature amounted to +10°C. In turn, the temperature of the top and lower plates was set at +20°C, and 0°C, respectively. The frame was placed in a thin bag to prevent moisture build-up in the tested material as well as the condensation on the cooler camera plate. Testing the samples involved placing them in a stabilising frame. The frame was made of extruded polystyrene and had the following dimensions: 600×600 mm (external), and 520×520 mm (centre cut) as well as the height of 45 mm. Before the test, the straw was dried at 50°C.

2.3. Analysed partitions

Partitions with the construction frame usually applied in this technology were accepted for calculations. The load-bearing structure is a timber frame placed in two rows – on the inner side and the outer side of the wall. The two adopted stud cross-section variants were 60×120 mm and 50×150 mm. In addition, a scenario with a double stud 50×150 mm was put to test. Five cases of axial spacing between studs were considered: 600, 650, 700, 750 and 800 mm. The heat distribution and the size of the thermal bridge were also checked in external corners made using timber elements with the adopted cross-sections and exemplary studs spacing of 700 mm. The adopted schemes are shown in figure 2.

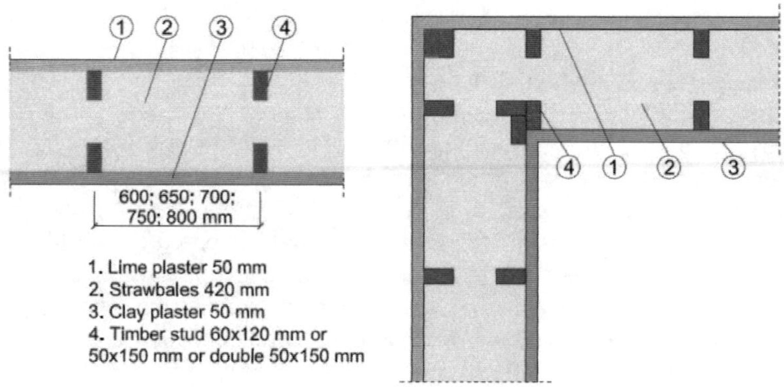

Figure 2. Schemes of analysed external wall and corner.

2.4. Thermal analysis

Temperature distribution was modelled by means of THERM 7.4 software (http://windows.lbl.gov), which can be used to conduct two-dimensional modelling of the determined heat flow in partitions, building elements and construction junctions [20, 21, 22]. It employs FEM for solving the heat flow equations. Junction modelling consists of the following steps [23]:

- model definition (geometry, materials and boundary conditions)
- mesh generation
- using Finite Element Analysis Solver to determine the temperature in nodes and heat streams
- reporting the analytical results (e.g. average heat transfer coefficient for a given node) and the processed results in a graphical form (e.g. isotherms)

The obtained results served to determine the linear heat transfer coefficient ψ [W/(m·K)] of a junction in accordance with ISO 10211 standard, given by the following formula (1):

$$\psi = L^{2D} - \sum_{i=1}^{j} U_i \cdot l_i \qquad (1)$$

where L^{2D} corresponds to linear thermal coupling coefficient obtained from numerical analysis as a product of the average heat transfer coefficient of a junction and its length [W/(m·K)]; U_i corresponds to the heat transfer coefficient of the i-the component of the junction [W/(m^2·K)] and l_i is the length assigned to the component with the heat transfer coefficient U_i [m].

The THERM software was employed for the generation of the U-factor; however, its values were approximated to the ones obtained from the calculations for the components of heterogeneous heat layers, in line with the ISO 6946 standard.

Table 1 summarises the thermal properties of individual materials (based on manufacturer data: www.tierrafino.com and ISO 10456 standard). Table 2 shows boundary conditions used in modelling, assuming the average external temperature for January in Lublin (Poland).

Table 1. Thermal conductivity of building materials and elements.

Building material/ element	Thermal conductivity [W/(m·K)]
Straw bale	0.047
Lime plaster	0.80
Clay plaster	0.91
Timber construction element	0.16

Table 2. Boundary conditions adopted in modelling.

Surface	Temperature [°C]	Surface resistance [(m²·K)/W]	Description
Internal	+20	0.13	Heat flow horizontal, simplified*
External	-2.6	0.04	Simplified*
Cut-off planes	–	–	Adiabatic

* the simplified model means that convective and radiative heat transfer is described by one common surface resistance

2.5. *Possibility of water vapour condensation*
Condensation of moisture on the internal surfaces of building partitions can arise in the locations where the temperature is below the dew point. Due to the drop in temperature, resulting from the increased heat transfer, the thermal bridge occurrence is to a considerable extent correlated with this phenomenon. On the basis of the ISO 13788 standard, the conditions that contribute to the occurrence of surface condensation were investigated. In order to avoid the mould growth, it is necessary to ensure that the temperature factor at the internal surface f_{Rsi} is greater than the design temperature $f_{Rsi,min}$ for the critical month (i.e. the one with the highest $f_{Rsi,min}$ value). In order to calculate both factors, it is necessary to use the increased thermal resistance at the initial surface (R_{si} = 0.25(m²·K)/W):

$$f_{0.25} = \frac{\theta_{si}-\theta_e}{\theta_i-\theta_e} \geq f_{0.25,min} = \frac{\theta_{si,min}-\theta_e}{\theta_i-\theta_e} \tag{2}$$

where θ_{si} is the surface temperature in the critical area [°C], $\theta_{si,min}$ is the minimum acceptable surface temperature [°C], θ_i is the internal temperature [°C], θ_e is the external temperature [°C].

The minimum surface temperature of the inner side of the wall occurs in the corner, therefore, only corners are taken into account for calculations.

3. Results and discussion

3.1. *Thermal conductivity*
Thermal conductivity and bulk density together with standard deviations are presented in table 3.

Table 3. Straw parameters.

Bulk density [kg/m³]	Standard deviation [kg/m³]	Thermal conductivity [W/(m·K)]	Standard deviation [W/(m·K)]
50.9	±4.26	0.0473	±0.0015

Munch [8] showed that placing straw bales "on flat" resulting in λ value about 0.063 W/(m·K) and "on edge" about 0.051 W/(m·K). Our tests on samples set "on edge" have shown a close fit with the literature data. The obtained smaller λ value may be associated with a lower density of straw bales because the degree of compaction of loose insulation materials affects their insulating properties [24]. Asdrubali [25] also reported that samples made with the stalks of straw perpendicular to the heat flow have better thermal insulation properties.

3.2. *Thermal analysis*
The graph below (figure 3) presents changes in the value of the averaged thermal transmittance coefficient of walls of different timber elements depending on the axial spacing of timber studs.

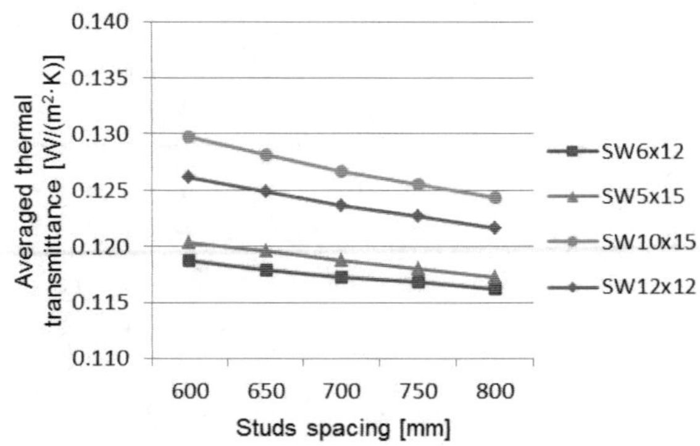

Figure 3. Averaged thermal transmittance coefficient of walls of different timber elements depending on the axial spacing of timber studs.

The considered external walls are characterised by the mean U-factor value from 0.117 to 0.130 W/(m^2·K), which is reduced as the stud spacing increases. Similar observations were reported in the analyses of walls made of hemp-lime composite with a stud spacing of 400-600 mm [26]. The results may be attributed to the increase in the straw bale area in the cross-section of the considered wall; this material exhibits superior insulating properties in relation to wood. The differences between the U-values characterising the walls with extreme stud spacing range only from 2.15 to 4.26%. Larger variations can be seen in the walls with doubled cross-sections of studs. Doubling the cross-section of studs results in an increase of the averaged thermal transmittance coefficient by 5.5 % and 6.8 %, respectively for columns 60×120 mm and 50×150 mm with cross-sections.

The graph in figure 4 presents the changes in the value of the linear thermal transmittance coefficient characterising walls of different timber elements, depending on the axial spacing of timber studs.

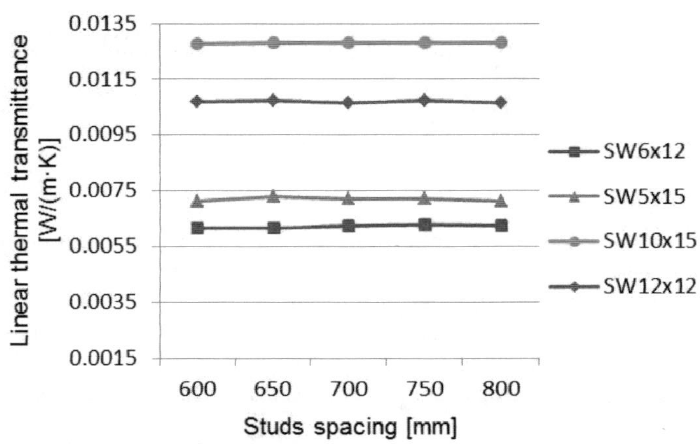

Figure 4. Linear thermal transmittance coefficient of walls of different timber elements depending on the axial spacing of timber studs.

Timber studs create linear thermal bridges, expressed as linear thermal transmittance coefficient ψ ranging from 0.0062 to 0.0128 W/(m·K). This coefficient increases along with the cross-section of timber studs. Their spacing has little effect on the linear thermal transmittance coefficient. Although certain small variations can be noticed, no clear relationship is observed. Similar observations are described in the literature [26]. Doubling the cross-section of studs increases the thermal bridges by 72% and 78%, for the columns with 60×120 mm and 50×150 mm cross-sections, respectively.

Table 4 presents the values of the averaged thermal transmittance and linear thermal transmittance coefficients for external corners with studs spacing of 700 mm.

Table 4. U and ψ value for external corners.

Symbol of corner	U [W/(m²·K)]	Ψ [W/(m·K)]
SC6×12	0.1005	-0.0434
SC5×15	0.1017	-0.0486

The corner with studs having a cross-section of 50×150 mm is characterised by an averaged thermal transmittance coefficient greater by 1.19% than in the case of the corner with studs of 60×120 mm and linear thermal transmittance coefficient lower by 12.1% than the corner with studs of 60×120 mm. Figures 8 and 9 show the graphical results corresponding to one of the considered types of external walls. The disruptions pertaining to the unidirectional heat flow at the area between thermal bridges and timber studs can be observed.

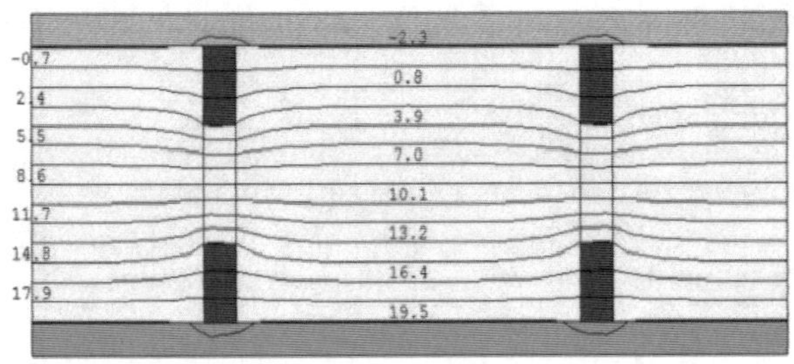

Figure 5. Isothermal distribution for an example analysed wall (stud (60×120 mm) spacing: 700 mm).

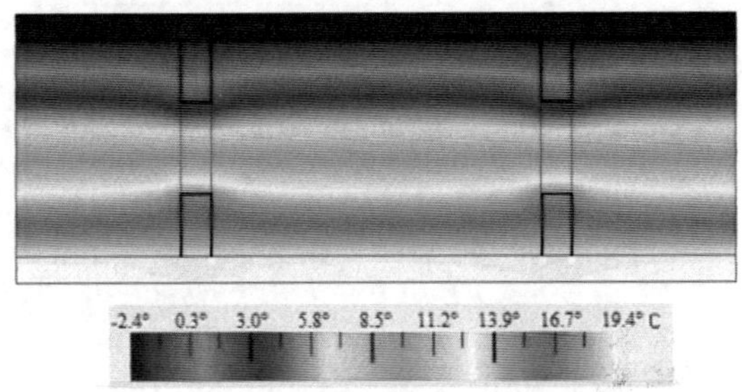

Figure 6. Temperature distribution for an example analysed wall (stud (60×120 mm) spacing: 700 mm).

3.3. Possibility of water vapour condensation

The temperature factors for the described corners are presented in table 5. The minimum temperature on the inner surface is the same in both cases.

Table 5. Temperature factors $f_{0.25}$ [°C].

Symbol of corner	$f_{0.25}$
SC6×12	0.92
SC5×15	0.92

National regulations in Poland allow for the use of the design temperature factor $f_{0.25.min}$ = 0.72 and the risk of surface condensation does not occur in any of the corners. Satisfactory thermal insulation parameters of the straw and the obtained U-value ensured the high temperature of the inner surface of the wall (18.2°C). Due to different diffusion resistance of wall layers and the presence of timber elements, it would be important to calculate the possibility of interstitial condensation in further studies.

4. Conclusions

Straw bales exhibit good thermal insulation properties, comparable with traditional insulating materials such as mineral wool and polystyrene. It would be reasonable to further investigate the thermal capacity of straw bales, as it can contribute to reducing heat loss under the actual conditions.

Although timber elements are characterised by relatively good insulating properties, they were shown to affect the local increase in heat flow within straw bale partitions, contributing to the creation of thermal bridges in walls. These linear heat transfer coefficients range from 0.0062 to 0.0128 W/(m·K); hence, they do not exceed the acceptable value recommended for energy-efficient buildings, i.e. 0.10 W/(m·K) [27]. The use of the studs with the largest cross-section and the smallest spacing resulted in an increase in averaged thermal transmittance coefficient by 10.6% compared to a wall with studs with the smallest cross-section and the largest spacing. The stud spacing to a limited degree contributes

IOP Conf. Series: Materials Science and Engineering **710** (2019) 012029 doi:10.1088/1757-899X/710/1/012029

to the changes in the linear transmittance coefficient values through the wall. Moreover, no relationship between its value and the stud spacing was observed. Surface condensation did not occur in the analysed corners.

The presented analyses prove that the timber frame configurations used in practice in straw bale walls are small thermal bridges. The results may also be helpful in designing as well as selection of this technology by investors. However, to assess the impact of thermal bridges more accurately, an energy analysis of the entire building should be performed.

5. References

[1] Khan T S, Mubeen U 2012 *Curr. Res. J. Biol. Sci.* **4(6)** 673–75

[2] Minke G and Mahlke F 2005 *Building with straw*. Birkhauser

[3] Atkinson, C 2016 *Building with straw bales*. Lecture notes, 2011, available at: (www.homegrownhome.co.uk) (retrieved on August 2016)

[4] Rahim M, Douzane O, Tran Le A D, Promis G and Langlet T 2016 *Constr. Build. Mater.* **102** 679-87

[5] Brzyski P and Łagód G 2018 *E3S Web of Conferences* **49** 00010

[6] Brzyski P and Widomski M 2017 *AIP Conference Proceedings* **1866** 040006

[7] Brzyski P, Barnat-Hunek D, Suchorab Z and Łagód G 2017 *Materials* **10(5)**

[8] Munch-Andersen J and Andersen B M 2016 *Straw Bale Houses -design and material properties. Danish Building and Urban Research*, Statens Byggeforskni-ngsinstitut; http://www.sbi.dk/download/pdf/jma_slides_halmhuse.pdf

[9] Schiavoni S, D'Alessandro F, Bianchi F and Asdrubali F 2016 *Renew Sust Energ Rev* **62** 988–1011.

[10] Chaussinand A, Scartezzini J L and Nik V 2015 *Energy Procedia* **78** 297-302

[11] Robinson J, Aoun H K and Davison M 2017 *Procedia Eng* **171** 1526-34

[12] Steen A S, Steen B, Bainbridge D and Eisenberg D 1994 *The Straw Bale House* (Vermont: Chelsea Green Publishing Company/ White River Junction)

[13] Lawrence M, Heath A and Walker P 2009 *Constr. Build. Mater.* **23** 2763-68

[14] Thomson A and Walker P 2014 *Constr. Build. Mater.* **68** 135-141

[15] Ashour T, Heiko G and Wu W 2011 *Energy Build.* **43(8)** 1960-67

[16] Lagod G, Suchorab Z, Guz L and Sobczuk H 2017 *AIP Conference Proceedings* **1866** UNSP 030002

[17] Majerek D, Guz L, Suchorab Z, Lagod G and Sobczuk H 2017 *AIP Conference Proceedings* **1866** UNSP 040024

[18] Fedorik F and Illikainen K 2013 *Int. J. Sustain. Built Environ.* **2/1** 19-26

[19] Viitanen H and Ojanen T 2007 Improved Model to Predict Mold Growth in Building Materials. Paper based on the VTT Projects "Building Biology" and "Integrated Prevention of Moisture and Mould Problems (Finland)

[20] Real S, Gomes M G, Rodrigues A M and Bogas J A 2016 *Constr. Build. Mater.***121** 460-70

[21] Murad C, Doshi H and Ramakrishnan R 2015 *Procedia Eng.* **118** 1030-37

[22] Stazi F, Tomassoni E, Bonfigli C and Di Perna C 2014 *Appl. Energy* **134** 176-196

[23] Mitchell R, Kohler C, Zhu L, Arasteh D, Carmody J, Huizenga C and Curcija D 2011 *Therm 6.3 / Window 6.3 NFRC Simulation Manual* (Lawrence Berkeley National Laboratory, California)

[24] Kosiński P, Brzyski P, Szewczyk A and Motacki W 2017 *J. Nat. Fibers* **15(5)** 717-30

[25] Asdrubali F, D'Alessandro F and Schiavoni S 2015 *SM&T* **4** 1–17

[26] Brzyski P, Duda S and Raczkowski A 2019 *MATEC Web of Conferences* **252** 05015

[27] Strom I, Joosten L and Boonstra C 2006 *Passive Houses Solutions* (Intelligent Energy Europe Promotion of European Passive House).

Modelling the monopedal robot

M Niechciał[1], D Rybarczyk[1] and J Buśkiewicz[2]
[1] Poznań University of Technology, Institute of Mechanical Technology, Piotrowo 3, Poznań, Poland
[2] Poznań University of Technology, Institute of Applied Mechanics, Piotrowo 3, Poznań, Poland

dominik.rybarczyk@put.poznan.pl

Abstract. The article describes modelling of the monopedal robot with the application of the numerical method for calculating kinematics. The first part presents a short review of the literature on monopedal robots and the theory of such designs. Due to the complexity of the natural solutions of the bipedal locomotion, the monopedal alternative simplifies the problem, by limiting its scope to the pulse movement in one axis– as required by jumping. In the subsequent part, basic mathematical and kinematic relations connected with a chosen designs. The model was based on a rigid quadrangle driven by a four-bar linkage. The effect was a set of vector equations, required to calculate its kinematics. Due to the complexity of the design, it could not be calculated using traditional analytical methods. The prepared kinematic structure was simulated using the numerical method to give the basis for the future experimental tests. The mathematical model was solved in the Wolphram Mathematica environment. Its main purpose was to analyse the movement of the final link that rests on the ground. The results from the simulation were compared to the movement parameters derived from a model designed in CAD software to verify its correctness. All values of links length and starting angular position, required to calculate the model, were taken from the said CAD project.

1. Introduction

One of the unresolved issues in modern robotics that limits the practical application of robots is a method of effective and smooth locomotion. The domination of mobile robots based on wheeled vehicle base is the outcome of simplicity in design and control. Due to its nature, however, the wheeled design is ineffective in most of the non-laboratory environments [1]. A possible progress in the area of locomotion problem is located in the bio mimic solution, reproducing animalistic limbs generated by the evolution. The most intuitive design human-wise is legged locomotion, which bypasses any problems connected with difficult terrain [2].

The most important advantages of the legged solution are [3]:

- adaptive to uneven terrain
- use of isolated footholds
- provide active suspension
- the environmental effects of legged vehicles are less than wheeled or tracked vehicles

Due to difficulties in control and supervision of bipedal designs [4] various projects have been started to devise monopedal jumping robots. The latter design requires dynamic balance and a special structure. The first model of a jumping robot was presented by Matsuoka in 1979, with the first solutions to the mentioned problems. The actuator was constructed of a pneumatic cylinder, restoring energy on impact; the balance in flight was controlled by rotational mass with the mass moment of inertia.

To perform a jump motion it needs a mechanical resonator that stores or modulates the power in any way. Most solutions to this problem apply a series of elastic actuators, in between the drive link and the motor, or another source of power. Its basic benefit is isolating and filtering all impacts delivered by dynamic jumping or falling, storing the energy of the motor for a period of time, until the launch, and the change in the definition of the control system. Due to the actual input delivered to the mechanism not being directly connected with the motor shaft, the real information about the force and speed is delivered by analysing relative rotary position between the elastic element and its elastic constant.

To ensure a correct motion, the jump itself must be repetitive or controlled in flight [5]. Because of a robust movement, most constructions also include the solution dealing with flight control [6].

1.1. Projects overview

One of the most important properties of all jumping robots is the ability to generate impulse of force that highly surpasses the dynamic possibilities of the drive. Owing to the selected method of storing and transferring energy, produced by the input link, the entire mechanism works similarly to a catapult. In the very basic example, given by nature, it is represented by the biological structure of frogs' legs, which are unable to maintain a process of walking, but due to modulating force, accumulated in crouching, enable jumping [7].

The catapult mechanism was represented in a simple flea-inspired mechanism size of 20 mm, acquiring extraordinary performance [8]. The combination of three spring actuators made of shape memory alloy (SMA) generates and stores energy during mechanical locking of the links. Due to its nature, further electrical heating of the springs enables an even greater force to be locked inside links. After the release, the mechanism jumps to the distance of up to 30 times its body size.

A design that also employs springs to accumulate energy is the Michigan robot [9]. The whole design is an enclosed kinematic chain consisting of 6 links creating a hexagon with four springs that force two opposed sides away from each other. Springs tension is generated by a rotor on one side connected with it, and with the link on the other. Its simplistic design with minimal influence of electronics in the drive causes that the mass could be minimised to achieve great effects in performance: with 25 grams of mass and almost 1 meter of jumping height [10]. Due to its robust flight phase, it was required to control its position with a tail [11], which is an offset mass connected with the body by a rotating link. It is a common solution based on animalistic tail [12]

A successful solution for jumping and control was presented by the Penn Jerboa robot [13]. The design is very much similar to the Michigan one, but for the purpose of maximum control and repetitive motion, its height was greatly reduced. The robot is driven by an enclosed quadrangle that is extended and compressed by an actuator with a spring on it. To bring more tension to the spring the whole robot is pressed to the ground by the momentum of the bouncing tail. Even though the height of the jump is only 0.16 meter, it is consistent, quick and predictable.

One of the designs that solves the problem of accumulating energy and controlling the flight trajectory is developed by the Berkeley team in a robot called Salto [14]. Its design is based on a rigid quadrangle driven by a four-bar linkage that moves a lever connected to the ground. Its greatest benefit is the modulation of the mechanical advantage that is the change of momentum to force ratio during the rotation of the drive. Thanks to low advantage at the starting position, the input energy of the motor is stored in a synthetic spring, which mediates in torque transmission, and moves the four-bar linkage in a slow manner. After reaching further angle the advantage raises allowing releasing momentum accumulated in the spring to acquire an impulse of the jump. Thanks to the construction of the robot, the final point of the mechanism moves in a linear manner aligned with the rotation axis of the motor perpendicularly to the ground. With this trait, the process of jumping can be in a straight line upwards without any oscillations in the air. The objective of this paper is to recreate the kinematic model of the construction using numerical methods, analyse its correctness with CAD project, composed for the calculation purpose, and perform prototype testing of its movement in work.

2. The kinematic model

A numerical simulation is required to perform geometric and kinematic analyses including determination of path, velocity and acceleration of mechanism points. The vector closed-loop method is utilised. The considered mechanism is of fourth class according to structural Assur classification [15].

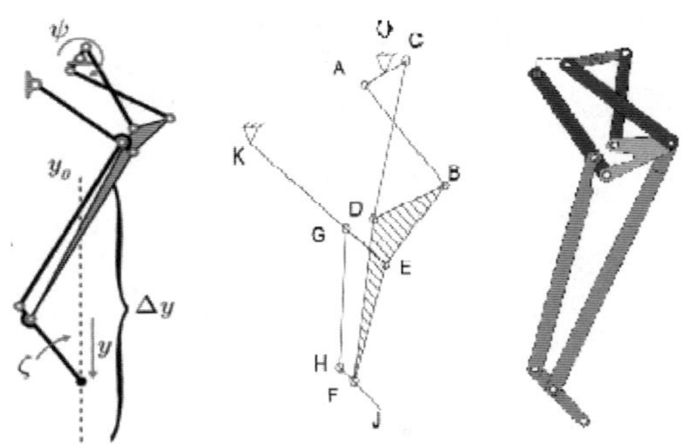

Figure 1. Models used in mathematical analyses. a) from the original Berkeley idea [14], b) the kinematic scheme on which further calculation was based on, c) model created in CAD environment.

The first step is to analyse the structure of the mechanism and compute its mobility [16]. The links GH and HJ are driven (passive group according to Assur classification) and do not change the mobility value, therefore, they are disregarded until the numerical method has been presented. The number of links (n) in the mechanism equals 7 and the number of the kinematic pairs (p) equals 10. Then, the mobility (w) of the mechanism equals to 1.

$$w = 3 * n - 2 * p \tag{1}$$

$$1 = 3 * 7 - 2 * 10 \tag{2}$$

All the kinematic equations have to include fixed revolute pivots K or O, which connect the mechanism to the frame. The first closed loop, which can be derived, is a modified four-link mechanism composed of joints OABCD.

The closed loop vector equation can be rewritten as two algebraic equations by equalling X and Y components [17]. The equations contain four variables (the indices at the angles correspond to the link lengths). Every angle is measured between familiar link and the horizontal line led from the joint connecting it with a previous link [18, 19].

Eq1:
$$\begin{aligned} L_1 \cos(\phi_1(t) + \pi) + L_3 \cos(\phi_3(t)) \\ = L_1 \cos(\phi_1(t)) + L_4 \cos(\phi_4(t)) \\ + L_5 \cos(\phi_5(t)) \end{aligned} \tag{3}$$

Eq2:
$$\begin{aligned} L_1 \sin(\phi_1(t) + \pi) + L_3 \sin(\phi_3(t)) \\ = L_1 \sin(\phi_1(t)) + L_4 \sin(\phi_4(t)) \\ + L_5 \sin(\phi_5(t)) \end{aligned} \tag{4}$$

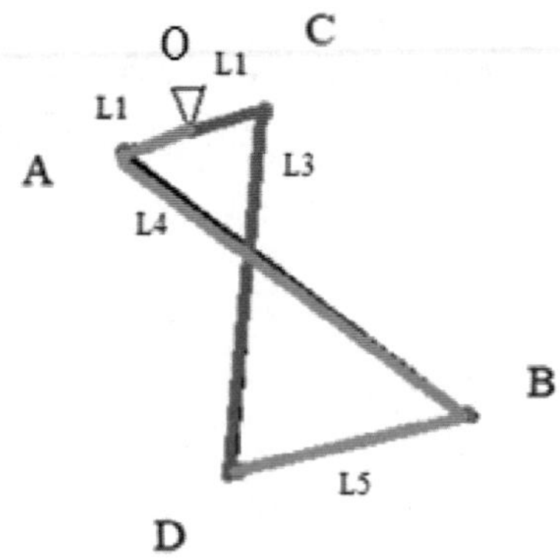

Figure 2. Visual presentation of the vector calculus providing the first set of equations.

The active link and joint K are connected through rigid quadrangle DBEF with defined angles. Rotations of all of these are dependent on a single one according to given geometric relationships.

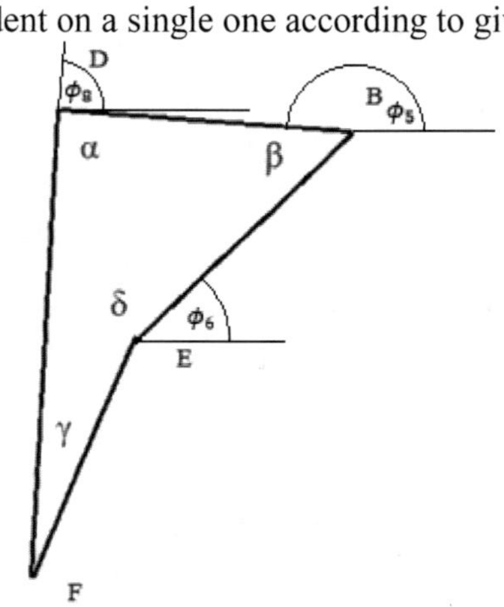

Figure 3. Geometric structure of the rigid quadrangle with its angles assigned.

$$\phi_6 = \phi_5 + \beta - \pi \tag{5}$$

$$\phi_8 = \phi_5 - \alpha \tag{6}$$

$$\phi_7 = \phi_5 - \alpha - \gamma \tag{7}$$

Providing formulas expressed by (Eq. 5-7) a kinematic equation can be given. The second loop consists of chains KEB and OCB. Lengths of links KE and EB are denoted as L_9 and L_6, respectively.

Eq3:
$$L_1 \cos(\phi_1(t)) + L_4 \cos(\phi_4(t)) = L_9 \cos(\phi_9(t)) + L_6 \cos(\phi_5(t) + \beta - \pi) + X \tag{8}$$

Eq4:
$$L_1 \sin(\phi_1(t)) + L_4 \sin(\phi_4(t)) = L_9 \sin(\phi_9(t)) + L_6 \sin(\phi_5(t) + \beta - \pi) + Y \tag{9}$$

The equations 8-9 are sufficient to describe the complete active kinematic chain of the mechanism. To numerically analyse the entire mechanism, in particular, the motion of point J, the equations 3, 4, 8 and 9 with another closed loop must be solved. This one is built on chains KGHF and OADF. Point F is built into the link HJ, therefore, the rotation angle of link HF provides information on the position of point J. Length L_{91} is corresponding to the link KG, L_{10} to GH, $L_{11}$to HF and L_{12} to DF.

Eq5:
$$L_{91} \cos(\phi_9(t)) + L_{10} \cos(\phi_{10}(t)) + L_{11} \cos(\phi_{11}(t)) = L_1 \cos(\phi_1(t) + \pi) + L_3 \cos(\phi_3(t)) + L_{12} \cos(\phi_5(t) + \pi - \alpha) \tag{10}$$

Eq6:
$$L_{91} \sin(\phi_9(t)) + L_{10} \sin(\phi_{10}(t)) + L_{11} \sin(\phi_{11}(t)) = L_1 \sin(\phi_1(t) + \pi) + L_3 \sin(\phi_3(t)) + L_{12} \sin(\phi_5(t) + \pi - \alpha) \tag{11}$$

This set of 6 equations 3, 4, 8, 9, 10, 11 was used to create the numerical simulation.

2.1. The numerical programme

To determine the mechanism angular positions starting from the known, prescribed first position, first and second derivatives of the closed-loop equations must be computed [20].

The matrix equation for angular velocities is then:

$$\mathbf{v} = \mathbf{b} * \mathbf{A}^{-1} \tag{12}$$

where:
v – angular velocities matrix,
b – column vector of prescribed data,
A – matrix of coefficients at unknown angular velocities.

The angular acceleration is computed from (where bp and Ap are the derivatives of b and A):

$$\mathbf{a} = (\mathbf{bp} - \mathbf{Ap} * \mathbf{v}) * \mathbf{A}^{-1} \tag{13}$$

Having given sets of derivatives and the initial position, an iterative method can be applied for every next instance.

$$p(t + \delta t) = p(t) + v * \delta t + \frac{1}{2} a * \delta t^2 \tag{14}$$

The programme using the numerical method for kinematic analysis was implemented in the environment of Wolfram Mathematica. The first step was to input all written equations into the script. Thanks to the character of the platform, which resembles a text editor, it was intuitional and self-explanatory.

The derivatives were computed and all the equations were transformed to the matrix form to change the format of selected equation arrays. For this purpose, the Mathematica function CoefficientArrays [,] was used with the given parameter of v array. The method segregates all variables in a selected array focusing on a given parameter. The output is a set of different objects (A, b and v matrices) that compose into the primary formula in equation 12. Because of the fact that all the next matrices are derivatives of the generated one, these were calculated using the D [,] function.

$$
\begin{aligned}
eq1 &= L1Cos[\phi1[t]+\pi] + L3Cos[\phi3[t]] = \\
&\quad = L1Cos[\phi1[t]] + L4Cos[\phi4[t]] + L5Cos[\phi5[t]]; \\
eq2 &= L1Sin[\phi1[t]+\pi] + L3Sin[\phi3[t]] = \\
&\quad = L1Sin[\phi1[t]] + L4Sin[\phi4[t]] + L5Sin[\phi5[t]]; \\
eq3 &= L1Cos[\phi1[t]] + L4Cos[\phi4[t]] == L9Cos[\phi9[t]] + L6Cos[\phi5[t]+\beta-\pi]; \\
eq4 &= L1Sin[\phi1[t]] + L4Sin[\phi4[t]] == L9Sin[\phi9[t]] + L6Sin[\phi5[t]+\beta-\pi]; \\
eq5 &= L91Cos[\phi9[t]] + L10Cos[\phi10[t]] + L11Cos[\phi11[t]] = \\
&\quad = L1Cos[\phi1[t]+\pi] + L3Cos[\phi3[t]] + L12Cos[\phi5[t]+\pi-\alpha]; \\
eq6 &= L91Sin[\phi9[t]] + L10Sin[\phi10[t]] + L11Sin[\phi11[t]] = \\
&\quad = L1Sin[\phi1[t]+\pi] + L3Sin[\phi3[t]] + L12Sin[\phi5[t]+\pi-\alpha]; \\
\phi &= \{\phi3[t], \phi4[t], \phi5[t], \phi9[t], \phi10[t], \phi11[t]\} \\
v &= \{\phi3'[t], \phi4'[t], \phi5'[t], \phi9'[t], \phi10'[t], \phi11'[t]\} \\
a &= \{\phi3''[t], \phi4''[t], \phi5''[t], \phi9''[t], \phi10''[t], \phi11''[t]\}
\end{aligned}
$$

Code 1. Definition of the equations.

After arranging all 18 equations from the mathematical model into matrix form, the input values are presented as follows:

$$
v: \quad \begin{bmatrix} \phi3'[t] \\ \phi4'[t] \\ \phi5'[t] \\ \phi9'[t] \\ \phi10'[t] \\ \phi11'[t] \end{bmatrix} \tag{15}
$$

$$
A: \quad \begin{bmatrix}
-L3Sin[\phi3[t]] & L4Sin[\phi4[t]] & L5Sin[\phi5[t]] & 0 & 0 & 0 \\
L3Cos[\phi3[t]] & -L4Cos[\phi4[t]] & -L5Cos[\phi5[t]] & 0 & 0 & 0 \\
0 & -L4Sin[\phi4[t]] & -L6Sin[\beta+\phi5[t]] & L9Sin[\phi9[t]] & 0 & 0 \\
0 & L4Cos[\phi4[t]] & L6Cos[\beta+\phi5[t]] & -L9Cos[\phi9[t]] & 0 & 0 \\
L3Sin[\phi3[t]] & 0 & L12Sin[\alpha-\phi5[t]] & -L91Sin[\phi9[t]] & -L10Sin[\phi10[t]] & -L11Sin[\phi11[t]] \\
-L3Cos[\phi3[t]] & 0 & L12Cos[\alpha-\phi5[t]] & L91Cos[\phi9[t]] & L10Cos[\phi10[t]] & L11Cos[\phi11[t]]
\end{bmatrix} \tag{16}
$$

$$
b: \quad \begin{bmatrix} 2L1Sin[\phi1[t]]\phi1'[t] \\ -2L1Cos[\phi1[t]]\phi1'[t] \\ -L1Sin[\phi1[t]]\phi1'[t] \\ L1Cos[\phi1[t]]\phi1'[t] \\ -L1Sin[\phi1[t]]\phi1'[t] \\ L1Cos[\phi1[t]]\phi1'[t] \end{bmatrix} \tag{17}
$$

$$a: \quad \begin{bmatrix} \phi 3''[t] \\ \phi 4''[t] \\ \phi 5''[t] \\ \phi 9''[t] \\ \phi 10''[t] \\ \phi 11''[t] \end{bmatrix} \tag{18}$$

$$Ap: \quad \begin{bmatrix} -L3Cos[\phi 3[t]]\phi 3'[t] & L4Cos[\phi 4[t]]\phi 4'[t] & L5Cos[\phi 5[t]]\phi 5'[t] & 0 & 0 & 0 \\ -L3Sin[\phi 3[t]]\phi 3'[t] & L4Sin[\phi 4[t]]\phi 4'[t] & L5Sin[\phi 5[t]]\phi 5'[t] & 0 & 0 & 0 \\ 0 & -L4Cos[\phi 4[t]]\phi 4'[t] & -L6Cos[\beta + \phi 5[t]]\phi 5'[t] & L9Cos[\phi 9[t]]\phi 9'[t] & 0 & 0 \\ 0 & -L4Sin[\phi 4[t]]\phi 4'[t] & -L6Sin[\beta + \phi 5[t]]\phi 5'[t] & L9Sin[\phi 9[t]]\phi 9'[t] & 0 & 0 \\ L3Cos[\phi 3[t]]\phi 3'[t] & 0 & -L12Cos[\alpha - \phi 5[t]]\phi 5'[t] & -L91Cos[\phi 9[t]]\phi 9'[t] & -L10Cos[\phi 10[t]]\phi 10'[t] & -L11Cos[\phi 11[t]]\phi 11'[\\ L3Sin[\phi 3[t]]\phi 3'[t] & 0 & L12Sin[\alpha - \phi 5[t]]\phi 5'[t] & -L91Sin[\phi 9[t]]\phi 9'[t] & -L10Sin[\phi 10[t]]\phi 10'[t] & -L11Sin[\phi 11[t]]\phi 11'[\end{bmatrix} \tag{19}$$

$$Bp: \quad \begin{bmatrix} 2L1Cos[\phi 1[t]]\phi 1'[t]^2 + 2L1Sin[\phi 1[t]]\phi 1''[t] \\ 2L1Sin[\phi 1[t]]\phi 1'[t]^2 - 2L1Cos[\phi 1[t]]\phi 1''[t] \\ -L1Cos[\phi 1[t]]\phi 1'[t]^2 - L1Sin[\phi 1[t]]\phi 1''[t] \\ -L1Sin[\phi 1[t]]\phi 1'[t]^2 + L1Cos[\phi 1[t]]\phi 1''[t] \\ -L1Cos[\phi 1[t]]\phi 1'[t]^2 - L1Sin[\phi 1[t]]\phi 1''[t] \\ -L1Sin[\phi 1[t]]\phi 1'[t]^2 + L1Cos[\phi 1[t]]\phi 1''[t] \end{bmatrix} \tag{20}$$

All values of length and angular position in time t = 0 were filled with those from the CAD model.

The algorithm for numerical calculation is based on filling given matrices with values, calculated in the previous step to determine the next position.

After all the angular positions were computed for one cycle of the active link motion, the position in XY plane of point J was determined using vector formula:

$$X_j = X_f + L_{91}Cos(\phi_9) + L_{10}Cos(\phi_{10}) + L_{110}Cos(\phi_{11}) \tag{21}$$

$$Y_j = Y_f + L_{91}Sin(\phi_9) + L_{10}Sin(\phi_{10}) + L_{110}Sin(\phi_{11}) \tag{22}$$

2.2. Simulation

The maximum and minimum distance between the drive and J point was 219.96 mm and 171.51 mm. All length values were provided from the CAD model. The simulation was performed with a set of given parameters:

- Number of iterations lp = 100
- Rotary speed of the motor n = 140 rpm
- Angular speed of the rotor w = $\pi/30 * n$
- Time of the single time step dt = $(\pi/2/w)/$lp
- Time of the full simulation tfull = $\pi/2/w$ = 0.107 s
- Starting angular position of the drive alpha = 0°
- Final angular position of the alpha drive F = 90°

The numerical solution gave a smooth, continuous movement without visible unpredictable spikes. The verification was performed by comparison of the final angular position of selected links with the one from the CAD model. The difference between the two readings amounted to less than 0.2°, thus classifying the method and the calculations as correct.

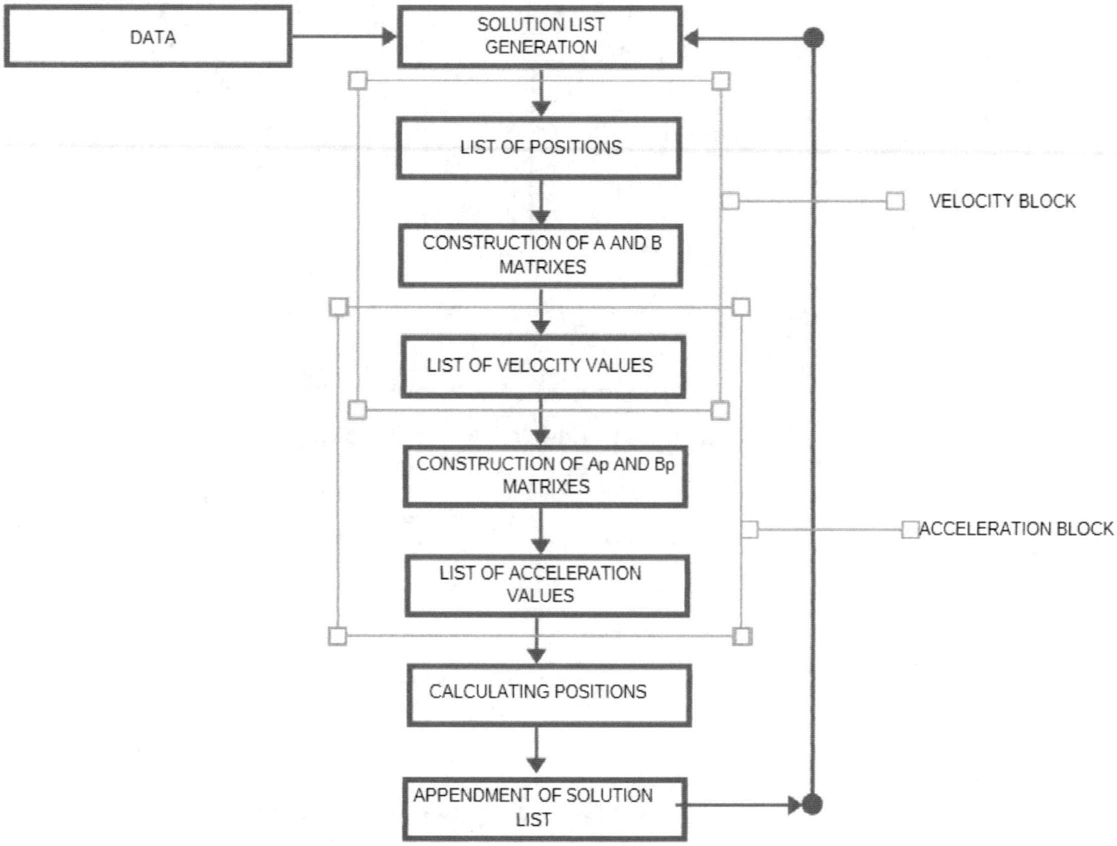

Figure 4. Algorithm of the numerical loop.

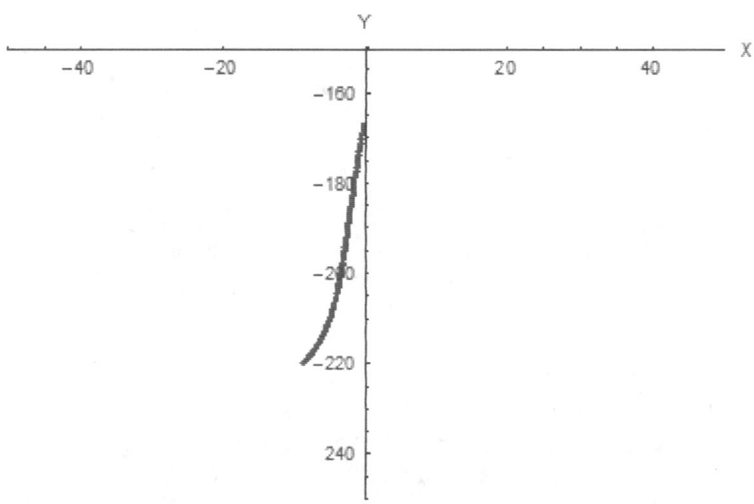

Figure 5. Movement of the final point on XY plane, in millimetres.

After calculating the movement of the final link on the XY-plane and placing it on the model in and adequate scale, it was visible that the movement was not entirely linear. The position of the final point was shifting to the back of the leg in its terminal phase. That should have a negative impact on the quality of the jump, because of the changing the orientation of the robot in the air. That gives an open field for study and testing with a view to further optimisation.

2.3. Experimental testing

To further analyse the given data, test the possibilities of the model and give the basis for further optimisation, the physical prototype was constructed.

Figure 6. Theoretical movement of the final point, in scale.

The links of the robot were created using a 3D printer. All links were connected with 3 mm brass rods locked in place with glued rings and placed in plain bearings, made by Igus company. The rotary force was generated by a torque spring and was transferred via rubber hose to the drive link.

Figure 7. Method of transferring momentum onto drive link.

The spring was primarily stretched to generate greater momentum. The mechanism was released and measured by the BNO055 accelerometer and gyroscope placed on top of the body, to check the performance. The model jumped to a height of approximately 150 mm that is roughly the size of the robot in its crouching position.

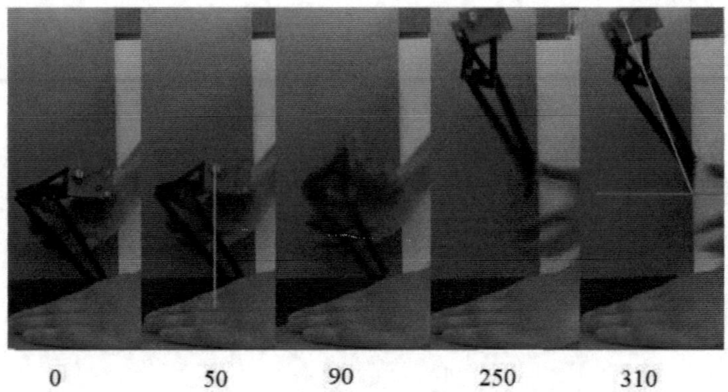

Figure 8. Process of jumping (time steps in milliseconds).

There was a visible alteration of the angular position of the body of the robot, as a consequence of the nonlinear character of the trajectory of the foot. Analysing the process of the jump by a gyroscope, there is a visible change in rotation, reaching the value of 97° per second.

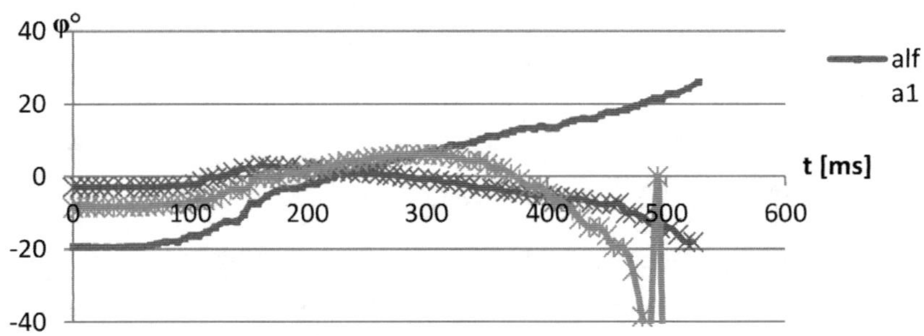

Figure 9. Chart of the angular position in the perpendicular axis. In the final phase, the change is inconsistent due to multiaxis rotation and impact on landing.

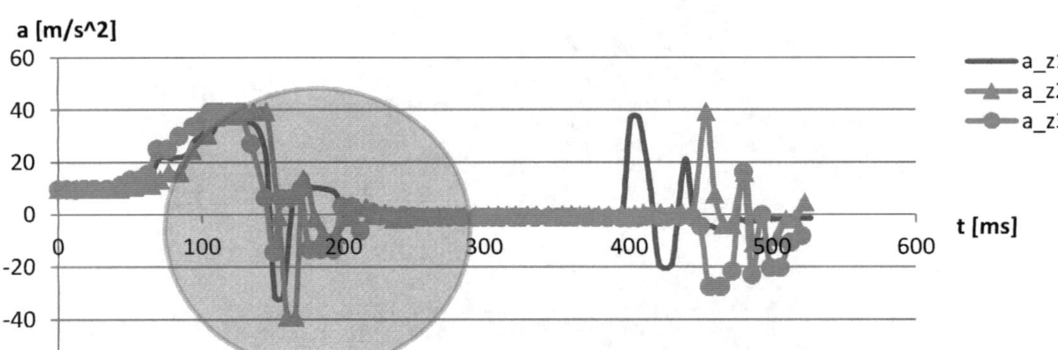

Figure 10. Chart of the acceleration in the Z-axis, shows a rise in acceleration just before the jump, shown in the red circle (positive values towards the ground). First high value is due to the method of stabilisation before the jump.

Data from the Z-axis resulted in a significant gain in acceleration shortly before the moment of the jump. The model is, therefore, correct in its basic principle: storing energy inside the series elastic element (hose) until the rise of efficiency.

3. References

[1] Fankhauser P, Bjelonic M, Bellicoso D, Miki T and Hutter M 2018 Robust Rough-Terrain Locomotion with a Quadrupedal Robot *2018 IEEE International Conference on Robotics and Automation (ICRA)* 1-8
[2] Burdick J and Fiorini P 2003 Minimalist jumping robots for celestial exploration *Int. J. Robot. Res* **22(7)** pp. 653–674
[3] Hardarson F 1997 *Locomotion for difficult terrain*
[4] Sayyad A, Seth B and Seshu P 2007 Single-legged hopping robotics research—A review *Robotica* **25(5)** pp. 587–613
[5] Schlegel M, Kovacˇ M, Zufferey J C and Floreano F 2010 Steerable miniature jumping robot *Auton Robot* **28** 295–306
[6] Johnson A, Libby T, Chang-Siu E, Tomizuka M, Full R and Koditschek D 2012 *Tail assisted dynamic self righting* University of Pennsylvania
[7] Astley H C and Roberts T J 2011 Evidence for a vertebrate catapult: elastic energy storage in the plantaris tendon during frog jumping *Biology Letters* **8(3)** pp. 386–389
[8] Minkyun N, Seung-Won K, Sungmin A, Je-Sung K and Kyu-Jin C 2012 Flea-inspired catapult mechanism for miniature jumping robots *IEEE Transactions on Robotics* **8(5)** pp. 1007–1018
[9] Zhao J 2013 MSU Jumper: A single-motor-actuated miniature steerable jumping robot *IEEE Transactions on Robotics* **29(3)** pp. 602–614
[10] Zhao J, Xi N, Cintron F J, Mutka M W and Xiao L 2012 A single motor actuated miniature steerable jumping robot *IEEE/RSJ Int. Conf. Intell. Robots Syst*. pp. 4274–4275.
[11] Zhao J, Zhao T, Xi N, Cintrón F, Mutka M and Xiao L 2015 *Controlling Aerial Maneuvering of a Miniature Jumping Robot Using Its Tail* arXiv:1502.05347
[12] Jusufi A, Kawano D, Libby T and Full R 2010 Righting and turning in mid-air using appendage inertia: Reptile tails, analytical models and bio-inspired robots *Bioinspiration & biomimetics*. **5**. 045001. 10.1088/1748-3182/5/4/045001
[13] De A and Koditschek D 2016 The Penn Jerboa: *A Platform for Exploring Parallel Composition of Templates Technical Report to Accompany*
[14] Haldane D, Plecnik M, Yim J and Fearing R 2016 A power modulating leg mechanism for monopedal hopping *2016 IEEE/RSJ International Conference on Intelligent Robots and Systems (IROS)* 4757-4764
[15] Parszewski Z 1978 *Teoria maszyn i mechanizmów* (Warszawa: Wydawnictwa Naukowo-Techniczne)
[16] Gronowicz A 2003 *Podstawy analizy układów kinematycznych* (Wrocław: Oficyna Wydawnicza Politechniki Wrocławskiej)
[17] Frączek J and Wojtyra M 2008 *Kinematyka układów wieloczłonowych : metody obliczeniowe* (Warszawa: Wydawnictwa Naukowo-Techniczne)
[18] Krzyżanowski P 2012 *Obliczenia inżynierskie i naukowe : szybkie, skuteczne, efektowne* (Warszawa: Wydawnictwo Naukowe PWN)
[19] Kurnik W 2012 *Wykłady z mechaniki ogólnej* (Warszawa: Oficyna Wydawnicza Politechniki Warszawskiej)
[20] Buśkiewicz J 2008 On the application of the mathematica environment for solving the problems of dynamics of mechanisms of third class *Vibrations in Physical Systems* **23** pp. 71–78

IOP Conf. Series: Materials Science and Engineering **710** (2019) 012031 doi:10.1088/1757-899X/710/1/012031

Finite element simulations of the bending resistance of square profiles with topological modifications

Q Estrada[1], D Szwedowicz[2], J Vergara-Vazquez[2], A Rodriguez-Mendez[3], J Silva-Aceves[1] and L C Wiebe[1]

[1] Universidad Autónoma de Ciudad Juárez (UACJ), Instituto de Ingeniería y Tecnología, Ciudad Juárez, Chihuahua, México
[2] Centro Nacional de Investigación y Desarrollo Tecnológico TecNM/Cenidet, Departamento de Ingeniería Mecánica, Cuernavaca, Morelos, Mexico
[3] University of California, Department of Mechanical Engineering, CA., USA

quirino.estrada@uacj.mx

Abstract. Bending resistance is one of the most important requirements in automobile structural members when lateral crashes occur. For this purpose, the current paper studies the bending and crashworthiness capacity of square profiles with topological modifications. Several finite element simulations were carried out using Abaqus/Explicit software. The topological modifications consisted in combining a square cross-section with polygonal and circular simple cross-sections. In all cases, the profiles evaluated were simulated with mechanical properties for aluminium alloy 6063-T5 with the same mass. The feasibility of our numerical study was validated by a three-point bending tests of a square tube using a Shimadzu universal test machine. Additional to the assessment of complex cross-sections, numerical results for simple cross-sections are presented. A better crashworthiness performance is obtained when the square section of a simple profile is reinforced with topological modifications. According to the simulation results, a complex cross-section can improve the crushing force efficiency up to 77.9% with respect to a single square profile. Regarding cross-sections with topological modifications, the best bending resistance and energy absorption capacity is obtained when the square section is combined with a square profile. This means an increase of 9.2 % of energy absorption (E_a) and 5.67% of CFE respect to the cross-section formed by square and triangular shapes.

1. Introduction

Safety of the passengers is one of the most important requirements when a new vehicle is designed. This priority is based on the annual deaths related to car crashes around the world [1,2]. Mexico is ranked in the ninth place of countries with the highest number of deaths. In this sense, 17,000 Mexican deaths associated with traffic accidents were reported in 2017 [3]. To counter these fatalities, the design of structural members considering the concept of crashworthiness is the fundamental target recognised by automobile manufacturers [4,5]. Crashworthiness refers to the ability of a structure to absorb a large amount of energy by plastic deformation and to minimise the injuries of the passenger during car collisions [6,7]. Thin-walled structures are widely used as crash boxes in frontal collisions or side-impact beams in lateral collisions [8]. Many parameters influence the energy absorption capacity of the thin-walled structures [9,10]. However, the cross-section is among the most important parameters and thus it is considered for topological studies [11,12]. Focusing on lateral crashes, an optimal design of the cross-section can reduce the mass of the components without sacrificing the strength and safety requirements. In this context, Tang et al. [13] carried out a numerical study of thin-walled structures with various cross-sections subjected to lateral impacts. In this study, simple geometries and composite complex profiles were evaluated by dynamic three-point bending tests. Complex profiles involve the combination of single profiles such as a circle, square, hat, ellipse and trapezoid profiles. According to the results, the use of complex cross-sections increases the crush force efficiency (CFE) by up to 0.87 relative to simple shapes. Wang [14] performed a numerical study of the bending resistance of multi-cell square arrangements by a three-point bending test. For this purpose, the square cross-section of simple profiles was reinforced by longitudinal and transversal plates. It was concluded that more partition plates do not directly translate improvement in the energy absorption capacity. However, this is strongly dependent on the relative position of the punch and partition plates. On the other hand, a similar behaviour was obtained by analysing aluminium circular profiles with stiffeners [15]. Special emphasis was set on the effect of damage evolution model on the energy absorption Ea. A better performance was obtained when the circular cross-section was reinforced in the longitudinal and transversal direction by 4 plates. A CFE value of 0.85 was obtained, which means an improvement of the energy absorption capacity of 34% relative to a single circular profile. In [16] a topological investigation of several profiles subjected to bending loads was carried out. In this study, single and multi-cell structures with triangular, square, hexagonal and circular shapes were numerically evaluated by a three-point bending test. According to their results, relative to single profiles, the effectiveness of multi-cell structures to improve the energy absorption performance was observed. After several simulations, the authors concluded that the best bending performance was for circular multi-cell structures with a CFE equal to 0.88. Lastly, many studies have been conducted in order to obtain the best cross-sectional shape. However, for the specific case of square profiles, it has been barely reported.

Thus, in the present article, a topological study of square thin-walled structures under bending load was performed numerically using Abaqus/Explicit FEM software. From a square base cross-section, several new complex cross-sections were proposed. In this sense, polygonal and circular shapes were combined with the square cross-section. The evaluation of the single and complex profiles was carried out by a three-point bending test.

1.1. Theoretical bending collapse for single square profiles

The bending resistance and energy absorption capacity of square profiles are directly influenced by the bending collapsed mode [17,18]. A torsional moment (M) at the end of the profile generates the formation of a central hinge line. This induces an inward fold at the compression flange [19]. As the inward fold gets deeper, new travel hinge lines GA, EA, KA and LA are formed. From these travel lines two outward folds at the adjacent flanges are formed (see figure 1).

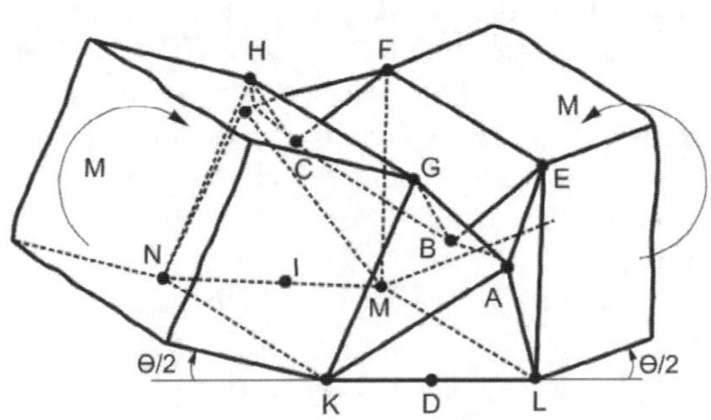

Figure 1. Theoretical bending collapse mode for a single square profile [19].

1.2. Energy absorption parameters

The assessment of simple and complex profiles regardless of manufacturing material can be carried out by dimensional and dimensionless parameters. In this way, the parameters listed in table 1 are obtained/calculated from bending force (F) versus displacement (δ) curves [20].

Table 1. Energy absorption parameters.

Peak load [P_{max}] kN	**Mean Force [P_m] kN**	**Energy absorbed [E_a] kJ**	**Crush force efficiency [CFE]**	**Specific energy absorption [SEA]**
Obtained from curve	$P_m = \frac{E_a}{\delta}$	$E_a = \int_0^{\delta} F.d\delta$	$CFE = \frac{P_m}{P_{max}}$	$SEA = \frac{E_a}{m}$

2. Validation of the three-point bending test model

To validate the feasibility of our numerical techniques, a square thin-walled profile was numerically and experimentally investigated in a three-point bending test (see figure 2). The discrete model was developed in Abaqus/explicit. The square profile has a side width (W) of 38.5 mm, length of 250 mm and thickness of 1.40 mm. The tube was modelled using shell elements (S4R) with elastoplastic properties for aluminium 6063-T5 alloy. An elastic modulus of E= 66,940 MPa, Poisson coefficient (υ) of 0.33, yield stress (σ_y) of 158.78 MPa and density of 2700 kg/m^3 were used. On the other hand, pin supports and a punch were modelled as rigid bodies with R3D4 elements. As boundaries conditions, the punch was unconstrained only in the y-direction to allow the bending process at 6 mm/min. During the contact interaction, a friction coefficient of 0.3 was used. Lastly, from a convergence analysis, a mesh with an average element size of 2 mm was implemented.

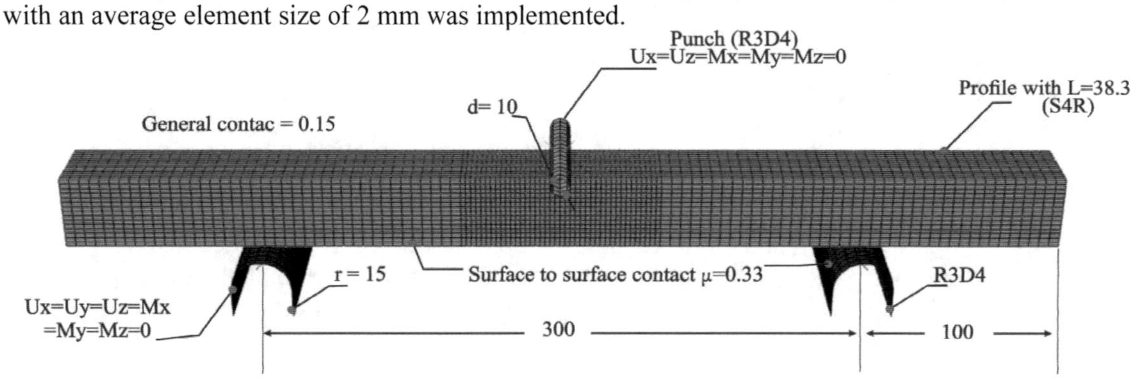

Figure 2. Discrete model of the three-point bending test of a square profile.

The numerical and experimental results are presented in figure 3. As expected, the square profile presented a typical bending force v displacement curve which is characterised by an initial maximum peak load (6 kN) followed by a drop of the bending force (see figure 3a). A good agreement between both models was calculated with differences in the order of 2%. On the other hand, to make a qualitative comparison of the bending collapse mode, it is observed that the discrete model captures correctly the deformation mode (see figure 3b). Thus, the discrete model is validated, and thus we continue with the topological studies of square profiles by numerical simulations.

3. Numerical topological study

In this study, a topological analysis of square profiles under bending load using Abaqus/explicit was conducted. The evaluated profiles were obtained by combining a square with polygonal and circular cross-sections. In order to obtain a comparison pattern, the study of single square profiles is also reported. In all cases, the thin-walled structures were made with aluminium alloy 66063-T5 and were evaluated by a quasi-static three-point bending test. Details of the evaluated profiles are displayed in table 2.

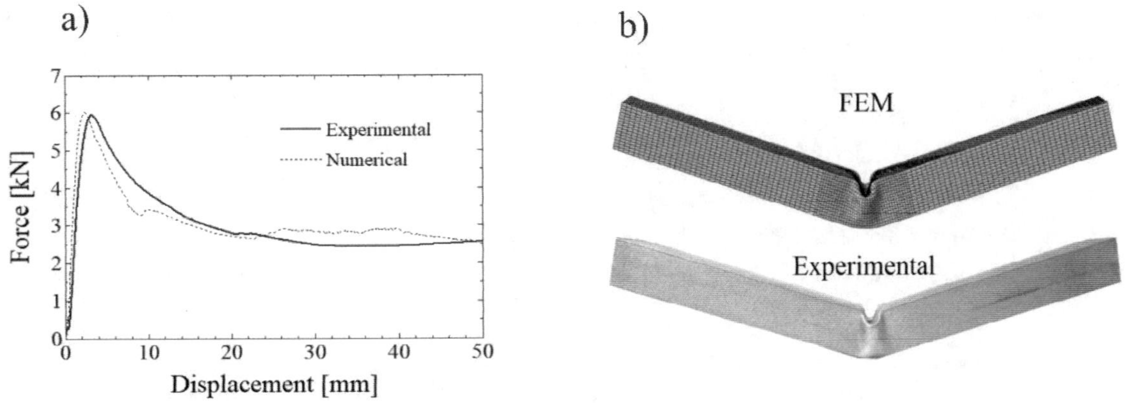

Figure 3. a) Force v displacement response of a square profile, and b) final deformation mode.

Table 2. Evaluated profiles with complex cross-sections, units in mm.

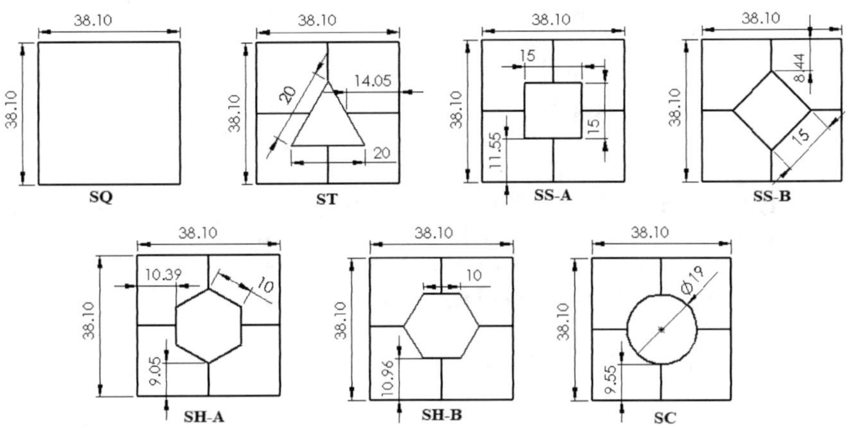

Specimen code	Geometrical base	Thickness [mm]	Length [mm]	Mass [kg]
SQ	Square	1.4	500	0.290
ST	Square, triangle	0.819	500	0.290
SS-A	Square, square	0.826	500	0.290
SS-B	Square, square	0.867	500	0.290
SH-A	Square, hexagonal	0.85	500	0.290
SH-B	Square, hexagonal	0.85	500	0.290
SC	Square, circular	0.853	500	0.290

4. Results and discussion

The bending and energy absorption performance of the evaluated structures was obtained by the study of force versus displacement curves. The curves show qualitative similarities with the curve for a typical square profile, which is characterised by an initial P_{max} at the bending process with a subsequent drop of force. However, the transition of P_{max} to Pm for complex profiles was smoothed. The opposite case is observed for the typical square profile where an abrupt drop of bending force is observed. According to Figure 4, as the number of edges of the internal cross-section increases, a convergence of the mechanical behaviour is observed. In this way, profiles with hexagonal (SH-A and SH-B) and circular (SC) reinforcement cross-section presented a similar bending response. Regarding the Pmax value, in all cases, the complex profiles presented lower Pmax values than a typical profile (SQ). In this sense, Pmax values in a range of 4.2 kN to 4.5 kN were obtained.

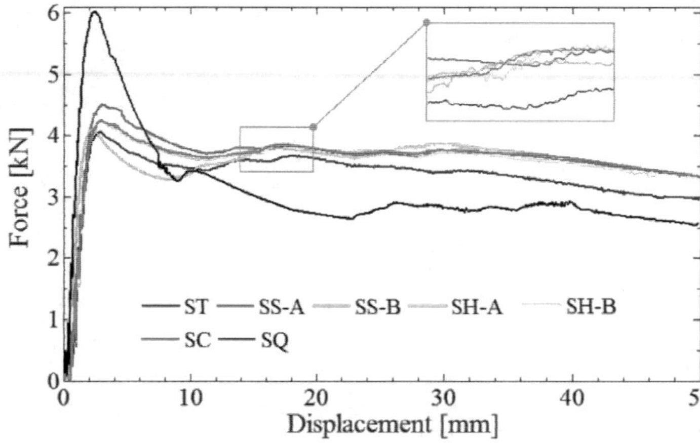

Figure 4. Force vs. displacement curves for the evaluated profiles.

From the area under the load v displacement curve, the energy absorption, including elastic and plastic deformation, is obtained (see figure 5). Three trends were clearly observed. As expected, the poorest performance was achieved for the single profile (SQ). Meanwhile, a reasonable performance was obtained by using a triangular reinforcement cross-section (ST). A third category is formed when the internal cross-section has an edge number equal to four or higher. For this group, similar energy absorption capacities are obtained in the order of 180 J. However, zooming at 50 mm of displacement, the importance of the topological analysis is visible. In this way, a greater number of edges contributed to a better stability of the structure. Thus, higher energy absorption was computed.

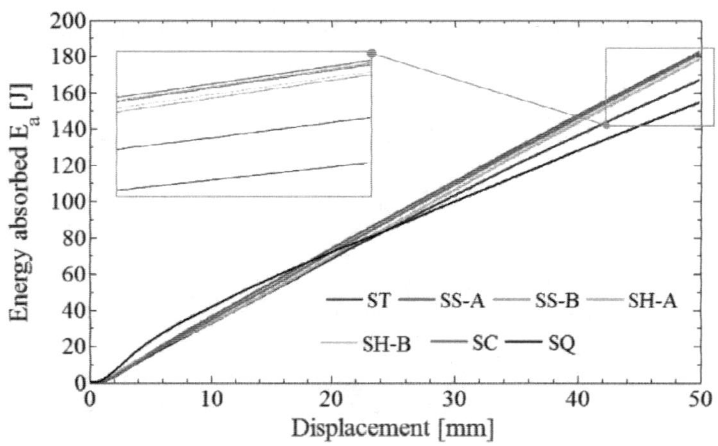

Figure 5. Comparison of energy absorption for the evaluated profiles.

Despite having the same mass, all complex cross-sections increased the stiffness of the square profile. Thus, the theoretical deformation mode, presented in figure 1, was modified. The final bending deformation at 50 mm is presented in Tables 3-4. From the cross-section cut view, as the reinforced internal profile tends to a circular shape, more advantageous plastic deformation is obtained. The opposite case was observed for the ST (triangular base) arrangement.

IOP Conf. Series: Materials Science and Engineering **710** (2019) 012031 doi:10.1088/1757-899X/710/1/012031

Table 3. Final deformation state for the evaluated complex arrangements.

Final deformation at 50 mm	Cross-section beneath punch
	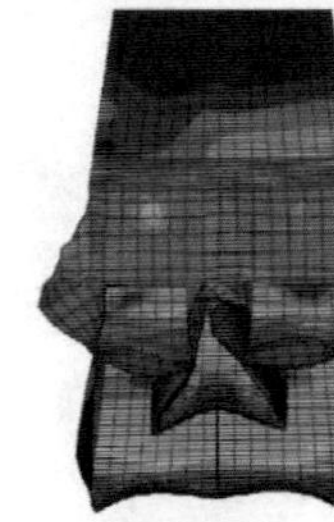
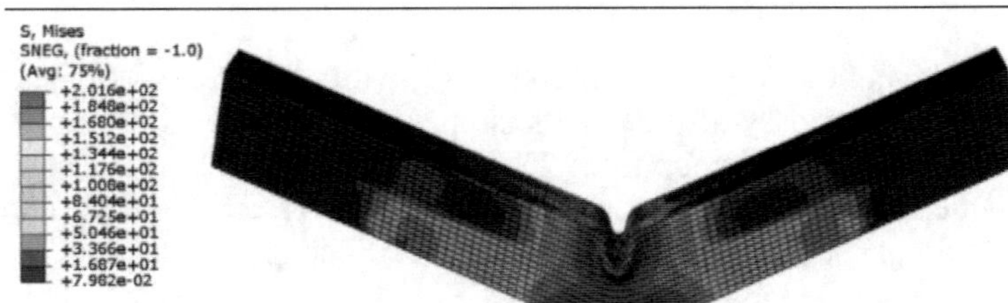	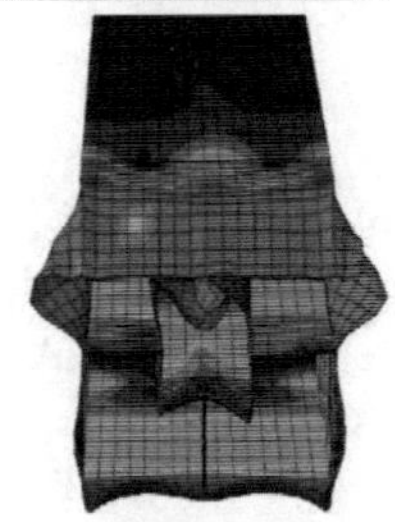
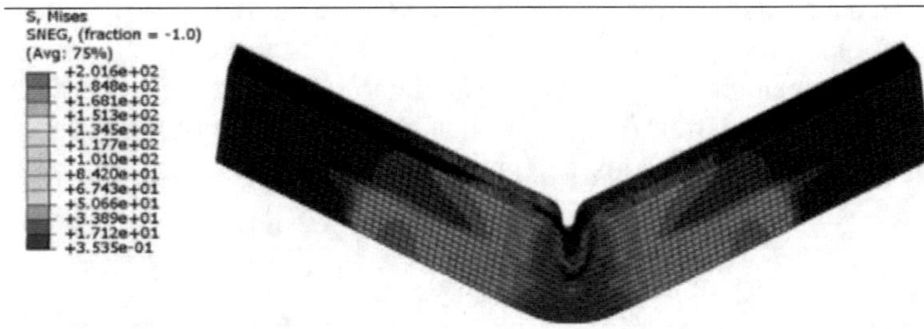	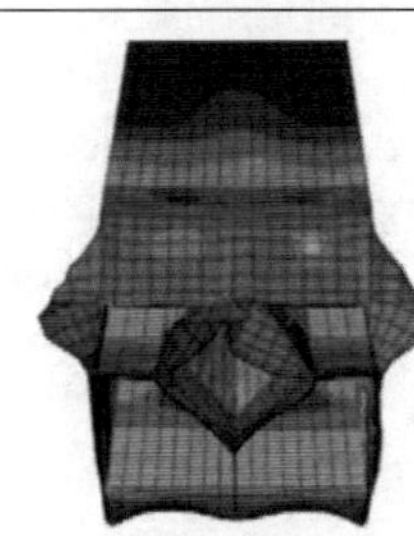

	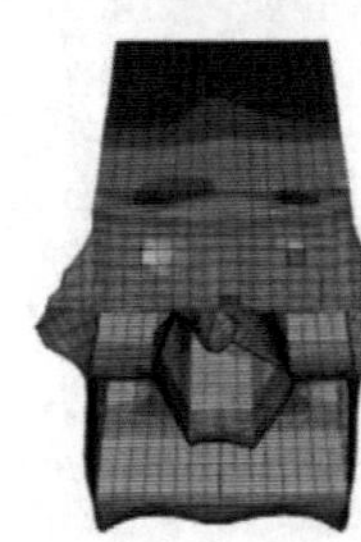

IOP Conf. Series: Materials Science and Engineering **710** (2019) 012031 doi:10.1088/1757-899X/710/1/012031

Table 4. Final deformation state for the SC profile.

Final deformation at 50 mm	**Cross-section beneath punch**
	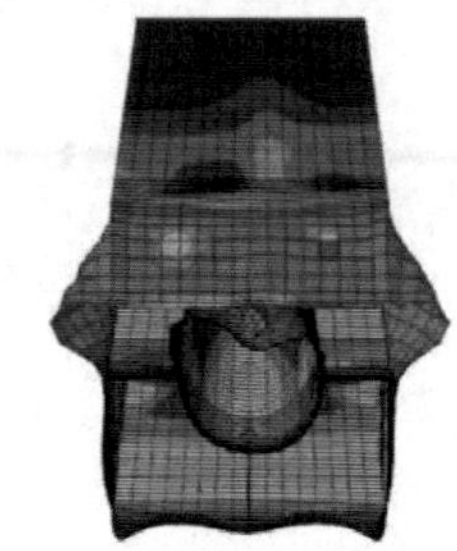

The assessment of the complex and single profile was carried out with the crashworthiness parameters showed in table 1. As a result, a summary of the results is presented in Table 5. Complex profiles allowed a decrease of peak load (P_{max}) in the range from 25.77% to 33.34% with respect to a single profile (SQ). The lowest P_{max} value equal to 4.06 kN was computed by the ST arrangement, and the highest, by the square profile with internal square reinforced, at 45° (SS-B). The effectiveness of the complex profile was demonstrated by calculating the energy absorption (E_a). An improvement of E_a capacity up to 21.34% was achieved by the SS-B profile. The highest values for mean force (P_m) and specific energy absorption (SEA) were 3.65 kN and 0.629 J/g, respectively. Despite obtaining the poorest E_a performance, the profile with triangular reinforcement (ST) presented an increase of 11.01% relative to the SQ profile. The poorest behaviour is attributed to the instability of the triangular shape during the bending process. According to our numerical results, it was determined that the E_a capacity of the structure is primarily dependent on the location of the internal profile followed by the cross-sectional shape. This explains the differences found for arrangements with denomination A and B, which have the same internal cross-section but at different positions.

Table 5. Summary of numerical results.

Code	P_{max} [kN]	P_m [kN]	E_a [J]	SEA [J/g]	CFE
SQ	6.09	3.01	150.42	0.518	0.4900
ST	4.06	3.34	166.99	0.575	0.8233
SS-A	4.10	3.57	178.64	0.616	0.8720
SS-B	4.52	3.65	182.52	0.629	0.8071
SH-A	4.25	3.63	181.75	0.627	0.8548
SH-B	4.23	3.58	179.44	0.618	0.8480
SC	4.25	3.62	181.33	0.625	0.8526

Energy absorption and peak load parameters are fundamental crashworthiness indicators. However, the optimal performance of the structure is obtained when the crush force efficiency (CFE) tends to unity. This means that a higher E_a value does not always represent the best crashworthiness behaviour. The best CFE performance was obtained for the SS-A structure, with 0.8720. In this way, a square internal reinforcement increased the stability of the structure during the bending process. Thus, better stress distribution was achieved. Additionally, the position of the internal square profile allowed a greater plastic deformation. Finally, from our numerical findings, the use of square reinforcement increases the crashworthiness capacity of square profiles. Thus, this kind of structure is highly recommendable to counter the harmful effects of lateral crashes.

IOP Conf. Series: Materials Science and Engineering **710** (2019) 012031 doi:10.1088/1757-899X/710/1/012031

5. Conclusions

A numerical study of crashworthiness and bending performance of square profiles with topological modifications was carried out. According to our study, the following conclusions were found.

1. Using the same mass value (0.29 kg), all proposed complex profiles showed a better energy absorption performance compared to a single square profile. An improvement in the range from 21.34 to 11.01% was computed.
2. The internal reinforcement cross-section plays an important role in the crashworthiness capacity of complex profiles. Polygonal cross-section with four or more edges presents a higher stability of the structure. Thus, the formation of travel and hinge plastic lines required greater external work.
3. Although the internal cross-section affects the crashworthiness performance of the structure, the position of the internal profile becomes more important. To compare complex profiles with a similar internal cross-section i.e SS-A, SS-B, SH-A, SH-B, a better performance was obtained when the load is transferred directly to the horizontal edge. In this way for equal internal cross-sections, differences of CFE between 0.80% and 7.44% were computed.
4. The best crashworthiness performance was obtained by a square profile with an internal square tube (SS-A). In this way, an increase of 21.34% and 77.9% for SEA and CFE were achieved. Therefore, this topology is an effective and low-cost alternative to decrease the injuries and fatalities during lateral car crashes.

6. References

[1] WHO 2019 *Global status report on road safety 2018* Available from: https://www.who.int/violence_injury_prevention/road_safety_status/2018/en/
[2] Nagel G M and Thambiratnam D P 2006 *Int. J. Impact. Eng.* **32(10)** 1595-620
[3] CONAPRA 2018 *Informe sobre la situación de la seguridad vial, México 2017.* Secretaria de Salud. Available from http://mapasin.org/wp-content/uploads/2018/10/Informe_SV_2017.pdf
[4] Tran T, Le D, and Baroutaji A. 2019 *Eng. Struct.* **182** 39-50
[5] Huang J P, and Liu R Q. 2018 IOP *Confer, ser. Mater.Sci.Eng.* **382(4)** 042012
[6] Lukaszewicz, D. H.-J. A 2013 Automotive Composite Structures for Crashworthiness. *In Advanced Composite Materials for Automotive Applications* ed. John Wiley & Sons Ltd (Chichester, UK) p 99–127
[7] Lozzi, A. 1981 *Int. J. Veh. Design.* **2(4)** 470-79
[8] Baroutaji A, Sajjia M, and Olabi A G. 2017 *Thin-Walled Struct.* **118** 137-63
[9] Huang Z, and Zhang X. 2019 *Thin-Walled Struct* **137** 231-50
[10] Huang Z, and Zhang, X. 2018 *Int. J. Mech. Sci.* **144** 461-79
[11] Li Z and Lu F 2015 *J Reinf Plast Comp.* **34(9)** 761–68
[12] Yin H, Xiao Y, Wen G, Qing Q, and Deng Y. 2015 *Thin-Walled Struct.* 94, 1-12
[13] Tang T, Zhang W, Yin H and Wang H 2016 *Thin-Walled Struct.* **102** 43–57
[14] Wang Z, Li Z and Zhang X 2016 *Thin-Walled* **107** 287–99
[15] Estrada Q, Szwedowicz D, Gutierrez-Wing E, Silva-Aceves J, Rodriguez-Mendez A, Elias-Espinosa M and Bedolla-Hernandez J 2019 *P. I. Mech. Eng. D-J Aut.* **233(8)** 2120–38
[16] Estrada Q, Szwedowicz D, Rodriguez-Mendez A, Silva-Aceves J, Wiebe L C, Castro J S, Castillo S A 2019 *MATEC Web of Conferences* **252** 07005
[17] Kecman D. 1983 *Int. J. Mech. Sci,* **25(9-10)** 623-36
[18] Elchalakani M, Zhao X L, and Grzebieta R H. 2002 *Int. J. Mech. Sci,* **44(6)** 1117-43
[19] Santosa S, Banhart J and Wierzbicki T 2001 *Acta Mech.* **148(1–4)** 199–213
[20] Sun G, Tian X, Fang J, Xu F, Li G, and Huang X. 2015 *Int. J. Impact. Eng.* **78** 128-37.

IOP Conf. Series: Materials Science and Engineering **710** (2019) 012032 doi:10.1088/1757-899X/710/1/012032

Modelling of oil tribotechnical data

J Kuben[1], P Račková[1], O Šimon[2] and M Zajac[3]

[1] University of Defence in Brno, Department of Mathematics and Physics, Kounicova 65, 662 10 Brno, Czech Republic
[2] University of Defence in Brno, Department of Communication Technologies, Electronic Warfare and Radiolocation, Kounicova 65, 662 10 Brno, Czech Republic
[3] University of Defence in Brno, Department of Informatics, Cybernetic Security and Robotics, Kounicova 65, 662 10 Brno, Czech Republic

jaromir.kuben@unob.cz

Abstract. The paper deals with statistical processing of field data gained from tribotechnical diagnostics. The data originate from an engine oil analysis. The oil was taken at various locations from several hundreds of medium lorries T810 used in the Czech army. The goal was to propose regression models describing dependence of these quantities on kilometres travelled. Such models can be used, among others, to estimate first hitting time for individual quantities indicating the necessity of the oil change. Two types of models have been used – linear regression models and nonlinear regression models – which seem to be the most appropriate choice for the description of processed data.

1. Introduction

Tribotechnical diagnostics provides us with non-destructive methods of finding out information about the state of various mechanisms. In this case, medium lorries T810 used in the Czech army were an object of interest. A lot of precious items of information can be gained from oil of these vehicles as it is assumed that the condition of a vehicle tightly corresponds to composition of its engine oil. A large amount of data originating from an engine oil analysis, involving over 30 components of various importance was at this research's disposal.

From the point of view of further processing, oil components can be split into three groups containing observed quantities of interest, see [1]:

- Physical and chemical properties of oil, e.g. kinematic viscosity, flash point, total base number (TBN), anti-oxidation and anti-wear particles (AOWP),
- Products of wearing, e.g. Fe, Cu, Pb,
- Outer contaminants, e.g. soot and Si.

In the analysis of oil, various methods were used: Atom Emission Spectrometry Fourier (AES) – the content of iron (Fe), lead (Pb) and other elements in [ppm] and Fourier Transformation Infra-Red (FTIR) – content of soot, anti-oxidant, and anti-wear additives (AOWP) in [%] and normed methods for stating physical and chemical parameters – kinematic viscosity, flash point, total base number etc.

Our goal was to propose appropriate regression models which could be used to determine first hitting time, that is instant when critical values of observed quantities are reached. The acquired outcomes will serve practical approaches such as i) operation optimisation, ii) maintenance planning and iii) rationalisation of life cycle costs.

Similar topics were studied by numerous authors using various mathematical methods for the investigation of oil tribotechnical data originating from various vehicles. In [2–4], the goal was to keep both preventive and corrective maintenance costs as low as possible; while in [5], the importance of preventive maintenance is demonstrated based on the comparison of data from vehicles with and without such a maintenance. In [6], the influence of soot in oil is studied for armoured personnel carriers (APC), heavy off-road trucked vehicles T-72/97 and medium lorry trucks T810. In [7], aggregation of field data and their assessment is discussed. Analytical methods can be applied to study particles from wear process and oil deterioration, see [8, 9]. In [10], mathematical methods like stochastic processes, diffusion models and Wiener process are applied. In [11, 12], regression models are used for Fe, Pb, AOWP, and soot. Results in [6–12, 19] and further papers referenced therein, indicate that oil change periods could be extended, and considerable expenses saved.

2. Linear regression models

Standard methods of regression analysis were used. The critical step was to propose an appropriate regression function which would fit data well. Extent of samples was relatively big, almost 300 observations. The explanatory variable was each time the distance travelled by the vehicle in [km]. Typically, the dependant variable is changing only mildly, most often it decreases or increases slowly. But there are exceptions. For example, kinematic viscosity can decrease when fuel gets into oil and can increase as a consequence of a primary or secondary pollution. Similarly, soot content decreases (soot dissolves) when the vehicle does not operate while it increases when the vehicle operates. With respect to the above mentioned facts, linear regression models $Y_i = \eta(x) + \varepsilon_i$, $\eta(x) = \beta_0 + \beta_1\varphi(x)$ were used, where error terms ε_i are assumed to be independent and normally distributed with $\varepsilon_i \sim$ Norm $(0, \sigma^2)$. Choices x, $\sqrt{x}$, $1/x$, arctan x and other functions for $\varphi(x)$ were tested. Using least squares method the following well-known formulae for point estimates b_0, b_1 of β_0, β_1 can be derived (n is the number of observations, $\overline{x}$ and $\overline{y}$ are sample means):

$$b_2 = \frac{\sum_{i=1}^{n}(x_i - \overline{x})y_i}{\sum_{i=1}^{n}(x_i - \overline{x})^2}, \qquad b_1 = \overline{y} - b_2\overline{x}, \quad \text{where } \overline{x} = \sum_{i=1}^{n} x_i, \qquad \overline{y} = \sum_{i=1}^{n} y_i. \tag{1}$$

Moreover, we also want to find confidence and prediction intervals for the response. A confidence interval for the mean response is

$$\left(\eta(x) - t^*\sigma\sqrt{\frac{1}{n} + \frac{(x-\overline{x})^2}{\sum_{i=1}^{n}(x_i - \overline{x})^2}},\ \eta(x) + t^*\sigma\sqrt{\frac{1}{n} + \frac{(x-\overline{x})^2}{\sum_{i=1}^{n}(x_i - \overline{x})^2}} \right). \tag{2}$$

where $t^* = t_{n-2}\left(1 - \frac{\alpha}{2}\right)$ is a quantile of Student distribution, α is a level of significance and an unknown parameter σ is estimated with the sample standard deviation

$$s = \sqrt{\frac{\sum_{i=1}^{n}(\eta(x_i) - y_i)^2}{n-2}}.$$

Replacing in (2) the quantile $t_{n-2}\left(1 - \frac{\alpha}{2}\right)$ with $\sqrt{F_{2,n-2}(1-\alpha)}$, where $F_{2,n-2}(1-\alpha)$ is a quantile of Fisher distribution we get a confidence band for the whole regression function.

Finally, a confidence interval for an individual response, which is usually called prediction interval, is

$$\left(\eta(x) - t^*\sigma\sqrt{1 + \frac{1}{n} + \frac{(x-\overline{x})^2}{\sum_{i=1}^{n}(x_i - \overline{x})^2}},\ \eta(x) + t^*\sigma\sqrt{1 + \frac{1}{n} + \frac{(x-\overline{x})^2}{\sum_{i=1}^{n}(x_i - \overline{x})^2}} \right). \tag{3}$$

For the derivation of the above formulae and the explanation of concepts confidence and prediction intervals and bands see, e.g., [13-16].

3. Evaluation of regression functions

As an example, we present the evaluation of regression functions for three oil components which are generally considered to be significant for the assessment of vehicles condition. The formulae were obtained using Computer Algebra System Maple 2018 and the results are listed in table 1.

Table 1. Linear regression models.

Component	Regression function
Iron (Fe)	$10.55 + 0.1599\sqrt{x}$
Anti-oxidant and anti-wear particles (AOWP)	$99.83 - 0.00054x$
Total base number (TBN)	$10.79 - 0.0101\sqrt{x}$

The option of $\varphi(x)$ was chosen to get smaller residual sum. In all cases a null hypothesis H_0: $\beta_1 = 0$ was tested and rejected even for very small level of significance α. Likewise, normality of error terms was tested using χ^2-test, Shapiro-Wilk's test, and d'Agostino et al. tests based on skewness and kurtosis coefficients but the results were negative. Nevertheless, as the number of observations is quite high this is not a serious problem due to the central limit theorem.

In figures 1, 2 and 3 graphs of residual functions are displayed including confidence bands for mean response, for regression function and prediction bands for individual response for $\alpha = 0.05$.

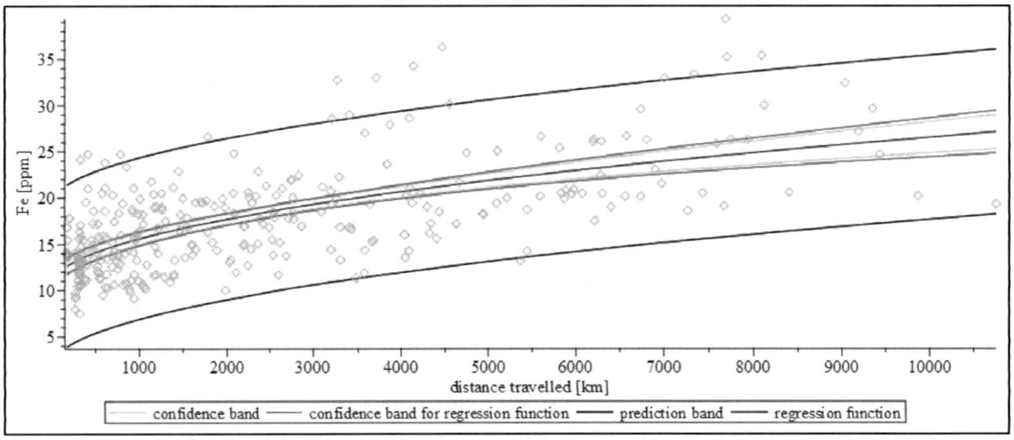

Figure 1. Amount of iron depending on distance travelled.

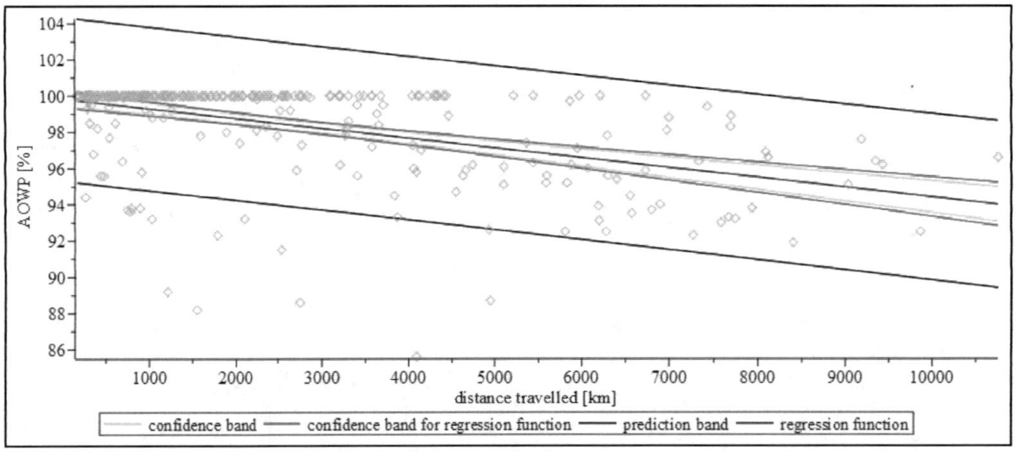

Figure 2. Amount of anti-oxidation and anti-wear particles depending on distance travelled.

4. Nonlinear regression models

Data for some oil components are considerably inhomogeneous and it is difficult to find a linear regression model that fits them well. Since the extrapolation to the right outside the range of observed values of an independent variable is required to estimate first hitting time (instants when critical values of observed variables are reached) we have also tested simple nonlinear models. We considered smooth piecewise linear or quadratic functions or their combinations with one break point. For such a type of nonlinear regression models see e.g. [17].

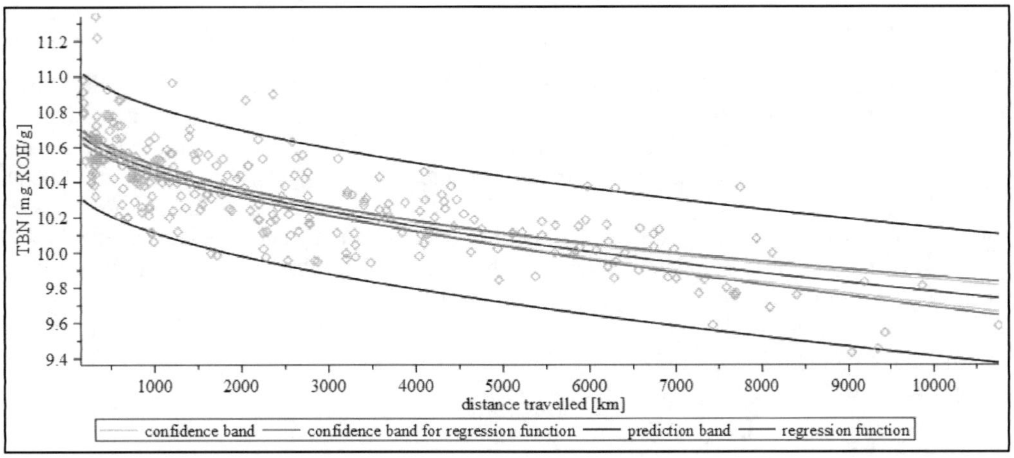

Figure 3. Value of total base number depending on distance travelled.

We concluded that the most satisfying piecewise polynomial functions of the considered types are the ones which are quadratic to the left of the break point and linear to the right. They can be described with a formula

$$\eta(x) = \begin{cases} a_1(x-x_0)^2 + a_2(x-x_0) + y_0 & \text{for } x \le x_0, \\ a_2(x-x_0) + y_0 & \text{for } x > x_0, \end{cases} \tag{4}$$

where x_0 is a break point. The parameters a_1, a_2, x_0, y_0, coefficients of the regression function, are to be determined using nonlinear least squares method.

As an example, let us present the results for total base number (TBN) data. Appropriate initial values for searched parameters must be set because the algorithm for numerical solution of nonlinear least squares method only guarantees to find local minima. We obtain:

$$a_1 = 3.161 \cdot 10^{-7}, \qquad a_2 = -8.175 \cdot 10^{-5}, \qquad x_0 = 1030, \qquad y_0 = 10.42.$$

The corresponding graph is displayed in figure 4. Residual sum of squares 8.991 is smaller than the one given by the linear model above, which is 9.387 [km^2].

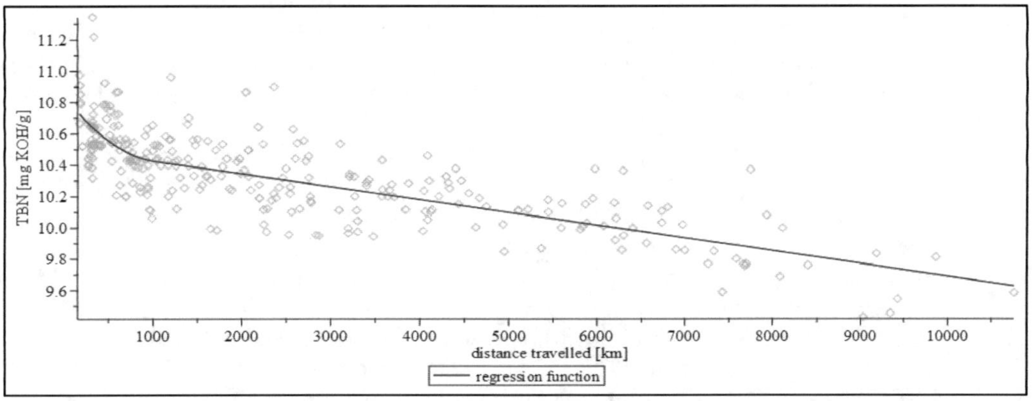

Figure 4. Value of total base number depending on distance travelled.

IOP Conf. Series: Materials Science and Engineering **710** (2019) 012032 doi:10.1088/1757-899X/710/1/012032

5. Conclusions

We will estimate first hitting times for the three considered oil components. Critical values for T810 requiring oil exchange are listed in table 2.

Table 2. Critical values.

Component	Critical value
Iron (Fe)	50 [ppm] (upper bound)
Anti-oxidant and anti-wear particles (AOWP)	20 [%] (lower bound)
Total base number (TBN)	6 [mg KOH/g] (lower bound)

Using regression functions from table 1 we evaluated that the estimates for first hitting time are 60 881, 147 649, 223 617 [km], respectively. Nonlinear regression function (piecewise polynomial function) described by (4) gives the value 55 133 [km]. These values are much bigger than those recommended for vehicles T810, which are prescribed (depending on the type of oil) to about 720 days or 200 Mh corresponding to 6 000 km.

Similar results were obtained for other oil components as well. This indicates that recommended oil change periods could be significantly extended, which would lead to substantial savings of oil expenses. This fact is very important from the economical point of view.

Recently we obtained new large data for this lorry involving even more detailed pieces of information. We plan to continue with this study and use further statistical methods as for example quantile regression, see [18].

Acknowledgement

This work was supported by the project of Specific University Research *Preparation, assessment and mathematical processing of tribotechnical diagnostics field data* SV17-215 financed by MŠMT of the Czech Republic.

6. References

[1] Stodola, J and Novotný P 2018 *Tribodiagnostika BSV* (Brno: Univerzita obrany) ISBN 978-80-7231-984-8 (in Czech)

[2] Glos J, Vališ D and Žák L 2014 Possibilities for optimizing maintenance of military vehicles on the results of tribological diagnostics In: *International Scientific – technical Conference trans&MOTOAUTO"14. Bulgaria*, Varna: Scientific-technical union of mechanical engineering pp. 17–19 ISSN 1310-3946

[3] Glos J, Žák L and Vališ D 2012 Possibilities of Tribo-diagnostics Regarding Maintenance Optimisation. In: *Intelligent technologies in logistics and mechatronics systems ITELMS'2012*, Kaunas: Kaunas University of Technology pp. 73–78 ISSN 2345-0088

[4] Glos J, Žák L and Vališ D 2013 Possibilities of mathematical modelling of tribo-diagnostics data In: *Transport Means Kaunas* (Lithuanian Republic: Kaunas Univerzity of Technology) pp. 216–219

[5] Stodola J and Stodola P 2009 Tribology´s Contribution to Efficient Maintenance of Military Engines In: *ICMT'09. Oprox*, Brno. 10 p. ISBN 978-80-7231-649-6

[6] Vališ D, Hasilová K and Žák L 2017 Combustion engine condition estimation based on indirect diagnostics data In: *Risk, Reliability and Safety: Innovating Theory and Practice* (London: Taylor & Francis Group) pp. 1930–1936 ISBN 978-1-138-02997-2 doi: 10.1201/9781315374987-293.

[7] Vališ D, Žák L, Glos J and Walek A 2012 Possible inputs to maintenance optimisation based on non-direct diagnostics In: *11th International Probabilistic Safety Assessment and Management Conference and the Annual European Safety and Reliability Conference 2012 (PSAM11 ESREL 2012)* (New York: Curran Associates, Inc) pp. 143–153

[8] Vališ D, Žák L, Walek A and Glos J 2013 On approaches of assessment of tribo data from heavy tracked vehicle *Applied Mechanics and Materials* **436** pp. 303–310

[9] Vališ D, Žák L and Chaloupka J 2013 Prediction of further Operation based on Vehicle Tribo Data. In: *Proceedings of the 2013 IEEE IEEM* (Bangkok: IEEE) ISBN 978-1-4799-0986-5

[10] Vališ D and Žák L 2017 Contribution to prediction of soft and hard failure occurrence in combustion engine using oil tribo data *Engineering Failure Analysis* **82(4)** 583–598

[11] Žák L and Vališ D 2015 Oil Additives Used as Indicator and Input for Preventive Maintenance Optimisation In: *International Conference on Military Technologies 2015 (ICMT)* (Brno: IEEE) pp. 1–6, ISBN 978-80-7231-976-3. doi: 10.1109/MILTECHS.2015.7153659

[12] Žák L and Vališ D 2016 Finding the optimal time for preventive maintenance using linear regression. In: *MENDEL* 2016, *22nd International Conference on Soft Computing* (Brno: University of Technolog) pp. 235–240

[13] Anděl J 2007 *Statistické metody 4th edition* (Praha: MATFYZPRESS) (in Czech)

[14] Faraway J J 2014 *Linear Models with R. 2nd edition* (New York: Boca Raton, CRC Press)

[15] Pruim R 2018 *Foundations and Applications of Statistics 2nd edition* (Providence, Rhode Island: American Mathematical Society)

[16] Zvára K 2008 *Regrese* Praha: MATFYZPRESS. 254 p. ISBN 978-80-7378-041-8. (in Czech)

[17] Seber G A F and Wild C J 2013 *Nonlinear regression. Hoboken* (New Jersey: John Wiley & Sons)

[18] Zhou K Q and Portnoy S L 1996 Direct use of regression quantiles to construct confidence sets in linear models *The Annals of Statistics* **24(1)** pp. 287–306

[19] Woch M, Kurdelski M and Matyjewski M 2015 Reliability at the checkpoints of an aircraft supporting structure *Eksploatacja i niezawodnosc-maintenance and reliability* **17(3)** pp. 457-462

IOP Conf. Series: Materials Science and Engineering **710** (2019) 012033 doi:10.1088/1757-899X/710/1/012033

Reduction of excessive heavy metals accumulation in drinking water with natural zeolites

J Kujawska and H Wasag

Lublin University of Technology, Faculty of Environmental Engineering, 38 Nadbystrzycka Str., 20-618 Lublin, Poland

j.kujawska@pollub.pl

Abstract. This paper reports on the study of heavy metal removal from groundwater intended for human consumption using natural aluminosilicates of the zeolite group. The natural zeolite material has been extracted from the Sokirnica deposit located in the Zakarpattia Region in Ukraine. The central focus of this study was the course of the heavy metal adsorption process itself, in which natural zeolites were applied for the removal of Cr, Cd, Ni and Pb ions from aqueous solutions in the presence of large amounts of Ca and Mg ions, competing for ion exchange sites. The capacity of zeolite as an adsorbent was assessed with the use of adsorption isotherms prepared for the heavy metals under tests, whereas the mathematical description of the process was provided by Freundlich and Langmuir isotherm equations. The results from the analysis confirm the high efficiency, and thus the applicability, of zeolite deposits in water filtration. The effective Cr, Cd, Ni and Pb removal rate from groundwater intended for human consumption was confirmed even when the content of these metals in water was exceeded by a factor of 10, compared to the permissible limits provided in legal regulations.

1. Introduction

There are over 800 naturally occurring silicate minerals, and the silicate and aluminosilicate mineral content in the rock structure amounts to approx. 75% by weight. Zeolites constitute a large group of aluminosilicate minerals of different chemical composition, properties and crystalline forms. They are typically found as hydrated aluminosilicates of Na and Ca, and less frequency of Ba, Sr, K, Mg and Mn. Derived from Greek etymology, the word "zeolite" comprises two parts "zeo" = cook and "lithos" = stone, which translates to "a boiling stone" [1]. The term was coined in 1756 by a Swedish chemist and mineralogist, Axel Fredrik Cronstedt, following an observation that during heating, the mineral was losing substantial amounts of water. Already in the nineteenth century, the minerals were known to be aluminosilicates, however, it was not until the following century that more accurate analyses and categorisation were carried out. It was then that zeolites were found, by Smith and Beck, to be crystalline, microporous aluminosilicates containing exchangeable cations characterised by their water adsorption/desorption properties. Liebau described these materials as tectonic-silicate minerals built of chambers or channels, which enable them to diffuse the previously adsorbed molecules [2].

Zeolites are crystalline, porous and hydrated metal aluminosilicates of group I and II of the periodic table (Na, K, Ba, Sr). Their simplified structure is given by the formula:

$$M_{x/n}\left[Al_xSi_yO_{(x+y)}\right]pH_2O \qquad (1)$$

where M stands for (Na, K, Li) and/or (Ca, Mg, Ba, Sr), n is cation charge; y/x = 1–6, p/x = 1–4. Other formulas also apply, e.g. using patterns of the oxide:

$$M_{2/nO} \cdot Al_2O_3 \times SiO_2 yH_2O \quad (2)$$

The zeolite skeleton is built by two basic structures, the SiO_4 and AlO_4 tetrahedrons, which are linked by shared oxygen atoms. The Al:Si ratio varies from 1:5 to 1:1. The replacement of the silicon-oxygen group with an aluminium-oxygen group results in the emergence of an excess negative charge in the network, which attracts positive cations Na^+, K^+, Li^+, Ca^{2+}, etc – as they can move freely. In its aluminosilicate skeletal framework, zeolites contain free spaces storing high-mobility water ions and molecules, which enable cation exchange and reversible dehydration [3].

The tetrahedron, consisting of four oxygen atoms arranged around the central atom of aluminium or any other element with a valence less than 4, is negatively charged and compensated by the off-network cations (figure 1).

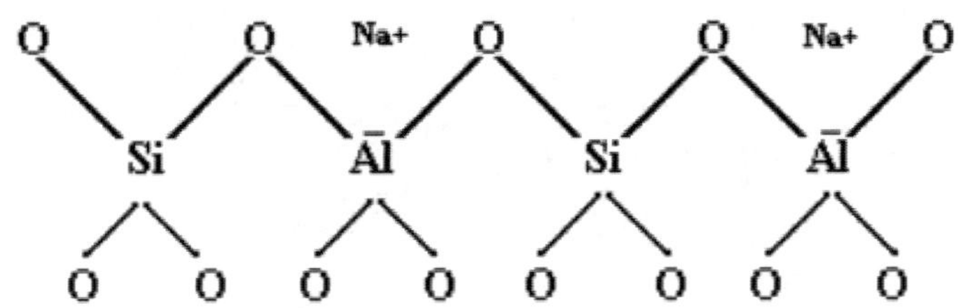

Figure 1. Schematic structure of the natural zeolite [2].

Currently, there are over 40 known types of natural zeolites, however, only seven of which are used commercially (clinoptilolite, chabazite, mordenite, erionite, ferrierite, analcime, phillipsite). Among the natural zeolites, it is clinoptilolite that is the most widely and frequently studied mineral. This results not only from its common occurrence in nature and the lowest price but also from its unique physicochemical properties [4]. The specific physicochemical properties of zeolites include: high ion exchange selectivity, reversible hydration and dehydration, high gas adsorption capacity, high thermal stability, resistance to aggressive media.

Their particular properties determine that zeolites provide a suitable solution in a wide range of environmental, scientific and industrial problems. Since their discovery in 1756, zeolites have been a considerably popular object of scientific studies. Axel Fredrik Cronstedt has found them to be highly effective adsorbents, ion exchangers and molecular sieves. In particular, this is their last application (molecular sieves) that is the most widespread in engineering. Zeolites are used to separate straight-chain hydrocarbons from branched ones [2], as chemical sensors [5] in the control of industrial processes, monitoring indoor air quality and environmental quality, wastewater control and in automatic flue gas treatment. The application of natural zeolites for water and wastewater treatment has been continuing and studied for years. With respect to their industrial application, zeolites were first implemented as water softeners in high-pressure boiler plants; on the other hand, their water softening properties in groundwater treatment have been studied to date [6]. However, in recent decades, natural zeolites have been under scrutiny considering their capacity for the removal of heavy metals from wastewater [7, 8, 9]. Numerous researches have been undertaken to improve the process by chemical or thermal activation of natural zeolites [10, 11]. Considering water treatment, the issues that are typically approached by researchers are the removal of ammonia and humic compounds [12], seawater desalination [13] or Mn and Fe filtering [12, 14]. In addition to the Mn removal, zeolites' purification efficiency has been confirmed in other heavy metals, such as Pb and Zn [15, 16, 17, 18].

Despite the technological progress, which has led to the partial replacement of natural zeolites with synthetic ion exchangers in the second half of the 20th century, the minerals of the zeolite group remain irreplaceable in a wide range of processes, *e.g.* removal of ammonia from water. The high-quality, easily available and cost-effective natural zeolite is a leading solution currently in use in water and wastewater treatment technologies, heavy metal filtration from wastewater and Fe, Mn and ammonia removal from water. In response to the progressive contamination of groundwater with heavy metals, this study attempts to investigate the use of natural zeolite as a fully ecological material for the removal of excessive heavy metal accumulation in drinking water.

The central aim of the presented investigation is the assessment of zeolite potential as a natural ion-exchange adsorbent to remove heavy metal excess (Cr, Cd, Ni and Pb) from groundwater intended for human consumption.

2. Materials and methods

The natural zeolite sample has been purchased from Andalusia Group (Warsaw, Poland) and originally extracted from the Sokirnica deposit located in the Zakarpattia Region in Ukraine. The main component of the zeolite material is clinoptilolite (70-75%), and its detailed composition is presented in Table 1.

Table 1. The chemical composition of the natural zeolite used in the study.

Chemical composition of natural zeolite	
Oxides	wt. % zeolite
SiO_2	67.7
Al_2O_3	12.6
Fe_2O_3	2.0
Na_2O	0.35
K_2O	2.8
CaO	2.6
MgO	1.0
SO_4^{2-}	0.43

The material subjected to testing was the commercial zeolite powder (particle size 1.5-2.0 mm, bulk density 1.04 g/cm^3). The tested groundwater was sampled from an individual water intake on a private property in the suburbs of Lublin, from a depth of approx. 45 m. The pH of the water was neutral, pH = 7.1, and it was characterised by a high content of Ca and Mg, 100 mg/dm^3 and 18 mg/dm^3, respectively. From the ICP-OES analysis, employed to determine the Cd, Cr, Ni and Pb content, it was shown that the content of metals in question is below the limit of detection (Cd < 2, Cr < 5, Ni < 10, Pb < 5 ppb).

The solutions with a metal content of 0.01 mol/dm^3 were prepared by dissolving their respective salts in double-distilled water: $Cr(NO_3)_2 \cdot 6H_2O$, $CdCl_2 \cdot 3H_2O$, $Ni(NO_3)_2 \cdot 6H_2O$, $Pb(NO_3)_2$. Pure reagents for analysis were used to prepare the working solutions of the analysed metal substrates.

The concentration of metals in all test samples was determined by means of the inductively coupled mass spectrometry in inductively coupled plasma ICP-OES JY 238 Ultrace (Jobin Yvon-Horriba France).

2.1. Adsorption isotherms

The adsorption of heavy metal cations (Cr, Cd, Ni, Pb) onto zeolite material is known to depend on the equilibrium concentrations of these metals in aqueous solutions; therefore, to obtain in-depth information on the relationship between the two, the *static method of separate sampled portions* was employed. The test procedure was as follows: 100 ml plastic flasks were filled with 95 ml of groundwater water (containing competitive adsorption Ca and Mg ions) with the addition of 1 g of zeolite and subjected to shaking for 12 hours. Subsequently, working solutions of the Cr, Cd, Ni and Pb were added to the mix.

The concentration of the solutions was 0.01 M in 0.1 to 1 cm^3. The flasks were shaken for 2 hours, upon which time, a 10 ml sample was taken for analysis and 0.2 ml of nitric acid (V) was added. Thus obtained test samples were, finally, shaken for 1 hour, after which the heavy metal concentration in the solution was determined. The amount of adsorbed metal was calculated from the difference between the introduced and the analysed amount of the adsorbate. The collected data were interpreted in the form of adsorption isotherms.

2.2. Heavy metal removal from water

The removal of heavy metal cations from water was performed under dynamic conditions by filtering the water at a speed of 2 m/h through a 75-cm-high zeolite bed. The filtration process was continued, and the heavy metal content (Cr, Cd, Ni and Pb) was increased at 24-hour steps. For the first day, the tested groundwater was filtered without Cr, Cd, Ni or Pb, and the following day, water with the addition of permissible heavy metal content levels for drinking water (Cr 50 μg/l, Cd 5 μg/l, Ni 20 μg/l, Pb 10 μg/l). Over the next three 24-hour filtration cycles, the heavy metal concentration was increased by 2, 5 and 10 times, respectively. In each filtration cycle, after 6, 12, 18 and 24 hours, 50 cm^3 of filtered water samples were collected. The 24-hour average sample was analysed for heavy metal content.

3. Results

The tests set out to determine the adsorption rate of Cr, Cd, Ni and Pb cations by natural zeolite depending on the heavy metal concentration in aqueous solutions, containing high levels of competitive Ca and Mg ions. The collected results provided data for the sorption isotherms presented in figures 1-4. The experimentally determined amounts of adsorbed heavy metals (q_e), depending on their equilibrium concentration in the solution (C_e), provided the data for Freundlich and Langmuir equations, thus delivering the description of the investigated adsorption process. The constants in the mathematical models, as well as the error of their determination, were established by the Gauss-Newton method for solving nonlinear least-squares regression problems, with the use of Statistica software package [19]. The values of adsorption isotherm coefficients of the tested heavy metals onto natural zeolite are presented in table 2.

Table 2. Coefficients of heavy metal adsorption isotherms onto natural zeolite.

Isotherm	Formula	Coefficient	Cd	Cr	Ni	Pb
		a	0.119	0.140	0.205	0.091
		k	1.073	0.631	0.547	1.036
Freundlich	$q_e = aC_e k$	R	0.923	0.823	0.892	0.926
		a	0.121	0.211	0.363	0.085
	qe =	k	-0.019	0.465	0.696	-0.016
Langmuir	aCe/(1+kCe)	R	0.922	0.836	0.901	0.929

From the table, it emerges that there is a rectilinear relationship between the adsorption rate and Cd and Pb ion content, hence, in practice, the adsorption rate of the said heavy metals onto zeolite rises proportionally to the increase in their concentration in the solution (figures 2, 5).

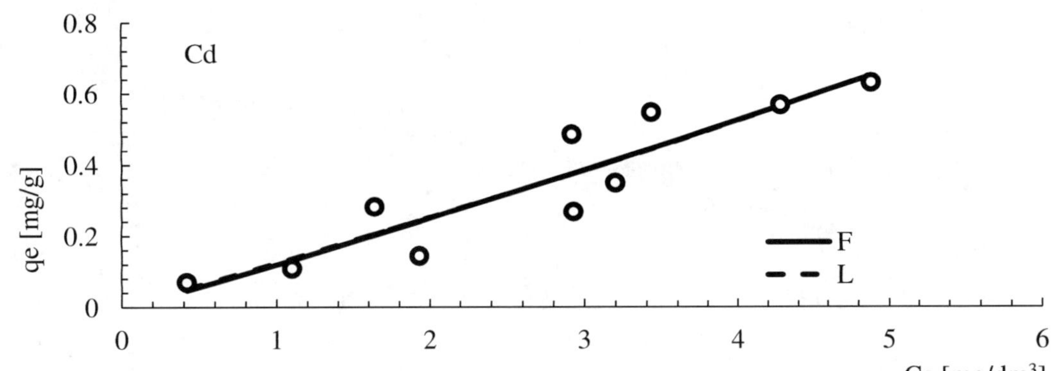

Figure 2. Relationship between the amount of adsorbed Cd onto zeolite bed and its equilibrium concentration in groundwater (model adsorption isotherms: F – the Freundlich isotherm, L – the Langmuir isotherm).

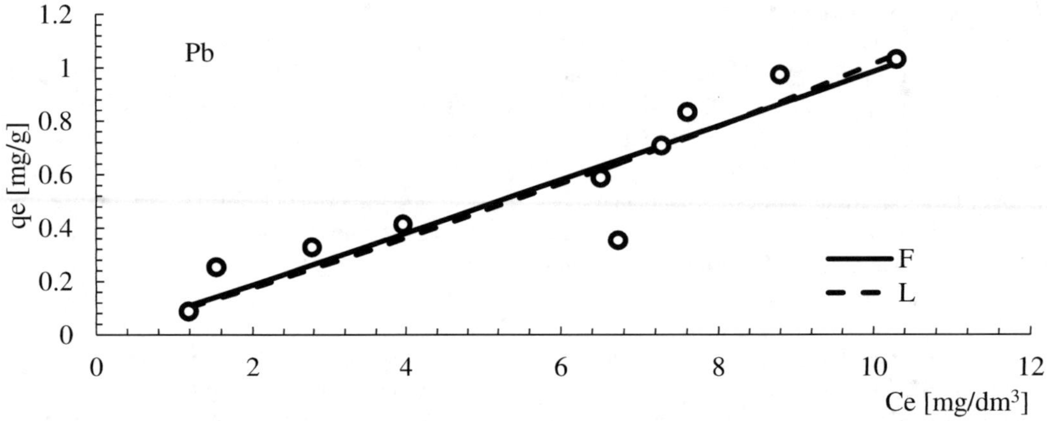

Figure 3. Relationship between the amount of adsorbed Pb onto zeolite bed and its equilibrium concentration in groundwater (model sorption isotherms: F – the Freundlich isotherm, L – the Langmuir isotherm).

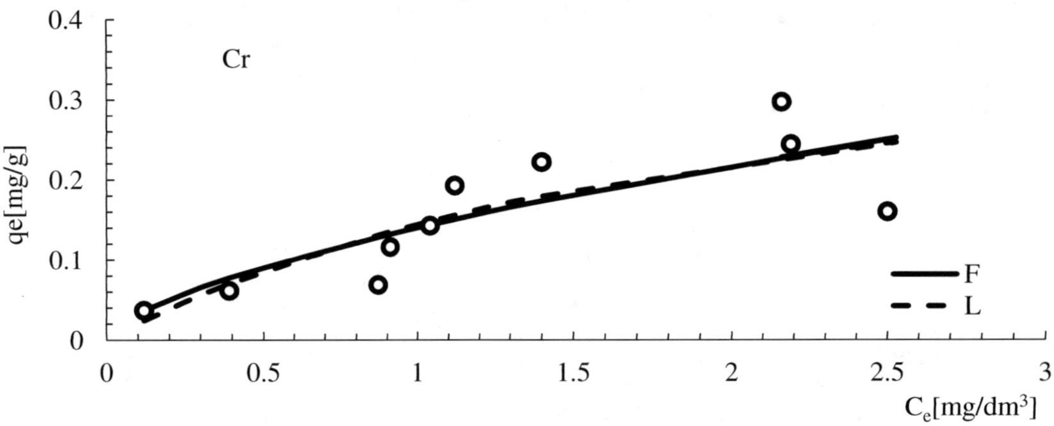

Figure 4. Relationship between the amount of adsorbed Cr onto zeolite bed and its equilibrium concentration in groundwater (model sorption isotherms: F – the Freundlich isotherm, L – the Langmuir isotherm).

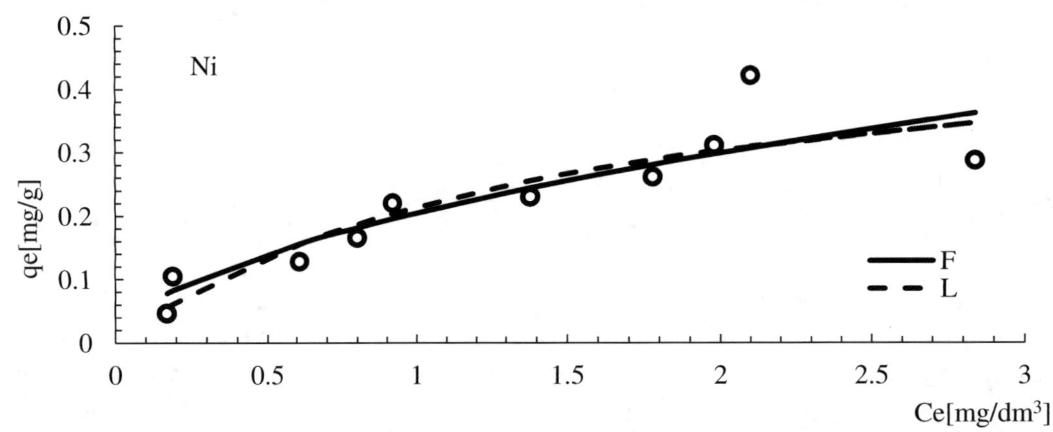

Figure 5. Relationship between the amount of adsorbed Ni onto zeolite bed and its equilibrium concentration in groundwater (model sorption isotherms: F – the Freundlich isotherm, L – the Langmuir isotherm).

In terms of Cr and Ni ions, the amounts of these metals adsorbed by natural zeolite decrease as their concentration in the water phase increases (figures 1, 4). The two applied mathematical models, i.e. Freundlich or Langmuir equations, describe the course of Cr, Cd, Ni and Pb adsorption on natural zeolite from Ca- and Mg-ion-rich aqueous solutions.

The error of equilibrium adsorption constants, determined with the use of Freundlich or Langmuir equations, are in the range of 0.82 to 0.93. Thus, it can be concluded that the heavy metal ion removal observed in our tests occurs as a result of two activities, the adsorption in micropores and the monolayer adsorption in the surface region [20]. Although similar studies report a considerably higher degree of compliance of the mathematical model with the experimental data, this applies to situations where adsorption of heavy metal ions on zeolite occurs from monometal ion solutions in the absence of competing ions [16].

The coefficient *a* in Freundlich and Langmuir adsorption isotherm equations, which expresses the heavy metal adsorption capacity of natural zeolite, indicates a comparable adsorption rate for Cd and Cr, and a notably higher zeolite affinity for Ni, and, finally, relatively low efficiency in Pb ion removal. Compared with the research data from other studies, it becomes clear that the filtration capacity of the investigated material emerging from our study is, admittedly, quite low. This discrepancy should be attributed to the fact that the reported studies investigated different sorption materials, which was, furthermore, analysed in test conditions excluding the impact of competitive ions, which decrease the sorption efficiency of zeolite [10, 17].

Furthermore, considering the same adsorbent, as shown by our previous studies, the presence of Ca and Mg ions in the solution has a profoundly negative effect on the natural zeolite's adsorption capacity in relation to the heavy metals in question. The adsorption capacity of the natural zeolite used in the study was exponentially higher in monometal solutions [21]. Therefore, competitive adsorption in the presence of considerable amounts of Ca and Mg ions substantially reduces the adsorption capacity of natural zeolite for Cr, Cd, Ni and Pb ions. However, the heavy metal filtration efficiency was proved sufficient for the zeolite material to be subjected to drinking groundwater purification tests. The results from the subsequent tests are summarised in table 3.

Table 3. Changes in heavy metal concentration in water over subsequent zeolite filtration cycles.

No. of 24-hour filtration cycle	Heavy metal content [μg/dm^3]							
	Raw water				Filtered water			
	Cd	Cr	Ni	Pb	Cd	Cr	Ni	Pb
1	< 2	< 5	< 10	< 5	< 2	< 5	< 10	< 5
2	5*	50*	20*	10*	< 2	6,1	< 10	< 5
3	10	100	40	20	< 2	11,2	< 10	< 5
4	25	250	100	50	2.7	27.6	14.2	9.4
5	50	500	200	100	3.6	44.3	15.7	12.2

< - heavy metal content below the limit of detection,
* - permissible heavy metal content levels for drinking water, as per the Regulation of the Minister of Health of 7 December 2017 on the quality of water intended for human consumption.

The most obvious finding emerging from the analysis is that the natural zeolite bed exhibits high efficiency in the removal of heavy metal ions from drinking water. At the lowest of the tested toxic heavy metal concentrations in water, i.e. not exceeding the permissible limits provided in the respective legal regulations, the filtration was shown to reduce their levels below the lower limit of detection, characteristic of the employed analytical method; it was only in the case of Cr that analytically detectable concentrations of this metal were recorded. Following filtration on zeolite, the Cr concentration in water showed a 10-fold decrease compared to the levels measured prior to the water treatment. Simultaneously, the increased content of tested heavy metals in raw water results in their higher concentration in the filtrated water. However, it was proved that even when the permissible norms are exceeded five times, the resulting Cr, Cd, Ni and Pb concentrations in the treated water are within the standard for drinking water. Even in extreme scenarios and the most severely polluted water, a 5th filtration cycle solves the heavy metal pollution problem, successfully reducing Cr, Cd and Ni concentrations below the acceptable threshold [22]. It was only in the case of Pb that the concentration in the filtered water was above the permissible limit, of 10 μg/dm^3, having reached 12.2 μg/dm^3. This stems from the adsorption properties of the tested zeolite, which shows low Pb adsorption capacity; furthermore, similar trends were also observed for other natural zeolites [16,17]. The experimental data suggest that the issue could be resolved by proper optimisation of the filtration process conditions, e.g. by increasing the filter bed height or reducing the filtration speed, although further work is planned to confirm these assumptions.

4. Conclusions

1. Adsorption isotherms for the Cr, Cd, Ni and Pb ion heavy metal adsorption in aqueous solutions, despite the presence of substantial concentrations of competitive Ca and Mg ions, confirm that the natural zeolite is an effective ion exchange adsorbent.
2. The course of ion-exchange adsorption of Cr, Cd, Ni and Pb onto natural zeolite is described at high accuracy by Freundlich and Langmuir isothermal models.
3. The values of the natural zeolite ion exchange capacity for Cr, Cd, Ni and Pb ions under competitive adsorption conditions obtained from the tests prove that natural zeolite exhibits sufficient effectiveness to be considered as a filtering bed material, which is further accentuated by the fact that this mineral does not require any chemical pre-treatment prior to its application as a filtration medium.
4. The tests conclude that natural zeolite shows high selectivity for Cr, Cd, Ni and Pb ions, which enables effective removal of heavy metal excess from water intended for human consumption.

5. References

[1] Polat E, Mehmet K and Onus Naci A 2004 J. *Fruit Ornam. Plant Res.* 12 183-87
[2] Auerbach S M, Carrado K A, Prabir K. and Dutta P K 2003 *Handbook of Zeolite Science and Technology* (United Kingdom: CRC Press)
[3] Wang S and Peng Y 2010 *Chem Eng J* **156** 1-24
[4] Hubicki Z, Rudaś M and Jakowicz A 2000 *Ecol Chem Eng S* **9**(5–6) 603-11
[5] Li D, Yao J and Wang H 2013 *Prog. Nat. Sci. Mater. Int.* 2**2** 684-9
[6] Hailua Y, Tilahuna E, Brhane A, Resky H and Sahu O 2019 Groundwater for Sustainable Development 8 457–67
[7] Fu F and Wang Q 2011 *J Environ Manage* **92** 407-18
[8] Al-Haj Ali A and El-Bishtawi R 1997 *J. Chem. Tech. Biotechnol.* **69** 27-34
[9] Shaheen S M, Derbalah A S and Moghanm F 2012 *Int. J. Environ. Sci. Dev* **3**(4) 362-67
[10] Wang Z, Tan K, Cai J, Hou S, Wang Y, Jiang P and Liang M 2019 Colloid Surface A 561 388–94
[11] Visa M 2016 *Powder Technol* **294** 338–47
[12] Margeta K, Zabukovec Logar N, Šiljeg M and Farkaš A 2013 *Natural zeolites in water treatment – How effective is their use* (United Kingdom: Intech)
[13] Wibowo E, Sutisna A, Rokhmat M, Murniati R, Khairurrijal A and Mikrajuddin A 2017 *Procedia Engineer* **170** 8 – 13
[14] Barloková D 2008 Slovac *Journal of Civil Engineering* **2** 8-12
[15] Obaid S S, Gaikwada D K, Sayyedc M I, AL-Rashdia K and Pawara P P 2018 *Mater Today-Proc* **5** 17930–34
[16] Perić J, Trgo M and Vukojević Medvidović N 2004 *Water Res* **38** 1893–99
[17] Dursun S and Pala A 2007 *Int. J. Appl. Environ. Sci.* **2**(1&2) 11-19
[18] Caputo D and Pepe F 2007 Micropor Mesopor Mat 105 222–31Wang S, Peng Y 2010 *Chem Eng J* **156** 1-24
[19] Ignatowicz K 2008 *Przem. Chem.* **87** 464-6

[20] Limousin G, Gaudet J-P, Charlet L, Szenknect S, Barthe V, Krimissa M and Kiyoku H 2007 *Appl Geochem.* **22** 249–75
[21] Borowski G, Kujawska J, Wasąg H 2019 Physicochem. Probl. Miner. Process 55(6) 1467-74
[22] The Regulation of the Minister of Health of 7 December 2017 on the quality of water intended for human consumption (Dz.U. 2017 poz. 2294).

Influence of IT outsourcing on selected groups of stakeholders (original research)

J Sloniec

Lublin University of Technology, Management Faculty, Department of Organisation of Enterprise, 38 Nadbystrzycka Str., 20-618 Lublin, Poland

j.sloniec@pollub.pl

Abstract. This paper reports on the study conducted to determine the stakeholders of information technology outsourcing (ITO) and to the impact that IT outsourcing exerts on them. Six groups of ITO stakeholders were defined: vendor, customer, employees of supplier and customer, national labour market, national economy. In the study, employees of 200 large Polish and foreign companies operating in Poland were surveyed. The results indicated both positive and negative impact of ITO on its stakeholders: positive and negative on the customer and positive for the customer employees, which were subsequently correlated in search for conclusions The originality of the paper consists in that an extensive original study has led to defining ITO stakeholders and determining the impact of IT outsourcing on them, clarifying the positive and negative impact of ITO on customer and employees, and the correlation between the factors of the positive/negative impact of ITO on the customer. The results demonstrate that with the increase in the benefits of IT outsourcing (factors of the positive impact of outsourcing on the customer), the risks associated with it (factors of the negative impact of outsourcing on the customer) grow. This is further confirmed by a high and very high positive correlation between these values.

1. Introduction

IT outsourcing (ITO) is a transfer of information technology to an external business entity to manage and implement activities, processes and functions. In today's understanding, outsourcing is a relatively new concept, which emerged in the second half of the twentieth century.

The dynamically developing global IT market is predicted to maintain a similar growth rate until 2020 [1]. Given the importance of IT outsourcing among other branches of the Polish and global economy, many of its aspects have not been well studied and therefore there emerges a need for conducting and continuing research in this area. The subject matter discussed in this article contributes to the study of a new, previously unexplored issue regarding IT outsourcing stakeholders from the Polish perspective. The aim of the article is to define information technology outsourcing (ITO) stakeholders and to determine the impact of IT outsourcing on them.

Therefore, based on the results of own survey, undertaken among the employees of 200 large companies operating in Poland that make use of IT outsourcing, the level of the positive and negative impact of ITO according to selected factors was evaluated for three stakeholders of this service. In addition, the research intended to determine the relationship between the positive and negative factors of ITO for the customer.

The use of IT outsourcing involves risks [2-4], which at the same time apply to all parties to the service and all stakeholders. For the benefit of the industry and the economy it seems then crucial to

IOP Conf. Series: Materials Science and Engineering **710** (2019) 012034 doi:10.1088/1757-899X/710/1/012034

identify the positive and negative impact of ITO on stakeholders and to weigh the relevance of significant impact factors in direct research. Establishing the direction and the extent of dependencies between these impact factors will enable to execute a more effective control and management of IT outsourcing, which will in turn directly translate into an economic profit of ITO stakeholders. As no similar research in Poland has yet been conducted, this study responds to the paucity of data to describe IT outsourcing stakeholders in our country.

Outsourcing appeared and developed in the twentieth century. Under the current name and meaning, the concept came into practice of management sciences in the 1980s, whereas the theoretical foundations were introduced 1990s [5-6].

All companies have suppliers and co-operators and, in part, these contacts take the form of outsourcing. The phenomenon of outsourcing spread during the development of craft production in the 18th century. At that time, a system of cooperative relations, which is currently referred to as outsourcing connections, was created. The industrial revolution of the late 19th and early 20th century contributed to the development of large industrial enterprises, simultaneously leading to diminishing the position of craft manufactories. The scale of outsourcing has been reduced. It was not until the 1960-1970s' emergence of new management concepts and the development of the existing ones that the concept of outsourcing revived. The precursor of the new wave of developments in outsourcing as a phenomenon, in IT outsourcing in particular is R. Perot, whose company offered paid implementation of IT functions to Frito-Lay enterprise, namely the management of the company's IT equipment. In the 1980s, the process of external procurement of parts for cars, *i.e.* outsourcing, was implemented by General Motors.

At the same time, it should be noted that from the 1960s and 1970s, the practice of sharing employees has been in force. In effect, shared IT employees would perform the same tasks in a number of companies. However, what distinguishes employee and outsourcing is the nature of the latter. While small companies have been benefitting from shared work, outsourcing is used by all enterprises, and IT outsourcing is also very much present in companies with large and well-developed IT departments [7]. Initially, IT outsourcing was used primarily in the areas of non-critical information, peripheral for the company, its loss did not cause a decline in competitiveness. Then, IT outsourcing partners have been granted access key information of critical importance for the competitiveness of the company or organisation [8]. Information on banking operations is considered to be of high-sensitivity and has previously been stored and processed exclusively internally. Since the 1990s, it is not. Financial, insurance, health, communication and retail information are among the most frequently data processed as part of IT outsourcing [9]. This phenomenon can be explained by changing direction and attitude to using and managing information, as well as the use of information technologies. Taking this view into account, outsourcing can be considered as an evolving data processing service, much like a telephone service or a power supply service. We are not interested in the infrastructure we use. We are interested in the possibility of conducting a conversation or switching on an electrical device and with respect to data or information, canal that matters is exchange of data or information. Therefore, the company does not waste its strength and resources on building an infrastructure while it can use it efficiently [7].

Information loses its strategic importance. It is caused by its universal availability as well as the effortlessness and cost of its acquisition [10].

Already in the 1990s, it was predicted that due to the widespread availability of information, it would cease to be a lasting competitive advantage [11].

The phenomenon of globalisation also affects the development of outsourcing through:

1. The trend among enterprises to implement production processes where production factors are cheaper. This is the phenomenon of so-called global outsourcing, which is when software is created and developed in countries with highly-qualified IT employees and where the remuneration for work is relatively low (such as India – the leader in the global outsourcing market) [12].

2. Increased acceptance of the Internet as a communication medium, which affects the phenomenon of application outsourcing – tendencies to create and use applications via the Internet, instead of installing them in the local environment [13].

The increase in the scope and size of IT outsourcing services was mentioned by many authors of research already in the 1990s [8; 14-16]. According to those early works, this increase would also take place in subsequent years, which has appeared to have been confirmed by the current results of the IT market analysis.

Outsourcing creates a certain bond between enterprises, having the character of partner relations. This relation influences the functioning of the whole enterprise. As part of the relationship, the service vendor takes over the task of cost rationalisation, while striving to secure the level of quality of the services provided.

There are many outsourcing classifications and extraction criteria. One of the criteria is the company's connections with the outsourcing partner [6]. There are:

1. Contract outsourcing, when the cooperation is carried out on market terms, the service is outsourced to a completely independent partner;
2. Capital outsourcing, when entities are related by capital but are organisationally independent. Often, in the process of enterprise restructuring, a capital company is separated to perform a given function.

The division into these two groups of outsourcing will be discussed in more detail in the later part of this report.

Other important categories of outsourcing are strategic and tactical outsourcing. The former is related to the company's development strategy and is permanent, the latter – relates to shorter time horizons.

It is also common to divide outsourcing into offshore outsourcing – relocation of tasks to distant countries (in the case of Poland e.g. to India), nearshore outsourcing – use of companies in neighbouring countries, onshore outsourcing – use of companies located in the same country as the customer.

Outsourcing may relate to a single function, implemented on the basis of self-employment, functions related to processes and functional areas. It can be based on market principles and partnership cooperation [17-18]. Each of these types of outsourcing has both advantages and disadvantages and is used by enterprises in specific situations, depending on their needs and capabilities.

2. IT outsourcing development in Poland

The development of outsourcing in Poland resembled the dynamics observed in other countries. The first company to implement ITO in Poland was the state-owned enterprise ZETO (Electronic Works of Computational Engineering) created by the chairman of the Science and Technology Committee in 1964 (it had headquarters in several large Polish cities). The company provided data collection and processing services to state institutions, organisations and enterprises. In the following years, ZETO has undergone transformations, some branches were closed. However, there are still branches that have been transformed and "under changed names" are currently providing IT outsourcing services. This proves that Poland was very early in the group of countries developing computational techniques, including IT outsourcing services.

IT outsourcing is a dynamically growing area of IT, also in Poland. According to specialist publications, the estimated growth in the value of the IT outsourcing market is around 7% per annum [19]. This results in a great interest in businesses in this field but also in the interest of enterprises to outsource IT-related activities. Companies providing IT services in Poland are currently employing over 140,000 people.

In 2013, IT services in Poland accounted for an estimated 1.7% of the global IT market, while the value of the IT services market in Poland accounted for 29% of the total value of the IT market. The number of companies providing services in the IT industry has increased by over a half in the last four years. Two-thirds of employees of large IT service companies are employed in foreign enterprises, the largest of which are American and French companies. The value of the IT services market in 2013 amounted to approximately USD 3.14 billion [19], the largest proportion of which is for design services, including systems integration in particular. The next locations are occupied by support services and outsourcing services, which constituted about 19% of the value of the IT services market. Analysts of the Polish IT market predicted that due to the rapid development of the market in 2018 the share of outsourcing services in the market of IT services would reach 22%. The most dynamic development would include infrastructure hosting services, customer application management, and especially the use of cloud solutions [20]. The predictions were confirmed by recent data [21].

3. Stakeholders of IT outsourcing

The concept of stakeholders was introduced by the Stanford Research Institute in the 1970s and meant initially "groups without whose support the organisation would cease to exist" [22]. It was then developed by R. Ackoff and R.E. Freeman – the scientists representing the system approach [23]. The term is used especially in strategic management and project management. In strategic management, stakeholders are individuals or institutions that have or can have an impact on the organisation. In project management, stakeholders are people who are not directly involved in the project, but the results of the project may have an impact on them, which is why these people have a vested interest in a successful or unfavourable course of the project.

According to the theory of stakeholders [24-25], the unit of analysis is the relationship between business and groups and individuals who may influence it. With regard to IT outsourcing as a service provided by a supplier to a customer, the scope of who stakeholders are is extended to include as persons, groups of people, organisations, the national economy and elements of the national economy – the labour market.

The characteristics of stakeholders in IT outsourcing were undertaken by several foreign researchers [26-32]. This article presents a different approach to the topic, focusing on two stakeholders – ITO customers and suppliers, employees of the customer, and selected impact factors. The choice of stakeholders and factors was caused by the availability of data from the previous ITO research carried out by the author in large organisations in Poland.

Considering the impact of ITO on these people, groups of people, organisations, elements of the national economy and the national economy, the following six groups of stakeholders were distinguished:

1. ITO services vendor is an organisation that performs IT services and receives payment for services performed.
2. ITO service customer is an organisation that outsources IT services for which these services are provided for remuneration.
3. Employee/employees of the company – ITO vendor is a person/ group of people employed in the supplier's company and realising the outsourced IT services.
4. Employee/employees of the ITO customer is/are a person/ group of people employed in the implementation of functions related to IT support for the customer company.
5. The national labour market, one of the markets functioning in the economy, where employers are purchasers and the sellers are employees.
6. The economy of Poland as a whole of economic activity carried out on the territory of the state.

With regard to IT outsourcing stakeholders, it would appear that the first four stakeholders (supplier and customer, supplier and customer employees) are undeniably important. Why are the national labour market and national economy included? These two elements must essentially be considered among the stakeholders, because IT outsourcing undeniably affects (positively/negatively) the state of the labour market and the condition of the national economy. That is why, Table 1 and 2 consider these stakeholders from the point of view of the Polish labour market and the national

IOP Conf. Series: Materials Science and Engineering **710** (2019) 012034 doi:10.1088/1757-899X/710/1/012034

economy of Poland. The approach to ITO stakeholders adopted in this study is, therefore, different in comparison with studies of foreign scientists [26-33].

IT outsourcing affects all of the enumerated stakeholders, and the impact that it makes can be specified. Tables 1 and 2 detail the positive and negative impact of ITO on individual stakeholders (by the example of Poland, especially in relation to the labour market and national economy).

Table 1. Impact of ITO on stakeholders – Poland (first part).

No	Stakeholder	Impact	
		Positive (generating profits)	Negative (generating costs)
1.	ITO supplier	• Income for ITO services • Possibility of company development (entering new markets, expanding the scope of activity, increasing competitiveness) • The possibility of specialising the company	• The need to hire employees with the highest qualifications and thus to pay high remuneration • The need to pay for the continuous improvement of IT staff qualifications • The need to create very good working conditions in order to keep employees and encourage new, with the highest qualifications • The need to follow technological novelties and the need to bear the costs of technological novelties
2.	ITO customer	• **Savings in personnel costs (reduction of employment in the IT department or liquidation of the IT department)** • **Cost-saving of hardware costs in case of IT department limitation or liquidation** • **Access to knowledge of highly qualified employees** • Savings in the cost of employee training • **Flexibility of external employees** • Cost savings for the entire project • Shorter implementation time of the project by specialised external employees • **Higher quality of project implementation by specialised external employees**	• **Incurring costs of providing outsourcing services** • **Loss of know-how** • **Security issues (data, information, procedures), hostile takeovers** • The need for additional time to introduce external employees into specific company issues • **Over-dependence on the supplier** • Danger of misunderstanding the company's needs and implementation of the project inconsistent with the demands • **Potential tensions among employees facing changes in the IT department** • Loss of the synergy effect associated with the existence of stable employee teams
3.	Employee of supplier	• High wages of IT employees • Shifting focus on specific issues • Flexibility of employment (flexible working hours, remote work) • Challenging work (variable problems, variable troubleshooting conditions)	• Lack of stable workspace • Insistence on project effects • Insistence on project duration • Insistence on cost • Possible problems in the future related to work instability (e.g. pension issues)
4.	Employee of customer	• **Possibility of focusing on specific issues in the case of leaving the IT department in the company – greater autonomy** • **Smaller work scope** • **Increased job satisfaction**	• Loss of employment in the event of reduction or liquidation of the IT department • Necessity to retrain • Possibilities to reduce the pay you receive

Note: the impact factors were marked by using Bold, which were then tested
Source: own work and [26-32]

Positive and negative impact factors for particular groups of IT outsourcing stakeholders are listed. For all ITO stakeholders, the number of positive impact factors is similar to the number of negative impact factors. So, you can talk about a similarly strong positive impact as the negative ITO on all groups of stakeholders.

Table 2. Impact of ITO on stakeholders –Poland (second part).

No	Stakeholder	Impact	
		Positive (generating profits)	Negative (generating costs)
1.	National labour market	• Increased pay of IT employees • Increasing employment in IT outsourcing in the case of the emergence of new ITO companies or extension of existing activities • Increased demand for highly qualified IT employees (emphasis on employee training) • The emergence of new professions and functions (related to data security, information, processes, related to big data analysis)	• Reduction in the total number of IT employees employed (reduction of employment in IT departments in particular companies) • Potential increase in unemployment • Possibility of employee precarisation [32] • Increased demand for highly qualified IT employees (inability of some employees to improve their qualifications) • The disappearance of certain professions and IT-related functions in companies
2.	The economy of Poland	• Increase in the number of companies providing IT outsourcing services • Increasing tax revenues from newly created and developing IT outsourcing companies • The possibility of becoming a regional or global tycoon that provides IT outsourcing services	• Reducing the number of jobs (possibilities to increase unemployment) • Reduction in tax revenues from domestic ITO companies when using offshoring • Increasing the number of employees from the precariat • Possibility of social unrest in the case of a significant enlargement of the precariat

Note: the impact factors were marked by using Bold, which were then tested
Source: Own elaboration and [26-32]

4. The impact of IT outsourcing on selected groups of stakeholders (based on own research)

4.1. Own research methodology

The study reported in this paper is a part of a larger study of big organisations operating in Poland and using IT outsourcing. The research group was homogeneous – large organisations using IT outsourcing. The survey was conducted in 2016 on a group of 200 large organisations, in the vast majority of enterprises, Polish or foreign, but having a branch in Poland. The average number of employees was 250, which was the main aspect for recognising a given business as belonging to a group of large organisations.

The study was carried out by means of the CATI (*Computer Assisted Telephone Interview*) method. It involves conducting telephone interviews and recording the responses using a special computer script. The selection of the research sample was intentional (only companies using ITO), however, on the other hand, it was a simple random selection because the companies and organisations participating in the study were randomly selected and their chances of participating in the study were equal. The questionnaire contained a control and primary question in relation to it. The consistency of answers to these questions was 97.6%. The choice of the questionnaire as a research tool was dictated by the area and scope of research (IT outsourcing in large organisations in Poland), as well as a previous literature query of similar research abroad. 77% of the respondents represented the managers or employees of IT departments, the remaining 23% were directors, managers and administration staff. The survey tool was a questionnaire. Answers to substantive questions were provided in the expanded seven-point Likert scale. The answer "1" in this scale meant "it is not important at all/it has not been achieved", the answer "7" meant "very important/has been fully achieved". The answer found on the enlarged scale in the middle of the scale (in the quoted studies answer "4") is the neutral answer "undecided" [34]. Standard statistical methods (percentage structure, correlation) were used to develop the test results.

Three of the examined values were derived from these studies: success factors, risk factors and the impact of IT outsourcing on the respondents' job position.

The study was intended to test the hypotheses regarding the extent of the positive and negative impact of IT outsourcing on the customer's company and the size of the positive impact of IT outsourcing on the customer's employees. It also aimed to address the hypotheses regarding the relationship between the positive and negative impact of ITO on the customer. The research questions regarding the hypothesis are formulated as follows:

1. Is there a relationship (measured by the correlation coefficient) between selected factors of the positive/negative impact of ITO on the customer?
2. If the correlation exists, is it positive or negative?
3. If the relationship exists, what is its correlation ratio size?

In search for the answers to these questions, an analytical tool was put to use, Statistica 13.1 software.

4.2. Brief description of the surveyed organisations

The organisations were included in the group of large organisations only due to the average number of employees – over 250 employees. The turnover was not considered because in some organisations this is sensitive or secret information. The organisations represented various industries: trade, services, logistics, education/ research/science. Their percentage share in the structure was similar, except for administration and non-profit organisations, which participated in the study at the level of 3% (Figure 1).

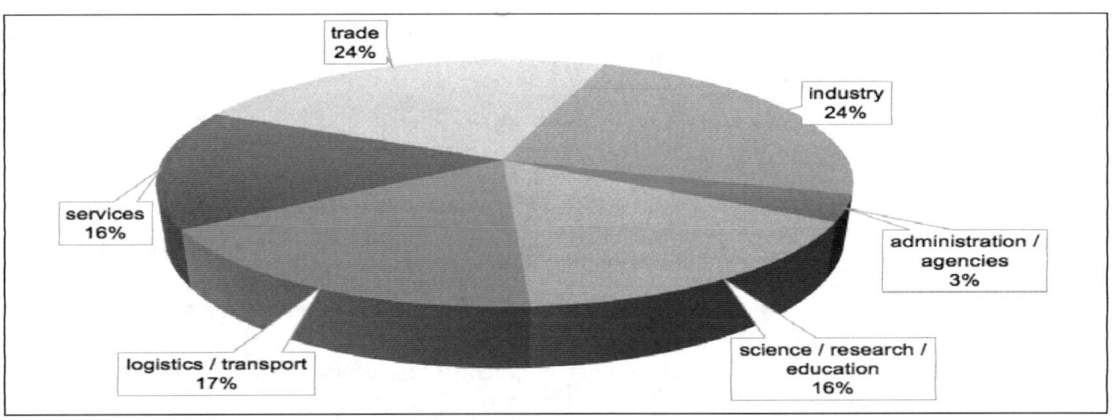

Figure 1. Industry structure of organisations participating in the survey, %
Source: own work

Almost half of the organisations surveyed were established in the years 1981-2000, i.e. during the economic transformation in Poland, around 15% were created after 2000 and 25% before 1981. The majority of the surveyed companies' capital was Polish 80%, while 7% was foreign, 6% mixed (Polish and foreign), and 5% of respondents did not answer the question. These were usually independent organisations that did not form part of the structure of other entities. 54% of the surveyed companies are located in the metropolis, 45% have a location outside the metropolis, whereas 1% has not indicated their location. Only 6% of enterprises are in special economic zones, while 90% outside these zones and 4% of respondents did not indicate where the head office is located.

Considering the characteristics of the surveyed organisations, it can be assumed that they were diverse.

4.3. Evaluating the impact of IT outsourcing on companies, customers and employees of the customer's organisation based on my own research

The positive factors listed in Tables 1 and 2 of the ITO's impact on the stakeholders correspond with the selected factors of benefits from ITO indicated in the conducted studies of large organisations using this service (Table 3).

Table 3. Positive impact factors on the ITO customer obtained from stakeholders and based on own study.

No	Positive factors of influence on the ITO customer company	The corresponding benefit factor from ITO (own study)
1	Savings in personnel costs (reduction of employment in the IT department or liquidation of the IT department)	Savings in personnel costs
2	Equipment cost savings in the case of IT department reduction or liquidation	Savings in the costs of using technology
3	Access to knowledge of highly qualified employees	Facilitating access to new technologies Reducing the risk of technological obsolescence

Source: own work

In order to assess the positive impact of IT outsourcing on organisations – service customers, the answers to the question about the benefits of using ITO were used. The positive and negative impact was assessed in the expanded seven-point Likert scale. Answer "1" meant "not reached at all", answer "7" meant "was fully achieved", then for drawing 2, 3, 4, the answers from "1" to "3" were added, meaning "it has not been achieved" and the answers from "4" to "7" have been summed up, meaning "has been achieved". Answer "4" was neutral. The percentage structure of the ITO benefits is shown in Figure 2.

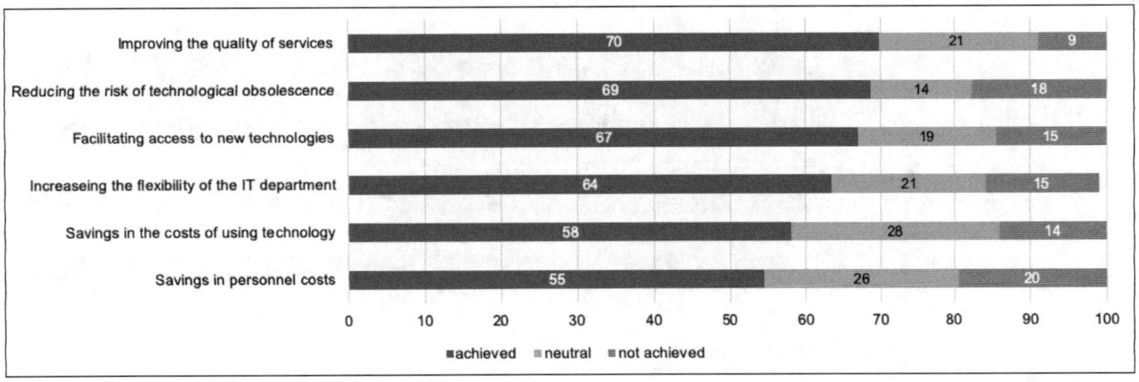

Figure 2. Structure of achieving positive factors of the impact of IT outsourcing has a customer company, %
Source: own work

Among the respondents – the 200 large organisations operating in Poland – the vast majority assess the above-mentioned factors of the positive impact of outsourcing as achieved, from 69.8% for the factor *improving the quality of services offered* to 54.5% for *the personal cost savings* factor. If the above-mentioned factors of positive impact on the customer company are assessed much more frequently as achieved, it means that the impact of ITO on the stakeholder – the customer company based on the conducted research – is indeed positive. This confirms the hypothesis that most of the surveyed organisations evaluate the beneficial effects of ITO on their company as achieved. It also confirms the advisability of using ITO by companies.

It should be noted that a percentage of the organisations (from 9% to 19.5%) assessed that the above-mentioned factors of the positive impact of IT outsourcing were not achieved.

Similarly, the negative factors listed in Tables 1 and 2 of the impact of ITO on its stakeholders correspond with the selected risk factors of ITO indicated in the conducted studies of large organisations using this service (Table 4).

Table 4. Negative impact factors on the ITO customer obtained from stakeholders and based on own study.

No	A negative factor affecting the ITO customer	Corresponding ITO risk factor (own study)
1	The cost of providing outsourcing services	Hidden contract costs
2	Loss of know-how	Loss of know-how
3	Security issues (data, information, procedures), hostile takeovers	Security issues
4	Excessive dependence on the supplier	Excessive dependence on the supplier
5	Opportunity to resist employees against changes in the IT department	Possible resistance of employees

Source: own work

In order to assess the negative impact of IT outsourcing on customers, they were asked a question about the risks associated with the use of ITO. The percentage structure of the validity of ITO risk factors is shown in Figure 3.

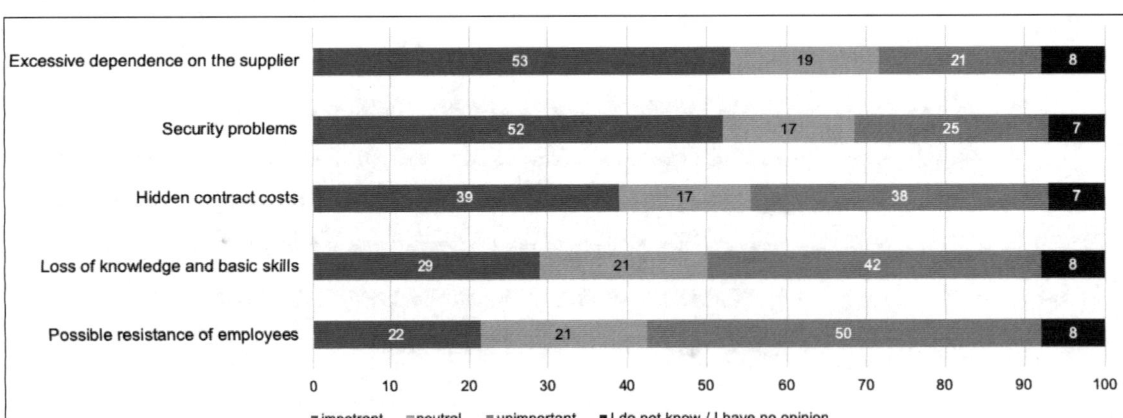

Figure 3. Structure of the importance of negative impacts of IT outsourcing has a customer company, %
Source: Own elaboration

Among the examined factors of the negative impact of ITO on the customer, two were assessed as definitely more important: *excessive dependence on the supplier* assessed as important by 53% of respondents and *security issues* rated as important by 52% of respondents. The *hidden contract costs* factor was rated by a similar percentage of the organisation as valid and invalid (39% and 37.5%).

Two other risk factors – *the loss of know-how* as well as *the resistance of employees* were assessed as invalid (42% and 49.5% of responses). IT outsourcing also has a negative impact on the customer company, which is presented in Tables 1 and 2 and confirmed in the conducted research.

However, it should be noted that the following factors have been recognised as critical: *over-reliance on the supplier* and *security issues*. Considering activities that could reduce the negative impact of these factors on the customer's services would be:

- Draw an ITO contract properly so as to minimise the impact of these factors on the customer;
- Adherence to the rules of contract performance by performing periodic inspections of contract realisation.

Existence of factors that negatively impact ITO on the customer company does not disqualify the ITO service. It allows recognising the risk of ITO and, as a consequence, carrying out activities that minimise the risk or eliminate its impact.

This is a partial confirmation of the assumption that the majority of the surveyed organisations assesses as an important impact risk of ITO (negative impact factors) on their company.

In the case with the ITO stakeholder – ITO customer, the selected factors of the positive and negative impact of ITO on the customer were examined. It is, therefore, possible to compare the impact of these factors. Before the comparison, it should be emphasised that the study was carried out on a group of organisations that employed ITO. They appreciate the advantages of the service, at the same time noticing its deficiencies. *Increasing the quality of services* assessed as being achieved by the largest percentage of organisations may involve incurring additional costs (one of the TQM principles is that higher quality is achieved through earlier expenses), but also with security issues. Security, on the other hand, was rated as important by more than half of the surveyed organisations. This example shows that the positive aspects of ITO using are correlated with negative aspects.

Table 5. Correlation between the risk factors and the benefits of using ITO.

		Benefits					
		Increased flexibility of the IT department	Improving the quality of services offered	Savings in personnel costs	Facilitating access to new technologies	Savings in the costs of using technology	Reducing the risk of technological obsolescence
Risk	Excessive dependence on the supplier	0.777	0.810	0.767	0.790	0.767	0.747
	Loss of know-how	0.674	0.697	0.695	0.693	0.682	0.708
	Hidden contract costs	0.691	0.698	0.666	0.670	0.659	0.661
	Possible resistance of employees	0.639	0.673	0.664	0.651	0.657	0.652
	Security issues	0.722	0.767	0.697	0.737	0.720	0.714

Note: correlation coefficients were significant with $p < 0.05$
Correlation $0.5 < |r| \leq 0.7$ is high and $0.7 < |r| \leq 0.9$ is very high [36].
Source: own work

While IT outsourcing is associated with substantial benefits, it can also prove to be a risky endeavour [35-36]. It is then interesting from the scientific point of view to confirm whether the increase in the benefits of ITO (factors of positive impact on the customer) is accompanied by an increase in risk (factors of negative impact on the customer) or not. It can be assumed that such a relationship exists, that is, that with the increase in the benefits of ITO, its risk increases. In order to test this hypothesis, the correlation between benefits (factors of positive impact on the stakeholder –

ITO customer) from the use of ITO and risk factors (factors of negative impact on the stakeholder – ITO customer) was examined (Table 5). It intended to determine whether the factors studied are statistically significant, what the strength of this relationship and its direction is. The factors of benefit and risk were measured in the extended 7-point Likert scale.

It was found that there is a high or very high positive correlation between all risk factors examined (these are factors of the negative impact of outsourcing on the customer) and benefits from outsourcing (they are factors of the positive impact of outsourcing on the customer). The existence of a high or very high positive correlation between the studied benefit and risk factors from the ITO indicates that the increase in the value of one variable (for example an excessive dependence on the supplier) corresponds to a linear increase in the value of the second variable (for example improvement of the quality of services offered). For the given example, the value of the correlation coefficient is the highest and amounts to 0.810. With the improvement of the quality of services offered the dependence on the supplier increases. The interpretation of the causes presented in dependencies is not easy. We can only intuitively assess that indeed a correlation between the aforementioned benefits and risks of ITO should exist. The presented analyses confirm the existence of this supposition, the greater the benefits, the greater the ITO risk. With respect to the stakeholder analysis, it can be concluded that the higher positive impact of ITO entails a higher negative impact on the customer.

For a graphical representation of the occurrences of correlation between variables, an exemplary scatter plot of the variables "improvement of the quality of the offered services" and "excessive dependence on the supplier" is given in Figure 4.

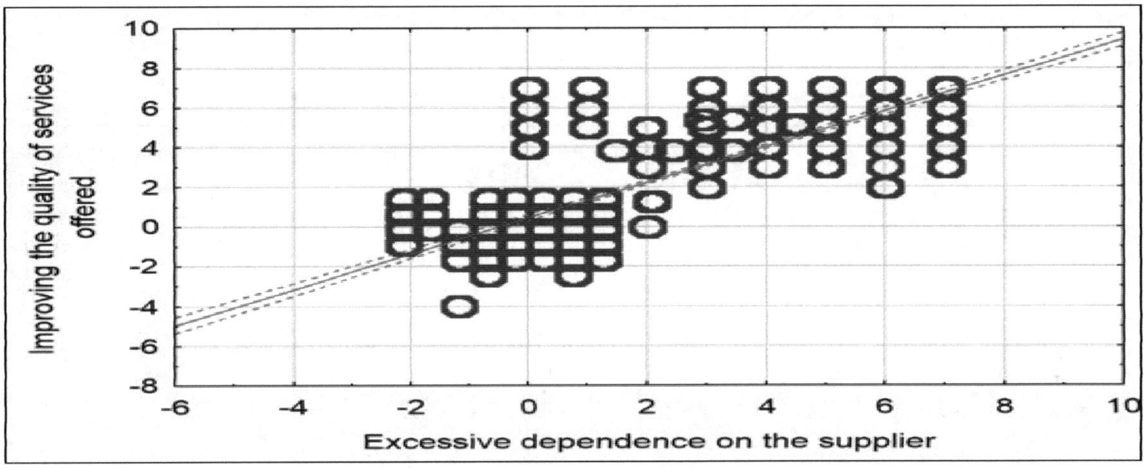

Figure 4. Scatter plot of the variables "to improve the quality of services" and "excessive dependence on suppliers" (correlation coefficient 0.810)
Source: own work

Regarding the central research questions, it should be noted that:

1. There exists a relationship between the selected factors of the positive and negative impact of ITO on the customer.
2. The direction of dependence is positive, i.e. the importance of the factors of the positive impact of outsourcing on the customer (benefits from ITO) is accompanied by the increase in the importance of the negative impact of outsourcing on the customer (ITO risk) and the increase in the importance of the negative impact of outsourcing on the customer (ITO risk) is accompanied by an increase in the importance of the factors of positive impact of outsourcing on the customer (ITO benefits).

3. The correlation coefficient assumes high and very high values, which means that there is a strong relationship between variables.

From the analysis of the reasons for the established relationships, a reference should be made to the specific benefits and risks of ITO under study. The higher the flexibility of the IT department is rated, the greater the dependence on the supplier. The higher the dependence on the supplier, the higher the risk of technological obsolescence is rated. IT outsourcing is associated with risk, however, it does bring measurable and immeasurable benefits to the organisation. By monitoring risk factors, outsourcing can be used in a rational way, increasing the organisation's benefits. The methods for monitoring the risk of outsourcing include:

- Quantitative and qualitative control of the fulfilment of the outsourcing contract
- Security monitoring (data, procedures, processes).

IT outsourcing brings significant benefits to the organisation provided it is used rationally and in a controlled manner.

Similarly, the factors listed in Tables 1 and 2 of ITO's impact on its stakeholders correspond with the selected ITO impact factors on the work of IT department employees, indicated in the conducted studies of large organisations (Table 6).

Table 6. Positive impact factors on the employee of the ITO customer obtained from stakeholders and based on own study.

No	A positive factor influencing the employee of the ITO customer	A corresponding factor of ITO influence on the work of the IT department (own research)
1	Possibility of focusing on specific issues in the case of leaving the IT department in the company – greater autonomy	Autonomy (the possibility of making decisions and its implementation with the minimum opposition)
2	Smaller work scope	Demand (on time, energy and expectations related to what is expected from the position)
3	Greater job satisfaction	Job satisfaction (joy of work, sense of fulfilment)

Source: own work

In order to assess the positive impact of IT outsourcing on employees of the customer company, the answers to the question about the impact of ITO on the position of an IT department employee were used. The percentage structure of the ITO impact is shown in Figure 5.

In the opinion of the surveyed organisations, three factors of the positive impact of ITO on the IT worker's workplace were rated significantly more positively (from 40.50% to 28%) than negatively (from 10.5% to 8.5%). The factor of *satisfaction with the work performed* was by far the highest. An evaluation of the positive impact of IT outsourcing on the work of IT employees of the customer company provides another confirmation of the positive effect of ITO.

The positive impact of ITO on employees of the customer concerns employees who continue their work given that IT outsourcing is typically associated with the earlier reduction of personnel numbers (the tasks to be expelled and full-time employees of the customer are taken over by employees of the ITO vendor). Loss of work by the employees of the ITO customer is not associated with a big threat of being out of work, because the IT area is currently an employee market in Poland (they can choose from many jobs offers). IT employees continuing their work after the introduction of ITO usually need to change the scope of their professional duties and retrain.

The working hypothesis that the majority of employees of the surveyed organisations positively assess the impact of ITO on their work position has been partially confirmed. Less than half of the surveyed employees do not assess the impact of ITO on their work position as positive. At the same time, it should be stated that almost half of the surveyed employees assessed this impact as neutral, and small percentage (around 10%) stated that the impact was negative.

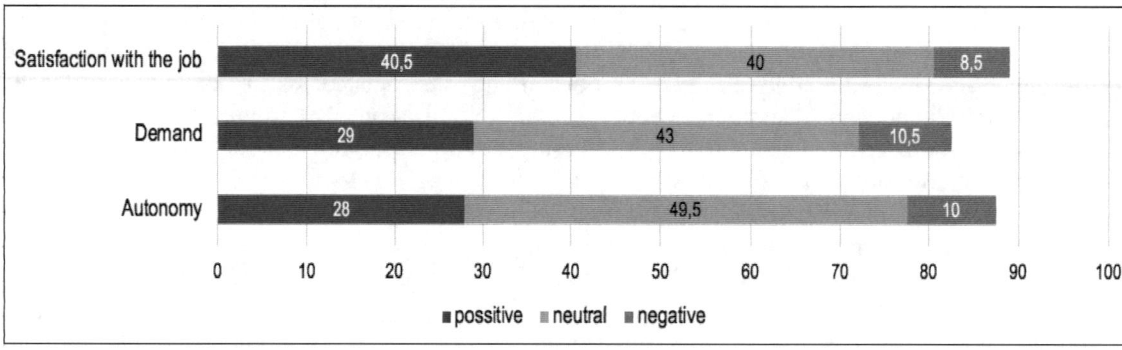

Figure 5. Structure of the impact of IT outsourcing on the IT employee's job position, %
Note: the structure does not add up to 100% due to the lack of response from some respondents (answers: *I do not know/I have no opinion*)
Source: own work

5. Summary

IT outsourcing is the everyday reality in modern enterprises. The share of IT services in the structure of the IT market is increasing, which translates into an increase in the value of the global IT outsourcing market. This trend is even more dynamic in Poland, where over the last five years the share of ITO in the market of IT services has increased by over 3% and continues to indicate an upward trend [37].

The purpose of the reported study was to define the stakeholders of this service and to evaluate the impact that IT outsourcing has on them. The paper defines six groups of IT outsourcing stakeholders: ITO supplier company, ITO customer company, supplier employees, customer employees, national (Polish) labour market, and the economy of Poland. The positive and negative impact of the ITO on each studied stakeholder group was determined. The original research presented here, resulted in determining the positive and negative ITO impact of selected factors on ITO customers and the positive impact of selected factors on customers' employees. An important positive impact (55%-70% of the responses) of the ITO on the customer company was noted due to all six factors studied. An important negative impact (52-53%) of two factors was noted: *over-reliance on suppliers* and *security issues*. The impact of other negative factors is lesser. IT outsourcing has a positive effect (28-41%) on the customer's IT employees rather than negative (8.5-10.5%).

The dependence between the positive and negative impact of ITO on the customer was identified and confirmed on the basis of own study. |It was found that there is a positive high or very high correlation between the studied benefit and risk factors of IT outsourcing. The higher the importance of ITO benefits, the higher the importance of risk factors, or alternatively, the higher the validity of risk factors, the higher the ITO benefits. Higher benefits are associated with higher risk factors. In order to maximise the benefits, the risk factors of the ITO organisation should be adequately monitored.

As a result of these investigations, suggestions were identified for future research: IT outsourcing has a positive effect on the ITO customer and IT department employees in this company. The hypothesis that most of the surveyed organisations evaluate the achieved beneficial effects of ITO on their company as achieved was confirmed. Partially, the assumption that most of the surveyed organisations assess the risk of impact (factors of negative impact) of ITO on their company as important has been confirmed as well. Also to a certain extent the results confirm the hypothesis assuming that most of the employees of the surveyed organisations positively assess the impact of ITO on their job position. The study has succeeded in showing the dependence between positive and negative factors influencing the ITO on the customer. A high or very high positive correlation between positive and negative ITO impact factors on the customer has been identified. It can be expected that there is a correlation between other factors that have not been accounted for in the presented work. However, to confirm this, a separate study should be carried out. The existence of correlation can be an indication for companies using ITO to continuously monitor the implementation of the outsourcing contract.

The present study was designed to determine the effect of ITO on stakeholders. The conclusions that emerged from this study are as follows: Both the positive and negative impact of IT outsourcing on all of the stakeholders have been identified. The task of the management of ITO supplier and ITO customer is to minimise the negative impact factors. The task of the managers of the national economy is to minimise the negative impact factors and maintain the balance of the industry development (steady development of ITO suppliers and customers, balance on the labour market, growth of the national economy).

This study was unable to encompass the entire group of stakeholder. The limitations resulted from the content of data derived from the author's research on IT outsourcing in large organisations in Poland. Future research directions should account for the unexamined stakeholder groups (ITO suppliers and employees of supplier companies). It is also possible to extend the research to other groups of stakeholders as identified in foreign surveys: sponsors, outsourcing project management on the part of the supplier and the customer, participants of the project team. An interesting research problem would also be to identify the relationship between stakeholders and factors of the positive and negative impact of ITO on individual stakeholders. An important research problem is also to determine the range of the level of balance between the amount of positive and negative impact of ITO on individual stakeholders. As mentioned earlier an effective ITO control and management translates directly into economic benefits for ITO stakeholders.

As it was stated earlier, the prospects for the development of the IT outsourcing market are highly promising. The key IT development trends (they also concern IT outsourcing) for the nearest future [38] are:

- Dynamic development of innovative technologies,
- Progressive process automation,
- Increased importance of Data Science,
- Further increase in the use of remote solutions and cloud solutions.

6. References

[1] Jadczak A 2017 Polski rynek IT W 2016 roku wart był 40 mld zł. *IT wiz*, 08.09.2017. http://itwiz.pl/polski-rynek-2016-roku-wart-byl-40-mld-zl// (accessed 01 May 2019)

[2] Gonzalez M R, Gasco J L, Llopis J 2015b Razones y riesgos del outsourcing de sistemas de información en las grandes empresas españolas *Revista Europea de Dirección y Economía de la Empresa* **24** 175-89

[3] Lacity M C, Khan S and Yan A 2016 Review of the Empirical Business Services Sourcing Literature: An Update and Future Directions *Journal of Information Technology* **31**(3) 269-328

[4] Słoniec J, Kaczorowska A, Motyka S 2016 Risk Assessment of IT Outsourcing in Enterprises Depending on Their Branch *Zeszyty Naukowe Politechniki Śląskiej Seria Organizacja i Zarządzanie* **99** 473-86

[5] Mullin R 1996 Managing the outsourced enterprise *Journal of Business Strategy* **17**(4) 28-36 doi:10.1108/eb039792

[6] Słoniec J 2018 Outsourcing w dużych organizacjach w Polsce Modelowanie z wykorzystaniem równań strukturalnych (Toruń: Wydawnictwo Dom Organizatora)

[7] Grover V, Cheon M J and Teng T C 1994 A Descriptive Study on the Outsourcing of Information Systems Functions *Information and Management* **27**(1) 36

[8] McLellan K, Marcolin B and Beamish P 1995 Financial and Strategic Motivations behind IS

Outsourcing *Journal of Information Technology* **10**(4) 310
[9] Gupta G and Gupta H 1992 Outsourcing the IS function Is it necessary for your organization? *Information Systems Management* **9**(3) 44
[10] Chen Q and Lin B 1998 Global Outsourcing and its Managerial Implications *Human Systems Management* **17**(2) 110
[11] Hopper M D 1990 Rattling SABRE – New Ways to Compete on Information *Harvard Business Review May-June 1990* 118-25
[12] Heeks R, Krishna S, Nicholson B and Sahay S 2001 Synching or Sinking: Global Software Outsourcing Relationships. *IEEE Software* **18**(2) 54-60
[13] Marchand N and Jacobsen H-A 2001 An Economic Model to Study Dependencies between Independent Software Vendors and Application Service Vendors *Electronic Commerce Research* **1**(3) 315-334
[14] Palvia P C 1995 A Dialectic View of Information Systems Outsourcing: Pros and Cons *Information and Management* **29**(5) 265
[15] Caldwell B 1996 The New Outsourcing Partnership *Information Week* **585** 51
[16] Lacity M C and Willcocks L P 1998 An Empirical Investigation of Information Technology Sourcing Practices: Lessons from Experience *MIS Quarterly* **22**(3) 363
[17] Gonzalez M R, Gasco J L and Llopis J 2006 Information systems outsourcing: A literature analysis. *Information and Management* **43**(7) 821-34
[18] Gonzalez M R, Gasco J L and Llopis J 2015a Outsourcing de sistemas de información: situación actual evolución y tendencias *Investigaciones Europeas de Dirección y Economía de la Empresa* **21** 93–99
[19] Kośnik Ł, and Majman S 2013 *IT@PL Rynek pracy IT w Polsce* file:///Users/jolantasloniec/Downloads/IT%20Report%20-%20Poland%20-%20Polish.pdf/ (accessed 15 May 2019)
[20] Czupryn P, Górecki J, Kapusta E, Kujawa A, Micek G and Śniadała G 2015 Rynek usług IT w Polsce http://absl.pl/wp-content/uploads/2016/10/raport_it_2015_PLv151210.pdf/ (accessed 15 May 2019)
[21] IT service market in Poland *SkyRes* 2017 http://www.skyres.eu/en/it-services-market-in-poland/ Accessed 15 May 2019
[22] Freeman R E and Moutchnik A 2013 Stakeholder management and CSR: questions and answers *Umwelt Wirtschafts Forum* **21**(1) 5-9 doi:10.1007/s00550-013-0266-3.
[23] Freeman R E, Phillips R and Sisodia R 2018 Tensions in Stakeholders Theory *Business and Society* 1-19 doi:10.1177/0007650318773750
[24] Parmar B L, Freeman R E, Harrison J S, Wicks A C, de Colle S and Purnell L 2010 *Stakeholder Theory: State of the Art* (Cambridge University Press)
[25] Phillips R 2003 *Stakeholder Theory and Organizational Ethics* (San Francisco: Berrett-Koehler Publishers)
[26] Alban D, Eynaud P, Malaurent J, Richet J L and Vitari C 2019 Information Systems Stakeholders *Information Systems Management* (Springer) doi: 10.1002/9781119332497.ch1
[27] Berger H 2010 Outsourcing and Communications, the Role of Stakeholder Analysis - A Case Study in Practice *AMCIS 2010 Proceedings* 66
[28] Bourne L 2002 The forgotten stakeholders: forming teams in an outsourced it environment https://mosaicprojects.com.au/PDF_Papers/P022_Forgotten_Stakeholders.pdf (accessed 02 October 2019)
[29] Lacity M C and Willcocks L P 2000 Relationship in IT outsourcing: a stakeholder perspective *Framing the Domains of IT Management. Projecting the Future Through the Past* ed R W Zmud (Cincinnati, OH: Pinnaflex Educational Resources, Inc) pp 355–84
[30] Robertson Ch J, Lamin A and Livanis G 2010 Stakeholder Perceptions of Offshoring and Outsourcing: The Role of Embedded Issues *Journal of Business Ethics* **95** 167-89 doi: 10.1007/s10551-009-0353-0

[31] Rajasekhar D, Banerjee P and Ponnam A 2017 Risks Perceived Regarding Recruitment Process Outsourcing: Stakeholder Concerns *Strategic Outsourcing an International Journal* **10(1)** 112-36 doi: 10.1108/JGOSS-07-2016-0023
[32] Tate W L, Ellram L M and Brown S W 2009 Offshore Outsourcing of Services: A Stakeholder Perspective *Journal of Service Research* **12** 56-72 doi: 10.1177/1094670509338617
[33] Standing G 2011 *Precariat: The New Dangerous Class* (Bloomsbury Academic)
[34] Likert R 1932 A Technique for the Measurement of Attitudes *Archives of Psychology* **140** 14
[35] Aubert B A, Dussault S, Patry M and Rivadr S 1999 Managing the Risk of IT Outsourcing *Proceedings of the 32nd Hawaii International Conference on System Sciences – 1999* doi: 10.1109/HICSS.1999.772972
[36] Bahli B and Rivard, S 2003 The information technology outsourcing risk: a transaction cost and agency theory-based perspective *Journal of Information Technology* **18** 211–21
[37] Watkins J C 2016 *An Introduction to the Science of Statistics: From Theory to Implementation* (USA: University of Arizona)
[38] Rynek pracy IT w 2018 https://nofluffjobs.com/static_vendor/raporty/rynek-pracy-it/assets/NFJ-Rynek-pracy-IT-2018.pdf/ (accessed 02 May 2019)

IOP Conf. Series: Materials Science and Engineering **710** (2019) 012035 doi:10.1088/1757-899X/710/1/012035

Shot peening intensity effect on bending fatigue strength of S235, S355 and P460 structural steels

W Macek[1,2], M Szala[3], M Kowalski[4], J Gargasas[5], A Rehmus-Forc[6] and A Deptuła[7]

[1] Energopiast sp. z o.o., Polna 12, 55-011 Siechnice, Poland
[2] Wroclaw School of Information Technology, ul. Ks. M. Lutra 4, 54-239 Wrocław, Poland
[2] Lublin University of Technology, Faculty of Mechanical Engineering, Department of Materials Engineering, 36 Nadbystrzycka Street, 20-618 Lublin, Poland
[4] Opole University of Technology, Faculty of Mechanical Engineering, Mikolajczyka 5, 45-271 Opole, Poland
[5] Vilnius Gediminas Technical University, Faculty of Mechanics, Department of Mechanical and Material Engineering, Saulėtekio al. 11, LT-10223 Vilnius, Lithuania
[6] State University of Applied Sciences in Elbląg, ul. Wojska Polskiego 1, 82-300 Elbląg, Poland
[7] Opole University of Technology, Faculty of Production Engineering and Logistics, Street Sosnkowskiego 31, 45-272 Opole, Poland

wojciech.macek@yahoo.com

Abstract. In this study, the effect of shot peening intensity of the structural steel specimens on the bending fatigue strength was investigated. Three structural steels S235JRG2, S355J2+N, P460NH were shot-peened with three levels of intensity. The results of fatigue tests were compared with non-peening processed samples. S-N curves in fatigue were determined in cyclic bending fatigue. The novel attempt to evaluate the bending fatigue, which relies on the 3D optical profiler measurement of the side-area of fractures, was proposed. All investigated steels present ferritic-pearlitic structure. Moreover, due to the peening process, refinement of the grains size was observed. Also, fractures were analysed with the 3D profiler. In the presented results of research, the highest level of peening gave the greater increase in fatigue life. For all types of investigated steels, shot-peening gave superior results of fatigue bending performance compared with unpeened specimens.

1. Introduction

Systematically increasing performance standards impose also more stringent safety standards for machine systems. This fact leads not only to the need to look for a new type of materials, but also to use material treatments to increase strength and durability of existing materials [1–9]. Many deterioration processes such as corrosion, abrasion, sliding, erosion, fatigue wear or even inaccurate maintenance of technological systems, shorten the operation time of industry components [10–18]. Commonly used in the industry are various processing methods for surface layer treatment. Apart from the conventional processes such as heat treatment, carburising or nitriding, the literature reports special treatment processes as laser surfacing [19], ion implantation [20], shot peening [21], thermal spraying [22], physical vapor deposition [23] and many more treatments that increase the operation lifetime of components. Despite the variety of treatments listed above, the shot peening technology seems the most cost-effective and promising for implementation into industry conditions. Scholtes and Vöhringer [24] characterises the shot peening process as a method that involves the mechanical interaction of hard particles of shot having sufficient kinetic energy with the surface of a workpiece of given geometry, defined chemical composition, and certain mechanical properties. Shot peening is usually carried out by the wheel method (shot is propelled by a motor-driven bladed wheel) or the air blast method (shot is introduced by an air blast machine, either by gravity or by direct pressure). The attainable near-surface changes in the material are dependent on the shot used, its parameters and type, and the condition of the workpiece [24].

The peening parameters have an influence on the surface layer performance. The metallic materials cause inhomogeneous plastic deformation of near-surface layers which finally result in compressive residual stresses, work-hardening and/or work-softening, or surface roughness development [24]. Shot peening with not accurate process parameters induces roughness on the surface that is linearly proportional to the velocity/intensity of the shot [25]. Thus, intensity of the peening on the operation properties is systematically investigated in the literature e.g. [21,26,27].

Shot peened steel is used for manufacturing the wide variety of components starting from simple structures, casings, machine components, and further for heavy industry and energetics industry. Shot peening is a process specifically designed to enhance fatigue strength of components which are subjected to high alternating stress [28]. The main advantage of the shot peening treatment is the possibility of imparting into the material compressive residual stresses useful from the fatigue life point of view [29–33]. Despite the wide range of research presented in the literature, materials after the treatment of shot peening sharps hard have not yet been adequately tested for fatigue phenomena depending on the process parameters. The type of work based on fatigue bending is especially rare but also needed [14,15,34].

Moreover, many methods for bending fatigue are employed in the industry. Most of them needs time-consuming tests. Therefore, Macek [35] proposed fractal analysis for bending-torsion samples or in the other work [36], presents the idea of fracture surfaces analysis. Thus, in a current paper an original method for evaluation of bending fatigue results is signalised. This method will be suitable for not-fractured samples and especially promising for fatigue samples tested with the failure criterion which is based on exceeding the critical amplitude of lever dislocation.

This study was subjected to untreated and shot-peened specimens made of three steel grades labelled S235JRG2, S355J2+N, P460NH. The main parameter which is guided during the research is the multiplicity of the re-process (process intensity) [27,37]. The material is subjected to a single, double, and threefold shot peening process of the surface.

The main aim of the study was to investigate the effect of shot peening intensity of S235, S355 and P460 steels on bending fatigue strength. Moreover, the new method for evaluation of bending fatigue strength with usage of computer methods that basis on an optical profiler results analysis, is presented.

2. Materials and methods

2.1. Materials

The materials used in this study were 12mm sheet metals made of three steel grades labelled as S235JRG2, S355J2+N, P460NH according to EN10028 and EN10025-1 standards [38]. In the following part of the work, in order to simplify the recording, the following designations were adopted: S235, S355 and P460. The mechanical properties of investigated materials are shown in Table 1 and the structure of the steel was observed on metallographic samples and quantitatively analysed with usage of Matlab software [38–40].

Table 1. Mechanical properties according to the manufacturers' certificates.

Material	Yield stress, Re (MPa)	Ultimate stress, Rm (MPa)	Young modulus, E (GPa)	Elongation, A (%)	Re/Rm	Hardness, HV
S235	291.0	424.5	210	30.0	0.69	132
S355	391.0	516.0	213	33.0	0.76	160
P460	616.5	717.0	205	24.6	0.86	212

2.2. Shot peening process and structure characterisation

The main parameter which is guided during the research is the multiplicity of the re-process (i.e. the process intensity). The material was subjected to a single, double, and threefold shot peening process. The steel-ball shot stream was propelled at 0.7 MPa. The peening process lasted about 4 minutes, shot granulation was 0.75-0.85 mm. The process was carried out under industrial conditions with comparable to the literature parameters [40–42]. The specimens were treated from four sides of the working part. The samples were subjected to a single, double, and threefold shot peening process of the surface. The surface quality of the peened samples was examined using optical microscope and optical profiler. The microstructure of the processed surface layer was observed on metallographic cross-sections with the usage of a light optical microscope.

2.3. Cyclic bending test

The fatigue research was carried out on the MZGS-100 fatigue test stand [39,43,44]. The tests were carried out under loading with the controlled force and loading frequency 28.8 Hz. The fatigue tests were performed under the stress ratio R = - 1 .The object of the study consisted of specimens with geometry and dimensions shown in Figure 1. Samples were cut from sheets 12 mm plate in thickness according to the rolling direction. 15 specimens of each material were prepared. Investigated specimens were subjected to different values of bending moment (cyclic bending test) [16,45,46]. The failure criterion was based on exceeding the critical amplitude of lever dislocation [47–49]. Statistical analysis of Stress-Life (S-N) fatigue data was performed according to ASTM standard [50].

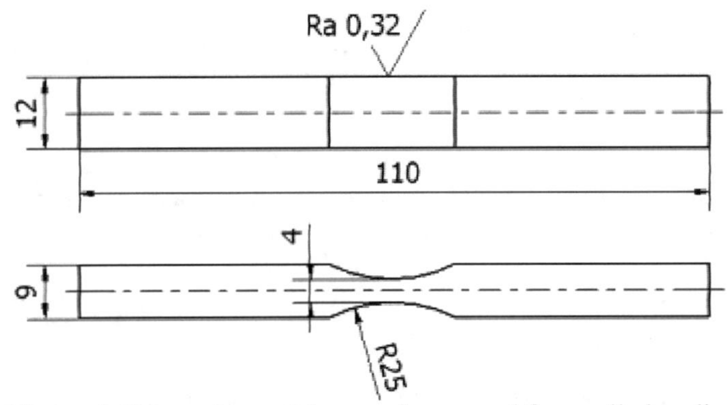

Figure 1. Dimensions of the specimen used for cyclic bending.

2.4. Characterisation of surface of peened samples with computer methods

Metallographic examinations of shot peened samples were made using the optical microscope [51,52]. The surface morphology surface analysis was performed using an optical 3D test stand Alicona G4 facilitating the acquisition of data sets at a high depth of focus according to the procedure described in [53]. The morphology of each steel peened with different peening intensity were comparatively analysed. The surface roughens parameters Sa, Sq, Ssk, Sku, Sp, Sv and Sz, calculated according to the ISO standard were compared on the basis of measurements done in specific areas of damage surfaces. Moreover, the facture analysis of the samples was conducted with usage of the optical profiler and special software which allows to conduct the analysis relating to surface morphology development.

IOP Conf. Series: Materials Science and Engineering **710** (2019) 012035 doi:10.1088/1757-899X/710/1/012035

3. Results and discussion

3.1. The effect of shot peening on structure and surface morphology

Three investigated types of structural steel, namely: S235, S355, and P460, presented ferritic-pearlitic structure however the microstructure of each steel differed in size of initial grain size visible in the core area in cross-sections of shot-peened in Figure 2. Also, quantitative metallographic results given in [39], confirmed decreasing volume content of ferrite (15,7%, 29,6% and 37,9%) and decreasing ferrite grains size (22,96%, 6,97% and 4,54%) for samples S235, S355 and P460, respectively. Analysis conducted after shot-peening process confirmed that the surface layer was deformed due to peening process, Figure 2., and refinement of the grains size was observed. The comparative analysis of the morphology of the shot peened surfaces indicated the influence of intensity on the surface layer structure Figure 3.

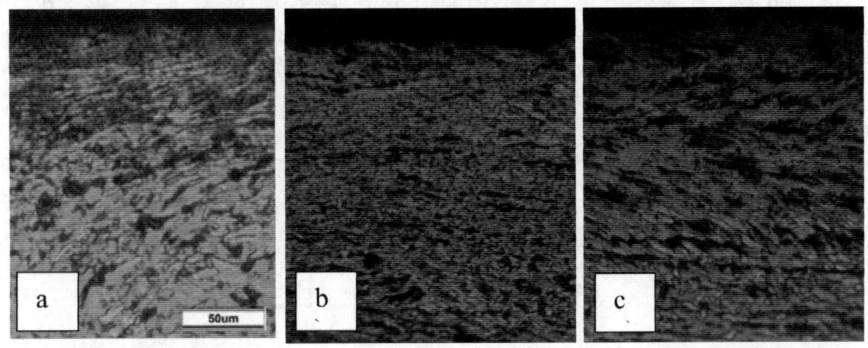

Figure 2. Metallographic cross-sections of specimens after shot peening a) S235, b) S355, c) P460.

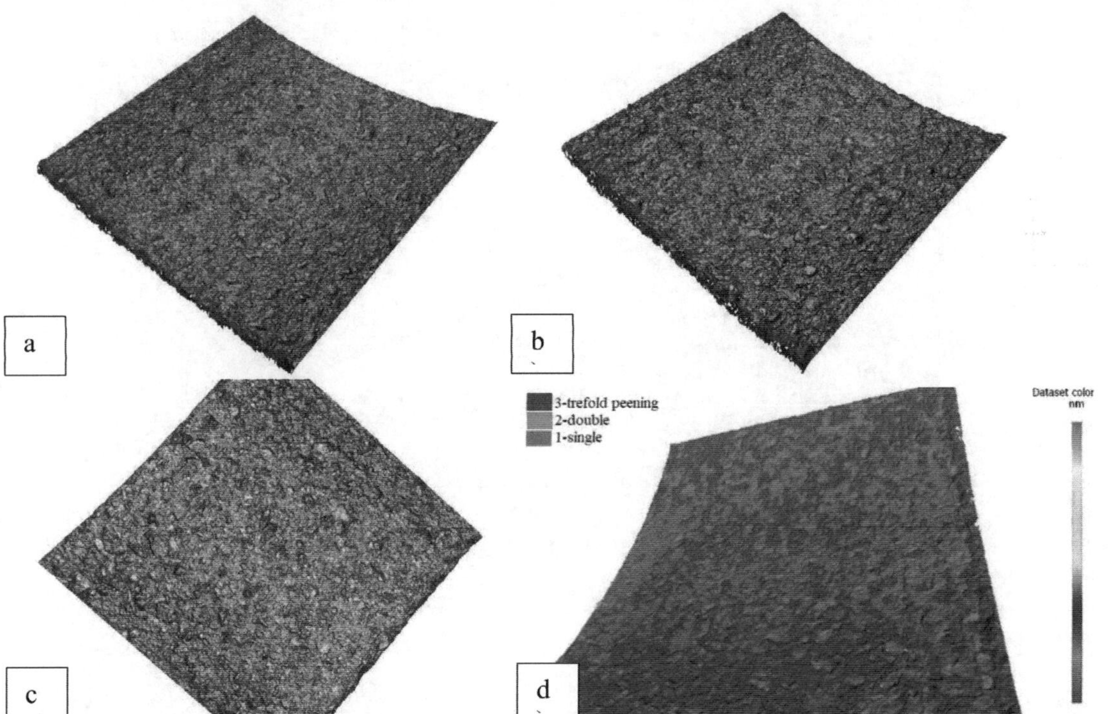

Figure 3. Characterisation of the surface plastic deformation due to intensity of the shot peening process of S235 steel. a) 1-single, b) 2-double c) 3-trefold peening d) comparison of surface morphology for various process parameters.

3.2. Fatigue tests results

Statistical analysis results of Stress-Life (S-N) fatigue data performed according to ASTM E739 standard, are presented in Figure 4. Thus, the plots compared peening intensity on the fatigue resistance.

General conclusion is that the S235 steel shot peening process increased the bending fatigue life of tested steel S235, S355 and P460 samples in reference to unpeened steel, Figure 4. However, due to the spread of the fatigue test results, it is difficult to clearly state the effect of peening process intensity on the bending fatigue. In the case of S355 steel, shot peening process increased the fatigue life of the material. In the case of two-times, there was a decrease in fatigue life compared to one and three-times. The reason for reduction of the fatigue characteristics after two-times peening may be the spread of results. Accuracy of constructed characteristics can improve the testing of more samples. The metallographic analysis of the material after two-times shot peening indicated occurrence of numerous microcracks caused by the selection of incorrect parameters of the process, which could have also affected the fatigue life of the specimens. Failure of shot peened steels under fatigue may be caused by crack propagation starting at locations where surface structural non-uniformities occurred, which were observed at the samples cross-sections. P460 steels showed a slight increase in fatigue life after the one-time process. After two and three-time shot peening, an increase in durability was observed, however, the indication of differences between durability for individual multiples of the process is difficult due to the dispersion of experimental studies.

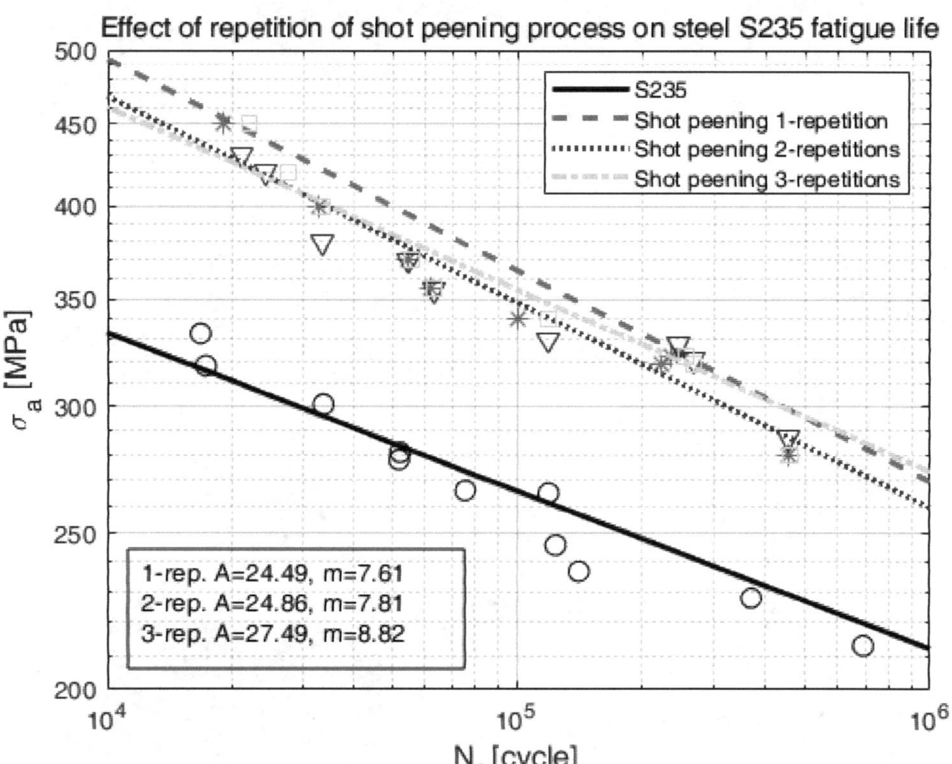

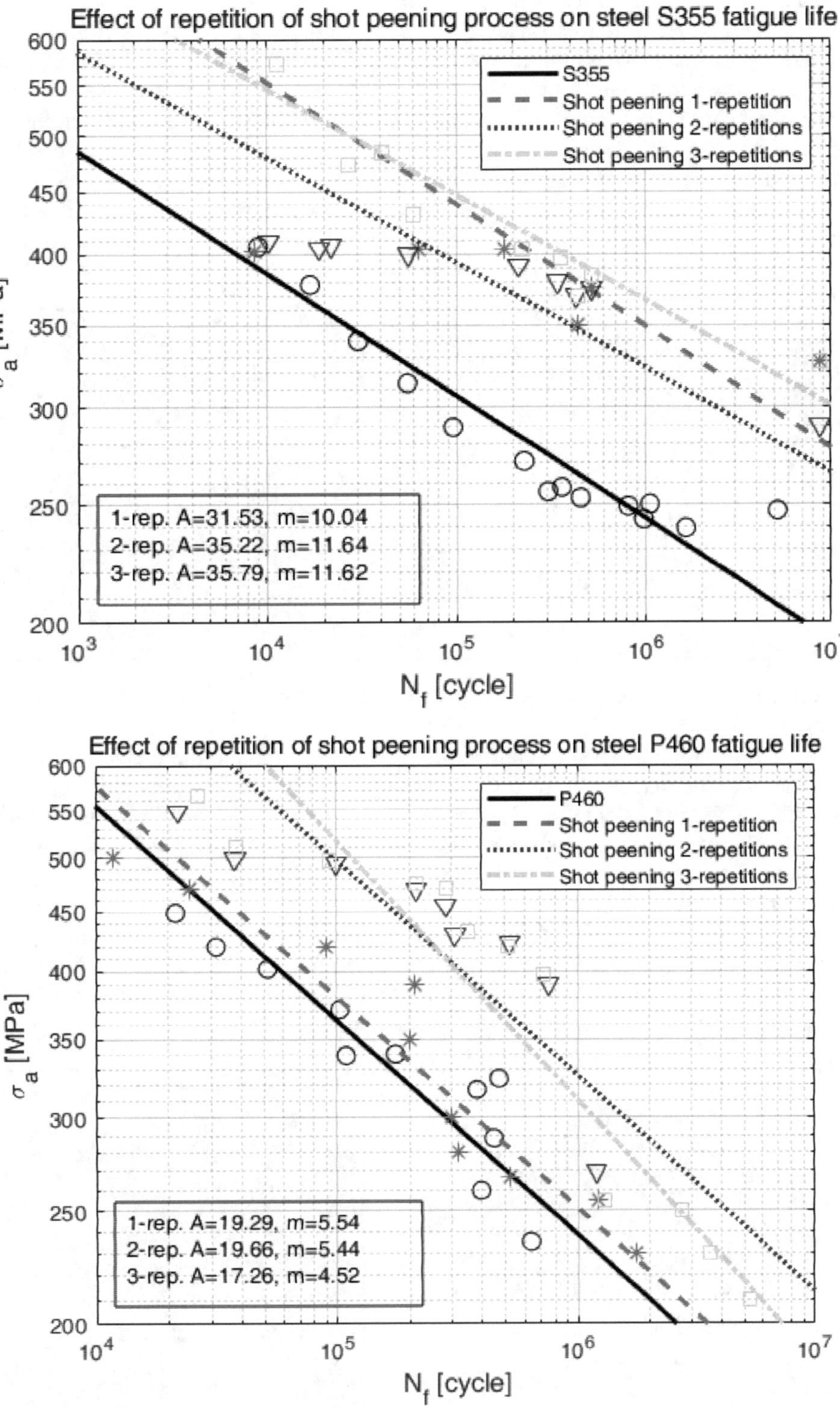

Figure 4. Fatigue characteristics of the untreated and shot peening processed steels S235, S355 and P460 with different process intensity.

IOP Conf. Series: Materials Science and Engineering **710** (2019) 012035 doi:10.1088/1757-899X/710/1/012035

3.3. Fractographic investigation with usage of an optical profilometer

After a series of fatigue tests, it was observed that the shot peening process improved the fatigue properties of the specimens. During fatigue tests, the materials cracked in narrow places, which allowed observation of cracks. All materials, both S235 and S355 as well as P460, after subsequent shot peening, cracked identically, i.e. the initiation of fatigue cracking occurred in the upper or lower part of the cross-section, which is presented in Figure 5. However, in the case of Steel P460 after a three-fold process shot peening in the range of a small number of cycles, i.e. 60-70 thousand samples were bursting analogously to those described earlier. But in the case of samples whose service life exceeded considerably one million cycles, the material cracked in the upper and lower part of surface of the tested specimen. It was a characteristic feature only for this material.

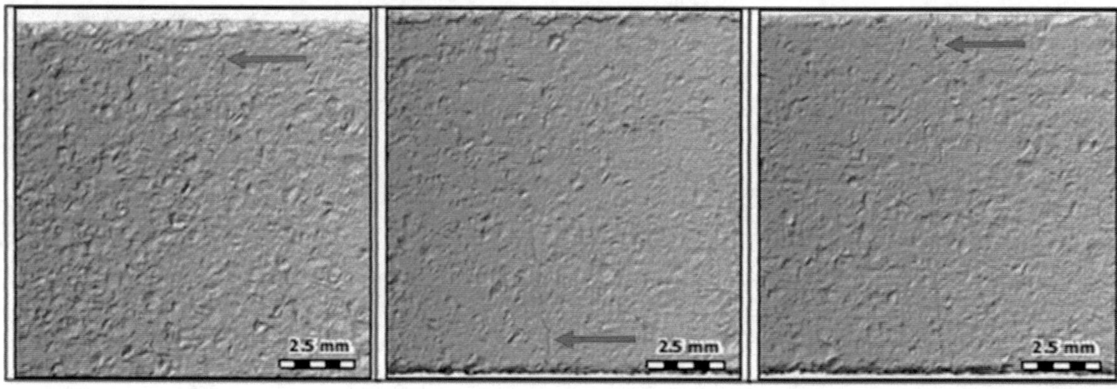

Figure 5. Surface fatigue-cracks (marked by arrows) after triple shot peening, S235, S355 and P460.

The 3D surface investigation allows to identify the crack and complete the crack analysis. In Figure 6 the exemplary sample side scan, see Figure 1, is presented. Surface topography measurement was made after bending test on the radius-side of the sample, see Figure 6. The profilometer technique clearly visualise cracks and area of the fatigue deformation. In Figure 7 the exemplary comparison of fracture areas is presented. The comparison of the centre area with the near crack area provides important information about the fracture mechanism and fracture resistance. The aim of this paper is to just signalise the idea and methodology of the investigation, thus it can be deduced from Figure 7 that the elaborated idea works and provides interceding information for fracture result interpretation especially for propagation mechanism. This quite unique attempt for bending fracture results analysis invented by Macek [35,36] is now being developed and will be published in our future work.

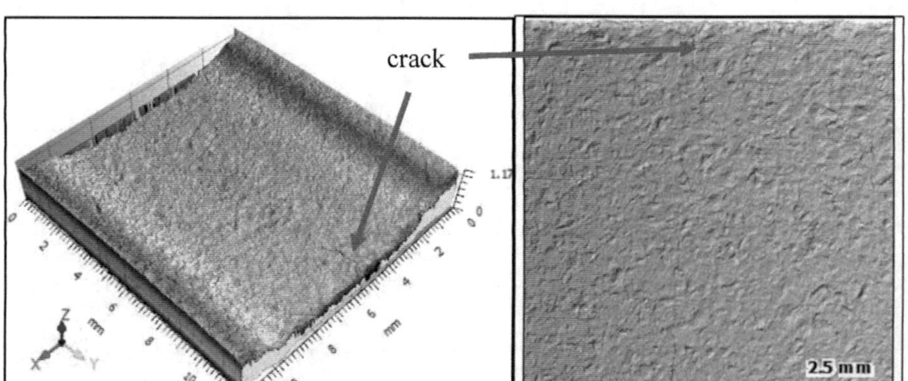

Figure 6. 3D view presenting the crack propagation after bend testing.

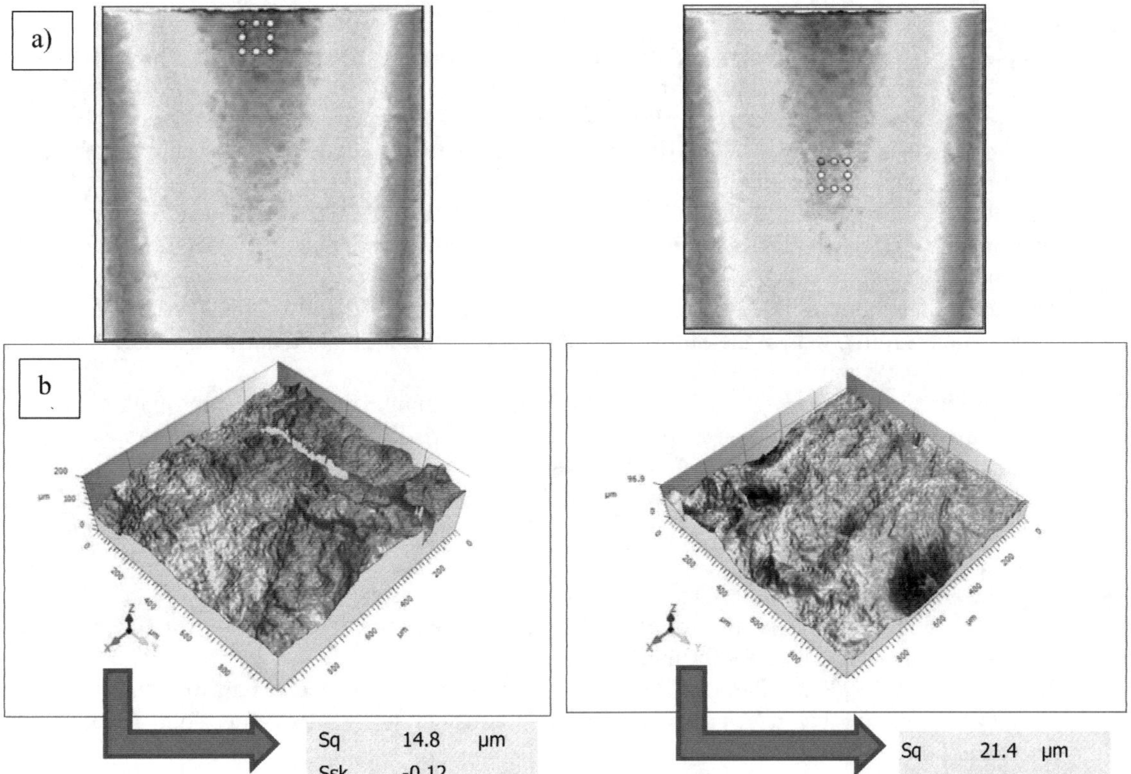

Sq	14.8	µm
Ssk	-0.12	
Sku	3.38	
Sp	54.2	µm
Sv	42.6	µm
Sz	96.9	µm
Sa	11.3	µm

Sq	21.4	µm
Ssk	-0.64	
Sku	5.71	
Sp	67.4	µm
Sv	138	µm
Sz	205	µm
Sa	16.0	µm

Figure 7. 3D view presenting the crack propagation after bend testing and roughness of marked areas: a) overview of the investigated area b) marked core area of the sample and roughness.

4. Conclusions

In the presented work, the effect of shot peening intensity of S235, S355 and P460 steels on bending fatigue strength was examined.

All examined steel grades an increase in fatigue life with respect to the reference (unpeened) material was noted. Shot peening for all tested steel grades had a positive effect in the case of low-cycle fatigue (LCF) and for single shot peening process.

In LCF range the highest level of peening gave the greater increase in fatigue life. In fatigue the intermediate level of peening gave results similar to those of the high level peening. These results are explained in terms of observed the surface damage.

The new method for evaluation of the bending fatigue strength with usage of the computer methods, that basis on the optical profilometer results analysis, was suggested.

5. References

[1] Wojciechowski S, Maruda R W, Barrans S, Nieslony P and Krolczyk G M 2017 Optimisation of machining parameters during ball end milling of hardened steel with various surface inclinations *Measurement* **111** 18–28

[2] Kovacı H, Hacısalihoğlu, Yetim A F and Çelik A 2019 Effects of shot peening pre-treatment and plasma nitriding parameters on the structural, mechanical and tribological properties of AISI 4140 low-alloy steel *Surface and Coatings Technology*

[3] Szala M, Dudek A, Maruszczyk A, Walczak M, Chmiel J and Kowal M 2019 Effect of atmospheric plasma sprayed TiO2-10% NiAl cermet coating thickness on cavitation erosion, sliding and abrasive wear resistance *Acta Phys. Pol. A* **136** 335–41

[4] Sanjurjo P, Rodríguez C, Peñuelas I, García T E and Belzunce F J 2014 Influence of the target material constitutive model on the numerical simulation of a shot peening process *Surface and Coatings Technology* **258** 822–31

[5] Deptuła A, Kunderman D, Osiński P, Radziwanowska U and Włostowski R 2016 Acoustic Diagnostics Applications in the Study of Technical Condition of Combustion Engine *Archives of Acoustics* 345--350

[6] Myoung S-W, Lee S-S, Kim H-S, Kim M-S, Jung Y-G, Jung S-I, Woo T-K and Paik U 2013 Effect of post heat treatment on thermal durability of thermal barrier coatings in thermal fatigue tests *Surface and Coatings Technology* **215** 46–51

[7] Gargasas J, Valiulis A, Gedzevicius I and Pokhmurska H 2015 The Research of Thermal Arc Sprayed Coatings Tribological Properties by Using Rubber Wheel Test *Solid State Phenomena* **220–221** 693–7

[8] GARGASAS J, Valiulis A, GEDZEVIČIUS I, Mikaliunas S, NAGURNAS S and Pokhmurska H 2016 Optimization of the Arc Spraying Process Parameters of the Fe–Base Mn-Si-Cr-Mo-Ni Coatings for the Best Wear Performance *Materials Science* **22**

[9] Deptuła A, Osiński P and Partyka M 2017 Identification of influence of part tolerances of 3PWR-SE pump on its total efficiency taking into consideration multi-valued logic trees *Polish Maritime Research* **nr 1**

[10] Szala M and Łukasik D 2018 Pitting Corrosion of the Resistance Welding Joints of Stainless Steel Ventilation Grille Operated in Swimming Pool Environment *International Journal of Corrosion* **2018**

[11] Hejwowski T, Marczewska-Boczkowska K and Kobayashi A 2013 A comparative study of electrochemical properties of metallic glasses and weld overlay coatings *Vacuum* **88** 118–23

[12] Fydrych D, Świerczyńska A, Rogalski G and Łabanowski J 2016 Temper Bead Welding of S420G2+M Steel in Water Environment *Advances in Materials Science* **16** 5–16

[13] Dybowski B, Szala M, Hejwowski T J and Kiełbus A 2015 Microstructural phenomena occurring during early stages of cavitation erosion of Al-Si aluminium casting alloys *Solid State Phenomena* **227** 255–8

[14] Lambert R D, Aylott C J and Shaw B A 2018 Evaluation of bending fatigue strength in automotive gear steel subjected to shot peening techniques. *Procedia Structural Integrity* **13** 1855–60

[15] Bianchetti C, Delbergue D, Bocher P, Lévesque M and Brochu M 2019 Analytical fatigue life prediction of shot peened AA 7050-T7451 *International Journal of Fatigue* **118** 271–81

[16] Owsiński R, Kamiński S, Niesłony A and Łagoda T 2015 Evaluation of fatigue life of steel using steel grain size: Evaluation der Ermüdung von Stahl in Abhängigkeit der Korngröße *Materialwissenschaft und Werkstofftechnik*

[17] Szala M, Szafran M, Macek W, Marchenko S and Hejwowski T 2019 Abrasion Resistance of S235, S355, C45, AISI 304 and Hardox 500 Steels with Usage of Garnet, Corundum and Carborundum Abrasives *Adv. Sci. Technol. Res. J.* **13**

[18] Szala M, Beer-Lech K and Walczak M 2017 A study on the corrosion of stainless steel floor drains in an indoor swimming pool *Engineering Failure Analysis* **77** 31–8

[19] Janicki D 2018 Microstructure and Sliding Wear Behaviour of In-Situ TiC-Reinforced Composite Surface Layers Fabricated on Ductile Cast Iron by Laser Alloying *Materials* **11** 75

[20] Kamiński M, Budzyński P, Szala M and Turek M 2018 Tribological properties of the Stellite 6 cobalt alloy implanted with manganese ions *IOP Conf. Ser.: Mater. Sci. Eng.* **421** 032012

[21] Żebrowski R and Walczak M 2019 Effect of the Shot Peening on Surface Properties and Tribological Performance of Ti-6Al-4V Alloy Produced by Means of DMLS Technology *Archives of Metallurgy and Materials* **Vol. 64, iss. 1**
[22] Łatka L, Niemiec A, Michalak M and Sokołowski P 2019 Tribological Properties of Al2O3 + TiO2 Coatings Manufactured by Plasma Spraying *Tribologia* **nr 1**
[23] Szala M, Walczak M, Pasierbiewicz K and Kamiński M 2019 Cavitation Erosion and Sliding Wear Mechanisms of AlTiN and TiAlN Films Deposited on Stainless Steel Substrate *Coatings* **9** 340
[24] Scholtes B and Vöhringer O 2016 Mechanical Surface Treatment *Reference Module in Materials Science and Materials Engineering* (Elsevier)
[25] Ismail S, Ahsan Q and Haseeb A S M A 2017 2.7 Recent Advances in Mechanical Surface Treatment *Comprehensive Materials Finishing* ed M Hashmi (Oxford: Elsevier) pp 171–9
[26] Żebrowski R and Walczak M 2018 The effect of shot peening on the corrosion behaviour of Ti-6Al-4V alloy made by DMLS *Advances in Materials Science* **18** 43–54
[27] Jamalian M and Field D P 2019 Effects of shot peening parameters on gradient microstructure and mechanical properties of TRC AZ31 *Materials Characterization* **148** 9–16
[28] Anon What is Shot Peening? Shot Peening Equipment - Wheelabrator
[29] Zhuang W Z and Halford G R 2002 Investigation of residual stress relaxation under cyclic load *International Journal of Fatigue*
[30] Wu D, Yao C and Zhang D 2018 Surface characterization and fatigue evaluation in GH4169 superalloy : Comparing results after finish turning; shot peening and surface polishing treatments *International Journal of Fatigue*
[31] Kamiński S, Kowalski M, Owsiński R and Łagoda T 2016 Influence of technological parameters of shot peening sharps hard to own stress level *Mechanik*
[32] Mahmoudi A H, Ghasemi A, Farrahi G H and Sherafatnia K 2016 A comprehensive experimental and numerical study on redistribution of residual stresses by shot peening *Materials & Design* **90** 478–87
[33] Zhou J, Sun Z, Kanouté P and Retraint D 2018 Reconstruction of residual stress and work hardening and their effects on the mechanical behaviour of a shot peened structure *Mechanics of Materials* **127** 100–11
[34] Maleki E, Unal O and Reza Kashyzadeh K 2018 Fatigue behavior prediction and analysis of shot peened mild carbon steels *International Journal of Fatigue* **116** 48–67
[35] Macek W 2019 Fractal analysis of the bending-torsion fatigue fracture of aluminium alloy *Engineering Failure Analysis* **99** 97–107
[36] Macek W 2019 Post-failure fracture surface analysis of notched steel specimens after bending-torsion fatigue *Engineering Failure Analysis* **105** 1154–71
[37] Chen M, Jiang C, Xu Z, Zhan K and Ji V 2019 Experimental study on macro- and microstress state, microstructural evolution of austenitic and ferritic steel processed by shot peening *Surface and Coatings Technology* **359** 511–9
[38] Kamiński S, Szymaniec M and \Lagoda T 2015 The influence of the ferrite and pearlite grain size on the SN fatigue characteristics of steel *Solid State Phenomena* vol 224 (Trans Tech Publ) pp 3–8
[39] Owsiński R, Kamiński S, Szymaniec M, Niesłony A and Łagoda T 2015 Evaluation of fatigue life of steel using steel grain size *Materialwissenschaft und Werkstofftechnik* **46** 1059–67
[40] Kamiński S 2017 *Fatigue durability of steel elementssubjected to shot blasting* (Opole, Poland: Opole University of Technology)
[41] Soyama H 2019 Comparison between the improvements made to the fatigue strength of stainless steel by cavitation peening, water jet peening, shot peening and laser peening *Journal of Materials Processing Technology* **269** 65–78

IOP Conf. Series: Materials Science and Engineering **710** (2019) 012035 doi:10.1088/1757-899X/710/1/012035

[42] Segurado E, Belzunce F J and Pariente I F 2018 Effects of low intensity shot peening treatments applied with different types of shots on the fatigue performance of a high-strength steel *Surface and Coatings Technology* **340** 25–35

[43] Macek W and Macha E 2015 The Control System Based on FPGA Technology For Fatigue Test Stand MZGS-100 PL *Archive of Mechanical Engineering* **62**

[44] Macek W and Macha E 2010 Energy-Saving Mechatronic System for Fatigue Tests of Materials under Variable-Amplitude Proportional Bending and Torsion *Solid State Phenomena*

[45] Macek W and Mucha N 2017 Evaluation of Fatigue Life Calculation Algorithm of the Multiaxial Stress-Based Concept Applied to S355 Steel under Bending and Torsion *Mechanics and Mechanical Engineering* **21** 935–51

[46] Macek W, Łagoda T and Mucha N 2017 Energy-based fatigue failure characteristics of materials under random bending loading in elastic-plastic range *Fatigue and Fracture of Engineering Materials and Structures*

[47] Springer M and Pettermann H E 2018 Fatigue life predictions of metal structures based on a low-cycle, multiaxial fatigue damage model *International Journal of Fatigue* **116** 355–65

[48] Pawliczek R and Prażmowski M 2015 Study on material property changes of mild steel S355 caused by block loads with varying mean stress *International Journal of Fatigue*

[49] Marciniak Z, Rozumek D and Macha E 2008 Fatigue lives of 18G2A and 10HNAP steels under variable amplitude and random non-proportional bending with torsion loading *International Journal of Fatigue*

[50] ASTM International 2012 *E739, Standard Practice for Statistical Analysis of Linear or Linearized Stress-Life (S-N) and Strain-Life (e-N) Fatigue Data*

[51] Nasiłowska B, Bogdanowicz Z and Wojucki M 2015 Shot peening effect on 904 L welds corrosion resistance *Journal of Constructional Steel Research* **115** 276–82

[52] González J, Bagherifard S, Guagliano M and Fernández Pariente I 2017 Influence of different shot peening treatments on surface state and fatigue behaviour of Al 6063 alloy *Engineering Fracture Mechanics* **185** 72–81

[53] Macek W and Wołczański T 2017 Analysis of fracture roughness parameters of S355J2 steel and EN AW-2017A-T4 aluminium alloy *ITM Web Conf.* **15** 06002

IOP Conf. Series: Materials Science and Engineering **710** (2019) 012036 doi:10.1088/1757-899X/710/1/012036

Information security risk assessment using the AHP method

J Zaburko and J Szulżyk-Cieplak

Lublin University of Technology, Faculty of Fundamentals of Technology, Nadbystrzycka Str. 38, 20-618 Lublin, Poland

jzaburko@gmail.com

Abstract. This paper is concerned with the issues of cybersecurity and it presents AHP multi-criteria analysis methodology applied for assessing the level of real threats arising from cyberspace that affect the information security. Based on the results of surveys conducted among employees of a selected state institution, we carried out an analysis of awareness of the dangers of cyberspace and ways employees respond to incidents in systems. Taking into account the research results, an appropriate mechanism for the provision of an adequate level of information security was proposed.

1. Introduction

The term security appears in numerous various contexts and is defined differently depending on the field of knowledge or the area of analysis. With the development of civilisation, the term has systematically expanded to include previously unknown challenges and threats, as well as new means and ways to eliminate them [1]. The modern broad understanding of security results from the equally broad understanding of its threats, which goes far beyond the classic military domain. In the 21st century, the humanity has faced threats of a completely new nature as a result of intensive civilisation and technological progress. The development and use of the achievements of science and technology has brought substantial benefits in everyday life, however, numerous dangers entailed. Natural resources are being increasingly exploited, and the natural environment is undergoing gradual degradation, which is particularly evident in relation to air quality, water resources and soil [2-5]. Industry and transport are the key sources of pollution, which has an adverse effect on the health and life of living organisms, is the cause of negative phenomena, both global and local, which in turn requires the use of comprehensive preventive measures and the use of sustainable and the most environmentally sound methods of waste management [6,7].

One of the key features of modern civilisation is the fact that the importance of information is continuously rising. A particularly relevant consequence of this distinction is systematic highlighting the role of information security. In the world dominated by electronics, almost every sphere of life is becoming more and more dependent on Information and Communication Technologies (ICT). ICTs come into an infrastructure which comprises cyberspace, containing data and key systems in many fields. Growing number of computer system security breaches and the safety of users of services offered by modern information and communication technologies means that maintaining security in cyberspace is currently one of the most critical problems at a national and international level. Threats to cyberspace security are becoming more and more advanced, and the scale of incidents is increasing every day. The problem of cybercrime related to illegal activities in cyberspace is also growing, including malware scams aimed at massive attacks on electronic banking accounts or identity theft [8, 9].

Cyberterrorism, which is a new form of crime in the virtual world, has become one of the greatest dangers. The reasons for the emergence of the phenomenon of cyberterrorism can be both political goals or purely for a profit.

Politically motivated cyberterrorism is inspired by the threat of destroying network infrastructure and intimidation or extortion of the government of a given country and citizens by far-reaching destabilising information activities in cyberspace (hacktivism) and physical attacks on ICT systems in order to obtain confidential data [10]. Cyberterrorists use virtual space as a tool for serving propaganda, spreading disinformation and recruiting by: hacking computers (hacking) and IT systems (cracking) to achieve material benefits accessing servers bypassing security (backdoor), eavesdropping on information transmitted between computers and intercepting passwords and logins (sniffing), impersonating another computer (IP spoofing), sending viruses and computer worms (malware), phishing confidential information (phishing) [11].

The typical targets of personalised attacks are:

- military systems storing information about the location of satellites, the location of military bases and long-range weapons as well as the communication systems used by the army of a given country;
- systems in institutions storing information important for the company's operations, such as information about reservations, clients, employed technologies, in which case the main perpetrators are most often employees or former employees who cooperate with competitors or take revenge for poor remuneration;
- systems included in the critical state infrastructure, i.e. systems of: state administration, armed forces and institutions dealing with national defence, energy, telecommunication, transport, services intended for emergency that store information important for national security [12].

Cybercrime is also a crime already known in the real world and is found in newer and newer forms. Identity theft, one of the most common crimes, consists in the deliberate use of another person's personal data, address of residence, PESEL number, most often to achieve financial benefit, which may be performed by setting up fake profiles on social networks, extorting bank loans through websites, using the personal data of another person against their will or using credit cards without its owner's consent [13].

Another example of cybercrime is computer fraud involving extortion of money and other tangible goods by entering into a binding purchase and sale agreement on the Internet, receiving the agreed amount and not sending the goods to the buyer. This phenomenon has been long known but it has intensified in recent years on a large scale.

When analysing the activities of cybercriminals, it may be noticed that they use manipulation techniques based on displaying fictitious goods on well-known sales platforms at prices substantially lower than those in stationary stores. In order for the transaction to be concluded, the buyer must pay the full amount for the ordered goods as well as for postal or courier services to the account indicated in the private message from the seller. The potential victim pays money to the account, the purchased goods never reach the buyer and no attempts to contact the seller bring any results.

Earning money has become a guiding principle for malware writers. Cybercriminals have focused their attention on the possibilities of encrypting data on computers that have received their viruses. In 2013, a virus called CryptoLocker appeared and ushered in the "ransomware" era. It was designed to block available user data and prevent access to the computer through encryption. Data recovery was only possible after making prompt payment of $300, which cybercriminals demanded by displaying information on the computer screen. A method called "asymmetric cryptography" was used to encrypt files [14]. This is a type of cryptography that uses sets of two related keys with a length of 330-660 characters for encryption, but one of the keys must be publicly made available to the recipient. The other key is private and known only to the creator of the virus so, in order to obtain it, one is required to make a payment to his account within a certain time. After ignoring the message and on the expiry of the time limit, the recovery and decryption of files becomes impossible. Opening a crafted, infected attachment received via e-mail usually caused virus infection of files on the disk. These files were usually orders to pay overdue invoices, underpayments of invoices from the mobile network operator or information from the delivery company about the inability to deliver a parcel.

In 2017, a more dangerous version of CryptoLocker appeared, known as WanaCrypt0r 2.0, which got into computers constituting the critical infrastructure of the attacked state. Due to a massive attack, the virus caused extensive damage by disabling traffic management systems, the work of hospitals, banks and stock exchanges as well as airports and train stations [15].

Hosting companies that make their disk space available for websites or online stores are also highly vulnerable to cyberattacks by using security breaches of crafted software that can break the security of servers storing sensitive data.

According to the website, niebezpiecznik.pl, on 18 December 2018, there was an unauthorised breach of security of the online store, morele.net, the data from whose customer database was obtained. The stolen database contained about 2 million records, including name, surname, e-mail address, password hash, and phone numbers. Breaking of passwords hashes by using "rainbow tables" or specially designed software turned out to be only a matter of time because after two days it was revealed that a criminal had already broken almost 350 thousand passwords. Customers of the online store were notified of a data leak in accordance with the GDPR Directive of 25 May 2018 and also informed the General Inspector for Personal Data Protection.

Company/Institution information resources protection is, therefore, a significant challenge for cybersecurity specialists. Cybersecurity consists in the implementation and proper management of the technical security of an ICT system [16].

Effective security of processed information according to PN-ISO / IEC 27001 should be ensured by:

- confidentiality - the information is available only to those users who have received appropriate permissions and have appropriate security information and declarations;
- integrity - available information is complete, reliable and has not been changed in any way;
- availability - authorised users have access to information at any time when they need it [17].

On 22 January 2015, the President of the Republic of Poland signed the document "Cybersecurity Doctrine of the Republic of Poland," which set out the strategic directions of actions aimed to provide a high level of cybersecurity of the Republic of Poland in the cyberworld. This document indicates the need for the development of state-controlled ICT systems and technologies while maintaining compliance with technologies offered by NATO. Regular security audits were also recommended.

The document of particular importance for the state's cybersecurity is the Act of 5 July 2018 on the National Cybersecurity System, which contains the provision on the creation of Computer Security Incident Response Teams (CSIRT MON, CSIRT NASK and CSIRT GOV). This is the first legal act of this kind in Poland where public administration and the telecommunications industries have also been included. This document imposes a degree of responsibility for maintaining the efficiency of the national critical infrastructure, established pursuant to earlier acts and legal acts, by government ministries responsible for the structure of the state [18,19].

One of the methods of efficient and secure exchange of information is the introduction of mandatory provisions regarding information processing. This document is called a security policy and is a normative document, constituting a set of regulations, rules and procedures according to which the institution manages information and provides IT resources [20-22]. It includes network infrastructure, hardware and software because they make it possible to process information quickly and securely. Lack of implemented procedures may result in excessive chaos in determining the quality and importance of processed information and affect the risks caused by incidents of legal or disciplinary consequences. It is usually said that the weakest link that constitutes a threat in the security chain, is a man because their careless action may lead to the disclosure of information. This may happen when we are dealing with people who are poorly acquainted with broadly understood computerisation. Therefore, it should be noted that when creating security policy rules, all regulations regarding management, exchange, storage and processing of information should have clearly defined rules to be followed by employees. It has been assumed that the above-mentioned document should be submitted for signature after having read it. Another task is to define the security method of the information processing system. The introduced security policies specify the rules for creating accounts, user identification, the principle of accountability of people logging into the system, requirements as to the length of passwords, its complexity and period of validity in the system. In order for the system to function properly and safely, the managers of the organisational units, in which the system elements are located, are responsible for maintaining the documentation that includes: a list of system users, a list of workstations, a list of active network devices and physical topology of the network.

The method of entering and outputting data is also important for proper system security. It should be determined whether users ought to have independent access to peripheral devices, such as printers and scanners, external or internal CD drives or flash drives used for recording and archiving data, but which are also the source of potential threats. Contingency plans should also be developed in the event of a power outage or physical IT storage medium damage. The antivirus policy for the system should also be specified.

The updated operating system and antivirus programme protect a workstation against infections originating from the network and which may get into the system while performing work duties. Attention should also be paid to crashes and other abnormal functioning of the system that may indicate an attack or a real security breach.

The organiser of the system is required to develop procedures for effective crisis management. Depending on the type and scope of system operation, you should additionally develop procedures for responding to: information system failure and unavailability of the service, malicious code - reporting an incident to CSRIT GOV, violation of confidentiality and data integrity, and misuse of the service in the system.

Procedures for responding to crisis should include provisions regarding: a correct description of the problem encountered, including its scope, identification of the cause of the incident, planning corrective actions to prevent the reoccurrence of the event and how to document the solution to the problem [21, 23].

Striving to eliminate losses or minimise them is an indispensable component of the proper functioning of the institution. In this aspect, it is necessary to properly manage security and manage risks. In order to ensure the appropriate level of security in an institution, it is necessary to conduct a detailed analysis of the resources at risk, taking into account potential threats. Risk estimation is necessary in the scope of detailed identification of threats to information security and an indication of weaknesses that may occur in a given institution. This approach allows both to identify types of risk based on potential threats and to learn techniques and methods to mitigate them, which will consequently result in the institutions' management implementing appropriate decisions as part of raising the level of information security in the institution

2. Materials and methods

As part of the study, an analysis of the information security system was carried out in one of the state administration sector institutions in Lublin. Identification of potential threats and estimation of risks of information loss in the institution was performed by means of the method of hierarchical analysis of the AHP problem. In the proposed solution, to determine the weights of individual criteria, the experience and knowledge of employees were used. The respondents in the survey determined the impact of individual threats on the risk of information loss. The used method also made it possible to carry out an analysis of the respondents' reliability, which makes the results more objective and reliable.

2.1. The Analytic Hierarchy Process (AHP)

The hierarchical problem analysis (AHP) method, also known as the Saaty method, is one of the best-known multi-criteria decision support methods in the world [24]. Saaty proposed this method for use in many areas to facilitate optimal choices when a decision-maker has more criteria to evaluate different decision options [25]. The advantages of the AHP method, such as flexibility, transparency and ease of use, objectivity of variant selection, the ability to compare both qualitative and quantitative factors, determine its high popularity and wide application, both in scientific research and solving decision problems [26], including risk analysis [27-29].

The analysis of the decision problem by the AHP method takes place in two phases. The first phase involves developing a hierarchical representation of the problem (Figure 2.). The main goal (problem description) is at the highest level of the hierarchy. At a lower level, the criteria are defined that have an impact on the solution of the analysed problem, and even lower are the sub-criteria that in turn affect the main criteria. The goal of phase two is to generate grades from a mutual comparison of criteria. The criteria are evaluated by a pairwise comparison of elements at each level of the hierarchical model and related to the element at a higher level. A nine-degree, bipolar scale of the importance criteria is used:

1-equal significance, 3-slight advantage, 5-strong advantage, 7-very strong advantage, 9-absolute advantage. The numbers 2, 4, 6, 8 are intermediate values and are used only when necessary. The determined comparisons form a square matrix of pairwise comparisons (*n* x *n*) A={a_(*ij*)}_(1≤*i,j*≤n), where a_(*ij*)∈{1,2,...,9} for 1≤*i*<*j*≤*n*, and *n* is the number of compared pairs of elements (Figure 1.).

In the matrix, *n(n-1)*/2 reversible pairwise comparisons are made, for which $a_{ij} = 1/a_{ji}$. A characteristic feature of this matrix is the diagonal, whose elements have the value of 1 [25].

$$A = \begin{bmatrix} 1 & a_{12} & \dots & a_{1n} \\ 1/a_{12} & 1 & \dots & a_{2n} \\ \dots & \dots & \dots & \dots \\ 1/a_{1n} & 1/a_{2n} & \dots & 1 \end{bmatrix}$$

Figure 1. Comparison matrix [30].

In the next step the pair comparison matrix A is transformed (normalised) into matrix $B=[b_{ij}]$, whose individual elements are equal to:

$$b_{ij} = \frac{a_{ij}}{\sum_{i=1}^{n} a_{ij}} \tag{1}$$

The weights of the evaluated elements are determined as arithmetic means of rows of the normalised matrix B:

$$w_i = \frac{1}{n}\sum_{j=1}^{n} b_{ij} \tag{2}$$

Ratings from experts are not always completely objective; in order to eliminate them the CR inconsistency factor has been introduced:

$$CR = \frac{CI}{RI}.100\% \tag{3}$$

where: *CI* – consistency (consequence) index, *RI* – random index

The CR coefficient determines to what extent the mutual comparisons of the validity of the characteristics are inconsistent (inconsequent). It is assumed that to consider the results as consistent, the value of the CR coefficient cannot exceed 0.10 (10%).

The basis for calculating the CR coefficient is the determination of the so-called largest eigenvalues (λ_{max}) of matrix A. They are calculated in a simplified way as the sum of the products of the sum of the values of comparisons in each column of matrix A and the appropriate weighting factor for a given element:

$$\lambda_{max} = \sum_{i=1}^{n} \quad \sum_{j=1}^{n} a_{ij} \cdot w_i \tag{4}$$

The consistency factor CI is then calculated:

$$CI = \frac{\lambda_{\max - n}}{n-1} \tag{5}$$

Tables of the RI index were generated by Saaty based on the simulations for 500,000 matrices and their values are presented in Table 1 [31].

Table 1. RI index values [31].

n	1	2	3	4	5	6	7	8	9	10	11	12	13	14	15
RI	0	0	0.52	0.89	1.11	1.25	1.35	1.4	1.45	1.49	1.51	1.54	1.56	1.57	1.58

2.2. Hierarchical model development

In order to determine the importance of individual information security threats in the analysed institution, a set of assessment criteria and sub-criteria was established. Based on the analysis of the literature on the subject [16,32], three main criteria were identified: human-dependent, technical and random threats. The hierarchy of the criteria for assessing the risk of information loss in the institution is shown in Figure 2.

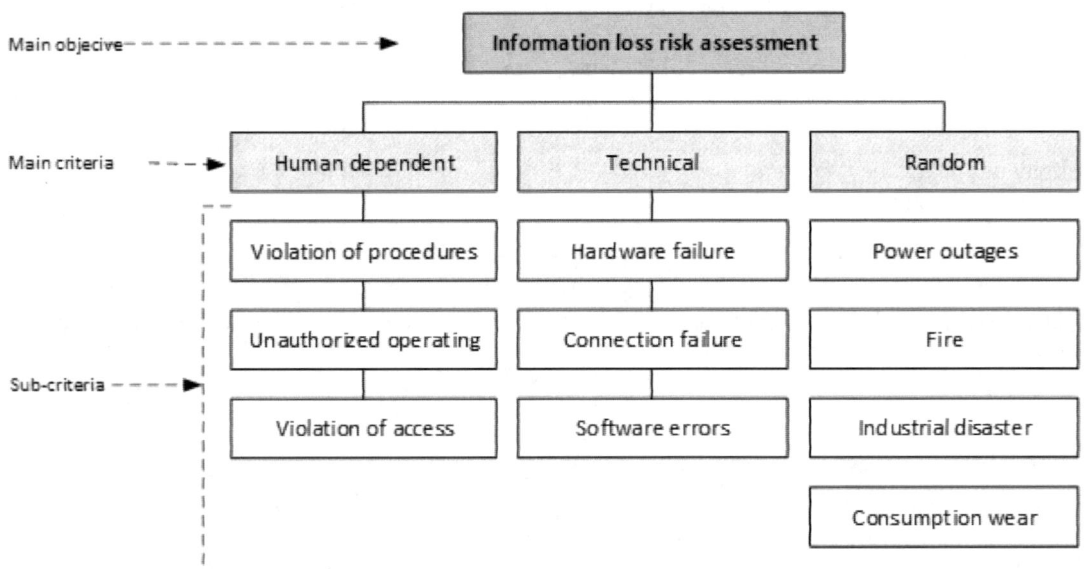

Figure 2. The hierarchical model of the problem (own work).

2.3. Course of research

By definition, the AHP method is based on the assessments of one decision-maker, or assessments of a brainstorming group. For the purposes of research, it was assumed that for correct comparisons, it is important to have theoretical or practical knowledge in the field of the analysed issues and to assess the weight of criteria, the assessments of experts, i.e. employees of institutions having daily contact with information processing, were used. To this end, a questionnaire was prepared which assessed the individual criteria, comparing them in pairs at each level of the hierarchy, giving importance to one of them according to a given scale. The research used a graphic presentation in the form of a table containing a bipolar verbal scale with a midpoint in a vertical arrangement. A fragment of the questionnaire and the method of responding are presented in Figure 3.

Violation of procedures (VP)		(VP) is definitely more important than (UO)
		(VP) is a lot more important than (UO)
	x	(VP) is much more important than (UO)
		(VP) is not much more important than (UO)
		Both factors are equally important
		(UO) is not much more important than (VP)
		(UO) is much more important than (VP)
		(UO) is a lot more important than (VP)
Unauthorised operation (UO)		(UO) is definitely more important than (VP)

Figure 3. Example of pairwise assessment of the selected criteria (own work).

3. Results and discussion

The study involved 8 employees who completed specialist training in information security, including 2 people who were appointed as Personal Data Protection Inspectors. Bearing in mind the course of conduct presented in the AHP method, weights of individual criteria (main criteria and sub-criteria) were calculated in relation to each of the respondents. Sample calculation results are presented in Table 2. and Table 3.

Table 2. Weight values for the main criteria and CR inconsistency coefficients for each respondent (own work).

	Employee 1	Employee 2	Employee 3	Employee 4	Employee 5	Employee 6	Employee 7	Employee 8
Human Dependent	0.55	0.39	0.59	0.63	0.33	0.46	0.29	0.22
Technical	0.37	0.50	0.32	0.32	0.53	0.48	0.50	0.67
Random	0.08	0.11	0.09	0.05	0.14	0.06	0.21	0.11
CR index	**0.092**	**0.071**	**0.085**	0.121	0.173	**0.062**	0.182	0.224

Table 3. Weight values for the "human-dependent" sub-criterion and CR inconsistency coefficients for each respondent (own work).

	Employee 1	Employee 2	Employee 3	Employee 4	Employee 5	Employee 6	Employee 7	Employee 8
Violation of procedures	0.62	0.35	0.55	0.45	0.51	0.39	0.41	0.29
Unauthorised Operating	0.19	0.31	0.22	0.44	0.33	0.42	0.37	0.42
Violation of Access	0.19	0.34	0.23	0.11	0.16	0.19	0.22	0.29
CR index	0.152	**0.061**	**0.085**	0.111	0.173	**0.052**	**0.072**	0.194

Using the calculated values of CR inconsistency coefficients, the level of reliability of individual respondents was determined. To assign weights to individual respondents, the AHP method was used, comparing CR indexes with each other and using the assessment scale presented in Table 4. [33]. The calculated weights of individual respondents are summarised in Table 5. Value 0 was assigned to respondents for whom CR> 0.10.

Table 4. The comparative scale of respondents' reliability [33].

Points	1	2	3	4	5	6	7	8	9
The r difference of CR indexes	$r<$ 0.02	0.02 $\leq r<$ 0.03	0.03 $\leq r<$ 0.04	0.04 $\leq r<$ 0.05	0.05 $\leq r<$ 0.06	0.06 $\leq r<$ 0.07	0.07 $\leq r<$ 0.08	0.08 $\leq r<$ 0.09	$r\geq$ 0.09

Table 5. Values of weights by individual respondents (own work).

	Employee 1	Employee 2	Employee 3	Employee 4	Employee 5	Employee 6	Employee 7	Employee 8
Main criterion	0.301	0.187	0.205	0	0	0.199	0.108	0
Sub-criterion „Human dependent"	0	0.232	0.177	0	0	0.252	0.339	0
Sub-criterion „Technical"	0.379	0	0	0.268	0.144	0	0	0.209
Sub-criterion „Random"	0.301	0	0	0.222	0.201	0	0.276	0

The total weights of the i-th criterion W_i were calculated using the formula [33]:

$$W_i = \sum w_{ij} \cdot w_{rj} \tag{6}$$

where: w_{ij} - weight of the i-th criterion according to the j-th respondent, w_{rj} - weight of the j-th respondent.

Based on formula 6, weights of the main criteria of information loss risk assessment in the analysed institution were calculated, which allowed estimating the significance of individual groups of information loss threats (Figure 4.).

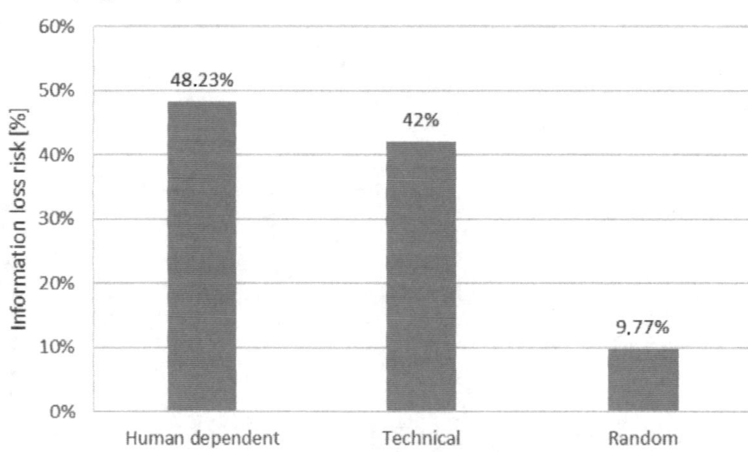

Figure 4. Significance of individual groups of information loss threats (own work).

Based on the results obtained from the computations, it can be concluded that the risks associated with the loss of information most often occur due to human factors (48.23%). Technical hazards are listed second (42%) and the least impact on the risk of information loss is random events that have reached significance at (9.77%).

Similarly, calculations were made to determine the weights of individual sub-criteria for each of the main criteria. Figure 5 shows the estimated risk of information loss for all sub-criteria adopted.

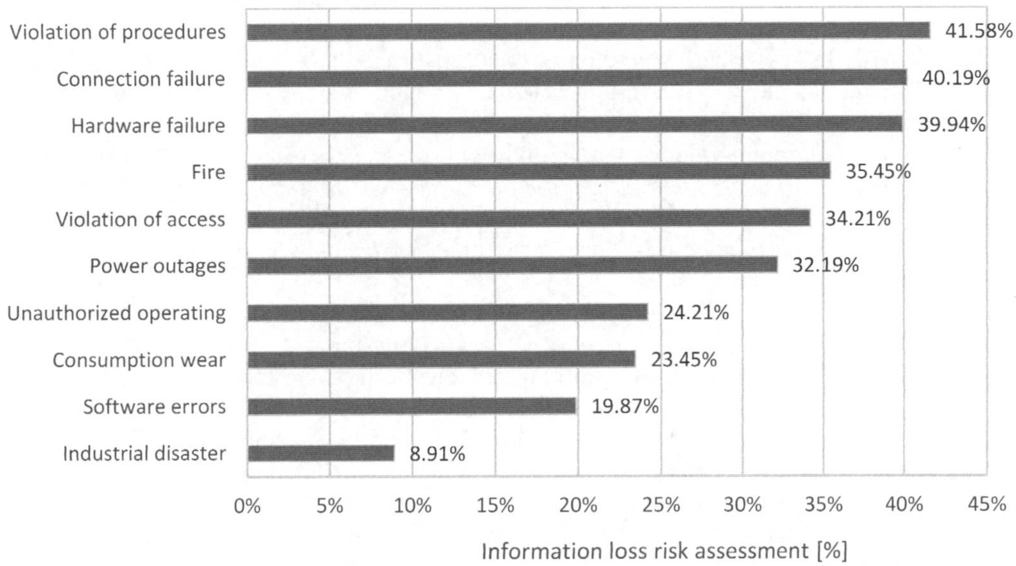

Figure 5. Risk of information loss in relation to individual threats (own work).

Thus obtained results served to determine which of the considered threats is the most real and which is the least. The most real human-dependent threat is the violation of the institution's procedures (41.58%), unauthorised operating (34.21%) and violation of access (24.21%). Random threats related to an industrial disaster have the least impact on the analysed risk of information loss. The risk of this phenomenon has been estimated at (8.91%).

4. Conclusion

The study conducted at the public institution enabled us to make a subjective assessment of the awareness of employees of a selected state institution regarding the threats of cyberspace. It should also be emphasised that the ability to measure non-compliance using this calculation procedure is seen as an advantage of the AHP method because it allows checking whether the experts participating in the survey answered the questions presented in a thoughtful and not accidental manner. According to the assessment of experts selected from the examined institution, information is usually lost due to human factors. Failure to comply with the guidelines contained in internal procedures may result in the violation of the information security policy, which clearly defines the rules related to access to data and work with information. Actions that should be taken to help employees understand the importance of cyberspace security include the intensification of training frequency and verification of knowledge obtained through tests conducted during training. Constant cooperation with IT staff and reporting observed irregularities in the operation of Information and Communication Technologies systems allows taking effective measures to prevent information disclosure or data leakage.

From the point of view of information protection, it is also recommended to prohibit the use of wireless technologies that pose a potential threat of being hacked into this network. It is also worth paying attention to the mobile solutions used in companies. These solutions guarantee the security of transmitted information in the form of encrypted connections, location of a lost or stolen device, information management on the phone from two levels - the user and the employee. Such a solution, although not cheap, as the final recipient of the technology is obliged to accept the offer of only one producer, is effective at the level of information security. It should be remembered, however, that the weakest link will always be a man who, whether consciously or not, can effectively deprive institutions of data and information.

5. References

[1] Fałdowski M 2018 Współczesny wymiar bezpieczeństwa *Zeszyty Naukowe SGSP **66(2)*** 109-22
[2] Bieganowski A, Józefaciuk G, Bandura L, Guz Ł, Łagód G and Franus W 2016 Evaluation of hydrocarbon soil pollution using E-nose *Sensors* ***18***(8) 2463
[3] Guz Ł, Łagód G, Jaromin-Gleń K, Guz E and Sobczuk H 2016 Assessment of batch bioreactor odour nuisance using an e-nose *Desalination and Water Treatment* ***57***(3) 1327-35
[4] Łagód G, Malicki J, Chomczyńska M and Montusiewicz A 2007 Interpretation of the results of wastewater quality biomonitoring using saprobes *Environmental Engineering Science* ***24***(7) 873-80
[5] Ratner B 2018 *Environmental security: dimensions and priorities. Scientific and Technical Advisory Panel to the Global Environment Facility* Washington DC
[6] Kamavisdar A 2010 Heavy metal pollution and their effects on environmental health *Indian Journal of Environmental Protection* ***30***(4) 335-40
[7] Widomski M, Gleń P and Łagód G 2017 Sustainable Landfilling as Final Step of Municipal Waste Management system *Problems of sustainable development* ***12***(1) 147-55
[8] Damphousse K and Smith B 1998 The Internet: A Terrorist Medium for the 21st Century," in Kushner, H. W. (ed.), *The Future of Terrorism: Violence in the New Millennium*, London: Sage 208–24
[9] Denning D E 2002 *Wojna Informacyjna i Bezpieczeństwo Informacji* (WNT Warszawa)
[10] Kołodziejczyk R 2017 Nowa odsłona terroryzmu - cyberterroryzm *Rocznik Bezpieczeństwa Międzynarodowego **17(2)*** 147-55
[11] Chen T, Jarvis L and Macdonald S 2014 *Cyberterrorism: Understanding, Assessment and Response* (Springer New York)
[12] Kowalewski J and Kowalewski M 2014 Cyberterroryzm szczególnym zagrożeniem bezpieczeństwa państwa *Telekomunikacja i techniki informacyjne **1-2*** 24-32
[13] Weijer van de S, Leukfeldt R and Bernasco W 2019 Determinants of reporting cybercrime: A comparison between identity theft, consumer fraud, and hacking *European Journal of Criminology* ***1***(4) 486-508
[14] Wobst R 2002 *Kryptologia. Budowa i Łamanie Zabezpieczeń* (Wydawnictwo RM Warszawa)
[15] Terlikowski M and Madej M 2009 *Bezpieczeństwo Teleinformatyczne Państwa* (Polski Instytut Spraw Międzynarodowych Warszawa)
[16] Górny P and Krawiec J 2016 Cyberbezpieczeństwo - podejście systemowe *Obronność – Zeszyty Naukowe Wydziału Zarządzania i Dowodzenia Akademii Sztuki Wojennej* ***2***(18) 78
[17] Kwiecińska M 2014 Wykorzystywanie mediów dla podniesienia bezpieczeństwa, potrzeby i sposoby realizacji *Obronność – Zeszyty Naukowe Wydziału Zarządzania i Dowodzenia Akademii Sztuki Wojennej* ***3***(11) 75-86
[18] *Doktryna Cyberbezpieczeństwa Rzeczypospolitej Polskiej* 2015 (BBN Warszawa)
[19] *Krajowe ramy polityki cyberbezpieczeństwa Rzeczypospolitej Polskiej na lata 2017–2022* 2017 (Ministerstwi Cyfryzacji Warszwa)
[20] PN-ISO/IEC 27001:2007, *Technika Informatyczna - Techniki bezpieczeństwa - Systemy Zarządzania Bezpieczeństwem Informacji - Wymagania* 2007 (PKN Warszawa)
[21] PN-ISO/IEC 27001:2014 - *Technika Informatyczna - Techniki Bezpieczeństwa - Systemy Zarządzania Bezpieczeństwem Informacji - Wymagania* 2014 (PKN Warszawa)

IOP Conf. Series: Materials Science and Engineering **710** (2019) 012036 doi:10.1088/1757-899X/710/1/012036

[22] PN-EN 50131-1 – *Systemy Alarmowe Systemy Sygnalizacji Włamania i Napadu Wymagania systemowe* 2009 (PKN Warszawa)
[23] Biernacki B and Kalman L 2018 *Systemy i Sieci Teleinformatyczne Sił Zbrojnych RP - Wielorakie Aspekty Bezpieczeństwa Cyberprzestrzeni* (Akademia Sztuki Wojennej Warszawa)
[24] Saaty T 1972 *An Eigenvalue Allocation Model for Prioritization and Planning* University of Pennsylvania 28-31
[25] Saaty T 1980 *The Analytic Hierarchy Process: Planning Priority Setting Resource Allocation* (New York International Book Company New York)
[26] Saaty T 2008 Decision making with the Analytic Hierarchy Process *International Journal of Services Sciences* ***1***(1) 83-98
[27] Janackovic G, Savic S and Stankovic I 2013 Selection and ranking of occupation al safety in dicators based on fuzzy AHP: A case study in road construction companies *South African Journal of Industrial Engineering* ***24***(3) 175-89
[28] Meng M and Enping L 2015 The Application Research of Information Security RiskAssessment Model Based on AHP *Methodournal of Advances in Information Technology* **6**(4) 201-206
[29] Ning X and Dong MZ 2011 The Research of Information Security RiskAssessment Method Based on AHP *Advanced Materials Research* ***187*** 575-80
[30] Adamus W and Gręda A 2005 Wspomaganie decyzji wielokryterialnych w rozwiązywaniu wybranych problemów organizacyjnych i menedżerskich. *Badania Operacyjne i Decyzje* ***15***(2) 10
[31] Alonso J and Lamata M 2005 *A Statistical Criterion of Consistency in the Analytic Hierarchy Process* (MDAI 2005 Springer-Verlag Berlin Heidelberg)
[32] Whitman M 2003 Enemy at the gates: Threats to information security *Comunication of the ACM* ***48***(8) 91-96
[33] Łękawa Z 2011 Metoda AHP w ocenie zdolności kredytowej jednostki samorządu terytorialnego *Annales Universitatis Maria Curie-Skłodowska* ***2*** 237

IOP Conf. Series: Materials Science and Engineering **710** (2019) 012037 doi:10.1088/1757-899X/710/1/012037

A preliminary study on the use of EEMD-RQA algorithms in the detection of degenerative changes in knee joints

J Jonak[1], R Karpinski[1], A Machrowska[1], P Krakowski[2] and M Maciejewski[3]

[1] Lublin University of Technology, Faculty of Mechanical Engineering, Department of Machine Design and Mechatronics, Nadbystrzycka 36, 20-618 Lublin, Poland
[2] Orthopaedic Department, Łęczna Hospital, Krasnystawska 52, 21-010 Łęczna, Poland
[3] Lublin University of Technology, Faculty of Electrical Engineering and Computer Science, Institute of Electronics and Information Technology, Nadbystrzycka 36, 20-618 Lublin, Poland

j.jonak@pollub.pl

Abstract. Degenerative changes, according to world literature, are one of the key reasons for disability, especially in the elderly population. Diagnosis and monitoring of the disease consist mainly in clinical examination, bedside interviews and imaging. However, during the healing process, there is currently no tool for fast, cheap, easily available and diagnostics that would be free from ionising radiation and that would enable evaluation of the course of the disease. Therefore, the scientific community is searching for new diagnostic methods, with the potential for wide application in medicine. Registration and analysis of knee joint vibration signals presents a chance for more accurate and faster diagnostics. The method is capable of detecting damage at an early stage, while specifying the selection of optimal treatment methods. Therefore, it seems crucial to develop methods of analysis appropriate for the nature of tested signals. The quality of low-frequency natural waveforms can be improved by filtration in selected bands, eliminating existing artefacts. This paper presents an application of the EEMD-RQA algorithm in the detection of degenerative changes in knee joints. Pre-processing in the form of filtration gives the opportunity to pre-test the usefulness of the algorithm RQA in the ability to create/subsequent development of indicators describing the condition of the joint surfaces examined without the need for surgical intervention.

1. Introduction

The knee joint is one of the biggest joints in the human body. Moreover, due to its unique location, it serves a vital role in a range of daily activities including work, leisure or sports. The main component of every synovial joint is the articular cartilage. The articular cartilage ensures smooth and painless movement between bones. Even though structures such as menisci, ligaments or synovial membrane protect the cartilage in motion, it is nonetheless subjected to high shearing and compressive forces, and its properties change during daily activities [1–3]. Moreover, if overused or after a traumatic episode, the forces in the cartilage may exceed its capacity and trigger an unstoppable chain reaction leading to osteoarthritis, which affects the biomechanical properties of the cartilage [4,5] and provokes pain and joint dysfunction. In 2010 osteoarthritis was the 11th most common cause of disability in the world, and research suggest that more and more people will be suffering from joint dysfunction due to osteoarthritis [6]. Also, the treatment of patients with degenerative changes generates substantial costs [7]. Total knee replacement is a gold standard in end-stage disease; however, every joint replacement has its limited

IOP Conf. Series: Materials Science and Engineering **710** (2019) 012037 doi:10.1088/1757-899X/710/1/012037

survival time. Especially young patients with osteoarthritis are susceptible to revision surgery, given that the 10-year survival rate of total knee replacement reaches 81%-92% [8]. One of the most important factors in prolonging joint endurance is the detection of early changes in the joint, which can induce osteoarthritis [9]. Orthopaedic surgeons have a variety of imaging modalities, which facilitate the detection of early osteoarthritis; however, each of these modalities has its weaknesses including radiation, examination waiting time or low sensitivity. None of the existing techniques such as computed tomography, conventional radiography, MRI or ultrasound, can be implemented on a larger scale as a screening evaluation. Requirements such as low costs, reproducibility, diagnostic accuracy or wide availability might be met by vibroacoustic analysis of signals emitted by joints during movement. With the advancement of imaging techniques, there has been progress in the field of medical signal [10,11] and image [12] processing methods. The method is non-invasive, simple, reproducible, cost-effective, and yet to date, it has not been implemented in medicine on a larger scale. This article presents the vibroacoustic analysis that can be a helpful tool supporting objective evaluation of arthrokinematic motion quality and detection of osteoarthritic changes in knee joints. Analysis of vibroacoustic processes generated by the knee joint in contrast to classical diagnostic methods provides detailed information on the tribological properties of the joint, associated, among others, with the state of the vitreous cartilage and rheological properties of the synovial fluid [4,13].

Vibroarthrography provides an opportunity to quantitatively and objectively evaluate the integrity and smoothness of the articular cartilage during joint motion. As shown before, vibroacoustic signals generated by a damaged articular cartilage might show specific properties that can be used in the detection of articular cartilage abnormalities [14,15]. Moreover, the signals generated by an abnormal cartilage have been shown to correspond with the degree of cartilage damage [16]. Some authors suggested that osteoarthritic changes in joints produce vibroacoustic signals of higher frequency, greater peaks and elongated duration in comparison to the healthy knee joint [17]. It is believed that evaluation of vibroacoustic emission adds insight into tribological properties of the joint, and reflects the articular cartilage and synovial fluid status [18], which in turn reflect the clinical appearance and the function of the articular cartilage, thus affecting the treatment regimen [19]. The vibroacoustic analysis is suggested as a complementary evaluating method in addition to imaging modalities such as MRI, ultrasound or conventional radiography [10]. However, to date no standards regarding equipment, testing routine or data processing have been established. Furthermore, the literature shows lack of consensus with respect to the examination protocol, whether it should be conducted in an open or a closed kinematic chain [20]. Therefore, further investigation of the subject is required in order to establish repeatable and reliable examination conditions, which could provide researchers with comparable results in order to set guidelines for further testing and preparation of commercial diagnostic devices.

The article presents the results of preliminary tests on the application of EEMD (Ensemble Empirical Mode Decomposition) and RQA (Recurrence Quantification Analysis) algorithms as potential tools in the detection of degenerative changes occurring in the knee joint.

2. Methods

The results were obtained from tests performed on two volunteers: one with healthy knee joints and the other one with diagnosed chondromalacia in both knees. The clinical appearance of the chondromalacia was evaluated with the visual analogue scale (VAS), which is a validated method of subjective pain assessment. The test session included slow knee extension in a sitting position (open kinetic chain) and rising up from chair (sit-to-stand) movements (closed kinetic chain). During the test session, both knees were independently tested. The acoustic signal was measured using an analogue contact microphone CM01b connected to a National Instruments data acquisition card. The signal was sampled at 1kS/second for 30 seconds. The analogue to digital converter resolution was set at 16 bits. At 5V reference voltage, this resulted in about 76 microvolts resolution. The recorded data have been subjected to signal analysis. The studies were connected with RQA (Recurrence Quantification Analysis) preceded by EEMD (Ensemble Empirical Mode Decomposition) filtration methods used for acquired bioacoustic signals. Four intrinsic mode functions (IMF) were extracted from the input signal to allow further

IOP Conf. Series: Materials Science and Engineering **710** (2019) 012037 doi:10.1088/1757-899X/710/1/012037

analysis, by removing irrelevant frequency components. The RQA tests were performed to compare bioacoustic signals generated by healthy and diseased joints. The signal processing graph of EEMD-RQA method is presented below.

Adaptive technique Empirical Mode Decomposition (EMD) is a tool in the time-frequency analyses of nonlinear and non-stationary signals [21–23]. The basic assumption of this algorithm is to decompose the signal into a finite number of mode oscillations imbedded in the input signal known as Intrinsic Mode Functions (IMFs). EMD analyses have found a wide range of applications [24,25].

EMD algorithm is based on the process of sifting the IMFs from given input data set. IMF acquisition is related to local extremes identification. The found local maxima are interpolated by cubic spline in order to build the upper signal envelope. The same procedure is followed for the determination of the lower envelope. Subsequently, at each point their mean $m_1(t)$ is calculated along with the difference between the input data x(t) and $m_1(t)$:

$$h_1(t) = x(t) - m_1(t) \tag{1}$$

Function $h_1(t)$ is an IMF when it satisfies two conditions [21]:
In the whole data set, the number of extremes and the number of zero crossings must be either equal or differ at most by one.
At any point, the mean value m1(t) of the envelope defined by the local maxima and the envelope defined by the local minima is zero.
Sifting procedure is repeated k times, where $h_k(t)$ is taken as input data:

$$h_n(t) = h_{n-1}(t) - m_n(t) \tag{2}$$

The process can be stopped when the condition of equality of the zero crossings and extremes number is fulfilled. Then it is determined a residual signal rn(t), described as:

$$r_n(t) = x(t) - \sum_{i=1}^{n} c_i(t) \tag{3}$$

where $c_i(t)$ is an i-th Intrinsic Mode Function and n expresses the largest number of mode oscillations possible to obtain.

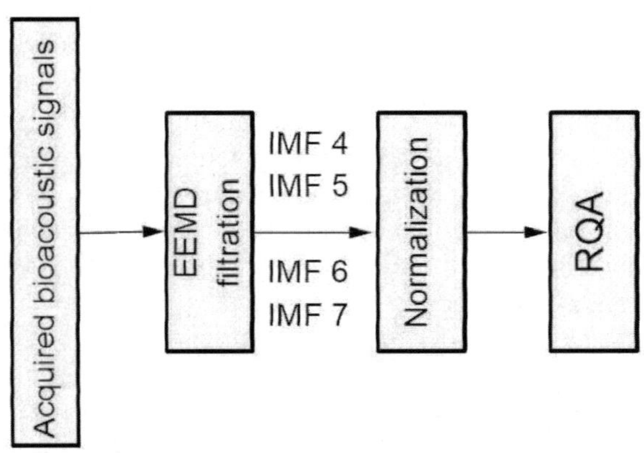

Figure 1. EEMD- RQA method.

The algorithm used in the study, EEMD, differs from EMD by white noise implementation. Additional white noise is helpful to force the ensemble to exhaust all possible solutions in the sifting process. In the case of the realised research, EEMD algorithm is used to place a signal portion in the IMF areas of the same scale.

Recurrence Quantification Analysis is one of the methods for analysing nonlinear dynamic systems. The inference on RQA coefficients is based on dependencies occurring at recurrence plots [26,27]. The recurrence rate (RR) parameter defines the density of recurrence points and the number of recurrences. RR informs about the probability of recurrence in a measured system. Determinism (DET) parameter

IOP Conf. Series: Materials Science and Engineering **710** (2019) 012037 doi:10.1088/1757-899X/710/1/012037

focuses on the number of recursive points forming linear segments parallel to the main diagonal line on RP (recurrence plot). It measures the degree of determinism of the analysed phenomenon. The coefficient describing the complexity of the deterministic structure in the system is known as Entropy (ENTR). Laminarity (LAM) is the measure of system behaviour stability. The last parameter used in the current analysis was trapping time (TT) parameter, which refers to the average length of vertical lines measuring the time scale of changes occurring in the studied process.

3. Results

Figures 2 and 3 show waveforms registered in opened and closed kinematic chains. There 5 movement cycles were executed for the healthy knees and for the dysfunctional ones.

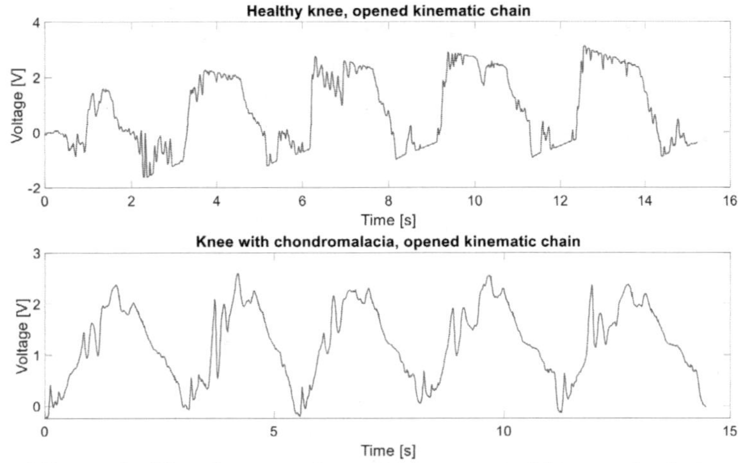

Figure 2. Waveforms registered in an opened kinematic chain.

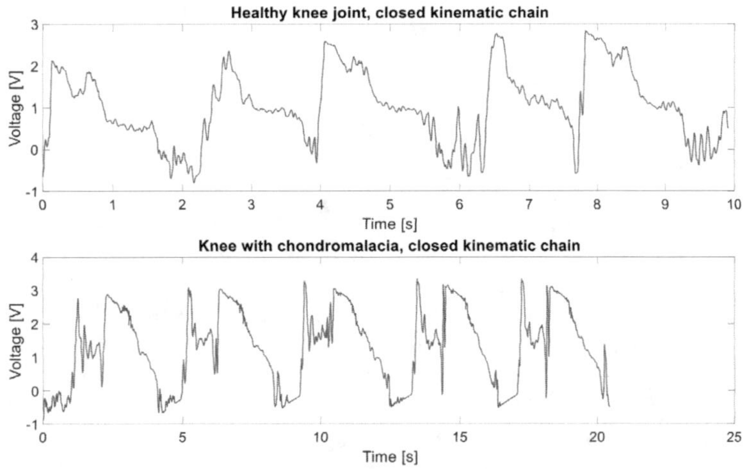

Figure 3. Waveforms registered in a closed kinematic chain.

Pre-processing included the EEMD filtration procedure. The results obtained for the opened kinematic chain are shown in figures 4 and 5.

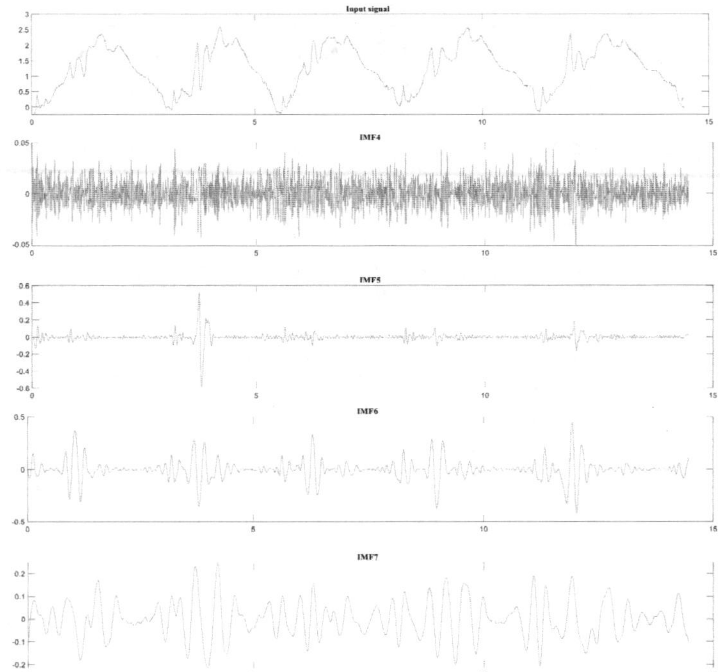

Figure 4. EEMD results: opened kinematic chain, joint with chondromalacia.

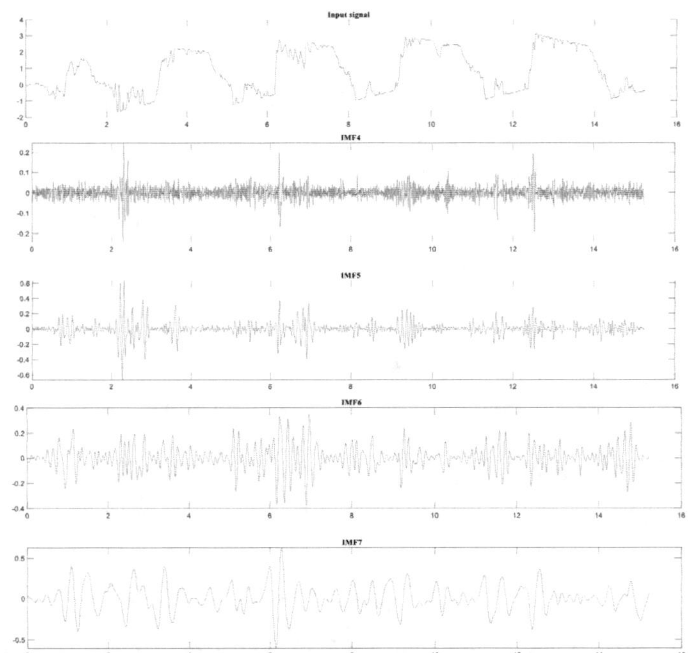

Figure 5. EEMD results: opened kinematic chain, healthy knee joint.

Closed kinematic chain results for healthy and damaged knee joint surfaces are placed in figures 6 and 7.

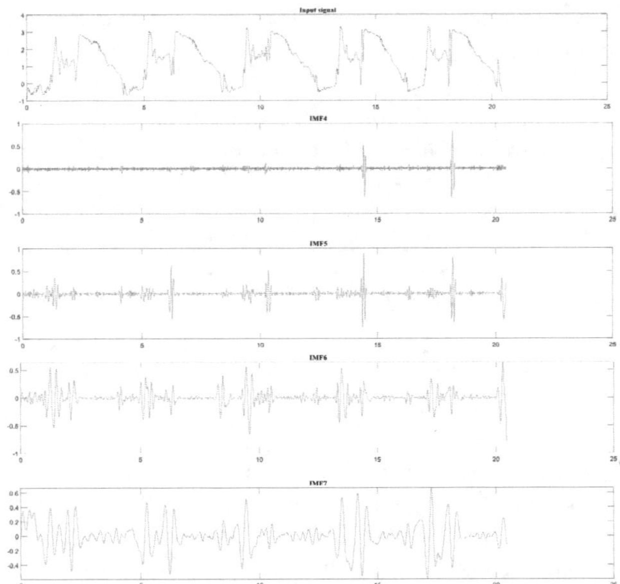

Figure 6. EEMD results: opened kinematic chain, knee joint with chondromalacia.

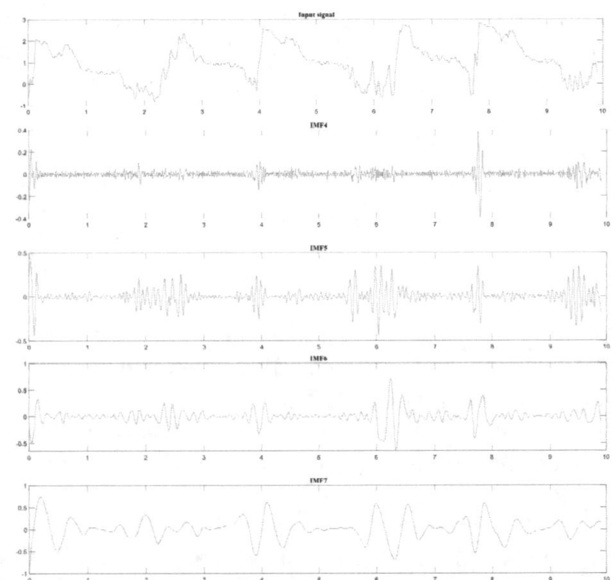

Figure 7. EEMD results: opened kinematic chain, healthy knee joint.

EEMD allowed obtaining 15 IMFs as a result of the sieving process. Low- and high-frequency components were separated, low-frequency signals were analysed more closely. Due to the lack of relevant diagnostic information that would provide the introduction of pathological changes contained in IMF 1-3, they were not included in further analyses [28].

In the case of degenerative changes for the open kinematic chain, higher amplitudes of oscillations at higher frequency ranges (IMF4) are observed. However, at lower frequencies, higher amplitude values are noted for healthy knee joints. It is connected with the occurrence of vibrations coming from the muscle tension and vibrations of the ligament apparatus.

Recurrence Quantification Analysis has shown significant differences between healthy and degenerated joints. The largest variations were observed in the case of the Trapping Time parameter.

For open kinematic chain movements, the discrepancies reach 35% and 30% for closed kinematic chain movements. RQA results are presented in table 1.

Table 1. Recurrence Quantification Analysis: ChO- chondromalacia, opened kinematic chain; HO- healthy, opened kinematic chain; ChC- chondromalacia, closed kinematic chain; HC- healthy, closed kinematic chain.

Type	RR	DET	ENTR	LAM	TT
ChO	127.500	99.996	8.159	99.997	196.306
HO	80.211	99.969	7.030	99.983	70.267
ChC	86.245	99.980	7.511	99.988	124.636
HC	65.757	99.967	6.946	99.982	43.623

4. Conclusions

The initial analysis has shown the existence of significant differences in the readings obtained from healthy and damaged knee joints. The obtained results clearly indicate the great importance of the TT parameter in RQA. Longer TT for joints with chondromalacia could point to the necessity of extending the movement time with a view to countering pain. The analysis of vibroacoustic processes generated by the knee joint is a high-precision objective tool for the assessment of the quality of the performed joint articulation . The non-invasive and low-cost character of the method enables the its application in monitoring the condition of joint structures. Bioacoustic signal analysis methods are a promising area for further research. Future investigations ought to engage a more numerous group of patients combined with an intraoperative assessment of the degree of damage to the articular cartilage.

Acknowledgments

This research project was financed in the framework of the Lublin University of Technology-Regional Excellence Initiative project, funded by the Polish Ministry of Science and Higher Education (contract no. 030/RID/2018/19).

5. References

[1] Gatti A A, Noseworthy M D, Stratford P W, Brenneman E C, Totterman S, Tamez-Peña J and Maly M R 2017 Acute changes in knee cartilage transverse relaxation time after running and bicycling *Journal of Biomechanics* **53** 171–7

[2] Karpiński R, Jaworski Ł and Zubrzycki J 2016 Structural analysis of articular cartilage of the hip joint using finite element method *Advances in Science and Technology Research Journ*al **10** 240–6

[3] Kępa P and Gajewski J 2019 The modelling and strength analysis of the endoprosthesis of the knee joint ed M Kulisz, M Szala, M Badurowicz, W Cel, M Chmielewska, Z Czyż, K Falkowicz, J Kujawska and T Tulwin MATEC Web of Conferences 252 07002

[4] Karpiński R, Jaworski Ł, Jonak J and Krakowski P 2019 Stress distribution in the knee joint in relation to tibiofemoral angle using the finite element method ed M Kulisz, M Szala, M Badurowicz, W Cel, M Chmielewska, Z Czyż, K Falkowicz, J Kujawska and T Tulwin MATEC Web of Conferences 252 07007

[5] Maier F, Lewis C G and Pierce D M 2019 The evolving large-strain shear responses of progressively osteoarthritic human cartilage Osteoarthritis and Cartilage

[6] Cross M, Smith E, Hoy D, Nolte S, Ackerman I, Fransen M, Bridgett L, Williams S, Guillemin F, Hill C L, Laslett L L, Jones G, Cicuttini F, Osborne R, Vos T, Buchbinder R, Woolf A and March L 2014 The global burden of hip and knee osteoarthritis: estimates from the Global Burden of Disease 2010 study Annals of the Rheumatic Diseases 73 1323–30

[7] Puig-Junoy J and Ruiz Zamora A 2015 Socio-economic costs of osteoarthritis: A systematic review of cost-of-illness studies Seminars in Arthritis and Rheumatism 44 531–41

[8] Castagnini F, Sudanese A, Bordini B, Tassinari E, Stea S and Toni A 2017 Total Knee Replacement in Young Patients: Survival and Causes of Revision in a Registry Population The Journal of Arthroplasty 32 3368–72
[9] Acevedo R J, Rivera-Vega A, Miranda G and Micheo W 2014 Anterior Cruciate Ligament Injury: Identification of Risk Factors and Prevention Strategies Current Sports Medicine Reports 13 186–91
[10] Jonak K, Krukow P, Jonak K E, Grochowski C and Karakuła-Juchnowicz H 2019 Quantitative and Qualitative Comparison of EEG-Based Neural Network Organization in Two Schizophrenia Groups Differing in the Duration of Illness and Disease Burden: Graph Analysis With Application of the Minimum Spanning Tree Clinical EEG and Neuroscience 50 231–41
[11] Krukow P, Jonak K, Karpiński R and Karakuła-Juchnowicz H 2019 Abnormalities in hubs location and nodes centrality predict cognitive slowing and increased performance variability in first-episode schizophrenia patients *Scientific Reports* 9
[12] Grochowski C, Krukow P, Jonak K, Stępniewski A, Wawrzycki K and Maciejewski R 2019 The assessment of lenticulostriate arteries originating from middle cerebral artery using ultra high-field magnetic resonance time-of-flight angiography *Journal of Clinical Neuroscience* 68 262–5
[13] Karpiński R, Machrowska, Anna and Maciejewski, Marcin 2019 Application of acoustic signal processing methods in detecting differences between open and closed kinematic chain movement for the knee joint *Applied Computer Science* 36–48
[14] Bączkowicz D and Majorczyk E 2016 Joint Motion Quality in Chondromalacia Progression Assessed by Vibroacoustic Signal Analysis PM&R 8 1065–71
[15] Shark L-K, Chen H and Goodacre J 2010 Knee Acoustic Emission: A Clue to Joint Ageing and Failure RHEUMATOLOGY 49 I79–I79
[16] Rangayyan R M, Oloumi F, Wu Y and Cai S 2013 Fractal analysis of knee-joint vibroarthrographic signals via power spectral analysis Biomedical Signal Processing and Control 8 23–9
[17] Shark L-K, Chen H and Goodacre J 2011 Knee acoustic emission: A potential biomarker for quantitative assessment of joint ageing and degeneration Medical Engineering & Physics 33 534–45
[18] Kim K S, Seo J H, Kang J U and Song C G 2009 An enhanced algorithm for knee joint sound classification using feature extraction based on time-frequency analysis Computer Methods and Programs in Biomedicine 94 198–206
[19] Shieh C-S, Tseng C-D, Chang L-Y, Lin W-C, Wu L-F, Wang H-Y, Chao P-J, Chiu C-L and Lee T-F 2016 Synthesis of vibroarthrographic signals in knee osteoarthritis diagnosis training BMC Res Notes 9 352
[20] Bączkowicz D, Kręcisz K and Borysiuk Z 2019 Analysis of patellofemoral arthrokinematic motion quality in open and closed kinetic chains using vibroarthrography BMC Musculoskeletal Disorders 20
[21] Du X, Li Z, Bi F, Zhang J, Wang X and Shao K 2012 Source separation of diesel engine vibration based on the empirical mode decomposition and independent component analysis Chinese Journal of Mechanical Engineering 25 557–63
[22] Jonak J, Machrowska A, Podgórski J and Bęc J 2016 Identification of the operating parameters for the mechanical systeIdentification system using EMD algorithm ed M Belhaq MATEC Web of Conferences 83 05001
[23] Nalband S, Prince A and Agrawal A 2018 Entropy-based feature extraction and classification of vibroarthographic signal using complete ensemble empirical mode decomposition with adaptive noise IET Science, Measurement & Technology 12 350–9
[24] Battista B M, Knapp C, McGee T and Goebel V 2007 Application of the empirical mode

decomposition and Hilbert-Huang transform to seismic reflection data GEOPHYSICS 72 H29–37

[25] Echeverría J C, Crowe J A, Woolfson M S and Hayes-Gill B R 2001 Application of empirical mode decomposition to heart rate variability analysis Medical & Biological Engineering & Computing 39 471–9

[26] Syta A, Jonak J, Jedliński Ł and Litak G 2012 Failure Diagnosis of a Gear Box by Recurrences Journal of Vibration and Acoustics 134 041006

[27] Webber C L and Zbilut J P 1994 Dynamical assessment of physiological systems and states using recurrence plot strategies Journal of Applied Physiology 76 965–73

[28] Wu Y, Yang S, Zheng F, Cai S, Lu M and Wu M 2014 Removal of artefacts in knee joint vibroarthrographic signals using ensemble empirical mode decomposition and detrended fluctuation analysis Physiological Measurement 35 429–39.

IOP Conf. Series: Materials Science and Engineering **710** (2019) 012038 doi:10.1088/1757-899X/710/1/012038

Life cycle assessment of the use of sewage sludge as Portland cement replacement

Z Pavlík[1], M Pavlíková[1], M Záleská[1], G Łagód[2], Z Suchorab[2] and L Guz[2]

[1] Czech Technical University in Prague, Faculty of Civil Engineering, Department of Materials Engineering and Chemistry, Thákurova 7, 166 29 Prague, Czech Republic
[2] Lublin University of Technology, Faculty of Environmental Engineering, Department of Water Supply and Sewage Disposal, Nadbystrzycka 38D, 20-618 Lublin, Poland

pavlikz@fsv.cvut.cz

Abstract. In this paper, we proposed an alternative thermal treatment of sewage sludge, aimed at its application in blended cements, which can reduce the energy demands and requires less technologically sophisticated processing. The life cycle assessment (LCA) was performed to quantify the environmental impact of the sewage sludge used as a partial cement replacement. The LCA was focused on the CO_2 emission and energy consumption. The functional unit of 1 m^3 of blended mortars composed of Portland cement CEM I 42.5, silica sand, and thermally treated sewage sludge, where the sludge was used as partial cement substitute in a dosage of 10, 20, and 30% by mass of cement, was analyzed. The positive effect of the use of sewage sludge as a partial cement replacement with respect to the energy consumption and GHG emission, considering the compressive strength of the analyzed composites, was quite apparent. The decrease of energy consumption necessary for the production of the analyzed blended binders was nearly 10% per each 10% of sewage sludge used as Portland cement replacement. The energy needed for the sludge thermal treatment at 700°C was 220 MJ/t. The emission of GHG related to the sludge thermal treatment was 40kg/t. The presented data were strongly affected by the applied distribution of electricity sources, with a high share of coal combustion plants.

1. Introduction

The sewage sludge originating from the municipal wastewater treatment plants is produced in large quantities all over the world. Due to the rapid progress in urbanization and industrialization, which is characteristic for the developing countries, a high increase in the production of sewage sludge could be observed over the last several decades [1]. Sewage sludge must be processed in an appropriate way, because it frequently poses a threat to the natural environment resulting, for instance, from high heavy metal content [2]. Deposition of sewage sludge in landfills and its application in agriculture were limited through the regulations issued by the European Union [3-6]. Currently, the most common method of sewage sludge disposal is still landfilling, and in some countries also disposal into the ocean. Moreover, it is partially applied in agriculture, for example as organic fertilizer and for the soil management [7]. However, in that case of disposing, it must always be considered that sludge contains pollutants and unstable pathogens which can present potential health and environmental hazards. Therefore, the sewage sludge landfilling will become inapplicable in a near future and other methods of sludge disposal must be searched. These alternative methods may consequently lead to the recovery of valuable raw materials from potentially dangerous materials, processing them in order to enable their use in agriculture, various branches of industry or heat and energy recovery.

Due to the high carbon footprint of cement production, different types of mineral admixtures were tested up to now as possible PC (Portland cement) substitutes. There are two aspects of cement production that result in emissions of CO_2 [8]. The first is the chemical reaction involved in the production of the main component of cement, clinker, as limestone is decomposed by the addition of heat [9]. According to recent estimation, these so-called "process" emissions contribute about 5% of total anthropogenic CO_2 emissions excluding land-use change. The second source of emissions is from the combustion of fossil fuels used for heating raw clinker ingredients up to melting temperature [10]. Total emissions from the cement industry could therefore constitute as much as 8% of global CO_2 emissions [11-14].

Supplementary cementitious materials became quite common in today's concrete industry [15], [16]. Typical examples are by-products of iron, silicon and ferrosilicon alloys production, and waste products from coal combustion in thermal power stations. Here, blast furnace slag, fly ash and micro silica are produced. At present, these products are commonly applied in concrete production, and their beneficial effects on concrete properties and performance are proven by many years of usage. However, there is still number of other pozzolana materials that are applied in blended binder and concrete production very rarely, and which contribution to concrete performance is not yet fully understood [17]. Therefore, we proposed an alternative thermal treatment of sewage sludge in this paper, aimed at its application in blended cements, which can reduce the energy demands and requires less technologically sophisticated processing.

2. Materials, characterization, conducted tests

2.1. Sewage sludge processing, composites preparation

The fermented and mechanically dewatered dried sewage sludge from wastewater treatment plant located in Lublin, Poland, was thermally treated for 2h at 700°C. The temperature of thermal treatment was chosen based on previous sludge TG/DSC analysis [18]. The produced sewage sludge ash (SSA) was mechanically activated by milling to the fineness similar to cement.

The composites (mortars) were prepared from OPC CEM I 42.5 R. As fine aggregate, silica sand of fraction 0–2 mm was used. It is a product of Filtrační písky, Ltd., Chlum u Doks, Czech Republic and it was mixed from three sand fractions. The weight ratio of sand fractions was 1:1:1. In mortars composition, OPC was partially substituted with SSA; its dosage was 10, 20, and 30% of cement mass. The water-binder ratio of 0.5 and the sand-binder (cement + SSA) ratio of 3.0 were constant for all composite mixes. Control mortar composed of OPC, sand, and batch water was also produced. The fresh mixes were casted into 40mm × 40mm × 160mm iron molds and after 1 day demolded, and cured in water at a temperature of (21 ± 2)°C. The characterization of mortar samples was conducted for samples matured for 28 days.

2.2. Experimental stage

The performed tests included measurement of basic structural properties of the developed mortar and their compressive strength. Among fundamental structural properties of researched mortars, bulk density, specific density, and total open porosity were measured. Before these tests, the specimens were dried in a vacuum drier at 60°C. The specific density ρ_s (kg/m^3) was measured using helium pycnometry. The bulk density ρ_b (kg/m^3) of mortar specimens was determined according to the EN 1015-10 (2000) [19] using a gravimetric method. The expanded combined uncertainty of the bulk density test was 1.2%. The total open porosity ψ (%) was calculated from the known bulk density and matrix density values with combined uncertainty of 1.7% [20]. The compressive strength test was conducted according to the EN 1015-11 standard (1999) [21]. The loading area was 40mm × 40mm. The relative expanded uncertainty of strength test was 1.4%.

3. LCA analysis

In order to determine the environmental impact of the use of SSA in blended mortars, two basic comparative criteria were chosen, namely carbon footprint and the amount of consumed energy. A functional unit of 1m^3 of final material and additional functional units as 1t of cement and 1t of SSA were used in calculations. The range of the analysis was reduced to the intermediate stage of material processing, i.e., the material production consisted basically of two main subsystems, namely production and processing of the binders (Portland cement and BASS) and mixing of the composite – mortar. The data of emissions and energy usage was compiled from the literature review, theoretical estimations, and information of cement producers. The emphasis was placed on the consideration of local factors, such as composition of energy mix or variety of used materials. Within the analysis, the Czech electricity mix, where the main portion of electricity comes from coal combustion and nuclear power plants, was used [18].

The total amount of emitted CO_2 and consumed energy during manufacturing of 1m^3 of studied composites with 10, 20, and 30% PC replacements with SSA were compared with the data calculated for the reference composite mix. For the intermediate stages, i.e., Portland cement and BASS manufacturing, the obtained values were related to 1t of a particular product.

4. Results and discussion

Basic structural parameters of studied composites, together with compressive strength data are presented in table 1. The bulk density slightly decreased with increasing dosage of SSA in mortar mix, but differences were mostly in the range of measuring uncertainty. Accordingly, the porosity of mortars with SSA was slightly higher as compared to that measured for control mortar mix. In agreement with structural parameters, mortar compressive strength remained almost unaffected by the SSA incorporation in mortar mix. It was a highly promising finding for the use of thermal treated sewage sludge in concrete and cement-based composited manufacturing.

Table 1. Structural and mechanical parameters of examined mortars.

Mortar	ρ_b (kg/m^3)	ρ_s (kg/m^3)	ψ (%)	f_c (MPa)
RM	2031	2251	9.8	52.4
SSA 10	2038	2267	10.1	52.3
SSA 20	2004	2234	10.3	51.9
SSA 30	2001	2235	10.5	51.8

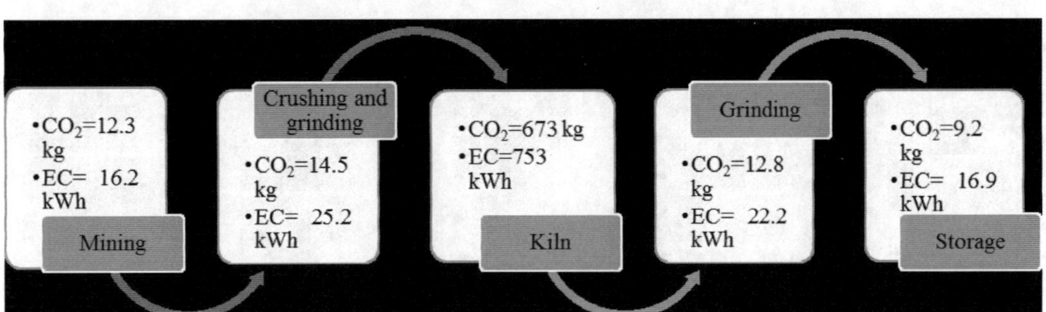

Figure 1. PC production – outputs and inputs (per 1t of cement).

IOP Conf. Series: Materials Science and Engineering **710** (2019) 012038 doi:10.1088/1757-899X/710/1/012038

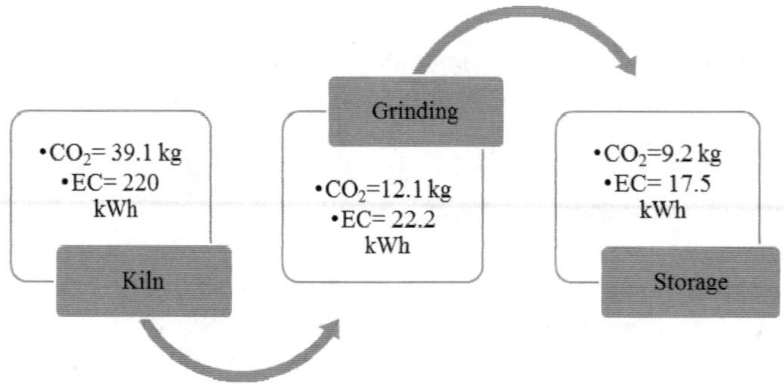

Figure 2. Sewage sludge processing and SSA production – outputs and inputs (per 1t of SSA).

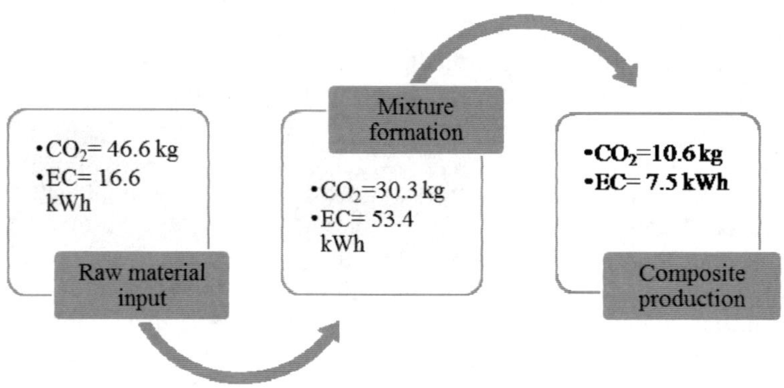

Figure 3. Mortars production – outputs and inputs (per 1m^3 of composite).

The partial life cycle processes included in the LCA analysis are given in figures 1-3. Here, EC stands for energy consumed. In figure 4, environmental impact of SSA use in mortars production is presented. The calculated data showed decreasing trends in both CO_2 production and energy consumption with the increasing dosage of SSA in mortar mix.

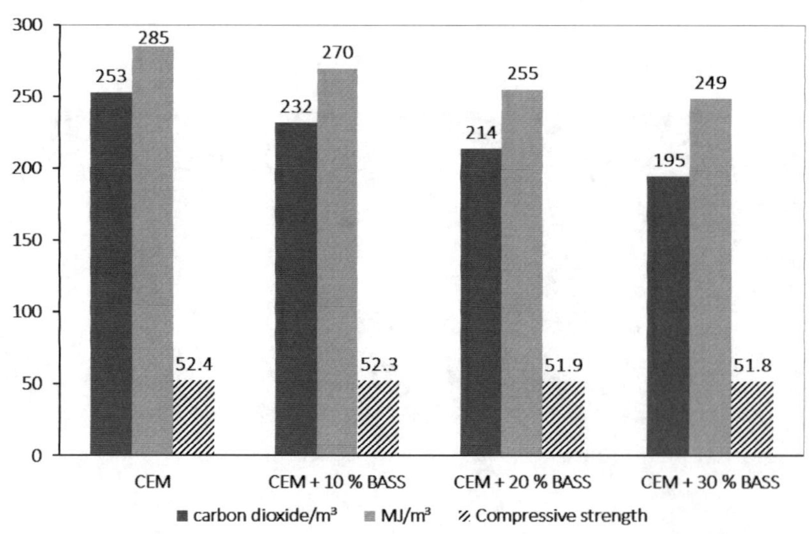

Figure 4. Carbon dioxide and EC for production of 1m^3 of composite.

Taking into consideration the functional properties of SSA mortars (table 1), which were close to the control material, the overall benefits of SSA use were quite obvious. The reduction of waste generation with minimization of material landfilling added to the profitability and efficiency of the sewage sludge application as well. The importance of obtained results is underlined by the fact that Czech Republic and Poland belong to major producers of sewage sludge in the EU. Moreover, sewage sludge recycling and reuse is in both countries still very limited.

5. Conclusions

Sewage sludge originating from the municipal wastewater treatment plant was thermally treated. The obtained ash was mechanically activated and used as partial OPC substitute in mortar composition. The positive effect of the use of sewage sludge ash as a partial cement replacement with respect to the energy consumption and GHG emission, considering the compressive strength of the analyzed mortars, was quite apparent. The decrease of energy consumption necessary for the production of the analyzed blended binders was nearly 10% per each 10% of sewage sludge ash used as Portland cement replacement. The energy needed for the sludge thermal treatment at 700°C was 220 MJ/t. The emission of GHG related to the sludge thermal treatment was nearly 40kg/t.

Based on LCA analysis and obtained structural and mechanical parameters of the developed mortars, the presented thermal treatment of sewage sludge and formed SSA can bring both environmental and economic benefits for concrete and cement industry.

Acknowledgment

The authors gratefully acknowledge the financial support received from the Czech Science Foundation, under project No 17-02815S, from the Ministry of Education, Youth and Sports of the Czech Republic under Czech/Polish bilateral project No 8JPL19027, and from the Polish Ministry of Science and Higher Education under project No NAWA PPN/BIL/2018/1/00045/00001.

6. References

[1] Yiming L, Shaoqi Z, Fuzhen L and Yixiao L 2012 Utilization of municipal sewage sludge as additives for production of eco-cement *J. Hazard. Mater.* **213-214** 457-65

[2] Montusiewicz A, Lebiocka M and Pawłowska M 2008 Characterization of the biomethanization process in selected waste mixtures *Arch. Environ. Prot.* **34** 49–61

[3] CECCEC 1986 *Commission of European Communities Council Directive 86/278/EEC of 4 July 1986 on the protection of the environment and in particular of the soil, when sewage sludge is used in agriculture* (Commission of the European Communities v Council of the European Communities)

[4] CECCEC 1991 *Commission of European Communities Council Directive 91/156/EEC on waste* (Commission of the European Communities v Council of the European Communities)

[5] CECCEC 1998 *Commission of European Communities Council Directive 91/271/EEC of March 1991 concerning urban waste–water treatment amended by the 98/15 EC of 27 February 1998* (Commission of the European Communities v Council of the European Communities)

[6] CECCEC 1999 *Commission of European Communities Council Directive 99/31/EC of 26 April 1999 on the landfill waste* (Commission of the European Communities v Council of the European Communities)

[7] Kujawska J, Pawłowska M, Wójcik K, Baran S, Żukowska G and Pawłowski A 2016 Application of exploratory waste and compost form municipal waste for the production of soil-like materials for reclamation of degraded area *Rocznik Ochrona Srodowiska* 18(2) 709-721

[8] Gao T, Shen L, Shen M, Chen F, Liu L and Gao L 2015 Analysis on differences of carbon dioxide emission from cement production and their major determinants *J. Clean. Prod.* **103** 160-70

[9] Andrew RM 2018 Global CO_2 emissions from cement production *Earth Syst. Sci. Data* **10** 195-217.

[10] Ishak S S and Hashim H 2015 Low carbon measures for cement plant – a review *J. Clean. Prod.*

IOP Conf. Series: Materials Science and Engineering **710** (2019) 012038 doi:10.1088/1757-899X/710/1/012038

103 260-74

[11] Ali M B, Saidur, R and Hossain M S (2011). A review on emission analysis in cement industries *Renew. Sustain. Energy Rev.* **15** 2252-61

[12] Le Quéré C, Andrew R M, Friedlingstein P, Sitch S, Pongratz J, Manning A C, Korsbakken J I, Peters G P, Canadell J G, Jackson R B, Boden T A, Tans P P, Andrews O D, Arora V K, Bakker D C E, Barbero L, Becker M, Betts R A, Bopp L, Chevallier F, Chini L P, Ciais P, Cosca C E, Cross J, Currie K, Gasser T, Harris I, Hauck J, Haverd V, Houghton R A, Hunt C W, Hurtt G, Ilyina T, Jain A K, Kato E, Kautz M, Keeling R F, Klein Goldewijk K, Körtzinger A, Landschützer P, Lefèvre N, Lenton A, Lienert S, Lima I, Lombardozzi D, Metzl N, Millero F, Monteiro, P M S, Munro D R, Nabel J E M S, Nakaoka S-I, Nojiri Y, Padin X A, Peregon A, Pfeil B, Pierrot D, Poulter B, Rehder G, Reimer J, Rödenbeck C, Schwinger J, Séférian R, Skjelvan I, Stocker BD, Tian H, Tilbrook B, Tubiello F N, van der Laan-Luijkx I T, van der Werf G R, van Heuven S, Viovy N, Vuichard N, Walker A P, Watson A J, Wiltshire A J, Zaehle S and Zhu D 2018 Global Carbon Budget 2017 *Earth Sys. Sci. Data* **10**, 405–48

[13] Meireles P D S, Pereira D S S, Melo M A F, Braga R M, Freitas J C O, Melo D M A and Silvestre F R S 2019 Technical evaluation of calcium sulphate α-hemihydrate in oilwell application: An alternative to reduce the environmental impacts of Portland cement *J. Clean. Prod.* **220**, 1215-21

[14] Rahhal V F, Trezza M A, Tironi A, Castellano CC , Pavlíková M, Pokorný J, Irassar E F, Jankovský O and Pavlík, Z 2019 Complex characterization and bahaviour of waste fired brick powder-Portland cement system *Materials* **12** 1650

[15] Meyer C 2009 The greening of the concrete industry *Cem. Concr.Compos.* **31** 601-5

[16] Záleská M, Pavlíková M, Pavlík Z, Jankovský O, Pokorný J, Tydlitát V, Svora P and Černý R Physical and chemical characterization of technogenic pozzolans for the application in blended cement *Constr. Build. Mater.* **160** 106-16

[17] Apprianti E, Shafigh P, Bahri S and Farahani J N 2015 Supplementary cementitious materials origin from agricultural wastes - a review *Constr. Build. Mater.* **74**, 176-87

[18] Záleská M, Pavlík Z, Pavlíková M, Scheinherrová L, Pokorný J, Trník A, Svora P, Jankovský O, Suchorab Z and Černý R 2018 Biomass ash-based mineral admixture prepared from municipal sewage sludge and its application in cement composites *Clean Technol. Environ. Policy* **20** 159-71

[19] EN 1015-10 2000 *Methods of Test for Mortar for Masonry—Part. 10: Determination of Dry Bulk Density of Hardened Mortar* (Brussels: European Committee for Standardization.

[20] Pavlíková M, Zemonová L, Pokorný J, Záleská M, Jankovský O, Lojka M, Sedmidubský D and Pavlík Z 2018 Valorization of wood chips ash as an eco-friendly mineral admixture in mortar mix design *Waste Manage.* **80**, 89-100

[21] EN 1015-11 1999 *Methods of Test for Mortar for Masonry—Part. 10: Determination of Flexural and Compressive Strength of Hardened Mortar* (Brussels: European Committee for Standardization).

IOP Conf. Series: Materials Science and Engineering **710** (2019) 012039 doi:10.1088/1757-899X/710/1/012039

Static response of FGM porous rhombic conoidal shell

M I Ansari[1], A Kumar[2], D Barnat-Hunek[3] and G Łagód[4]

[1] Jamia Millia Islamia, Department of Architecture, New Delhi-110025, India
[2] National Institute of Technology Patna, Department of Civil Engineering, Patna-800005, India
[3] Lublin University of Technology, Faculty of Civil Engineering and Architecture, Nadbystrzycka 40, 20-618 Lublin, Poland
[4] Lublin University of Technology, Faculty of Environmental Engineering, Nadbystrzycka 40B, 20-618 Lublin, Poland

er.mdirfan2k9@gmail.com

Abstract. The porosity effect on static analysis of functionally graded porous doubly curved rhombic conoids was investigated. Porosities inside functionally graded materials can occur due to a technical issue during the fabrication process and lead to the occurrence of micro-voids in the materials. The porosity effect was accounted for using a modified power law. The presented mathematical formulation includes normal curvatures in displacement field, as well as cross curvature in strain-displacement field. The improvement in the 2D mathematical model enables to solve problems of moderately thick FGM porous conoids. The feature that distinguishes the presented shell from other is that the maximum transverse deflexion does not occur at its centre. The improved mathematical model was executed in finite element code. The obtained numerical results were compared with the results available in the literature. Once validated, the current model was employed to study the effect of porosity, skew angle, boundary condition, volume fraction index, loading pattern and other geometric parameters.

1. Introduction

In the last two decades, shell structures made of functionally graded materials (FGM) have been broadly used by civil, mechanical, aeronautical and marine engineering. The FGM is an inhomogeneous material, composed of two (or more) materials, organised with a smooth gradation in the desired direction. However, at the time of fabrication, porosities are infused in the FGM which creates a major issue. Apart from this, the FGM manufactured using sintering process retain porosities because of different solidification rate of material constituents. Koizumi used FGM in advanced engineering structures experiencing elevated temperatures [1]. The bending analysis of FGM cylindrical shells using the element-free kp-Ritz method was analysed by Zhao [2]. An elastic solution for a sandwich panel with isotropic skins was presented by Kashtalyan and Menshykova [3]. The stresses in functionally graded doubly curved shells were calculated by incorporating the differential quadrature method in the first order shear deformation theory by Tornabene and Viola [4]. Viola used unconstrained third-order shear deformation theory for the static analysis of moderately thick functionally graded conical shells [5]. Dai and Dai developed an analytical solution based upon the classical shell theory for FGM cylindrical shell under thermomechanical loading [6]. Xiang and Liu used a meshless global collocation method with nth-order shear deformation theory for the static analysis of FGM sandwich plate [7].

IOP Conf. Series: Materials Science and Engineering **710** (2019) 012039 doi:10.1088/1757-899X/710/1/012039

Gutierrez Rivera and Reddy studied the large deformation behaviour of FGM shells by using seven-parameter continuum shell formulation [8].

Since casting and fabrication of conoids is easy due to its singly ruled surface, it is favoured in the construction industry. Conoids are structurally stiff, aesthetically appealing, and are used to cover the column-free large area in industrial structures, aircraft hangars and exhibition halls. The first study on simply supported and clamped conoids was carried out by Hadid [9]. A combined variational approach was used for the bending analysis of elastic conoids for both types of boundary condition. Finite difference method was carried out on the conoidal shell by Das and Bandyopadhyay for both experimental, as well as theoretical investigation [10]. Ghosh and Bandyopadhyay engaged their own formulation to examine the influence of cut-outs on the static analysis of conoidal shells [11]. The bending analysis of stiffened conoids was studied by Das and Chakravorty using a three-noded beam element [12]. Kumari and Chakravorty used FSDT for the study on the bending response of delaminated conoids [13]. Malekzadeh Fard and Baghestani explored the free vibration behaviour of moderately thick doubly curved shell based on FSDT with classic boundary conditions [14]. The bending and free vibration analysis of functionally graded porous plate was presented by Akbaş [15]. He used FSDT model and Navier solution for solving the problem. An analytical model was developed by Al Rjoub and Hamad to study the effect of porosity on free vibration analysis of porous beam [16]. Kiran et al. presented the effect of porosity on the static response of a functionally graded skew magneto-electro-elastic plate [17]. Gupta and Talha explored the influence of porosities on the flexure and free vibration response of graded plate using higher deformation theory [18,19].

The literature review reveals that no results for the static analysis of doubly curved FGM porous rhombic conoids are available. Therefore, an attempt was made in the present paper to study the bending behaviour of FGM porous conoidal shell with the help of an improved mathematical model. The FE coding was done by using a C0 nine-noded FE with seven nodal unknowns at each node for the presented improved mathematical model developed by the authors. C1 continuity requirement associated with the presented model is suitably circumvented. The present study facilitates the bending analysis by finite element (FE) modelling, keeping in mind the processing time using a computer and the simplicity of the approach. This study is the first step towards enhancing our understanding of the bending problem of FGM porous rhombic conoids.

2. Formulation

2.1. Porosity inclusion

The FGM porous conoidal shell of sides a, b and thickness h are depicted in Figure 1. As presented in Figure 1(b), the even type of porosity distributions along the thickness coordinate is considered in the present work. Using the modified power law for the inclusion of porosity effect, the effective material properties of the FGM porous conoidal shell at any point in thickness coordinate (z) can be stated as:

$$P(z) = P_c V_c(z) + P_m V_m(z) - (P_c + P_m)\frac{\beta}{2}$$

where, $$V_c(z) = \left(\frac{1}{2} + \frac{z}{h}\right)^n, \quad (0 \le n \le \infty), V_c(z) + V_m(z) = 1 \qquad (1)$$

where β (β << 1) is the volume fraction of porosities. P(z) implies the material properties like the Young's modulus of elasticity (E), material density (ρ) and Poisson's ratio (v) of the FGM porous conoidal shell. Subscripts c and m stand for ceramic and metallic constituents, respectively.

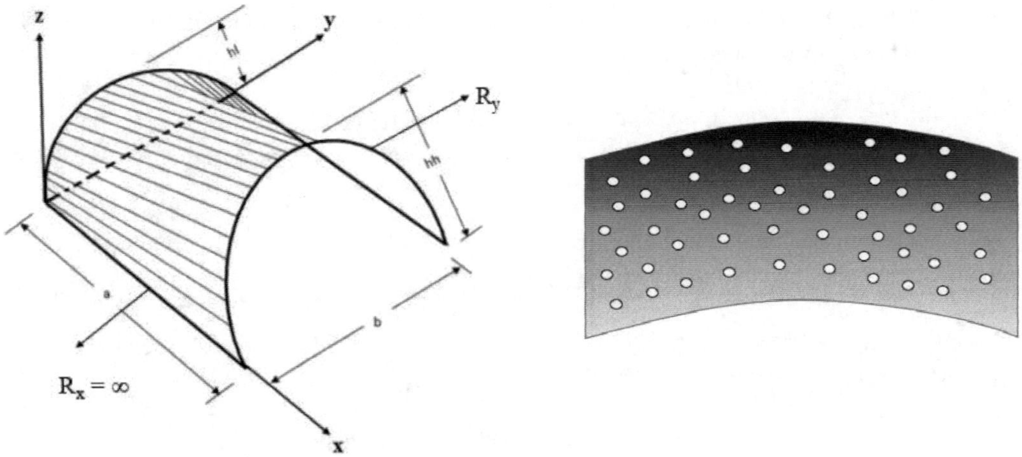

(a) (b)

Figure 1. (a) Geometry of conoidal shell and (b) porosities models for even porosity in thickness coordinate.

2.2. *Displacement field and strains*

In order to develop the mathematical model, the displacement fields (u, v, w) along the (x, y, z) coordinates for FGM conoidal shell are considered on the basis of the third order shell theory and are as follows:

$$u(x,y,z)=\left(1+\frac{z}{R_x}\right)u_0+z\theta_x+z^2\xi_x+z^3\zeta_x$$
$$v(x,y,z)=\left(1+\frac{z}{R_y}\right)v_0+z\theta_y+z^2\xi_y+z^3\zeta_y \qquad (2)$$
$$w(x,y,z)=w_0$$

where (u_0, v_0, w_0) are the corresponding displacements of any point on the mid-plane along the (x, y, z) coordinates. (θ_x, θ_y) are the rotations of normal to the mid-plane about the y and x-axes, respectively. The functions (ξ_x, ξ_y, ζ_x, ζ_y) are higher order terms of Taylor's series expansion at the mid-plane of the conoidal shell. By imposing the boundary condition (zero transverse shear strain at top and bottom) in equation (2), the function ξ_x, ξ_y, ζ_x and ζ_y are calculated as

$$u(x,y,z)=\left(1+\frac{z}{R_x}\right)u_0+\theta_x\left(z-\frac{4z^3}{3h^2}\right)-\frac{\partial w_0}{\partial x}\left(\frac{4z^3}{3h^2}\right)+v_0\left(\frac{4z^3}{3h^2R_{xy}}\right)$$
$$v(x,y,z)=\left(1+\frac{z}{R_y}\right)v_0+\theta_y\left(z-\frac{4z^3}{3h^2}\right)-\frac{\partial w_0}{\partial y}\left(\frac{4z^3}{3h^2}\right)+u_0\left(\frac{4z^3}{3h^2R_{xy}}\right)$$
$$w(x,y,z)=w_0 \qquad (3)$$

For omitting C^1 continuity problem associated with TSDT, the out of plane derivatives are replaced by the following relations

$$\psi_x=\frac{\partial w_0}{\partial x},\ \psi_y=\frac{\partial w_0}{\partial y} \qquad (4)$$

Hence, the field variables per node used in the present mathematical model are u_0, v_0, w_0, θ_x, θ_y, ψ_x and ψ_y. Mathematically, it may be expressed as

$$\{\delta\} = \{u_0, v_0, w_0, \theta_x, \theta_y, \psi_x, \psi_y\}^T \tag{5}$$

where $\{\delta\}$ is termed as a displacement vector.

The strain vector may be expressed as

$$\{\varepsilon\} = \{\varepsilon_x, \varepsilon_y, \gamma_{xy}, \gamma_{xz}, \gamma_{yz}\}^T \tag{6}$$

Further, the strain vector $\{\varepsilon\}$ can be associated with global displacement vector $\{X\}$ by the subsequent relationship

$$\{\varepsilon\} = [B]\{X\} \tag{7}$$

Here, matrix $[B]$ is known as strain-displacement matrix that contains the derivatives of shape functions.

The in-plane and transverse shear strains are

$$\begin{aligned}
\varepsilon_x &= \frac{\partial u}{\partial x} + \frac{w}{R_x} \\
\gamma_{xy} &= \frac{\partial v}{\partial x} + \frac{\partial u}{\partial y} + \frac{2w}{R_{xy}} \\
\gamma_{xz} &= \frac{\partial u}{\partial z} + \frac{\partial w}{\partial x} - \frac{u_0}{R_x} - \frac{v_0}{R_{xy}} \\
\gamma_{yz} &= \frac{\partial v}{\partial z} + \frac{\partial w}{\partial y} - \frac{v_0}{R_y} - \frac{u_0}{R_{xy}}
\end{aligned} \tag{8}$$

The strain relationship can be written as:

$$\begin{Bmatrix} \varepsilon_x \\ \varepsilon_y \\ \gamma_{xy} \end{Bmatrix} = \begin{Bmatrix} \varepsilon_x^0 \\ \varepsilon_y^0 \\ \gamma_{xy}^0 \end{Bmatrix} + z \begin{Bmatrix} k_1^1 \\ k_2^1 \\ k_6^1 \end{Bmatrix} - \frac{4z^3}{3h^2} \begin{Bmatrix} k_1^3 \\ k_2^3 \\ k_6^3 \end{Bmatrix}$$

$$\begin{Bmatrix} \gamma_{xz} \\ \gamma_{yz} \end{Bmatrix} = \begin{Bmatrix} \gamma_{xz}^0 \\ \gamma_{yz}^0 \end{Bmatrix} - \frac{4z^2}{h^2} \begin{Bmatrix} k_4^2 \\ k_5^2 \end{Bmatrix} \tag{9}$$

where

$$\gamma_{yz}^0 = \left(\frac{\partial w_0}{\partial y} + \theta_y\right) - \frac{u_0}{R_{xy}}, \quad \gamma_{xz}^0 = \left(\frac{\partial w_0}{\partial x} + \theta_x\right) - \frac{v_0}{R_{xy}}$$

$$k_1^1 = \frac{\partial \theta_x}{\partial x} + \frac{\partial u_0}{\partial x}\frac{1}{R_x}, \quad k_2^1 = \frac{\partial \theta_y}{\partial y} + \frac{\partial v_0}{\partial y}\frac{1}{R_y}$$

$$k_6^1 = \left(\frac{\partial\theta_y}{\partial x} + \frac{\partial\theta_x}{\partial y}\right) + \left(\frac{\partial u_0}{\partial y}\frac{1}{R_x} + \frac{\partial v_0}{\partial x}\frac{1}{R_y}\right)$$

$$k_1^3 = \left(\frac{\partial\theta_x}{\partial x} + \frac{\partial\psi_x}{\partial x}\right) - \frac{\partial v_0}{\partial x}\frac{1}{R_{xy}}$$

$$k_2^3 = \left(\frac{\partial\theta_y}{\partial y} + \frac{\partial\psi_y}{\partial y}\right) - \frac{\partial u_0}{\partial y}\frac{1}{R_{xy}}$$

$$k_6^3 = \left(\frac{\partial\theta_y}{\partial x} + \frac{\partial\psi_y}{\partial x}\right) + \left(\frac{\partial\theta_x}{\partial y} + \frac{\partial\psi_x}{\partial y}\right) + \left(\frac{\partial u_0}{\partial x} + \frac{\partial v_0}{\partial y}\right)\frac{1}{R_{xy}}$$

$$k_4^2 = (\theta_x + \psi_x) - \frac{v_0}{R_{xy}},\ k_5^2 = (\theta_y + \psi_y) - \frac{u_0}{R_{xy}} \tag{10}$$

2.3. Constitutive relationship
The linear stress-strain constitutive correlation for the FGM conoids are

$$\{\sigma\} = [Q]\{\varepsilon\} \tag{11}$$

where the constitutive matrix

$$[Q] = \begin{bmatrix} Q_{11} & Q_{12} & 0 & 0 & 0 \\ Q_{21} & Q_{22} & 0 & 0 & 0 \\ 0 & 0 & Q_{66} & 0 & 0 \\ 0 & 0 & 0 & Q_{44} & 0 \\ 0 & 0 & 0 & 0 & Q_{55} \end{bmatrix} \tag{12}$$

where

$$Q_{11} = Q_{22} = \frac{E(z)}{1-\nu^2},\quad Q_{12} = Q_{21} = \frac{\nu E(z)}{1-\nu^2}$$

$$Q_{66} = Q_{44} = Q_{55} = \frac{E(z)}{2(1+\nu)} \tag{13}$$

3. Finite element modelling

3.1. Introduction
For the present C^0 finite element (FE) model, nine-noded isoparametric Lagrangian element with seven degrees of freedom at each node was utilised in the present investigation.

3.2. Transformation of skew boundary of conoids
The plan of rhombic conoid is shown in Figure 2. In order to transform the global axes to local axes, transformation matrix [*T*] was used.

IOP Conf. Series: Materials Science and Engineering **710** (2019) 012039 doi:10.1088/1757-899X/710/1/012039

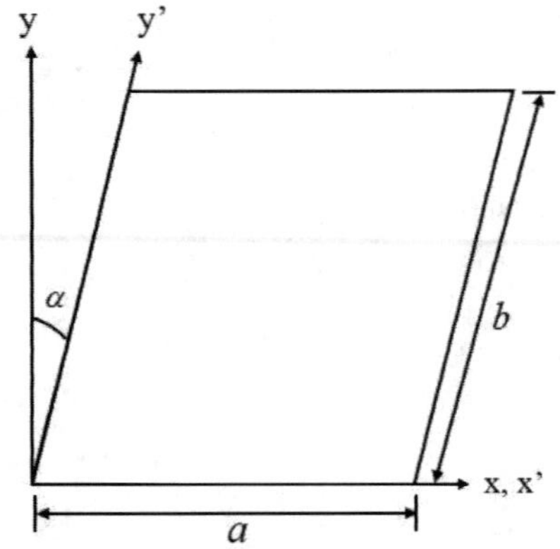

Figure 2. Plan of the rhombic conoidal shell.

$$[T]=\begin{pmatrix} c & -s & 0 & 0 & 0 & 0 & 0 \\ s & c & 0 & 0 & 0 & 0 & 0 \\ 0 & 0 & 1 & 0 & 0 & 0 & 0 \\ 0 & 0 & 0 & c & -s & 0 & 0 \\ 0 & 0 & 0 & s & c & 0 & 0 \\ 0 & 0 & 0 & 0 & 0 & c & -s \\ 0 & 0 & 0 & 0 & 0 & s & c \end{pmatrix} \tag{14}$$

where s = sinα, c = cosα and α is the skew angle of the conoidal shell.

3.3. Governing equation

The strain energy may be expressed as

$$U=\frac{1}{2}\iiint\{\varepsilon\}^T\{\sigma\}\mathrm{dxdydz} \tag{15}$$

By using the equation (11), the expression above can be represented as

$$U=\frac{1}{2}\iint\{\varepsilon\}^T[D]\{\varepsilon\}\mathrm{dxdy} \tag{16}$$

where, $[D]=\int[H]^T[Q][H]\mathrm{dx}_3$ in which [*H*] is the matrix that contains terms involving z and h.

By utilising equation (7) the stiffness matrix [*K*] is written as

$$[K]=\iint[B]^T[D][B]\mathrm{dxdy} \tag{17}$$

4. Numerical results and discussion

The bending analysis of FGM porous rhombic conoidal shells was performed under various types of mechanical loading. Parameters such as thickness ratio, aspect ratio, *hl/hh* ratio, porosity volume fraction and volume fraction index are also accomplished for numerical results. Unless stated otherwise, the following non-dimensional factors are utilised in the upcoming examples.

Non-dimensional quantities used are

$$\bar{w}=100\frac{wh^3E_c}{q_0a^4},\quad \bar{\sigma}_x=\sigma_x\left(\frac{a}{2},\frac{b}{2},\frac{h}{2}\right)100\frac{h^2}{q_0a^2}$$

Loading patterns used are

$$q = q_0,\ q = q_0 \sin\left(\frac{\pi x}{a}\right)\sin\left(\frac{\pi y}{b}\right)$$
$$q = q_0 \cos\left(\frac{\pi x}{a}\right)\sin\left(\frac{\pi y}{b}\right),\ q = q_0 \cos\left(\frac{\pi x}{a}\right)\cos\left(\frac{\pi y}{b}\right)$$

The details of some of the boundary condition (BC) used are as follows:

1. Simply supported (SSSS):

At $x = 0, a \quad v = w = \theta_y = \psi_y = 0$
At $y = 0, b \quad u = w = \theta_x = \psi_x = 0$

2. Clamped and simply supported (CCSS):

At $x = 0, a \quad u = v = w = \theta_x = \theta_y = \psi_x = \psi_y = 0$
At $y = 0, b \quad u = w = \theta_x = \psi_x = 0$

4.1. Comparison and convergence study

Three appropriate examples have been solved to check consistency and stability of the presented FE results. No results for FGM porous rhombic conoidal shell were found in the literature; hence, effectiveness of the present formulation has been tested by relating dimensionless deflection for FGM porous plate and isotropic conoids. The properties of the FGM constituents at room temperature used in the present study are as follows:

Al/ZrO_2: $E_c = 151$ GPa, $E_m = 70$ GPa, $\nu_c = \nu_m = 0.3$, $\rho_c = 3000$ kg/m^3, $\rho_m = 2707$ kg/m^3

Example 1. The convergence study of dimensionless maximum deflexion of FGM (Al/ZrO_2) plate was conducted for four power law indices ($n = 0, 0.5, 1, \infty$) and presented in Table 1. For this particular example, we used $hl = 0$ and $hh = 0$ in the present FE code of the FGM-1 conoidal shell for converting it into FGM plate. It was found that at 16 × 16 mesh, the results converge for the present nine-noded isoparametric elements. From Table 1, it was noted that our numerical results are consistent with the ones obtained by Ferreira et al. [20].

Example 2. Figure 3 represents the comparison of dimensionless maximum transverse displacement of simply supported FGM porous plate. The bottom layer of FGM porous plate is metallic and the top layer is ceramic. For various values of volume fraction indices, the results were compared with Akbaş and again it showed a decent agreement [15].

Example 3. No results for FGM porous conoidal shell exists in the literature; therefore, the presented FE result of FGM porous conoidal shell were compared with numerical result of isotropic conoidal shell. The Poisson's ratio $\nu = 0.15$, $hl/hh = 0.50$ and side-to-thickness ratio $a/h = 19$ was used to validate the presented result. Table 2 shows the validation of the presented FE formulation with Hadid and Bakshi and Chakravorty. The validation study confirms the improvement of the present obtained FE result over Bakshi and Chakravorty as the current numerical results are closer to Hadid [9,21].

Table 1. Validation study of dimensionless maximum deflexion of FGM simply supported plate.

Mesh	n			
	0	0.5	1	∞
6x6	0.02476	0.03326	0.03680	0.05348
8x8	0.02480	0.03331	0.03685	0.05352
12x12	0.02482	0.03333	0.03688	0.05353
16x16	0.02482	0.03333	0.03688	0.05353
Ferreira et al. [20]	0.0248	0.0330	0.0368	0.0536

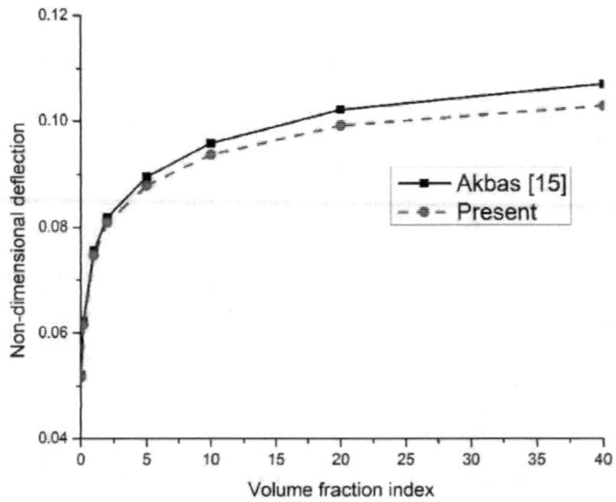

Figure 3. Comparison of dimensionless deflexion of FGM porous plate subjected to uniform loading.

Table 2. Comparison of deflexion (x10-2) of isotropic conoid along y/b = 0.50 subjected to uniformly distributed load.

x/a	Bakshi and Chakravorty [21]	Hadid [9]	Present
0.10	2.3680	2.5231	2.5621
0.40	5.1557	4.6333	4.7609
0.60	4.0510	3.5448	3.8575
0.70	3.4129	3.3149	3.2970
0.80	2.4809	2.5544	2.5949

4.2. Results and discussion

In order to analyse FGM porous conoids under various type of transverse loading, different composition of material constituents, different combination of boundary condition, numerous values of volume fraction index, side-to-thickness ratio, aspect ratio and *hl/hh* ratio were considered. Table 3 shows dimensionless maximum deflexion and their location subjected to uniform loading for the FGM porous rhombic conoids, respectively. The numerical results were computed for $a/h = 10$, $a/b = 1$, $hl/hh = 0.25$ ($hl = 0.05$, $hh = 0.2$), and $\alpha = 15°, 30°, 45°$ and $60°$. It can be seen that as we move from $n = 0$ to $n = \infty$, dimensionless deflexion of the porous conoidal shell is increased and it may be attributed to the higher volume fraction index which leads to a lesser ceramic content, thus reducing its stiffness. Interestingly, around 1.1 times increase in maximum dimensionless deflexion was noticed for all skew angle as the volume fraction index changed from 0 to ∞ for perfect conoidal shell, 1.4 times for porous conoidal shell with porosity volume fraction $\beta = 0.1$ and 1.7 times for porous shell with porosity volume $\beta = 0.2$.

Table 4 depicts dimensionless maximum deflexion of FGM porous rhombic conoids with skew angle for CCCC, CCSS, CSCS, CCFF and CFCF boundary condition. It was noted that the deflexion of the porous conoidal shell increased along with the porosity volume and it may be attributed to the decrease in the equivalent modulus of elasticity. The maximum dimension-less deflexion decreases along with skew angle. Due to the fact that an increase in skew results in the reduction of length of shorter diagonal and shortening of diagonal leads to improvement in stiffness of the rhombic conoids; thus, deflexion reduces.

Interestingly, highest deflexion was found for CFCF type end condition while CCCC retains lowest deflexion values among all types of end conditions. The effect of loading pattern on dimensionless maximum deflexion of simply supported FGM porous rhombic conoids is presented in Table 5. For all considered types of loading pattern, the non-dimensional value of deflexion decreases with an increase in skew angle. It can be noted that the effect of the skew angle is greater for cos-sin loading. The numerical results of dimensionless deflexion of FGM porous rhombic conoids with a/h ratio are plotted in Figure 4. These numerical results highlighted that non-dimensional deflexion value decreases with an increase in the a/h ratio. The dimensionless deflexion decreases up to a/h = 50; furthermore, no significant changes in dimensionless deflexion were noticed.

Table 3. Variation of dimensionless deflexion of FGM porous rhombic conoids subjected to uniform loading.

n	β	Skew angle			
		15°	30°	45°	60°
0	0	0.9342 (0.52,0.46)	0.5291 (0.60,0.39)	0.2214 (0.63,0.28)	0.0567 (0.63,0.17)
	0.1	1.0238 (0.52,0.46)	0.5808 (0.60,0.39)	0.2435 (0.63,0.28)	0.0623 (0.63,0.17)
	0.2	1.1269 (0.52,0.46)	0.6405 (0.60,0.39)	0.2691 (0.63,0.28)	0.0689 (0.63,0.17)
0.2	0	1.0225 (0.52,0.46)	0.5785 (0.60,0.39)	0.2410 (0.63,0.28)	0.0613 (0.63,0.17)
	0.1	1.1307 (0.52,0.46)	0.6406 (0.60,0.39)	0.2672 (0.63,0.28)	0.0679 (0.63,0.17)
	0.2	1.2577 (0.52,0.46)	0.7136 (0.60,0.39)	0.2980 (0.63,0.28)	0.0756 (0.63,0.17)
0.5	0	1.1042 (0.52,0.46)	0.6249 (0.60,0.39)	0.2597 (0.63,0.28)	0.0659 (0.63,0.17)
	0.1	1.2313 (0.52,0.46)	0.6977 (0.60,0.39)	0.2901 (0.63,0.28)	0.0735 (0.63,0.17)
	0.2	1.3837 (0.52,0.46)	0.7851 (0.60,0.39)	0.3267 (0.63,0.28)	0.0826 (0.65,0.19)
1	0	1.2062 (0.52,0.46)	0.6832 (0.60,0.39)	0.2834 (0.65,0.30)	0.0719 (0.65,0.19)
	0.1	1.3592 (0.52,0.46)	0.7707 (0.60,0.39)	0.3199 (0.50,0.30)	0.0810 (0.65,0.19)
	0.2	1.5479 (0.52,0.46)	0.8789 (0.60,0.39)	0.3650 (0.65,0.30)	0.0923 (0.65,0.19)
∞	0	2.0151 (0.52,0.46)	1.1413 (0.60,0.39)	0.4776 (0.63,0.28)	0.1222 (0.63,0.17)
	0.1	2.4366 (0.52,0.46)	1.3818 (0.60,0.39)	0.5798 (0.63,0.28)	0.1485 (0.63,0.17)
	0.2	3.0326 (0.52,0.46)	1.7235 (0.60,0.39)	0.7242 (0.63,0.28)	0.1854 (0.63,0.17)

Table 4. Variation of dimensionless deflexion of FGM porous rhombic conoids subjected to uniform loading.

BC	β	Skew angle			
		15°	30°	45°	60°
CCCC	0	0.9544 (0.58,0.48)	0.5615 (0.68,0.41)	0.2367 (0.74,0.32)	0.0604 (0.77,0.21)
	0.1	1.0814 (0.58,0.48)	0.6361 (0.68,0.41)	0.2681 (0.74,0.32)	0.0684 (0.77,0.21)
	0.2	1.2404 (0.58,0.48)	0.7291 (0.68,0.41)	0.3072 (0.74,0.32)	0.0783 (0.77,0.21)
CCSS	0	0.9970 (0.57,0.46)	0.5773 (0.67,0.39)	0.2433 (0.71,0.28)	0.0622 (0.72,0.19)
	0.1	1.1236 (0.57,0.46)	0.6508 (0.67,0.39)	0.2743 (0.71,0.28)	0.0701 (0.72,0.19)
	0.2	1.2807	0.7420	0.3129	0.0798

		(0.57,0.46)	(0.67,0.39)	(0.71,0.28)	(0.72,0.19)
CSCS	0	1.0140	0.5739	0.2344	0.0594
		(0.61,0.51)	(0.70,0.43)	(0.76,0.33)	(0.77,0.21)
	0.1	1.1460	0.6495	0.2658	0.0673
		(0.61,0.51)	(0.70,0.43)	(0.76,0.33)	(0.77,0.21)
	0.2	1.3102	0.7438	0.3049	0.0772
		(0.61,0.51)	(0.70,0.43)	(0.76,0.33)	(0.77,0.21)
CCFF	0	3.4178	2.7350	1.7582	0.7245
		(0.50,0.0)	(0.52,0.0)	(0.57,0.0)	(0.60,0.0)
	0.1	3.8095	3.0519	1.9652	0.8134
		(0.50,0.0)	(0.52,0.0)	(0.57,0.0)	(0.60,0.0)
	0.2	4.3110	3.4556	2.2288	0.9268
		(0.50,0.0)	(0.52,0.0)	(0.57,0.0)	(0.60,0.0)
CFCF	0	37.5010	29.2043	20.5013	10.7686
		(0.12,0.96)	(0.14,0.78)	(0.16,0.58)	(0.16,0.37)
	0.1	41.8795	32.6406	22.9386	12.0373
		(0.12,0.96)	(0.12,0.96)	(0.16,0.58)	(0.16,0.37)
	0.2	47.4397	36.9941	26.0191	13.6365
		(0.12,0.96)	(0.12,0.96)	(0.16,0.58)	(0.16,0.37)

Table 5. Variation of dimensionless deflexion of FGM porous rhombic conoids subjected to uniform and trigonometrical loading.

α	B	Loading pattern			
		uniform	sin-sin	cos-sin	cos-cos
15°	0	0.9544	0.7191	0.1070	0.1830
		(0.58,0.48)	(0.55,0.48)	(0.29,0.34)	(0.84,0.72)
	0.1	1.0814	0.8146	0.1200	0.2064
		(0.58,0.48)	(0.55,0.48)	(0.29,0.34)	(0.84,0.72)
	0.2	1.2404	0.9347	0.1366	0.2360
		(0.58,0.48)	(0.55,0.48)	(0.29,0.34)	(0.84,0.72)
30°	0	0.5615	0.4043	0.0432	0.1612
		(0.69,0.41)	(0.61,0.41)	(0.31,0.24)	(0.92,0.65)
	0.1	0.6361	0.4573	0.0480	0.1815
		(0.69,0.41)	(0.61,0.41)	(0.31,0.24)	(0.92,0.65)
	0.2	0.7291	0.5239	0.0542	0.2073
		(0.69,0.41)	(0.61,0.41)	(0.31,0.24)	(0.92,0.65)
45°	0	0.2367	0.1655	0.0180	0.0938
		(0.74,0.32)	(0.65,0.30)	(0.35,0.18)	(0.95,0.53)
	0.1	0.2681	0.1867	0.0199	0.1054
		(0.74,0.32)	(0.65,0.30)	(0.35,0.18)	(0.95,0.53)
	0.2	0.3072	0.2133	0.0223	0.1200
		(0.74,0.32)	(0.65,0.30)	(0.35,0.18)	(0.95,0.53)
60°	0	0.0604	0.0438	0.0059	0.0381
		(0.77,0.21)	(0.67,0.20)	(0.37,0.11)	(0.97,0.37)
	0.1	0.0684	0.0493	0.0065	0.0425
		(0.77,0.21)	(0.67,0.20)	(0.37,0.11)	(0.97,0.37)
	0.2	0.0783	0.0561	0.0072	0.0480
		(0.77,0.21)	(0.67,0.20)	(0.37,0.11)	(0.97,0.37)

Table 6. Effect of hl/hh ratio on dimensionless maximum deflexion of FGM porous rhombic conoids subjected to uniform loading.

hl/hh	*β*	α			
		15°	30°	45°	60°
0.20	0	1.2599 (0.52,0.46)	0.7067 (0.60,0.39)	0.2918 (0.63,0.28)	0.0741 (0.63,0.17)
	0.1	1.4195 (0.52,0.46)	0.7970 (0.60,0.39)	0.3291 (0.63,0.28)	0.0834 (0.63,0.17)
	0.2	1.6162 (0.52,0.46)	0.9087 (0.60,0.39)	0.3754 (0.63,0.28)	0.0950 (0.63,0.17)
0.15	0	1.3157 (0.52,0.46)	0.7309 (0.60,0.39)	0.3004 (0.63,0.28)	0.0764 (0.63,0.17)
	0.1	1.4821 (0.52,0.46)	0.8242 (0.60,0.39)	0.3388 (0.63,0.28)	0.0860 (0.63,0.17)
	0.2	1.6873 (0.52,0.46)	0.9395 (0.60,0.39)	0.3861 (0.63,0.28)	0.0978 (0.63,0.17)
0.10	0	1.3735 (0.52,0.46)	0.7559 (0.60,0.39)	0.3093 (0.63,0.28)	0.0787 (0.63,0.17)
	0.1	1.5471 (0.52,0.46)	0.8522 (0.60,0.39)	0.3487 (0.63,0.28)	0.0886 (0.63,0.17)
	0.2	1.7610 (0.52,0.46)	0.9712 (0.60,0.39)	0.3973 (0.63,0.28)	0.1007 (0.63,0.17)
0.05	0	1.4336 (0.52,0.46)	0.7816 (0.60,0.39)	0.3184 (0.63,0.28)	0.0812 (0.63,0.17)
	0.1	1.6146 (0.52,0.46)	0.8810 (0.60,0.39)	0.3589 (0.63,0.28)	0.0913 (0.63,0.17)
	0.2	1.8378 (0.52,0.46)	1.0039 (0.60,0.39)	0.4089 (0.63,0.28)	0.1038 (0.63,0.17)
0.0	0	1.4955 (0.52,0.46)	0.8080 (0.60,0.39)	0.3277 (0.63,0.28)	0.0837 (0.63,0.17)
	0.1	1.6841 (0.52,0.46)	0.9107 (0.60,0.39)	0.3693 (0.63,0.28)	0.0941 (0.63,0.17)
	0.2	1.9167 (0.52,0.46)	1.0375 (0.60,0.39)	0.4207 (0.63,0.28)	0.1069 (0.63,0.17)

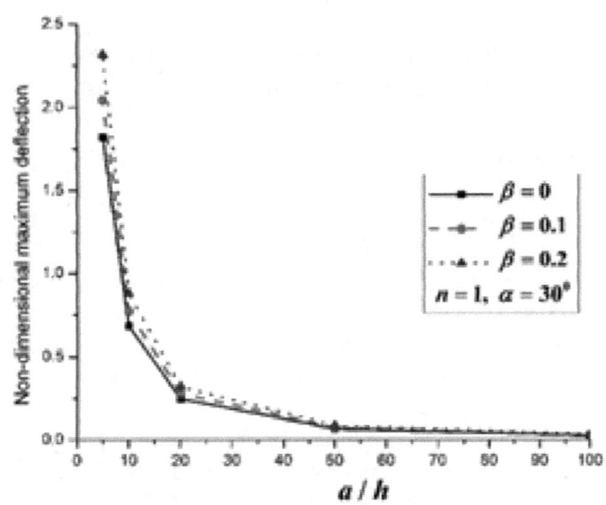

Figure 4. Variation of dimensionless maximum deflexion of FGM porous rhombic conoids subjected to uniform loading.

The correlation between *hl/hh* ratio and dimensionless deflexion of FGM porous rhombic conoids subjected to uniform loading was tested in Table 6. Interestingly, for higher value of *hl/hh* ratio, lower value of dimensionless deflexion was noticed. The decrease in *hl/hh* ratio reduces the curvature of the lower end (*hl*) of the conoidal shell. Thus, the stiffness of shell reduces, while deflexion increased. The results were calculated for $a/b = 1$ and $a/h = 10$. Figure 5 shows the dimensionless value of axial stresses along with thickness coordinate for FGM porous rhombic conoids having $\alpha = 15°$ and $\alpha = 30°$.

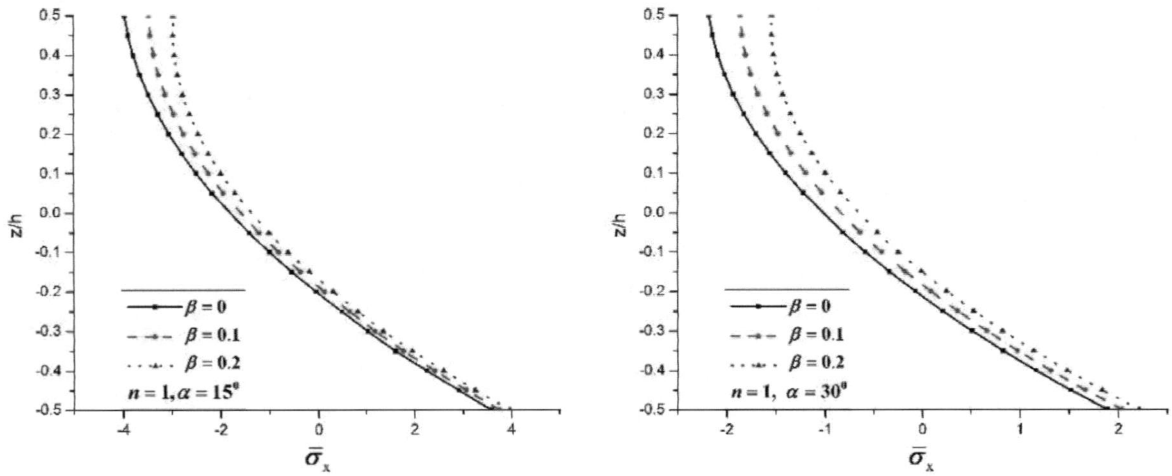

(a) (b)

Figure 5. Effect of porosity on dimensionless axial stress of FGM porous rhombic conoids subjected to uniform loading.

5. Conclusion

This paper focuses on the influence of porosity on the static analysis of FGM porous rhombic conoids based on improved TSDT subjected to the various types of loads using an efficient C^0 FE model. The subsequent outcomes of the present study were written below for the volume fraction indices, skew angles, thickness ratios, *hl/hh* ratios and various types of end support.

- The dimensionless deflexion of the FGM porous conoidal shell increases along with the porosity volume fraction and volume fraction index.
- With an increase in skew angle, dimensionless deflexion and normal stress decreases.
- The inclusion of porosity brings up deflexion in comparison with the perfect conoidal shell.
- The dimensionless deflection increases along with thickness of conoids.
- For all considered skew angles, a decrease in *hl/hh* ratio results in an increase in dimensionless deflexion.

6. References

[1] Koizumi M 1997 FGM activities in Japan *Compos Part B* **28B** 1-4 https://doi.org/10.1016/S1359-83689600016-9
[2] Zhao X, Lee Y and Liew K 2009 Thermoelastic and vibration analysis of functionally graded cylindrical shells *Int J Mech Sci* **519-10** 694-707 https://doi.org/10.1016/j.ijmecsci.2009.08.001
[3] Kashtalyan M and Menshykova M 2009 Three-dimensional elasticity solution for sandwich panels with a functionally graded core *Compos Struct.* **871** 36-43 https://doi.org/10.1016/j.compstruct.2007.12.003
[4] Tornabene F and Viola E 2013 Static analysis of functionally graded doubly-curved shells and panels of revolution *Meccanica* **48** 901-30. https://doi.org/10.1007/s11012-012-9643-1

[5] Viola E, Rossetti L, Fantuzzi N and Tornabene F 2014 Static analysis of functionally graded conical shells and panels using the generalized unconstrained third order theory coupled with the stress recovery *Compos Struct.* **112** 44-65. https://doi.org/10.1016/ j.compstruct. 2014.01.039

[6] Dai H and Dai T 2014 Analysis for the thermoelastic bending of a functionally graded material cylindrical shell *Meccanica* **49** 1069-81 https://doi.org/10.1007/s11012-013-9853-1

[7] Xiang S and Liu Y 2016 An nth-order shear deformation theory for static analysis of functionally graded sandwich plates *J. Sandw. Struct. Mater.* **185** 579-96 https://doi.org/10.1177/1099636216647928

[8] Gutierrez Rivera M and Reddy J 2016 Stress analysis of functionally graded shells using a 7-parameter shell element *Mech. Res. Commun.* **78** 60-70 https://doi.org/10.1016/ j.mechrescom. 2016.02.009

[9] Hadid H 1964 An analytical and experimental investigation into the bending theory of elastic conoidal shell. PhD Dissertation, University of Southampton

[10] Das A K and Bandyopadhyay J N 1993 Theoretical and experimental studies on conoidal shells *Computers and Structures* **493** 531-36 https://doi.org/10.1016/0045-79499390054-H

[11] Ghosh B and Bandyopadhyay J 1994 Bending Analysis of Conoidal Shells With cut-outs *Computers and Structures* **531** 9-18

[12] Das H S and Chakravorty D 2009 Composite full conoidal shell roofs under free vibration *ADV VIB ENG* **84** 321-28

[13] Kumari S and Chakravorty D 2010 Finite element bending behaviour of discretely delaminated composite conoidal shell roofs under concentrated load *Int. J. Eng. Sci.* **24** 54-70

[14] Malekzadeh Fard K and Baghestani A 2017 Free vibration analysis of deep doubly curved open shells using the Ritz method *Aerosp Sci Technol* **69** 136-48 https://doi.org/10.1016/j.ast.2017.06.021

[15] Akbas S D 2017 Vibration and Static Analysis of Functionally Graded Porous Plates *Journal of Applied and Computational Mechanics* **33** 199-207 https://dx.doi.org/10.22055/ jacm.2017.21540.1107

[16] Al Rjoub Y S and Hamad A G 2017 Free vibration of functionally Euler-Bernoulli and Timoshenko graded porous beams using the transfer matrix method *KSCE Int. J. Civ. Eng.* **213** 792-806. https://doi.org/10.1007/s12205-016-0149-6

[17] Kiran M, Kattimani S and Vinyas M 2018 Porosity influence on structural behaviour of skew functionally graded magneto-electro-elastic plate *Compos Struct.* **191** 36-77 https://doi.org/ 10.1016 / j.compstruct.2018.02.023

[18] Gupta A and Talha M 2017 Influence of porosity on the flexural and vibration response of gradient plate using nonpolynomial higher-order shear and normal deformation theory *INT J MECH MATER DES* **14** https://doi.org/10.1007/s10999-017-9369-2

[19] Gupta A and Talha M 2018 Influence of Porosity on the Flexural and Free Vibration Responses of Functionally Graded Plates in Thermal **18** https://doi.org/10.1142/S021945541850013X

[20] Ferreira A, Roque C, Jorge R, Fasshauer G and Batra R 2007 Analysis of Functionally Graded Plates by a Robust Meshless Method *MECH ADV MATER STRUC* **148** 577-87 https://doi.org/10.1080/15376490701672732

[21] Bakshi K and Chakravorty D 2014 First ply failure study of thin composite conoidal shells subjected to uniformly distributed load Thin-Walled *Structures* **76** 1-7 https://doi.org/10.1016/j.tws.2013.10.021

IOP Conf. Series: Materials Science and Engineering **710** (2019) 012040 doi:10.1088/1757-899X/710/1/012040

Life Cycle Assessment as a tool for the selection of solar hot water system

A Żelazna and J Gołębiowska

Lublin University of Technology, Faculty of Environmental Engineering, Nadbystrzycka 40B, 20-618 Lublin, Poland

a.zelazna@pollub.pl

Abstract. During the past decades, solar hot water systems became a popular solution, enabling utilisation of renewable energy by both, single users and enterprises. Maximisation of the energy gains is one of the factors influencing the environmental burdens connected with solar systems. The second important issue is a kind of elements of system, their production, transportation, usage and disposal. Therefore, as the method for estimation of the scale of environmental indicators Life Cycle Assessment should be used. In this study, LCA was applied for the selection of the individual design on the basis of two criteria (water tank capacity and hot water demand) considered in three system scenarios. The calculated indicators included greenhouse gas emission (Global Warming Potential) and IMPACT2002+ method. The results show the importance of proper designing in the case of environmental impact reduction.

1. Introduction

Modern society with its high energy demands faces a challenge of climate changes and a threat of depletion of natural resources. Those challenges led us to a concept of sustainability, which nowadays has become a common interest of various disciplines. This common interest is expressed through the performance of the sustainable development in our economy, society and environment, which includes the introduction of sustainable energy [1].

Sustainable development in the energy branch means using systems that are environmentally responsible and resource-efficient [2]. Responsible designing of energy systems is focused on activities which aim at environmental improvement of product and improvement of system efficiency. Environmental improvement is based on application of alternative processes, selection of materials with lower environmental effect, development in logistics and final disposal, etc. [2].

The answer to the question of sustainable energy seems to be energy from renewable sources. However, the introduction of such technologies need to be accompanied by an investigation of their environmental burdens. Minimisation of such burdens can be performed within the use of several methods, including the environmental Life Cycle Assessment.

In the case of small scale solar installations, one of the most important problems resulting from the economic issues is the lack of proper, individual design. This problem can be clearly visible on the example of mass-scale installations of the same system during performance of governmental programmes. In the mentioned case, it can be observed that the slope of collectors and the dimensions of installation is not individually designed, therefore issues such as maximisation of energy gains are neglected. Another important design parameters, hot water tank capacity and water demand, are the variables in the presented study.

2. Life Cycle Assessment of solar hot water systems

The subject of studies on the environmental Life Cycle Assessment of devices and energy systems was described in many papers, where environmental assessment of renewable energy sources is one of the most important issues [3-6]. Life Cycle Assessment is a popular tool for the analyses of renewable technologies since it allows the implementation of specified life cycle model and related environmental burdens of various types, like emissions, material etc. Concerning the topic of this paper, it is necessary to note that most of the analyses concern the ecological balance of SHW systems operating in Southern Europe, or only include Life Cycle Assessment at the stage of system components production. The selected examples of LCA studies are presented below.

Life Cycle Assessment method was used by Kalogirou to determine the environmental impact of flat plate collectors. The construction phase was analysed from the point of view of embedded energy. In the first paper, a system integrated with hot water storage tank was analysed. The consumption of primary energy for the production of integrated system was estimated as 2.7 GJ [7]. In another paper, Kalogirou described LCA for a system with two media: polypropylene glycol and water and calculated the primary energy consumption as 3.5 GJ [8].

A broad approach to the LCA for solar hot water system is presented in the paper of Alberti et al [9]. The authors compared environmental performance of two scenarios: a solar thermal system for providing domestic hot water (DHW) used in conjunction with a traditional natural gas heating system, and the natural gas heating system on its own. In addition, the authors calculated the national Greenhouse Gas emission reduction potential of a wider use of domestic solar hot water systems (DSHW) in China's and Spain's built environment by the use of five displacement methods. In general, some impact categories show a markedly better performance of the solar system (-65% in the case of greenhouse gas emission). However, weak points in the solar solution have been identified as there is an increase of impacts in categories such as acidification (+6%) and eutrophication (+61%), mostly due to used metals. Reduction of national emissions by promoting DSHW depends on the actual displaced technology and, in the case when 30% of the DHW were covered with solar sources, would place between 0.38% and 0.50% in Spain and between 0.12% and 0.63% in China.

Another example of LCA applied for the integrated solar system is the paper of Battisti and Corrado [10]. The use of primary energy in the life cycle of the tested system was calculated as 3.1 GJ, and greenhouse gas emissions equalled 219.4 kgCO2eq. The production phase of the system was responsible for 97.8% of the total value of indicator. On the other hand, its operational phase has been omitted because of the specific design of the system (a capacitive collector as a heat tank), which did not require a pump [18].

In the paper of Soulions et al. [11], the study of two advanced solar water heating systems, integrated with the facades and the roof of a social house building located in Nicosia, Cyprus and Athens, Greece, was presented. The first one comprises collector and storage tank in a single unit, called Integrated Collector Storage solar water heater; the second one combines a photovoltaic and a thermal collector called Hybrid Photovoltaic/Thermal device. The authors considered the integration of the studied solar systems on building's envelope, focusing on the effect of these systems in terms of building's thermal load along with their capacity to cover the energy needs, especially in terms of hot water and electricity consumption.

Computer simulations of energy load of building, as well as the energy output of solar systems, were carried out for two mentioned locations within South, West and East orientation modes. The implementation of the described solar systems into the building envelope allowed to diminish the thermal energy needs up to 10% in the case of heating and cooling purposes and up to 80% in the case of hot water and electricity needs. The method used for investigation of energy contribution was Life Cycle Assessment, allowing the evaluation of environmental impacts during the entire life cycle of the considered systems.

According to the nowadays state of knowledge, most of renewable sources can be treated as a source of clean, green energy. Though, one should remember about the need to individually consider the issues of their location. Local climate conditions are crucial for adaptation of the design solutions to the individual needs of the recipients, as well as for possible energy output of the designed systems [12].

3. Computer simulations as a tool for solar system designing

3.1. General assumptions

The presented study of system for solar hot water preparation was carried out on the basis of typical solution for a single-family house with four inhabitants. The conducted analysis included three versions of system with changing capacity of the hot water tank:

- 200 dm^3 in system no. 1 (S1),
- 250 dm^3 in system no. 2 (S2),
- 300 dm^3 in system no. 3 (S3).

In each system scenario hot water consumption was treated as variable between minor and higher water demand and equalled:

- 160 dm^3/d,
- 200 dm^3/d,
- 240 dm^3/d.

The selected location in central Poland (Łódź) is characterised by favourable insolation conditions, with yearly solar irradiance assumed in the study as 1147.93 kWh/m^2.

As solar thermal collector, 4.34 m^2 flat-plate collector with Al-Cu absorber was tested, characterised by the following data:

- optical efficiency n0=0.761 [-],
- 1st order heat loss coefficient a1=3.438 [W/m^2],
- 2nd order heat loss coefficient a2=0.008 [W/m^2].

3.2. Materials and methods

For a widespread implementation of computer modelling into designing process, two applications (GetSolar and SimaPro) were used to incorporate the variable design conditions and environmental aspects into the analysed model. The first program was used to estimate the possible energy output with the use of mentioned variable data on water consumption and storage volume for the selected collector. The second application, SimaPro, was used for forming of the system model and its life cycle.

As first, a life cycle impact assessment method IMPACT2002+ was used. In this damage-oriented method, the calculations concerning total environmental load expressed in Points are conducted on the basis of detailed life cycle inventory. The higher the calculated indicator, the more significant the environmental impact. As the most of damage-oriented indicators, Points calculated by the use of IMPACT2002+ are used mostly for comparisons of products or systems.

The calculations are performed throughout the midpoint level, including four damage categories (Human Health, Ecosystem Quality, Climate Change and Resources) and seventeen impact categories (Human toxicity, Respiratory effects, Ionizing radiation, Ozone layer depletion, Photochemical oxidation, Aquatic ecotoxicity, Terrestrial ecotoxicity, Aquatic acidification, Aquatic eutrophication, Land occupation, Terrestrial acidification/nutrification, Water turbined, Global Warming, Non-renewable energy, Mineral extraction, Water withdrawal, Water consumption). In the particular method, midpoint characteristics are presented in comparative way within the use of specific standard substances. The impact is considered within the time framework up to 500 years [13].

As second, a life cycle impact assessment method Global Warming Potential in 500 years perspective was used. This method is simpler than the previously described one and includes only the climate change category, based on carbon dioxide as a comparative substance. Selected emissions of substances to the air are compared with CO_2 on the basis of potential to cause the same effect as emission of carbon dioxide [14].

All the operations connected with environmental data modelling in this study, as well as the calculations of indicators were conducted in SimaPro v. 8.

IOP Conf. Series: Materials Science and Engineering **710** (2019) 012040 doi:10.1088/1757-899X/710/1/012040

3.3. Results of GetSolar modelling

GetSolar is the popular application enabling easy and quick modelling of solar systems power output.

The necessary parameters to be implemented into the model include:

- solar collector data:
  - optical efficiency n0 [-],
  - 1st order heat loss coefficient a1 [W/m²],
  - 2nd order heat loss coefficient a2 [W/m²],
  - collector area [m²].
- climate data including solar irradiance on collector plane [W/m²] and temperature.
- type of solar system (single, combined, etc.)
- type of water storage tank,
- capacity of water storage tank [dm³].
- orientation (0 – 360°)
- angle of inclination [°].

Data were assumed after the producers of collectors (technical data), and as typical design parameters. Two of the variables (water consumption and storage tank capacity) were examined to provide information about possible energy output, efficiency and the coverage of hot water demand during one year of operation.

The results of possible energy output modelling are presented in Table 1.

Table 1. The possible energy output, efficiency and hot water demand coverage modelled for systems 1, 2 and 3 with variable water demand.

System	Water demand, dm^3/d	Capacity of water tank, dm^3	Efficiency of system, %	Coverage of hot water demand, %	Possible energy output, kWh/yr	Average gain from collector, $kWh/yr/m^2$
S1	160	200	33	53	1630	376
	200		36	47	1814	418
	240		39	42	1949	449
S2	160	250	33	53	1652	381
	200		37	47	1834	423
	240		40	43	1978	456
S3	160	300	33	53	1661	383
	200		37	48	1847	426
	240		40	43	1994	459

The results presented in Table 1 show potential dependence between energy output and investigated variables. The examined parameters were assumed as typical for the selected type of system, with the consideration of changing water demand and storage volume. The increase in possible energy output reaches 22% yearly between the minimum and maximum of analysed variations. However, considering only one of the variables, the change in energy output exceeds 20% for water consumption and 2% for storage tank volume.

3.4. Results of SimaPro modelling

SimaPro v8 was used to enable the calculation of Life Cycle Impact Assessment indicators, particularly IMPACT2002+ (Figure 1) and GWP100a (Figure 2). This enabled the comparison between the assumed life cycle of System 1, System 2 and System 3 in three water consumption modes, considering the related environmental burdens. The detailed inventory, including production, operation and final disposal of System 1, System 2 and System 3, allowed the environmental impact assessment using the mentioned methods.

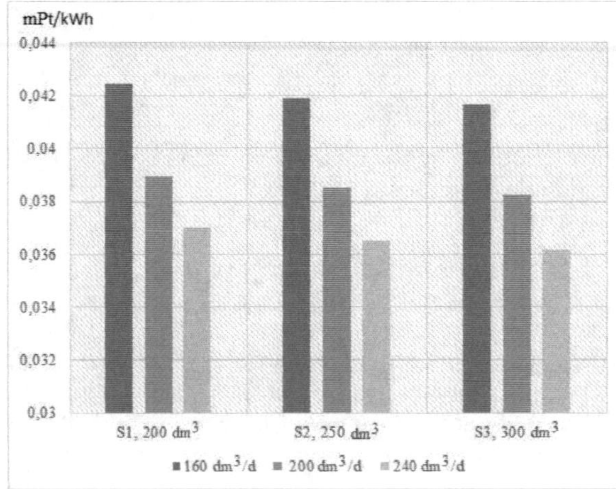

Figure 1. IMPACT2002+ results for System 1, System 2 and System 3 per functional unit.

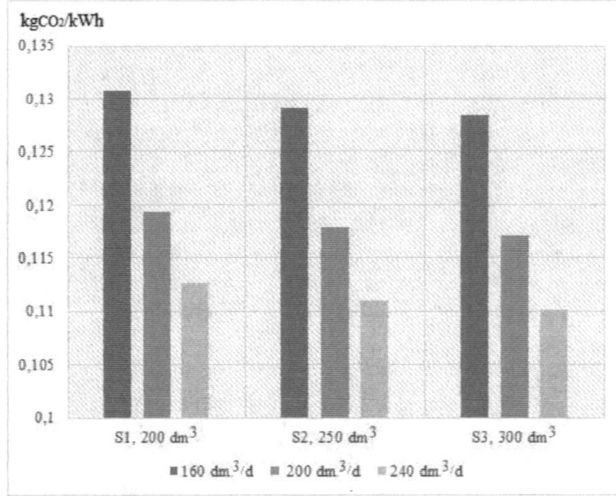

Figure 2. IPCC GWP100a results for System 1, System 2 and System 3 per functional unit.

The results presented in Figure 1 and Figure 2 are calculated for the functional unit, which included the energy possibly generated by solar hot water system, according to the previously presented calculations performed in GetSolar. This allowed to select the most environmentally-friendly version of system, considering the calculated indicators with the lowest value. In the described computer simulation, this was System 3 with the highest water consumption. It should be mentioned that high values of GWP indicator result from the kind of electricity used in the SimaPro model, since the operational phase is responsible for the most of calculated indicators.

In Figure 3, total IMPACT2002+ scores for System 1, System 2 and System 3 are presented within the damage categories. In this particular case, the share of impact categories in total value of indicator are similar for analysed systems since most of the environmental impact is connected with the use of electricity (polish energy mix). Therefore, issues such as conventional fuels, carbon dioxide emission and particulate matter emission are the most meaningful aspects, together with the use of metals for production of elements of the system.

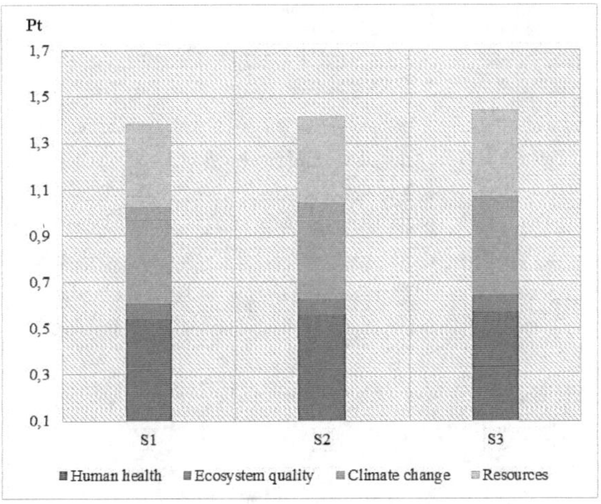

Figure 3. IMPACT2002+ damage assessment results for System 1, System 2 and System 3.

4. Conclusions

Life Cycle Assessment is an important tool that enables extending knowledge (including information about environmental burdens) of energy system designers and investors; thus, it helps in a decision-making process.

In this paper, the LCA methodology was used to assess the typical solar hot water preparation system applied to a single-family house with four inhabitants. The Global Warming Potential and IMPACT2002+ methods were used to calculate environmental indicators.

In the conducted analysis three system scenarios were considered, including different values of water tank capacity: 200 dm^3, 250 dm^3, 300 dm^3. Additionally, each scenario comprised three values of daily hot water consumption: 160 dm^3/d, 200 dm^3/d, 240 dm^3/d. The first three options influenced both the material and energy balance of examined systems, while the daily hot water consumption affected only the energy output. Therefore, three systems (System 1, System 2 and System 3) were indicated on the basis of water tank capacity, and further calculations were the options of water consumption for the mentioned systems.

The conducted analysis showed that the possible annual energy output from the different variants of examined system varies from 1630 kWh (tank capacity 200 dm^3, hot water demand 160 dm^3/d) to 1994 kWh (tank capacity 300 dm^3, water demand 240 dm^3/d). In general, in conducted analysis of systems, energy output increases with increasing tank capacity and water demand values. The difference between the highest and the lowest ones reaches 22%.

The analysis based on Life Cycle Assessment showed that System 3 (tank capacity 300 dm^3) with the highest water consumption of 240 dm^3/d is characterised by the lowest values of Global Warming Potential and IMPACT2002+ indicators. The tendency for total IMPACT2002+ score and GWP is similar for all considered scenarios since environmental impact is connected mainly with electricity consumption. For that reason, issues such as fossil fuels consumption, carbon dioxide and particulate matter emission are the biggest environmental problems.

Acknowledgement
This research work was supported from statutory founds of Faculty of Environmental Engineering, Lublin University of Technology.

5. References

[1] Pawłowski A *Probl. Ekorozw.* **4**(2) 9-12
[2] Zbicinski I, Stavenuiter J, Kozlowska B and van de Coevering HPM 2006 *Product Design and Life Cycle Assessment* (Uppsala, The Baltic University Press)

[3] Zambolin E and Del Col D 2010 *Sol. En.* **84**(8) 1382-96
[4] Gaidajis G and Angelakoglou K 2011 *Jour. of Eng. Sc. and Techn. Rev.* **4**(1) 92-5
[5] Varun A, Bhat IK and Prakash R 2009 *Renew. Sust. Energ. Rev.* **13** 1067
[6] Wild-Scholten MJ and Alsema EA 2006 *Mater. Res. Soc. Symp. Proc.*, 895-904
[7] Kalogirou S 2004 *Energ. Convers. Manage.* **45** 3075-92
[8] Kalogirou S, Tripanagnostopoulos Y and Souliotis M 2005 *Energ. Buildings* **37**(8) 824-35
[9] Alberti J, Raigosa J, Raugei M, Assiego R, Ribas-Tur J, Garrido-Soriano M, Zhang LH, Song GB, Hernandez P and Fullana-i-Palmer P 2019 *Sustain. Cities Soc.* **47** 101467
[10] Battisti RG and Corrado A 2005 *J. Clean. Prod* **13** 1295–300
[11] Soulions M, Panaras G, Fokaides P, Papaefthimiou S and Kalogirou S 2018 *Energ Build.* **169** 157-71
[12] Żukowski M 2017 *E3S Web of Conf.* **22** 00209
[13] Joilet O, Margni M, Charles R, Humbert S, Payet J, Rebitzer G and Rosenbau R 2003 *Int J LCA* **10**(6) 324-30
[14] Żelazna A 2013 *Rocznik Ochrona Środowiska* **15**(1) 258-71.

The comparison of automatic traffic counting and manual traffic counting

J Paľo[1], J Caban[2], M Kiktová[1] and Ľ Černický[1]

[1] University of Zilina, Faculty of Operation and Economics of Transport and Communications Department of Road and Urban transport, Univerzitná 1, 010 26 Žilina, Slovak Republic
[2] University of Life Sciences in Lublin, Faculty of Production Engineering, Głęboka 28, 20-612 Lublin, Poland

jacek.caban@up.lublin.pl

Abstract. The main purpose of traffic surveys is to know current state of transport and for design and solution of transport in the future. For collecting traffic data, traffic surveys are carried out which consist in counting the number of vehicles of different categories. There are many ways of collecting traffic data. This paper deals with manual counting and with automatic vehicle counter. Two traffic surveys were made and results from manual counting and automatic counting were compared. The aim is to find and describe advantages and disadvantages of selected types of traffic counting.

1. Introduction

The transport sector is influenced by a wide range of external social and economic factors such as demographics, living standards of the population, urban planning, organisation of production, structural changes in society and accessibility to transport infrastructure [1]. The main task of transport sector is to satisfy the customers' requirements for quality, flexibility, fastness and safety in transportation of people and goods [2]. Transport is an area with obvious and perceived problems such as noise, air pollution, traffic congestion and health problems [3]. Car travel is related to energy costs (fuel consumption), climate change (greenhouse gas), consumption on fossil fuels and traffic congestion [4, 5, 6, 7]. Congestion in urban areas is currently one of the most pressing problems in transport [8]. For the above reasons, traffic survey is a summary of activities designed to get traffic information. Traffic surveys mainly serve as a basis for solving and assessing suitability and quality of transport, for solving and designing an optimum outlook arrangement in traffic, for analysing the current traffic situation and actually solving each engineering work. The purpose of traffic surveys is to know the current state of transport [9].

Traffic surveys aim to capture data that accurately reflects the real-world traffic situation in the area. It may be counting the number of vehicles on a road or collecting journey time information for example, but there are many other types of data that traffic surveys collect.

In recent years, the manual approach has been largely replaced by new modern types of data collecting. This paper deals with automatic traffic counters, another way is recording traffic using video cameras, and then analysing the video footage later in the office. But this way of counting is excessively time-consuming [10].

IOP Conf. Series: Materials Science and Engineering **710** (2019) 012041 doi:10.1088/1757-899X/710/1/012041

2. Vehicle counter

The Sierzega SR4 vehicle counter is a statistical traffic data device that permits an unbiased, accurate and straightforward acquisition of traffic information for analysis. The radar device's inconspicuous appearance allows it to record vehicles without affecting normal flow of traffic. The radar device is powered through one or two 6-volt 12Ah batteries; each battery has enough power to last approximately one week depending on vehicle flow.

The Sierzega SR4 radar is primarily a speed measuring tool. Due to the way in which the data is gathered it gives volume, length and gap information. The data is recorded on per vehicle basis or VBV for short, each data record holds information for one vehicle, this data is made up of the following; vehicle speed in MPH (km/h), vehicle length in decimetres, gap in seconds, time and date. The data is compressed and stored in flash memory; the very latest radar devices are capable of recording 430,000 individual data records before it fills the onboard memory [11].

Once the data has been recorded and stored in memory, it is easily transferred directly to an existing PC for analysis using the standard serial port and SRA analysis software. The data can also be downloaded via a Pocket PC using pre-programmed terminal software and then synchronised to a desktop PC ready for analysis using SRA software.

There is also an option to export data to text files and then process them in another software, such as Microsoft Excel.

SRA software for data processing has predefined various diagrams which are easy to create, but also make classification of vehicles into four categories according to length of vehicles. Predefined borders for vehicles' classification are:

- 0 – 20 dm for Category 1,
- 20 – 60 dm for Category 2,
- 60 – 120 dm for Category 3,
- 120 – 255 dm for Category 4.

Also, every data has direction + or – according to direction of vehicles. (+ means oncoming vehicles). Then the raw data from the counter are: Date, Time, Length of vehicles, Speed of vehicles, Category, Gap, and Direction.

Date	Time	dm	km/h	Cat.	Gap	Dir.
2019-04-25	06:50:48	156	71	4	2,3	+
2019-04-25	06:50:50	126	69	4	0,7	+
2019-04-25	06:50:55	13	57	1	5,1	+

Figure 1. Example of data collected by the traffic counter.

2.1. Guidelines for installations

There are a few guidelines that need to be adhered to for ensuring that the data produced by the radar unit is both consistent and accurate. When considering a monitoring location if possible, try and find a pole that is between 0.5 meters and 2 meters parallel to the carriageway. The optimum mounting height of the radar unit should be at 1 meter above road level, however if the location is not suitable for the optimum mounting height due to footpaths or walkways then it can be installed at a height of up to 2.2 meters.

The automatic counter Sierzega offers various pole clamps according to country specific needs. The pole width should be between 0.06 m and 0.2 m. It is also advisable to avoid, where possible, dips, bends, and big masses of metal. If two radar units are to be deployed for better lane accuracy a distance of 50 m must be kept between radar beams.

2.2. Accuracy of traffic counter Sierzega SR4

According to producer, accuracy of the vehicle counter Sierzega SR4 is +/- 3% in speed, and +/- 20 % in length of vehicle. As the counter is primarily a speed measuring tool, there is no note about accuracy in number of counted vehicles. Classification of vehicles is carried out according to only one measured data, which is length of vehicles [12].

3. Vehicle counter

The manual counting means that people are standing by the side of roads and recording their observations on paper pads. The recording of vehicles makes to pre-prepared census sheets by comma method. In the case of high traffic, Arabic numerals there could be used. The crossing of vehicles is recorded at 15-minute intervals. Vehicles are divided into six categories: passenger car, motorcycle, truck, bus, heavy truck, bicycle. After the survey, these data are processed into Excel for the next evaluation [13, 14].

3.1. Accuracy of manual traffic counting

Manual counting usually refers to the practice of counting classified traffic in a 'manual fashion'. Some examples of traffic counting include vehicle counts at intersections, estimation of average daily traffic and annual average daily traffic [15].

Manual counting and classification can be carried out on the site or alternatively from video recordings. Counting and classification are simply based on visual examination and judgments by individual observers. The data is usually recorded using tally sheets or mechanical counters. After data have been collected for an interval (e.g. 1 min), totals are calculated and registered on a data sheet which can be inputted into computer later. It is usually taken for granted that errors in manual counts are small and can be ignored.

The investigation based on the manual counts derived from video recordings found out errors:

- counting errors are small, usually less than 1%,
- classification errors are significant, with an average between 4-5%.

The main errors are classification errors, reflection of difficulties in judging vehicles by a length criterion of 5.2 m from video recordings [16].

4. Comparison of results from traffic counts

In this paper, there are compared results from manual traffic counts and traffic counts from device Sierzega SR04. As there is quite different methodology of vehicles' classification, where vehicle counter carries out classification according to the length of vehicles and its accuracy is claimed to be +/- 20%, there is compared only total number of counted vehicles in this paper.

Supposing that errors in manual traffic counting are small [17] less than 1%, data from manual counts will be considered as the right number of vehicles. The data of manual traffic counts were collected in 15 minutes intervals.

Two surveys were conducted to compare the results of the automatic traffic count and the manual traffic count. The first survey was performed on April 25, 2019 near the town of Košice and the second survey on May 15, 2019 in the village of Palárikovo.

4.1. First survey

First survey was carried out at the selected crossing of roads R2 and III/3,401 near the town of Košice. Manually counted were all vehicles entering to the intersection, automatically counted were vehicles in two lanes at the entrance no. 3, see Figure 2.

Figure 2. The location of automatic vehicle counter - first survey.

The total number of vehicles that were manually counted was 10,120. Surprisingly the vehicle counter counted 11,149 of vehicles. This means, that automatic traffic counter counted 1,029 (10.2 %) more vehicles than it was counted manually. Looking closer to Figure 3, it can be seen that nearly at all 15 minutes intervals the vehicle counter counted more vehicles than manual counts. The greatest difference is in the morning in the time interval 6:30 – 6:45, in which vehicle counter counted 112 vehicles more (30.9%) than manual counts.

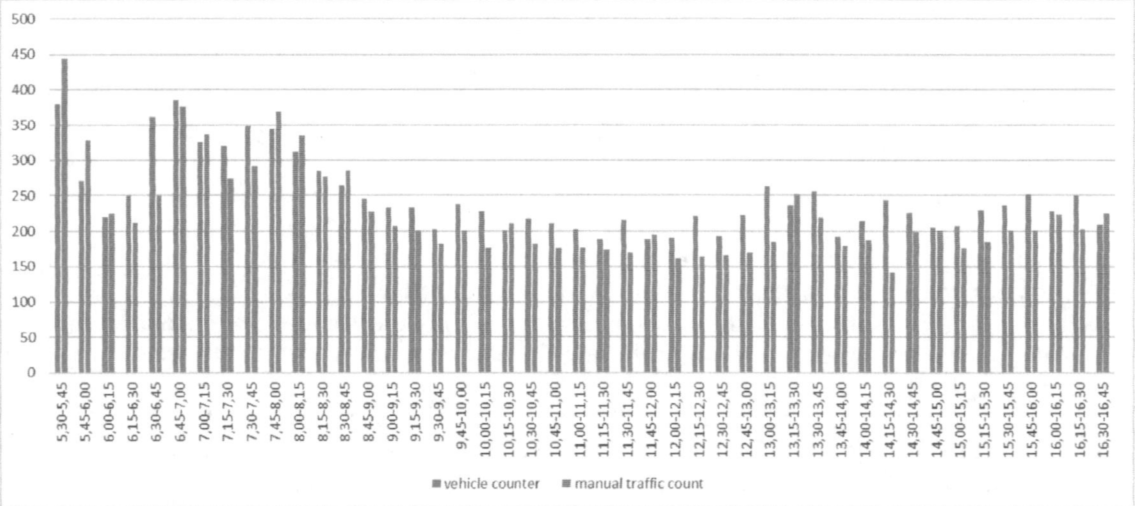

Figure 3. Difference in number of vehicles according to traffic volume –first survey.

The relation between traffic volume and numbers of errors was also investigated. As it can be seen in Figure 4, differences in number of counted vehicles are random. There was found no correlation between traffic volume and number of errors. The average difference from the manual traffic count calculated for intervals is 13.7% and the standard deviation is 12.8%.

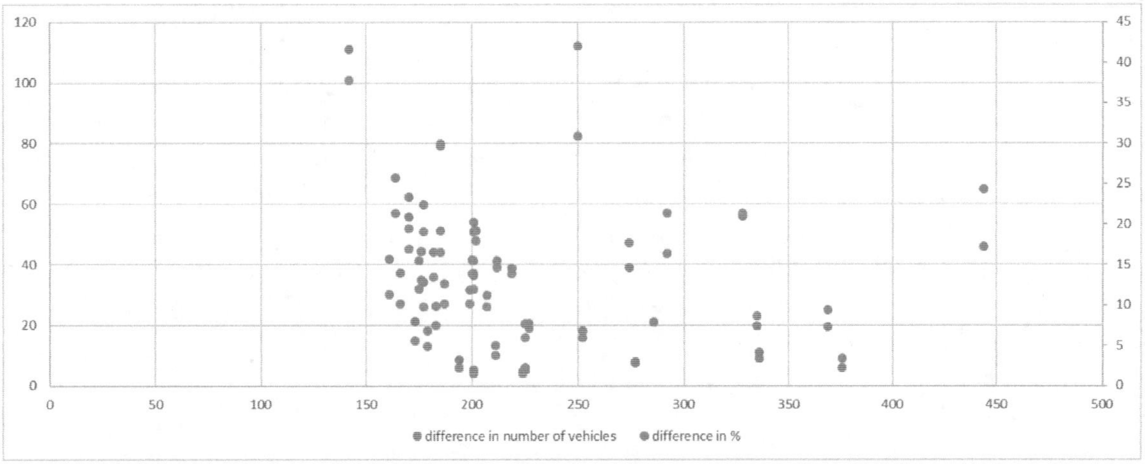

Figure 4. Difference in number of vehicles according to traffic volume – first survey.

4.2. Second survey
Second survey was carried out at the selected crossing on the road III 1,497 in the village Palárikovo. Manually counted were all vehicles entering to the intersection, automatically counted were vehicles at the entrance no. 3, see Figure 5.

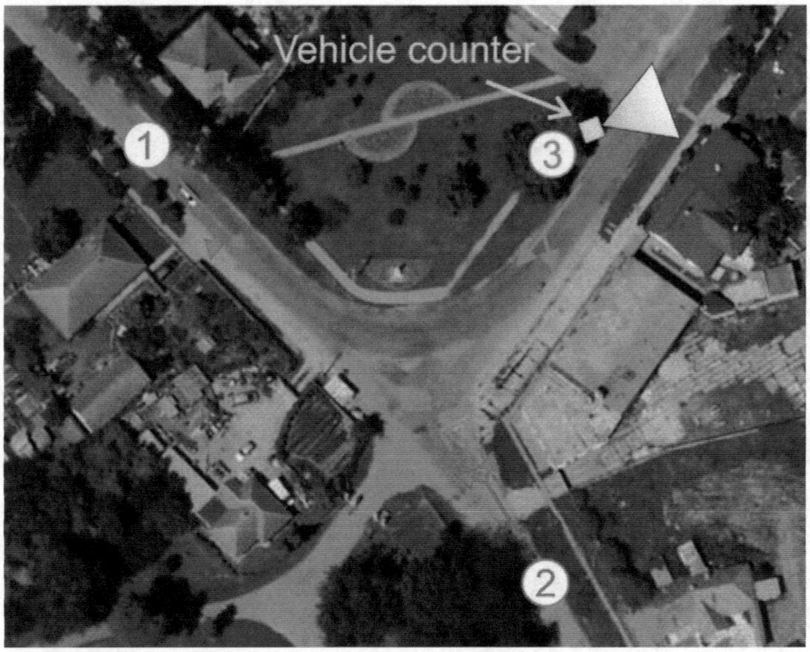

Figure 5. The location of automatic vehicle counter - second survey.

The total number of vehicles that were manually counted was 1,361. The vehicle counter counted more vehicles that it was counted manually again. The counter counted 1,439 of vehicles, that means that the counter counted 78 (5.7%) more vehicles than it was counted manually. Looking closer to the Figure 6, there can be seen differences in 15 minutes intervals; the greater difference is at the interval 14:15 – 14:30, in which the traffic counter counted 16 more vehicles than it was counted manually.

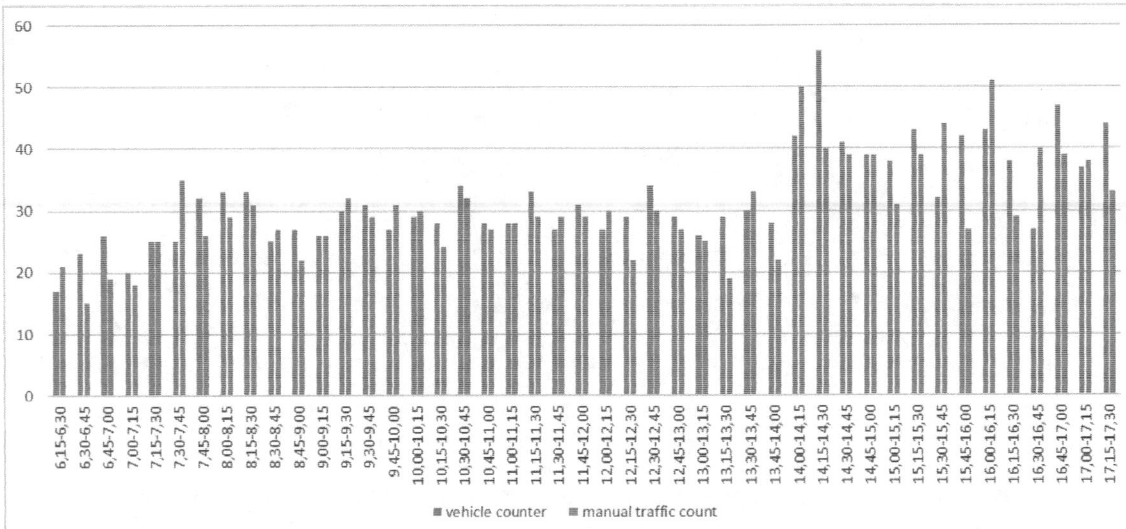

Figure 6. The comparison of manual traffic counting and automatic traffic counting – second survey.

Also, the relation between traffic volume and a number of errors was investigated. Again, differences in number of counted vehicles are random (Figure 7). There was found no correlation between traffic volume and the number of errors. The average difference from the manual traffic count calculated for intervals is 15.3 % and the standard deviation is 19%.

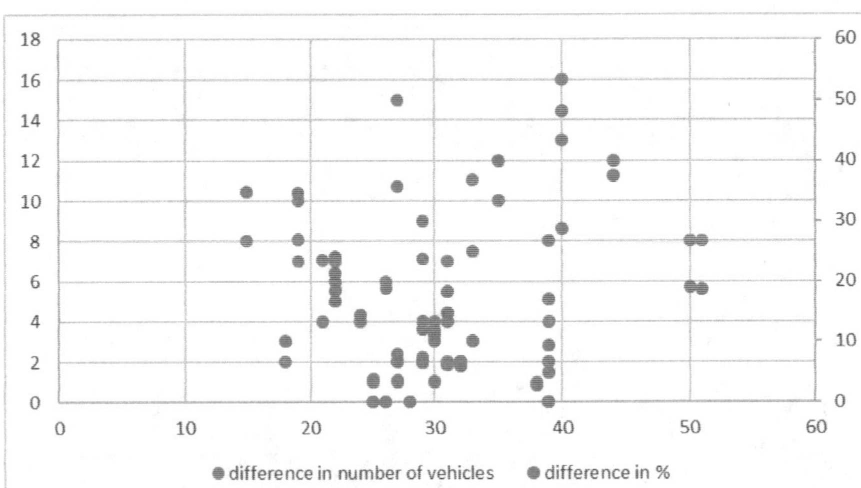

Figure 7. Difference in number of vehicles according to traffic volume –second survey.

Traffic counting with automatic traffic counter Sierzega gives more statistical outputs about traffic flow. While the manual traffic counts provide information about traffic volume and vehicles direction at intersection. Automatic counter Sierzega gives information about number of vehicles, speed of vehicles, their length and gap between vehicles. But automatic counter Sierzega does not give information about vehicles trajectory (direction) at intersection. Also, there is no way to find out this information with this vehicle counter [18].

The main advantage of using this traffic counter is its easy installation and data collection. According to authors' opinion this device is suitable for informative traffic counting, for long term traffic counting, and for obtaining additional traffic data for traffic modelling; e.g. vehicle's speed is very important parameter for traffic model calibration [19, 20, 21].

IOP Conf. Series: Materials Science and Engineering **710** (2019) 012041 doi:10.1088/1757-899X/710/1/012041

5. Conclusions

To assess the operating conditions used a number of approaches, due to the fact that in practice there are different circumstances allowing (or not) to use certain methods [22]. There are differences in the number of vehicles counted when comparing manual traffic counts and traffic counts carried out by automatic counter Sierzega.

If the total number of vehicles which passed the traffic counter was considered only, the difference is not so vast. During first traffic survey in Košice it was 12,000 to 13,000 of vehicles, which means difference of 2%. During the second traffic survey in Palarikovo it was 1,200 – 1,300 vehicles, which means difference of 3%. This number can be claimed as very good accuracy of traffic counts. But if we look closer to the number of vehicles in 15 minutes intervals, the differences were much greater. The maximum difference was up to 120 vehicles (40%).

This comparison also showed, that automatic counter Sierzega can be used also for traffic counts in two lines road. The average difference in traffic count at two-line road was 5% and the average difference in traffic counts at 1-line road was 6%.

Acknowledgement
Contribution has been prepared on the basis of the grant: VEGA no. 1/0436/18 - Externalities in road transport, an origin, causes and economic impacts of transport measures.

6. References

[1] Veternik M and Gogola M 2017. Examining of correlation between demographic development of population and their travel behaviour. *Proc. Engin.* **192**, 929-34

[2] Rybicka I, Droździel P, Stopka O and Luptak V 2018. Methodology to propose a regional transport organization within specific integrated transport system: A case study. *Transport Problems* **13,** 4, 115-125

[3] Moravčíková K and Križanová A 2016. Corporate social responsibility focusing on the protection of the environment through the cycling transport. Transport Means – *Proc. of the Inter. Conf.* 604-8

[4] Hurtova I, Sejkorova M, Verner J and Kucera A 2018. Preference and area coordination of public transport in modern city. *17th International Scientific Conference Engineering for Rural Development, ERD* **17**, 2085–91

[5] Rimkus A, Matijošius J, Bogdevicius M, Bereczky A and Torok A 2018. An investigation of the efficiency of using O2 and H2 (hydrooxile gas -HHO) gas additives in a ci engine operating on diesel fuel and biodiesel. *Energy* **152**, 640–51

[6] Skrucany T, Ponicky J, Kendra M and Gnap J 2016. Comparison of railway and road passenger transport in energy consumption and GHG production. *Proc. of the 3th Inter. Conf. on Traffic and Transport Engin. (ICTTE)* 744–49

[7] Sarkan B, Semanova S, Harantova V, Stopka O, Chovancova M and Szala M 2019. Vehicle fuel consumption prediction based on the data record obtained from an engine control unit. *III Inter. Conf. of Computational Methods in Engin. Science (CMES'18), MATEC Web of Conferences*, 252, 06009

[8] Lizbetin J and Bartuśka L 2017. The influence of human factor on congestion formation on urban roads. *10th International Scientific Conference Transbaltica 2017, Transportation Science and Technology, Proc. Engin.* **187** 206–211

[9] Cheluszka P and Mann R 2019. Vibration identification of the road header cutting head using high-speed cameras. MATEC Web Conf.*III International Conference of Computational Methods in Engineering Science (CMES'18).* **252** https://doi.org/10.1051/matecconf/201925203018

[10] Czerwinski D and Milosz M 2017. An inexpensive environmental monitoring system with IoT agents, ITM Web Conf., *II International Conference of Computational Methods in Engineering Science (CMES'17)* **15** https://doi.org/10.1051/itmconf/20171501001

[11] Sierzega, Using Manual, available at: https://www.sierzega.com/en-us/Products/Product-Viewer/sierzega-sr4-verkehrserfassungsgeraet

IOP Conf. Series: Materials Science and Engineering **710** (2019) 012041 doi:10.1088/1757-899X/710/1/012041

[12] Maćkowski M, Sawiski M and Walczysyn W 2019. Mobile system for road sign detection and recognition with template matching *MATEC Web Conf., 252. III Intern. Conf. of Computational Methods in Engin. Science (CMES'18).* https://doi.org/10.1051/matecconf/201925203014

[13] Ondruš J and Karon G 2017. Video system as a psychological aspect of traffic safety increase *Smart solutions in today's transport, Springer*, 167-77

[14] Palúch J and Kalašová A 2018. Traffic modelling on the roundabout in the town of Hlohovec and using the information from the traffic counter. *In: Management perspective for transport telematics, selected papers. Cham, Springer Inter. Publishing AG,* 2018. pp. 129-141

[15] Adebisi O 1987. Improving Manual Counts on Turning Traffic Volumes at Road Junctions. *Journal of Transportation Engineering-ASCE* **113,** 3, 256-67

[16] Zheng P and Mc Donald M 2012. An Investigation on the Manual Traffic Count Accuracy. *Pro. - Social and Behavioral Sciences* **43,** 226-31. doi.org/10.1016/j.sbspro.2012.04.095

[17] Zhao M, Garrick NW and Achenie LEK 1998. Data reconciliation-based traffic count analysis system. *Transportation Research Record*, **1625**, 12-17

[18] Badurowicz M and Cieplak T 2019. Real-Time Road Quality Assessment Using Smartphones and Cloud Lambda Architecture. Mobile system for road sign detection and recognition with template matching. MATEC Web Conf.,*III International Conference of Computational Methods in Engineering Science (CMES'18).* **252** https://doi.org/10.1051/matecconf/201925203011

[19] Kalašová A, Kubíková S and Palúch J 2018. Microscopic simulation of new traffic organisation in the city of Lučenec. *Intelligent transport systems - from research and development to the market uptake, proceedings. Cham: Springer International Publishing AG*, 2018. 102-10

[20] Gnap J, Kupčuljaková J and Semanová Š 2018. Determination of time savings for passengers by applying the public passenger transport preference. *Communications - scientific letters of the University of Žilina.* **20** 3, 3-8

[21] Findley DJ 2011. Comparison of mobile and manual data collection for roadway components, *Transportation Research Part C*, **19**, 3, 521-540

[22] Marczuk A, Caban J, Savinykh P, Turubanov N and Zyryanov D 2017. Maintenance research of a horizontal ribbon mixer. *Eksploatacja i Niezawodnosc – Maintenance and Reliability*, **19**, 1, 121–5

IOP Conf. Series: Materials Science and Engineering **710** (2019) 012042 doi:10.1088/1757-899X/710/1/012042

Diagnostics of welding process based on thermovision images using convolutional neural network

G Piecuch, M Madera and T Żabiński

Department of Computer and Control Engineering, Rzeszow University of Technology, al. Powstańców Warszawy 12, Rzeszów, Poland

gpiecuch@kia.prz.edu.pl

Abstract. Arc welding used at automated workstations in large-scale production systems requires continuous assessment of welded joints quality. There are known classical methods and diagnostic systems based on the observation of welding current or arc voltage, while along with the development of deep learning methods, the interest in diagnostics by the use of images is increasing. The article presents results of research conducted for the process of joining two stainless steel materials (AISI 304 and AISI 316L) of various thicknesses by means of a fillet weld, aimed at developing a method of diagnosing the welding process using a convolutional neural network. Infrared images recorded using two thermovision cameras mounted on a test stand were used to diagnose the process. EWM Tetrix 351 welding machine operating in TIG technology was used as an executive element. Welds were made at different currents and arc welding voltages, as well as at different welding speeds, which had a direct impact on its quality. The solution for binary classification of welded joints (correct or incorrect) with accuracy above 98% was achieved.

1. Introduction

In the era of digitisation and the fourth industrial revolution (Industry 4.0), there is a possibility to apply modern technology that allows automatization and high repeatability of the production processes, as well as automatic quality control at the manufacturing stage. The advantages of using artificial intelligence at manufacturing plants and Predictive Maintenance strategy were noticed by managers who pay more and more attention to maximizing profit while maintaining or improving the quality of manufactured goods. Advanced supervisory systems should therefore operate in real-time, to diagnose states of the machines, as well as technological processes [1]. In case of machine operation anomalies or process errors, intelligent supervisory system can automatically stop the production process and prevent defective components production [2]. Such systems can be used for various types of processes or machines parts e.g. cold forging [3], imbalance of the CNC spindle [4].

The research presented in this paper describes diagnosis of the welding process using images from thermovision cameras. In the literature research examples in which the welding process was studied using typical time signals [5-8] or thermovision [9-13] can be found. CNN networks are also increasingly used to diagnose the welding process [14-20]. Nevertheless, there is still a need for intensive scientific research to develop process diagnostics methods that exhibit intelligence, robustness and adaptation to disturbances, as well as environment changes. This article is entirely devoted to image-related method and automatic diagnosis of the quality of the welded joint by using a convolutional neural network.

IOP Conf. Series: Materials Science and Engineering **710** (2019) 012042 doi:10.1088/1757-899X/710/1/012042

2. Test stand and experiments

2.1. Test stand and parameters of the samples and process

The research test stand (scheme on figure 1) was divided into two sections: executive and measuring. The measurement part consisted of two thermal imaging cameras: Fluke TiS45 and Fluke Ti200, while the executive one was an EWM Tetrix 351 welding machine and a trolley moving in one plane, to which the welding torch was mounted (figure 2).

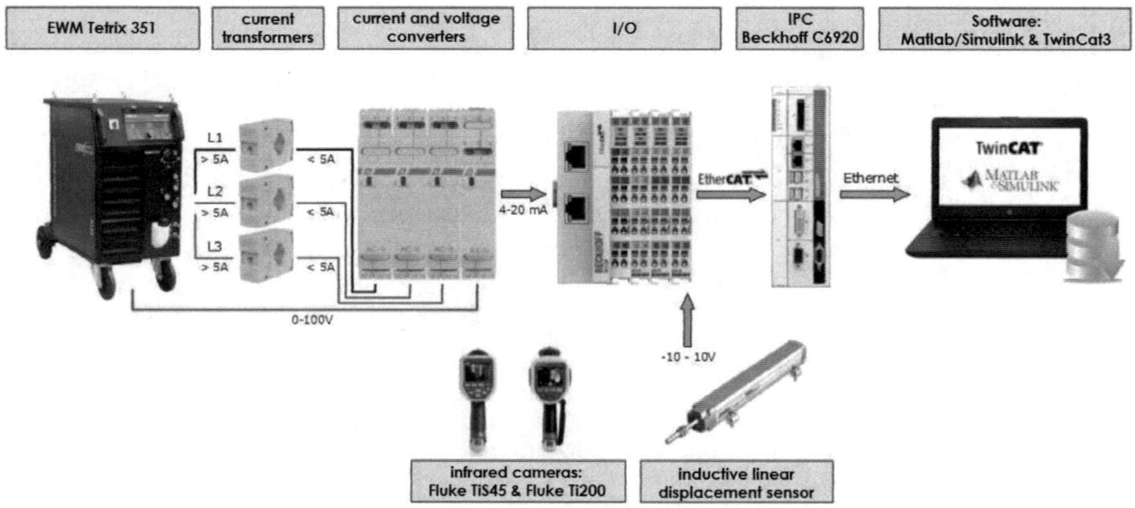

Figure 1. Scheme of test stand.

Figure 2. Test stand – trolley with welding torch mounted.

The tests were carried out for the process of joining two stainless steel materials: AISI 304 and AISI 316L with fillet welded joint. The tests were conducted on the basis of experiments carried out for an external company, which made it possible to connect only this type of elements (steel grade, thickness) and only with 60 degree deflexion. Before performing the tests using thermovision cameras, experiments were carried out for various welding parameters. After analysis of non-destructive tests (type of test: VT) results by an external company, optimal parameters for obtaining the correct welded joint were selected. For a well-made seam, the welding current was assumed: 100 A, and the speed of the trolley with the TIG holder: 200 mm/min. These values were modified during the welding process to introduce

IOP Conf. Series: Materials Science and Engineering **710** (2019) 012042 doi:10.1088/1757-899X/710/1/012042

anomalies in the process. Other process parameters that were constant for all experiments: argon as a shielding gas with a flow of 10 l/min, tungsten electrode with a diameter of 2.3 mm, chromium-nickel wire, which was fed at a speed of 800 mm/min. Exemplary samples are presented in figure 3, and the basic parameters of the samples are described in table 1.

Table 1. Base samples parameters.

Parameter	Value
Length of the bottom plate (base)	255 mm
Length of the upper plate	240 mm
Thickness of the bottom plate (base)	6 mm (AISI 316L)
Thickness of the upper plate	2 mm (AISI 304)
Angle of deflexion of plates	60° / 120° (from the side of the welded joint)

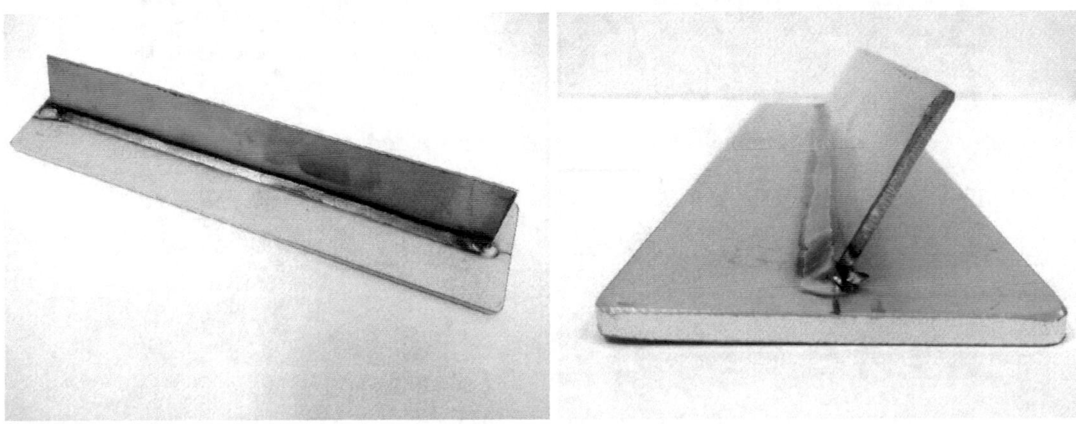

Figure 3. View of sample.

2.2. Experiments
In order to ensure uniform measurement conditions, the experiments were conducted according to the following procedure:

1. Marking the sample with a consecutive number
2. Setting the sample in the target place on the test stand ensuring repeatable positioning
3. Setting the position of the welding torch (in 3 axes) at appropriate distances from the sample
4. Starting image recording on both thermal imaging cameras
5. Igniting the arc
6. Starting the trolley move
7. Welding
   - If necessary (e.g. due to curvature of the welded material) correction of the TIG holder adjustment position
   - When performing incorrect samples – change welding parameters (feed speed of the trolley, welding current)
8. Turning off the welding
9. Stopping the trolley
10. Stopping recording images
11. Assessing of welded joint quality by an expert

As a result of the experiments: 18 correct and 15 incorrect samples were obtained (table 2), exemplary samples are shown in figure 4-6.

Table 2. Collection of samples.

Sample type	Quantity
Correct samples	18
Incorrect samples (obtained during a process performed with optimal parameters)	4
Incorrect samples – change of speed of the trolley from 200 mm/s to 100 mm/s	2
Incorrect samples – change of speed of the trolley from 200 mm/s to 400 mm/s	4
Incorrect samples – change in welding current from 100 A to 80 A	3
Incorrect samples – change in welding current from 100 A to 70 A	2

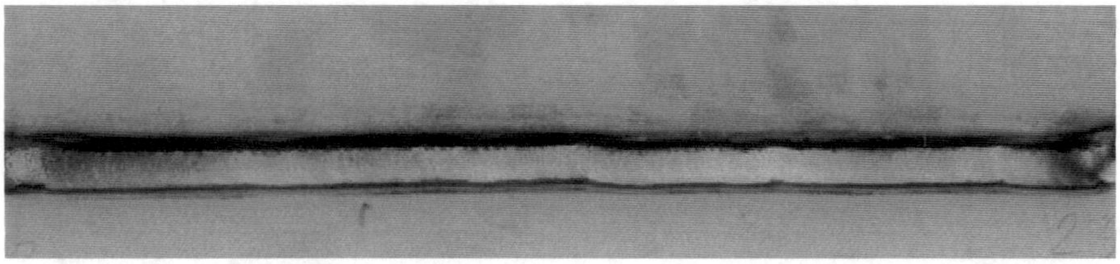

Figure 4. Sample with correct welded joint.

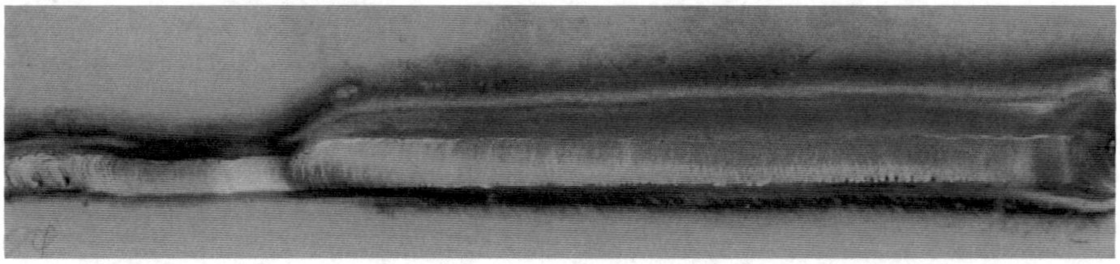

Figure 5. Sample with process anomaly – change of speed of the trolley from 200 mm/s to 100 mm/s (visible overheating of the sample).

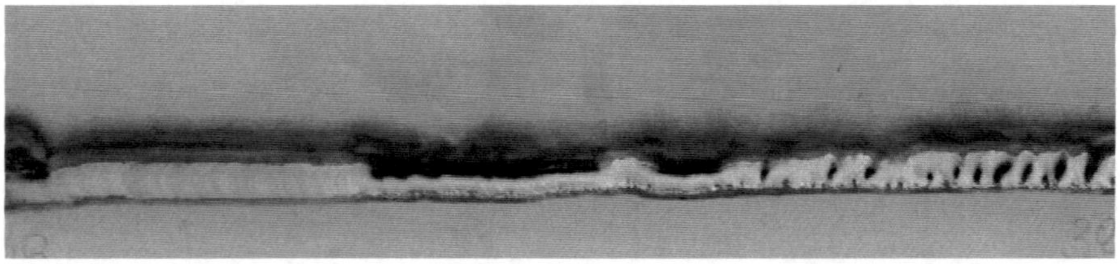

Figure 6. Sample with process anomaly – change in welding current from 100 A to 70 A.

3. Classification of the thermovision images

3.1. Data preparation

Even though there are different types of anomalies in the process (related to change of speed, change of current, etc.) the collected image data are split into two classes: correct and incorrect. Welding images were reviewed and labelled manually by an expert considering that only a part of the sample can be incorrect. Based on this information, related thermovision camera pictures were appropriately split to general correct and incorrect classes. This approach provided high quality and reliable data, so the classification model performance is not impacted by improper labels assignment.

For Fluke TiS45 there were 637 samples (images) collected with 2 classes: 190 correct samples like presented in Figure 7, and 447 incorrect samples like presented in Figure 8. Images were stored as BMP images with dimensions: 320 x 240 pixels and bit depth: 32.

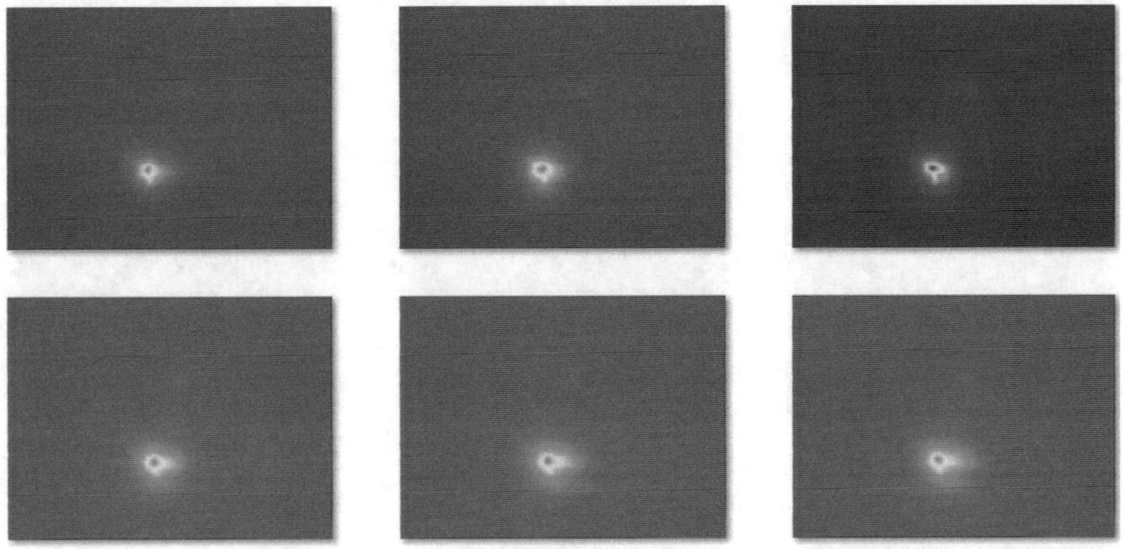

Figure 7. Example Fluke TiS45 thermovision images of correct welding (images for one sample).

Figure 8. Example Fluke TiS45 thermovision images of incorrect welding (images for one sample).

For Fluke Ti200 there were 13 158 samples (images) collected with 2 classes: 7005 correct samples like presented in figure 9, and 6153 incorrect samples like presented in figure 10. Images were stored as BMP images with dimensions: 280 x 210 pixels and bit depth: 24. The difference in the number of images recorded by both cameras was due to the fact that the Fluke TiS45 only recorded images (photos) with an interval every 3 seconds, while the Fluke TiS200 camera as a more extensive model, allowed for recording movies (frames from films were used).

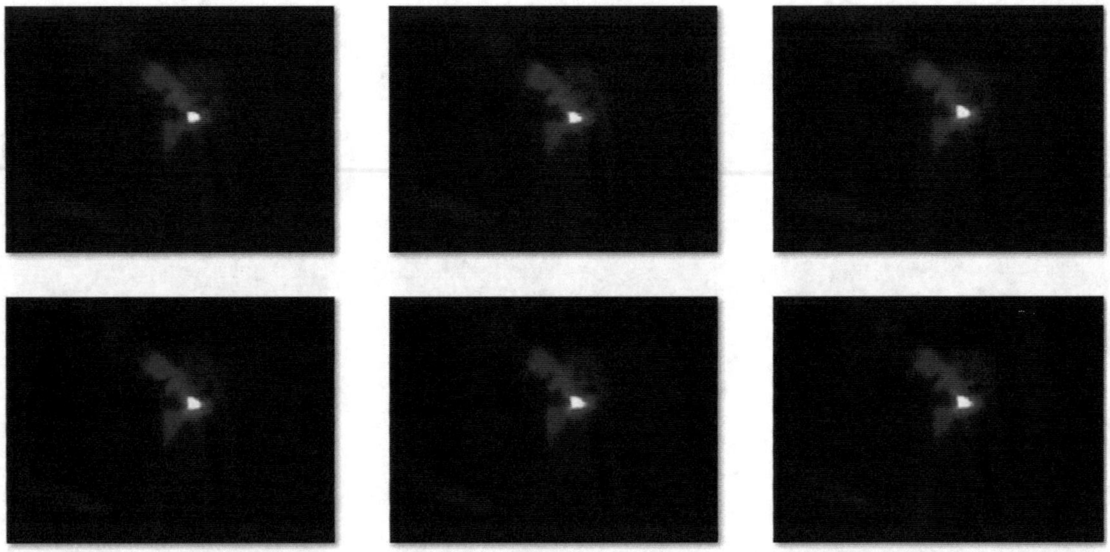

Figure 9. Example Fluke Ti200 thermovision images of correct welding (images for one sample).

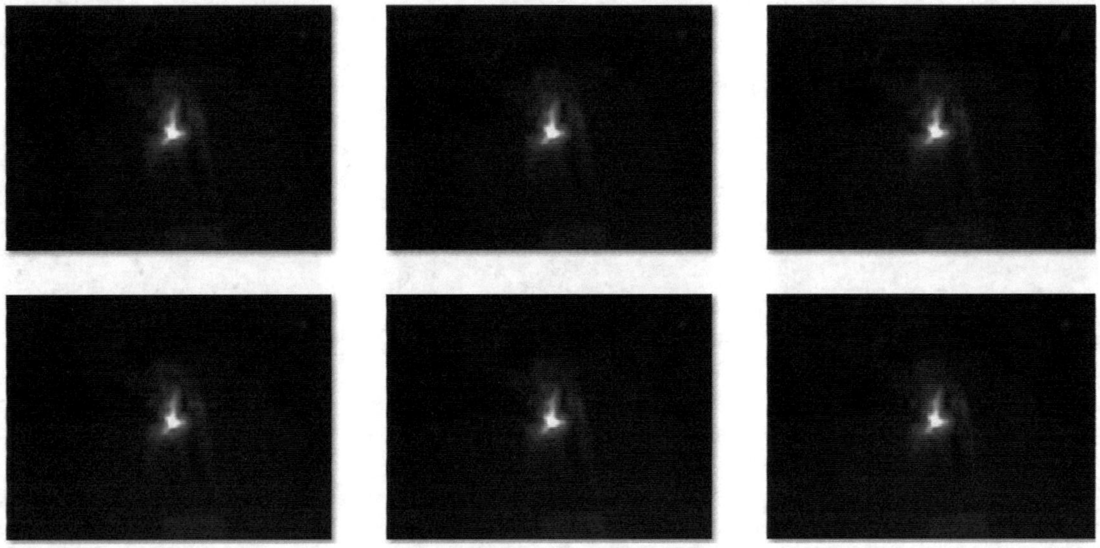

Figure 10. Example Fluke Ti200 thermovision images of incorrect welding (images for one sample).

3.2. Convolutional Neural Network architecture
The network architecture showed in figure 12 was proposed for Fluke TiS45 images learning and classification:

- 3 convolution layers, each with parameters:
  - Kernel size 32 x 32 (convolution window size)
  - Strides size 3 x 3 (strides of the convolution)
  - Activation function ReLU (rectified linear unit)
  - Max pooling operation with pool size 2x2 (factors by which to downscale)
- 2 densely-connected NN layers, each with parameters:
  - Dimension 128 units
  - Activation function ReLU

IOP Conf. Series: Materials Science and Engineering **710** (2019) 012042 doi:10.1088/1757-899X/710/1/012042

- One dropout operation after last hidden layer output with rate 0.3
- Last (output) layer with Sigmoid activation function
- Adam optimiser algorithm used to control learning rate with Binary Cross Entropy as objective function.

To obtain better classification performance, network structure for Fluke Ti200 (figure 11) has been simplified in comparison to the previous one:

- 2 convolution layers, each with parameters:
  - Kernel size 32 x 32 (convolution window size)
  - Strides size 3 x 3 (strides of the convolution)
  - Activation function ReLU (rectified linear unit)
  - Max pooling operation with pool size 2x2 (factors by which to downscale)
- One densely-connected NN layer with parameters:
  - Dimension 128 units
  - Activation function ReLU
- One dropout operation after last hidden layer output with rate 0.2
- Last (output) layer with Sigmoid activation function
- Stochastic Gradient Descent optimiser algorithm used to control learning rate with Binary Cross Entropy as objective function.

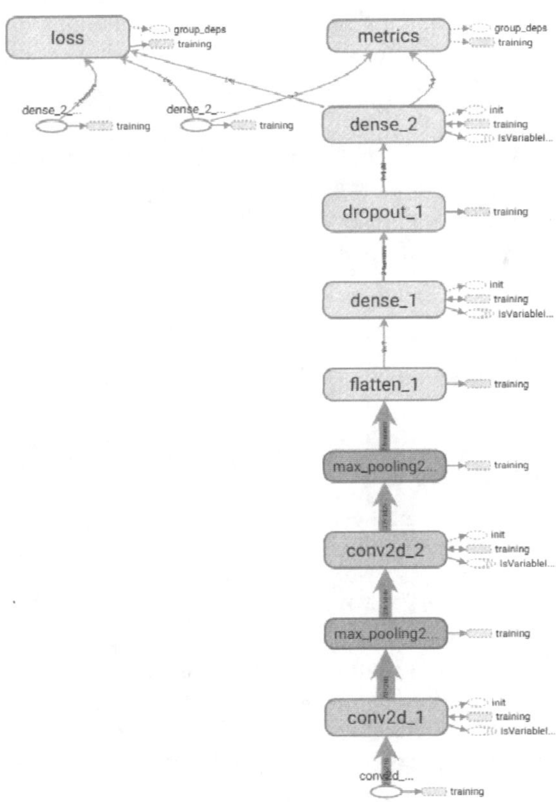

Figure 11. Network architecture graphs for Fluke Ti200.

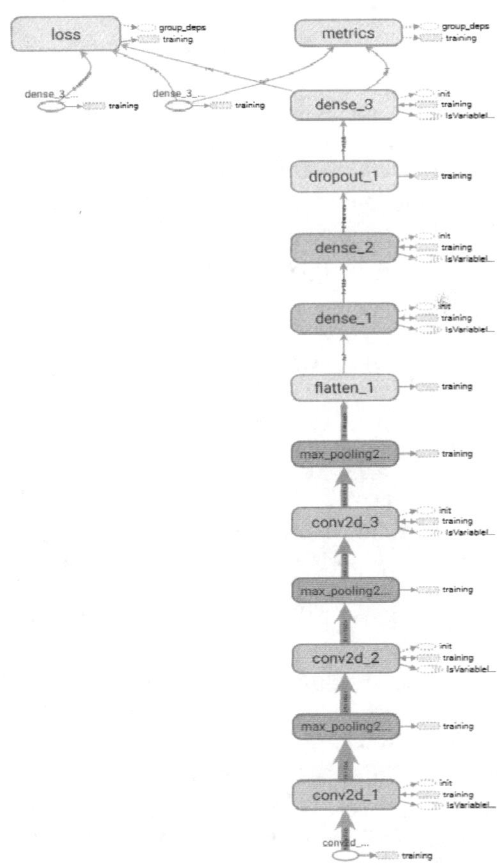

Figure 12. Network architecture graphs for Fluke TiS45

3.3. *Classification*

Classification model was built with correct samples treated as a negative example (N) and incorrect samples as a positive example (P) as presented in table 2. Accuracy was calculated as Acc=(TP+TN)/(TP+TN+FP+FN)·100%, sensitivity was obtained from the formula: Sen=TP/(TP+FN)·100% and false alarm rate value was defined as FAR=FP/(FP+TN)·100% (TP – true positive, TN – true negative, FP – false positive, FN – false negative).

Network architectures and parameters has been selected empirically. For Fluke TiS45 learning was performed with 100 epochs and it took approximately 16 minutes to reach reasonable accuracy level. Model learning and testing was performed with 10-fold cross validation method (10-FCV). Each fold accuracy per epoch is presented in figure 13. Accuracy for 10-FCV reached by the model was Acc=**95.13%**, sensitivity equal Sen=**97.06%**, and false alarm rate FAR=**9.23%**.

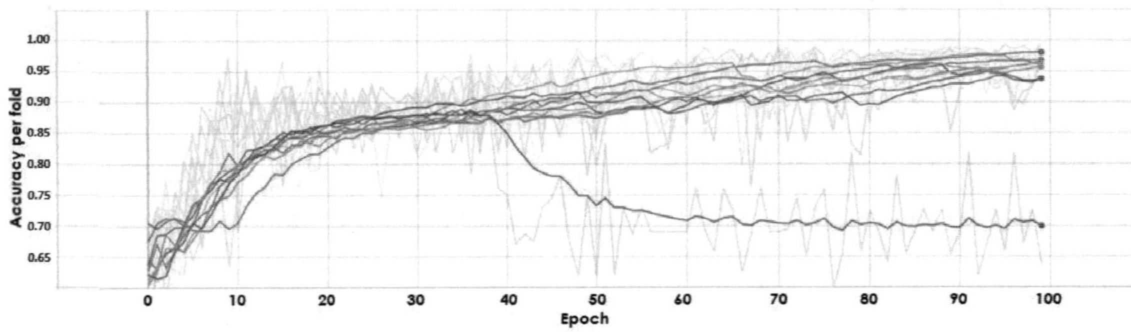

Figure 13. Accuracy for Fluke TiS45 each fold in function of epochs.

For Fluke Ti200 dataset, learning was performed with 100 epochs and it took approximately 4 hours to reach reasonable accuracy level. Model learning and testing was modelled with 10-FCV. Each fold accuracy per epoch is presented in figure 14. Accuracy for 10-FCV reached by the model was Acc=**98.59%**, sensitivity equal Sen=**98.87%**, and false alarm rate FAR=**1.64%**.

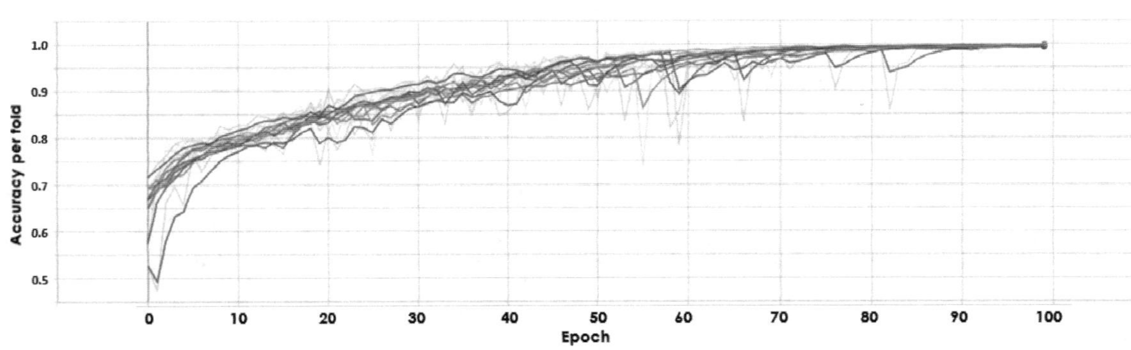

Figure 14. Accuracy for Fluke Ti200 each fold in function of epochs.

4. Conclusions

In this paper the classification of fillet welded joint quality between sheets made of AISI 304 and AISI 316L alloys based on images from thermal imaging cameras, using a convolutional neural network was described. The methodology of the conducted research and the results of the binary classification (correct or incorrect welded joint) are presented. Accuracy of classification at the level of 95.13% for images from the Fluke TiS45 camera and 98.59% for images from the Fluke Ti200 camera was achieved. The differences in the accuracy of classification of photos taken by TiS45 and Ti200 are mainly due to the varying number of samples and the different location of the cameras. The obtained results are similar to other parallel methods described in the literature [14, 15, 18, 19]. Due to the long learning time of the algorithms, for real-time process monitoring tasks it is recommended to prepare and learn networks offline and additional re-train online. High classification accuracy predestines the presented solution for use in quality control systems. Video supervision approach is not new, although it is usually related to review of the product after welding.

The objective of this work is to develop diagnostic method which can be used in the future as an element of the real-time welding supervision system which has been developed in Rzeszów University of Technology (Department of Computer and Control Engineering) [1].

Acknowledgments
This project is financed by the Minister of Science and Higher Education of the Republic of Poland within the "Regional Initiative of Excellence" program for years 2019 – 2022. Project number 027/RID/2018/19, amount granted 11 999 900 PLN.

5. References

[1] Żabiński T, Mączka T and Kluska J 2017 Industrial Platform for Rapid Prototyping of Intelligent Diagnostic Systems. In Mitkowski W, Kacprzyk J, Oprzędkiewicz K, Skruch P (Eds.), Trends in Advanced Intelligent Control, Optimization and Automation. Polish Control Conference (Kraków, Poland) pp. 712-21. doi: 10.1007/978-3-319-60699-6_69

[2] Żabiński T, Mączka T, Kluska J, Madera M and Sęp J 2019. Condition monitoring in Industry 4.0 production systems - the idea of computational intelligence methods application. In Teti R and Doriana M D'Addona (Eds.), *Proc. CIRP, Vol. 72. Proc. of the 12th CIRP Conference on Intelligent Computation in Manufacturing Engineering* (Naples, Italy) pp. 63-67. doi: 10.1016/j.procir.2019.02.012

[3] Żabiński T, Mączka T, Kluska J, Kusy M, Hajduk Z and Prucnal S 2014 Failures Prediction in the Cold Forging Process Using Machine Learning Methods. In: Rutkowski L, Korytkowski M, Scherer R, Tadeusiewicz R, Zadeh L A and Zurada J M (eds.) *ICAISC 2014, Part I. LNCS(LNAI), vol. 8467* (Heidelberg: Springer) pp. 622–33. doi:10.1007/978-3-319-07173-2_53

[4] Żabiński T, Mączka T, Kluska J, Kusy M., Gierlak P, Hanus R, Prucnal S and Sęp J 2015 CNC Milling Tool Head Imbalance Prediction Using Computational Intelligence Methods. In Rutkowski L, Korytkowski M, Scherer R, Tadeusiewicz R, Zadeh L and Zurada J (Eds.), *Artificial Intelligence and Soft Computing. Proc. of the 14th International Conference on Artificial Intelligence and Soft Computing ICAISC* (Zakopane, Poland) pp. 503-514. doi: 10.1007/978-3-319-19324-3_45

[5] Deyong Y, Xiangdong G and Seiji K 2014 Monitoring of high-power laser welding using high-speed photographing and image processing. *Mechanical Systems and Signal Processing, Vol. 49, Issues 1–2*, pp. 39-52. doi: 10.1016/j.ymssp.2013.10.024

[6] Fidali M and Jamrozik W 2013 Application of data fusion for welding process diagnostics. *Pomiary Automatyka Robotyka 12/2013*, pp. 90-95.

[7] Kovacevic R, Zhang Y M and Li L 1996 Monitoring of weld joint penetration based on weld pool geometrical appearance. *Welding Journal, Vol. 75*, pp. 317-329.

[8] Zhang Z, Ch H, Xu Y, Zhong J, Lv N and Ch S 2015 Multisensor-based real-time quality monitoring by means of feature extraction, selection and modeling for Al alloy in arc welding. *Mechanical Systems and Signal Processing, Vol. 60–61*, pp. 151-65. doi: 10.1016/j.ymssp.2014.12.021

[9] Fidali M and Jamrozik W 2013 Diagnostic method of welding process based on fused infrared and vision images. *Infrared Physics & Technology, Vol. 61*, pp. 241-53. doi: 10.1016/j.infrared.2013.08.010

[10] Nagarajan S, Banerjee P, Chen W and Chin B A 1992 Control of the welding process using infrared sensors. *IEEE Transactions on Robotics and Automation, Vol. 8*, pp. 86-93. doi:

10.1109/70.127242

[11] Pu Z, Li K, Wenzhong L, Jingjing Ch and Kaibo Z 2008 Real-time monitoring of laser welding based on multiple sensors. *Chinese Control and Decision Conference*. doi: 10.1109/CCDC.2008.4597620

[12] Sreedhar U, Krishnamurthy C V, Balasubramaniam K, Raghupathy V D and Ravisankar S 2012 Automatic defect identification using thermal image analysis for online weld quality monitoring. *Journal of Materials Processing Technology, Vol. 212*, pp. 1557-66. doi: 10.1016/j.jmatprotec.2012.03.002

[13] Górka J, Janicki D, Fidali M and Jamrozik 2017 Thermographic Assessment of the HAZ Properties and Structure of Thermomechanically Treated Steel, *International Journal of Thermophysic.* doi: 10.1007/s10765-017-2320-9

[14] Khumaidi A, Yuniarno E M and Purnomo M H 2017 Welding defect classification based on convolution neural network (CNN) and Gaussian kernel. *2017 International Seminar on Intelligent Technology and Its Applications (ISITIA) Surabaya, Indonesia.* doi: 10.1109/ISITIA.2017.8124091

[15] Zhang Z, Wen G and Chen S 2019 Weld image deep learning-based on-line defects detection using convolutional neural networks for Al alloy in robotic arc welding. *Journal of Manufacturing Processes Vol. 45,* pp. 208-16. doi: 10.1016/j.jmapro.2019.06.023

[16] Sizyakin R, Voronin V, Gapon N, Zelensky A and Pižurica A 2019 Automatic detection of welding defects using the convolutional neural network, *Proc. Vol. 11061, Automated Visual Inspection and Machine Vision III.* doi: 10.1117/12.2525643

[17] Park J K, An W H and Kang D J 2019 Convolutional Neural Network Based Surface Inspection System for Non-patterned Welding Defects, *International Journal of Precision Engineering and Manufacturing Vol. 20,* pp. 363-374. doi: 10.1007/s12541-019-00074-4

[18] Zhu H, Ge W and Liu Z 2019 Deep Learning-Based Classification of Weld Surface Defects, *Applied Sciences* MDPI. doi: 10.3390/app9163312

[19] Liu T, Bao J, Wang J and Zhang Y 2018 A Hybrid CNN–LSTM Algorithm for Online Defect Recognition of CO2 Welding, *Sensors* MDPI. doi: 10.3390/s18124369

[20] Tang S, Wang G, Zhang H and Wang R 2017 An online surface defects detection system for AWAM based on deep learning, *Solid Freeform Fabrication 2017: Proc. of the 28th Annual International Solid Freeform Fabrication Symposium – An Additive Manufacturing Conference, pp. 1965-81.*

IOP Conf. Series: Materials Science and Engineering **710** (2019) 012043 doi:10.1088/1757-899X/710/1/012043

Milling process diagnosis using computational intelligence methods

T Żabiński[1], G Piecuch[1], R Żyła[2] and S Prucnal[3]

[1] Rzeszow University of Technology, Department of Computer and Control Engineering, al. Powstańców Warszawy 12, Rzeszow, Poland
[2] Żbik Sp. z. o.o., ul. Emilii Plater 7, Rzeszow, Poland
[3] Rzeszow University of Technology, Department of Manufacturing and Production Engineering, al. Powstańców Warszawy 12, Rzeszow, Poland

tomz@kia.prz.edu.pl

Abstract. The paper presents the results of research aimed at developing a method for hard-to-machine metal alloy milling process diagnosis using computational intelligence methods. To diagnose the process, a signal from an accelerometer mounted on the spindle of a CNC machine was used. The data were recorded during milling of Inconel 625 alloy workpieces, performed by sharp and blunt cutters. The acceleration signal metrics, both in the time and frequency domains were used to develop the classifiers. Based on the experiments, it has been demonstrated that it is possible to effectively diagnose Inconel alloy workpieces milling process using shallow computational intelligence methods (decision trees, k-NN and linear support vector machines). Python was used for data processing and visualisation as well as classifiers development and testing.

1. Introduction

Despite new developments in manufacturing techniques (e.g. additive), machining still plays a significant role in the production practice. The concept of Industry 4.0 assumes automation and robotisation of production processes as well as limited participation of machine operators in technological processes supervision [1-2]. Therefore, there is a need for intensive scientific research to develop processes diagnostic methods that exhibit intelligence, robustness and adaptation to disturbances as well as environment changes [3-4]. The main purpose of the system monitoring the intelligent technological process conditions is to perform actions to prevent the production of out-of-specification components [5-6]. Intelligent computer systems that perform the diagnosis of technological processes in real-time enhances the efficiency, productivity and accuracy of manufacturing processes as well as increases the cutting tool life especially involved in processing new and demanding materials.

Numerous diagnostic methods are found in the specialist literature for various technological processes [7-8], CNC machines parts [9] and machine tools [10-11]. In this work, a method for Inconel 625 workpieces milling process diagnosis is developed as a continuation of our former research activity in the field [2,8,9,12]. The Inconel 625 is a highly demanding material used in critical applications, such as aerospace, chemical, and oil production industries [10].

The objective of this work is to develop a diagnostic method which can be used in the future as an element of the real-time milling supervision system, which has been developed in Rzeszów University

IOP Conf. Series: Materials Science and Engineering **710** (2019) 012043 doi:10.1088/1757-899X/710/1/012043

of Technology (Department of Computer and Control Engineering) [12]. To diagnose the process, a signal from the accelerometer mounted on the lower bearing of the CNC machine spindle was used. The data for the diagnostic method development was recorded during milling of Inconel alloy workpieces, performed by the use of sharp and blunt cutters, and subsequently subjected to analysis by means of shallow computational intelligence (CI) methods. Python programming environment was used to perform data processing and visualisation as well as classifiers development and testing.

This paper is composed of the following Sections. Section 2, describes the testbed and experiments. Section 3 is devoted to the milling process diagnosis using computational intelligence methods. In Section 4, the results are presented. In Section 5, the conclusions are formulated.

2. Testbed and experiments description

2.1. Testbed and data acquisition system

The experiments were conducted in the laboratory testbed consisting of Haas VM-3 CNC machine equipped with an inline direct-drive spindle (Figure 1) and a set of sensors: 7 accelerometers (sensitivity 100 mV/g) integrated with temperature sensors, 1 acoustic emission sensor (sensitivity 53 mV/Pa) and 1 force and torque sensor (3-axis). Two accelerometers are mounted on the lower bearing of the spindle, the next two are mounted on the higher bearing of the spindle, additional two accelerometers are mounted on Z-axis and the last one is mounted on the workpiece. The acoustic emission sensor is mounted in the machine cabin. The force and torque sensor is mounted in the chuck. In this study, a signal from the accelerometer mounted on the lower bearing of the CNC machine spindle was used (Figure 1).

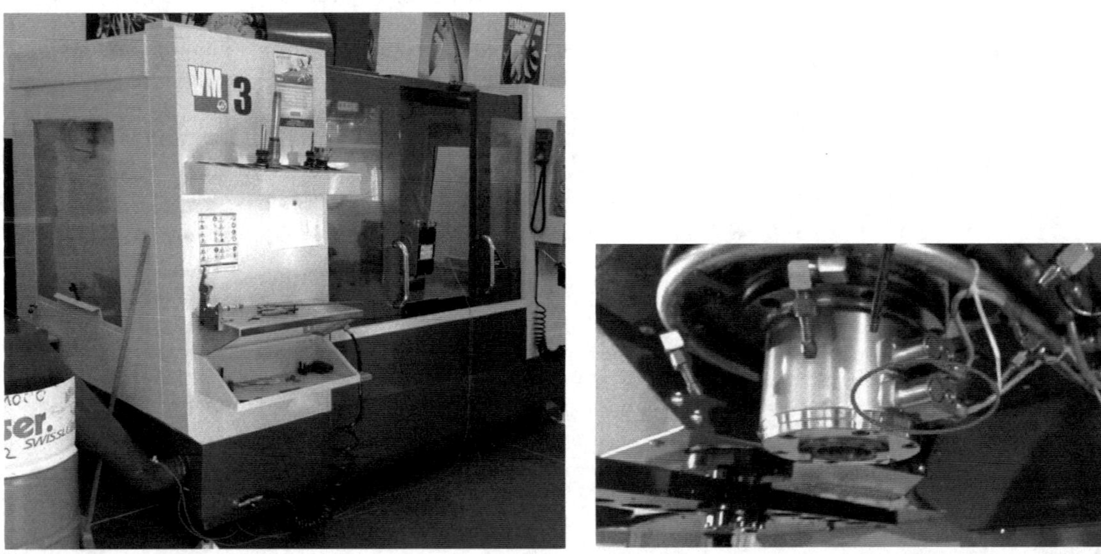

Figure 1. Haas VM-3 CNC machine (left picture) and CNC spindle with acceleration sensors mounted on lower bearing (right picture).

During the experiments, the data from the sensors were collected using a platform for rapid prototyping of intelligent diagnostic systems [12] which was composed of Beckhoff Industrial PC (IPC C6920), TwinCAT 3, MATLAB/Simulink projects and distributed input/output system based on EtherCAT protocol. IPC was used for signal acquisition and communication with a PC computer equipped with MATLAB/Simulink system, which performed data collection in External Mode. The PC computer received data from IPC and stored it as binary MATLAB files (mat) on the hard drive. The signal duration stored in one file and the sampling interval were equal to 640 ms buffer (16 000 samples) and 40 μs, respectively (25 kHz sampling frequency). The real-time PLC module for data acquisition operated with the main sampling interval of 2 ms. In the analog input modules EL3632 from Beckhoff, used to connect acceleration sensors, the oversampling factor (defined as the number of probes per one main sampling interval) was set to 50.

2.2. Experiments

During each experiment, one Inconel disc (Figure 2) was machined with a four-teeth milling cutter at the spindle speed of 862 rpm (revolutions per minute), which corresponds to 14.36 Hz. During each experiment one complete circular milling trajectory was performed at the edge of a disc using the same tool. Two types of milling cutters were tested: sharp and blunt. In the case of sharp tools, the resulting surface roughness was 0.55 µm and for blunt tools 1.05 µm.

Figure 2. Inconel disc with the trace of circular trajectory performed at the edge of the disc with different tools (sharp or blunt) for each disc quarter.

In this study, the data from 11 discs machining processes were used (7 executed with sharp tools and 4 with blunt tools). Each data buffer (640 ms, 16 000 samples) was used to calculate one set of time and frequency metrics. Data pre-processing phase defined two sets of labelled data (class) (Table 1).

Table 1. Data set.

Class	Cutter	Surface roughness	Number of buffers
Sharp	Sharp	0.55 µm	1085
Blunt	Blunt	1.05 µm	624

2.3. Measured and transformed signal

Figure 3 shows the acceleration signal plot for exemplary buffers of Sharp and Blunt classes.

In order to calculate frequency domain metrics, Fast Fourier Transform (FFT) was performed for each time data buffer. The length of FFT is equal to N=16384, taking into account 25 kHz signal sampling frequency, it leads to 1.56 Hz resolution of the transformed signal. Figure 4 shows the acceleration signal in the frequency domain for the exemplary data buffer.

IOP Conf. Series: Materials Science and Engineering **710** (2019) 012043 doi:10.1088/1757-899X/710/1/012043

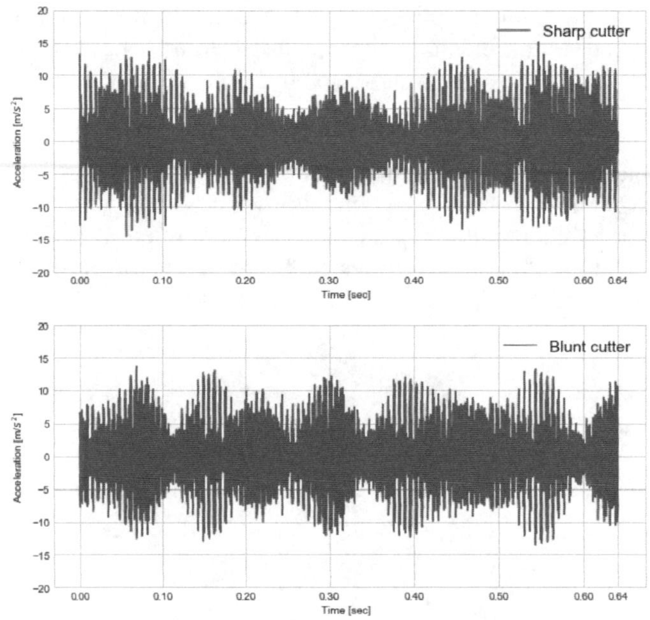

Figure 3. Plot of the acceleration time signal for one buffer of Sharp class and one of Blunt class.

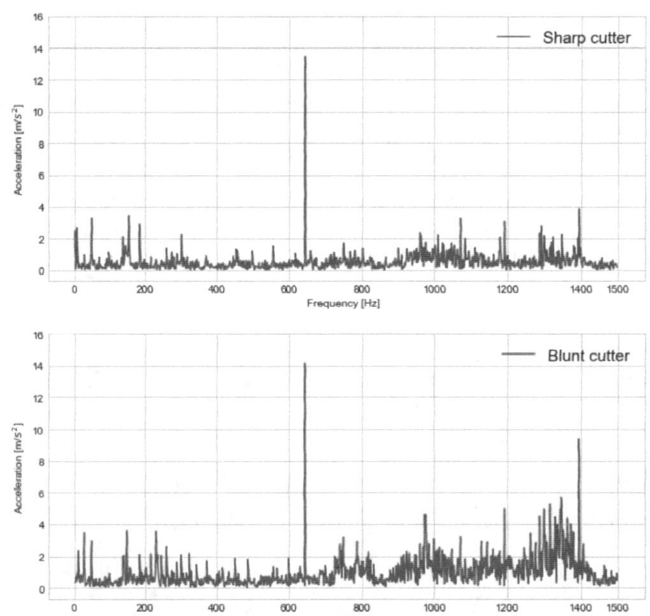

Figure 4. Plot of the acceleration signal in the frequency domain – FFT.

3. Milling diagnosis using computational intelligence methods

3.1. Programming tools used in the study

Data processing and classification experiments were carried out in Anaconda environment with Jupyter Notebook (Python version: 3.6.6). All classifiers were developed via Scikit-learn library, which provides the most commonly used CI algorithms (e.g. generalised linear models, support vector machines, decision trees). Scikit-learn tools for measuring the performance of classifiers, such as classification metrics and cross-validation methods, were used in the study. Pandas and NumPy libraries were applied for data processing, while Matplotlib with Seaborn for the visualisation of the signals and extracted features.

3.2. Features calculation
Various features were calculated in the time and frequency domains from each 640 ms data buffer, that is 22 in the time domain and 10 in the frequency domain. The time signal metrics used in the study are mentioned below [7]:

- Maximum
- Minimum
- Peak to peak – the difference between the maximum and the minimum value of the signal
- Median
- Maximum of the absolute value of the signal
- Mean
- Mean of the absolute value of the signal
- Variance
- Root mean square
- Standard deviation
- Energy
- Energy of the centered signal
- Kurtosis
- Skewness
- Moment order i (i = 5:10)
- Shannon entropy
- Signal rate

Features of the acceleration signal extracted in the frequency domain are mentioned below [9]:

- RMS amplitude spectrum in the frequency ranges: 0 Hz – 500 Hz, 500 Hz – 1000 Hz, 1000 Hz – 1500 Hz, 1500 Hz – 2000 Hz, 2000 Hz – 2500 Hz
- Area under the amplitude spectrum in the frequency ranges: 0 Hz – 500 Hz, 500 Hz – 1000 Hz, 1000 Hz – 1500 Hz, 1500 Hz – 2000 Hz, 2000 Hz – 2500 Hz

3.3. Features normalisation and selection
The calculated features were in different scales and units, which is why a normalisation procedure was performed to obtain better convergence of CI methods, especially those optimised by gradient descent [13-15]. K-nearest neighbors algorithm used in the study is also highly sensitive to normalisation because features with a larger range will have a higher influence impact on the calculated distance [16]. In our case, the features were scaled to the range between 0 and 1 with the following formula:

$$z = \frac{x - \min(x)}{\max(x) - \min(x)} \tag{1}$$

where min(x) and max(x) are minimum and maximum values of a feature, respectively, x is the value of the feature before normalisation and z is the value of the feature after normalisation.

Features selection method has been used to obtain high classifier performance with the smallest number of features. It helps reduce the computing time of classifier predictions and features calculation, especially with respect to real-time supervision systems. Logistic regression with L1 regularisation was used in this study [17-19]. Linear models with L1 norm are highly likely to estimate coefficients as zero, thus leading to sparse solutions with the most relevant features, which can be further processed by logistic regression itself or by other classifiers. During the process of feature selection, C is the most important parameter, responsible for the strength of regularisation and when it

IOP Conf. Series: Materials Science and Engineering **710** (2019) 012043 doi:10.1088/1757-899X/710/1/012043

is lower, and then regularisation is stronger. In the experiments, parameter C in logistic regression was set to 0.1 and only two features were selected:

- F1: Area under the amplitude spectrum in the frequency range 0 Hz – 500 Hz
- F2: Area under the amplitude spectrum in the frequency range 1000 Hz – 1500 Hz.

3.4. Computational intelligence methods
In this study, three different (one linear and two nonlinear) CI methods were examined:

- Single decision tree (SDT)
- K-nearest neighbor (K-NN)
- Linear support vector machine (Linear SVM)

In scikit-learn SDT implementation, CART algorithm is used. Gini impurity was the criterion for the information gain. Moreover, the decision tree classifier has not been cropped with max depth parameter set to none, as a result, there is a low risk of overfitting, given that only two features were used.

K-NN model was implemented with 5 neighbors, uniform weights and Euclidean distance as metrics. Linear SVM has been implemented with L2 regularisation, squared hinge as loss and penalty parameter C set to 1. All of these algorithms are well described in the literature, e.g. in [20], therefore they were not further elaborated in this article.

4. Results
The performance of each classifier was measured with the 10-fold cross-validation with three indicators: accuracy, sensitivity and false alarm rate. In this work, class Sharp is treated as a negative example (N) and Blunt is treated as a positive example (P). Accuracy is calculated as Acc=(TP+TN)/(TP+TN+FP+FN)·100%, sensitivity is obtained from the formula: Sen=TP/(TP+FN)·100% and false alarm rate value is defined as FAR=FP/(FP+TN)·100% (TP - true positive, TN – true negative, FP – false positive, FN – false negative).

Only 2 most relevant features (i.e. F1, F2) provided by logistic regression with L1 norm used as feature selector were used for each classifier. Results for all methods are shown in Table 2 (mean value and standard deviation obtained for the 10-fold cross-validation).

Table 2. Performance of classifiers developed in the study (mean and standard deviation values).

Algorithms	Acc [%]	Sen [%]	FAR [%]
SDT	99.94+/-0.18	99.84+/-0.49	0.0+/-0.0
KNN	99.94+/-0.18	100+/-0.0	0.09+/-0.27
Linear SVM	99.94+/-0.18	100+/-0.0	0.09+/-0.27

As it can be seen from Table 2, all the classifiers managed to achieve nearly perfect scores with only two best features, which were extracted in the frequency domain. It seems that the patterns located in the area under the amplitude spectrum in frequency ranges: 0 Hz - 500 Hz and 1000 Hz – 1500 Hz, found by logistic regression, are enough to develop a classifier that can separate two classes: Sharp and Blunt. Along with the performance, the time for making predictions and calculating features is another important factor for practical implementations of milling process supervision systems. Methods that do not require high computational power for making predictions (e.g. SDT) and use only two features are exhibit a great prospect for practical purposes.

5. Conclusions and future work
In this paper, the diagnosis problem for a milling process of Inconel 625 workpiece performed with sharp and blunt tools was considered. Two classes were examined, i.e. Sharp and Blunt. The data set consisted of 1709 records with 32 features in the time and frequency domains. The feature selection procedure was applied in order to extract the most significant attributes. As a result, 2 out of 32

attributes were selected and used for classifiers development. The area under the amplitude spectrum was selected as the most significant predictor calculated for two frequency ranges: 0 Hz – 500 Hz and 1000 Hz – 1500 Hz.

Three well known shallow CI methods were applied, i.e.: a single decision tree, the k-nearest neighbor and a linear support vector machine. The accuracy, sensitivity and false alarm rate were used to evaluate the performance of the algorithms. The values of these indicators were very high for all the methods for the 10-fold cross-validation testing method.

It was experimentally proved that it is possible to diagnose, with highly satisfactory performance, the milling process of Inconel 625 by means of CI methods. It is worth to emphasise that judging by its efficiency presented in the study, the implementation of the developed diagnostic methods in real-time operation seems to be possible. The result of the feature selection procedure shows that only 2 attributes of acceleration signal are sufficient for accurate diagnostics. For real-time purposes, the method with the lowest computational power requirements (e.g. SDT) can be used for the industrial implementations of milling process supervision systems. However, further research is required in this field.

Acknowledgments
This project is financed by the Minister of Science and Higher Education of the Republic of Poland within the "Regional Initiative of Excellence" program for years 2019 – 2022. Project number 027/RID/2018/19, amount granted 11 999 900 PLN.

6. References

[1] Hermann M, Pentek T and Otto B 2016 Design Principles for Industrie 4.0 Scenarios *Proc. of the 49th Hawaii International Conference on System Sciences* (Koloa, USA) pp 3928-37 doi:10.1109/HICSS.2016.488

[2] Żabiński T, Mączka T, Kluska J, Madera M and Sęp J 2019 Condition monitoring in Industry 4.0 production systems - the idea of computational intelligence methods application *Proc CIRP, Vol 72 Proc of the 12th CIRP Conference on Intelligent Computation in Manufacturing Engineering* eds R Teti and Doriana M D'Addona (Naples, Italy) pp 63-67 doi: 10.1016/j.procir.2019.02.012

[3] Institute for Prospective Technological Studies 2003 *Technical report: The future of manufacturing in Europe 2015-2020 - The challenge for sustainability* European Commission's Joint Research Centre

[4] National Research Council 1998 *Visionary manufacturing challenges for 2020* The National Academy Press (Washington, USA) doi: 10.17226/6314

[5] Ahmed N, Day J A, Victory L J, Zeall L and Young B 2012 Condition Monitoring in the Management of Maintenance in a Large Scale Precision CNC Machining Manufacturing Facility *IEEE International Conference on Condition Monitoring and Diagnosis* (Bali, Indonesia) pp 842-5 doi:10.1109/CMD.2012.6416281

[6] Henshall E and Campean F 2009 Implementing Failure Mode Avoidance *SAE Technical Paper 2009-01-0990* pp 309-320 doi:10.4271/2009-01-0990

[7] Codjo L, Jaafar M, Makich H, Knittel D and Nouari M 2018 Milling diagnosis using machine learning techniques toward Industry 4.0 *DX 2018 Principles of Diagnosis Proc of the 29th International Workshop on Principles of Diagnosis co-located with 10th IFAC Symposium on Fault Detection, Supervision and Safety for Technical Processes (SAFEPROCESS 2018)* eds L Travé-Massuyès and A Sztyber (Warsaw, Poland)

[8] Żabiński T, Mączka T, Kluska J, Kusy M, Hajduk Z and Prucnal S 2014 Failures Prediction in the Cold Forging Process Using Machine Learning Methods *ICAISC 2014, Part I LNCS(LNAI), vol 8467* eds L Rutkowski, M Korytkowski, R Scherer, R Tadeusiewicz, L A Zadeh and J M Zurada (Heidelberg: Springer) pp 622–33 doi:10.1007/978-3-319-07173-2_53

[9] Żabiński T, Mączka T, Kluska J, Kusy M., Gierlak P, Hanus R, Prucnal S and Sęp J 2015 CNC Milling Tool Head Imbalance Prediction Using Computational Intelligence Methods *Artificial Intelligence and Soft Computing Proc of the 14th International Conference on Artificial Intelligence and Soft Computing ICAISC* eds L Rutkowski, M Korytkowski, R Scherer, R Tadeusiewicz, L Zadeh and J Zurada J (Zakopane, Poland) pp 503-514 doi: 10.1007/978-3-319-19324-3_45

[10] Corne R, Nath C, El Mansori M and Kurfess T 2017 Study of spindle power data with neural network for predictingreal-time tool wear/breakage during inconel drilling *J Intell Manuf* **43**(2) 287-95 doi:10.1016/j.jmsy.2017.01.004

[11] Wang G and Cui Y 2013 On line tool wear monitoring based on auto associative neural network *J Intell Manuf* **24**(6) 1085-1094 doi:10.1007/s10845-012-0636-7

[12] Żabiński T, Mączka T and Kluska J 2017 Industrial Platform for Rapid Prototyping of Intelligent Diagnostic Systems *Trends in Advanced Intelligent Control, Optimization and Automation Polish Control Conference* eds W Mitkowski, J Kacprzyk, K Oprzędkiewicz, P Skruch (Kraków, Poland) pp 712-21 doi: 10.1007/978-3-319-60699-6_69

[13] Jayalakshmi T and Santhakumaran A 2011 Statistical normalization and back propagation for classification *International Journal of Computer Theory and Engineering* **3**(1) 89-93

[14] Al Shalabi L, Shaaban Z and Kasasbeh B 2006 Data mining: A preprocessing engine *J Compt Sci* **2**(9) 735-9 doi:10.3844/jcssp.2006.735.739

[15] Graf A B and Borer S 2001 Normalization in support vector machines *Joint Pattern Recognition Symposium* (Berlin, Heidelberg: Springer) pp 277-82 doi:10.1007/3-540-45404-7_37

[16] Ma Ch, Yang W and Cheng B W 2014 How the Parameters of K-nearest Neighbor Algorithm Impact on the Best Classification Accuracy: In Case of Parkinson Dataset *J Appl Sci* **14**(2) 171-6 doi:10.3923/jas.2014.171.176

[17] Demir-Kavuk O, Kamada M, Akutsu T and Knapp E 2011 Prediction using step-wise L1, L2 regularization and feature selection for small data sets with large number of features *BMC Bioinformatics* **12** 412 doi:10.1186/1471-2105-12-412

[18] Tang J, Alelyani S and Liu H 2014 Feature selection for classification: A review *Data classification: Algorithms and applications* **37** doi:10.1201/b17320-3

[19] Shi J, Yin W, Osher S and Sajda P 2010 A fast hybrid algorithm for large-scale l1-regularized logistic regression *J Mach Learn Res* **11** 713-41 doi:10.1145/1756006.1756029

[20] Wu X, Kumar V, Quinlan, J., et al. 2008 Top 10 algorithms in data mining *Knowl Inf Syst* **14**(1) 1-37 doi:10.1007/s10115-007-0114-2

IOP Conf. Series: Materials Science and Engineering **710** (2019) 012044 doi:10.1088/1757-899X/710/1/012044

Study of buoyant force acting on different fluids moving over a horizontal plate due to forced convection

J Singh[1] and C Singh[2]

[1] Mechanical Engineering Department, Thapar Institute of Engineering and Technology, Patiala 147004, Punjab, India
[2] Department of Mechanical Engineering, University College of Engineering, Punjabi University Patiala, 147002, Punjab, India

jashanpreet.singh@thapar.edu

Abstract. In this paper, the effect of buoyancy force on heat transfer over flat plate due to forced convection is investigated numerically. The governing equations namely continuity, Navier-stokes and energy equations were discretized by using an explicit finite difference technique. Code generation has been done in FORTRAN-95 programming language. The testing of FORTRAN code has been done for different fluids at different grid points. The length of horizontal flat plate (x) was taken as 0.5 m. The inlet velocity of fluid (u) at the edge of plate was taken as 3 m/s. The kinematic viscosity (υ) of air and water at 323 K were taken as 17×10^{-6} m^2/s and 66×10^{-7} m^2/s respectively. The surface temperature (T_s) of plate and flow temperature (To) of fluid were taken as 300 K and 323 K respectively. The grid was discretized by continuous method into 1001×101 ($m\times n$) mesh points. The results show the boundary layer thickness, velocity and temperature evolutions, and heat transfer coefficient formed by different fluids. Also, the heat transfer coefficient to surrounding fluid across flat plate was found higher for air as compared to water. Results showed that buoyancy force acts from downward position so it was observed maximum at lower distance of plate. However, buoyancy effect reduces as the distance of x increases.

1. Introduction

In present study, the computational fluid dynamics (CFD) was employed to solve the heat transfer due to forced convection across flat plate. Forced convection heat transfer across flat plate is universally involved in many engineering applications such as cooling coils, radiators, cooling circuits, refrigerators, steam generators, etc. [1]. CFD is globally accepted to mathematically solve the heat transfer problems [2]. In fluid dynamics, the governing equations such as Navier-Stokes equation, continuity equation and additional conservation equations (like example energy or species concentrations), are solved by using CFD tools/software packages or programming languages [3]. Buoyancy plays a significant role in heat transfer problems related to fluid flow. Buoyancy is generated due to the temperature difference between a fluid and surroundings. Buoyancy creates a significant effect in movement of fluid flow which contributes in transverse velocity that has a large effect on the heat transfer process [4].

In early days, Sakiadis [5] performed a numerical study to investigate the effect of skin friction on boundary layer thickness over a horizontal plate of finite length. They used integral method to numerically solve the governing equations by applying the appropriate boundary conditions. They reported that the integral method was effective in solving the turbulent boundary layer on a moving continuous flat surface. Peters [6] performed a numerical study to investigate the laminar hydrogen-oxygen diffusion flame on a flat plate. They investigated the non-equilibrium and nearly-equilibrium structure at various positions in the boundary layer. They used 15 elementary reactions between 8 species and investigated the effect of variable properties including thermo diffusion. Results showed that no chemical reactions occurred at larger distances from the starting edge of plate due to the low temperatures. Janssen and Armfield [7] investigated the boundary layer stability of steady state cavity convection flow. They applied artificial perturbation to generate the start-up waves and analyzed the stability character of the boundary layers in fully developed flow. During the experiments, they kept the top and bottom cavities as insulated. They found that heated and cooled walls exhibit travelling wave instabilities in the thermal boundary layers that form on the walls. They also found that these waves were observed at start-up for the water (Pr=7.5) at Ra=6×10^8. Additionally, the thermal boundary layers in the fully developed flow have approximately the same stability character as the start-up flow. In 21st century, CFD has received an extensive attention amongst the community of researchers/investigators to solve the engineering problems numerically by using computers. Mahgoub [8] performed a numerical study to investigate the heat transfer due to forced convection across a flat plate in a porous media. They found that the local heat transfer coefficient starts decreasing as the axial distance increases on the flat plate.

Various researchers have done a lot of work on boundary layer analysis. The various methods and techniques to solve the boundary layer are Approximate Generalized Method [9], Newton-Raphson method [9], Homotopy analysis method [10], Crank-Nicolson method [11], Integral method [12], etc. From literature survey, it is found that these methods are very tedious and complex. Moreover, these methods are difficult to unproductive to obtain approximate solutions of fluid flow problems. On the other hand, Finite difference method and Gauss-Seidel iteration are very simple techniques to analyze laminar boundary layer for flow over a flat plate. In this context, present study was carried out to analyze the velocity and temperature profiles of boundary layer over a flat plate using finite difference method using Gauss-Seidel iteration. Air and water were considered as the flowing fluids.

2. Mathematical modelling

2.1. Mathematical equations

Three different governing equations were used to solve the forced convection around flat plate. These equations are explained below [1]:

Continuity Equation:

$$\frac{du}{dx}+\frac{dv}{dy}=0 \tag{1}$$

Steady-state Navier-Stokes Equation:

$$u\frac{du}{dx}+v\frac{du}{dy}=v\left[\frac{d^2u}{dx^2}+\frac{d^2u}{dy^2}\right] \tag{2}$$

Energy equation:

$$u\frac{dT}{dx}+v\frac{dT}{dy}=\alpha\left[\frac{d^2T}{dy^2}\right] \tag{3}$$

2.2. Finite difference method

Discretization of governing equation was done by using explicit scheme of Finite difference method. Discretization of Navier-Stokes equation was done to find x-component of velocity (u) whereas the discretization of the continuity equation was done to find the velocity (v). However, the discretization the energy equation of thermal boundary layer was done to find out the value of temperature. Discretization of governing equations is as follows [13, 14]:

- *Continuity equation:*

$$v_{i+1,j+1} = v_{i+1,j-1} - \frac{2.\Delta x}{\Delta x}\left[u_{i+1,j} + u_{i,j}\right] \tag{4}$$

- *Navier-Stokes equation:*

$$u_{i+1,j} = u_{i,j} - \frac{\Delta x.v_{i,j}}{2\Delta y.u_{i,j}}\left[u_{i,j+1} - u_{i,j-1}\right] + \frac{v.\Delta x}{\Delta y^2.u_{i,j}\left[u_{i,j+1} + u_{i,j-1} - 2u_{i,j}\right]} \tag{5}$$

- *Energy equation:*

$$T_{i+1,j} = T_{i,j} - \frac{\Delta x.v_{i,j}}{2\Delta y.u_{i,j}}\left[T_{i,j+1} - T_{i,j-1}\right] + \frac{v.\Delta x}{\Delta y^2.u_{i,j}\left[T_{i,j+1} + T_{i,j-1} - 2T_{i,j}\right]} \tag{6}$$

3. Numerical model & boundary conditions

3.1. Physical model of problem

The physical model of problem consists of a horizontal flat plate placed on a smooth surface, schematically illustrated as Figure 1. The length (*alen*) of plate was taken as 0.5 m. The thickness (*t*) of plate was taken as 0.1 m.

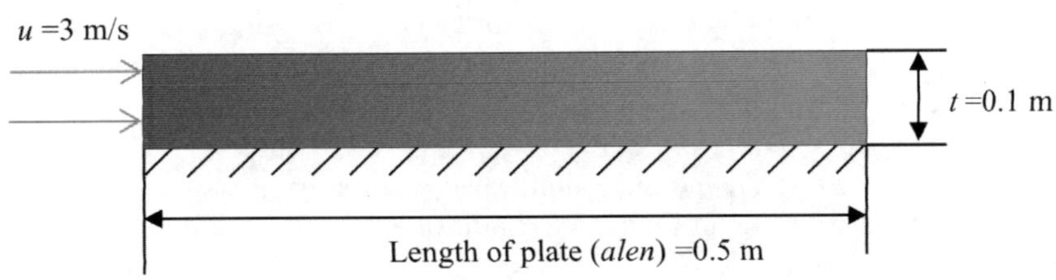

Figure 1. Physical model of problem.

3.2. Solution domain

The solution domain was divided into *m* and *n* number of cells. Number of grid points in *x* and *y* direction were designated as *i* and *j* respectively. The schematic diagram of solution domain is shown in Fig 2. The spacing between grid points along *x*-direction was solved by using the following correlation:

$$delx = \frac{alen}{m-1} \tag{7}$$

Where, the symbol *alen* represents the length of plate on which fluid flows, and *m* is the value of number of grid points along *x*-direction (i=1, 2, 3.........m+1) whose value was taken as 1001. However, the spacing between grid points along *y*-direction was solved by using the following correlation:

$$delx = \frac{t}{n-1} \tag{8}$$

Where, *t* is the thickness of plate = 0.1 m, and *n* is the number of grid points along *y* direction (j=1, 2, 3.........n+1) whose value was taken as 51.

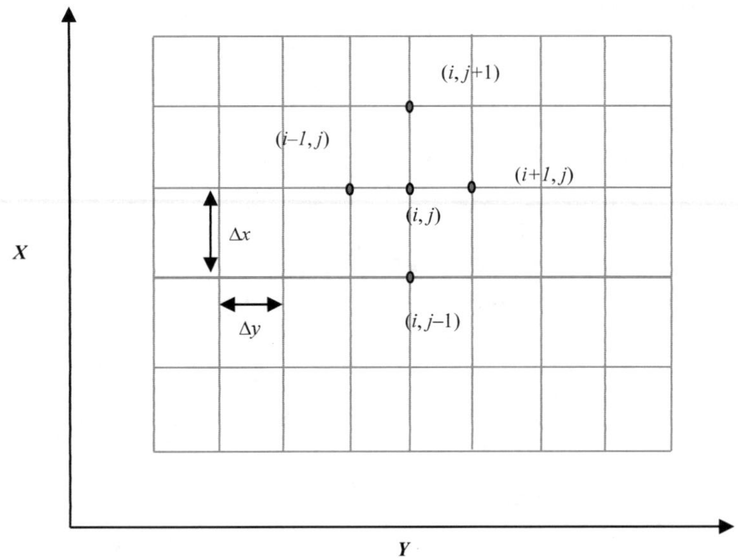

Figure 2. Discrete grid points in the mesh.

3.3. Boundary conditions

Various boundary conditions to solve momentum equation, continuity equation and thermal energy equation are defined. Four different boundary conditions (namely no slip, free slip, inflow, and outflow conditions) were applied to solve the discretized governing equations. The boundary conditions are written as:

- *No Slip boundary conditions:* The no-slip boundary condition for viscous fluids states that the fluid achieves zero velocity (i.e. hydrodynamic condition) and temperature gradient (i.e. thermal condition) at a solid boundary. In present study, the solid boundary was the surface of horizontal flat plate. So, the no slip boundary conditions were applied as follows:

  Hydrodynamic,
  $$u\,(i,1) = v(i,1) = 0 \qquad \text{for } 1 < i < m \tag{9}$$
  Thermal,
  $$u\,(i,1) = v(i,1) = 0, \text{ and } T(i,1) = T_w \qquad \text{for } 1 < i < m \tag{10}$$

- *Inflow boundary conditions:* Inflow boundary conditions were applied for the inlet of fluid (in Figure 2, at the left hand side of mesh i.e. i= 1). Hydrodynamic as well as thermal boundary conditions were applied as follows:

  Hydrodynamic,
  $$u(1,j) = u_o\ ,\ v(1,j) = 0 \qquad \text{for } 2 \le j \le n \tag{11}$$
  Thermal,
  $$u(1,j) = u_o\ ,\ v(1,j) = 0,\ T(1,j) = T_o \qquad \text{for } 2 \le j \le n \tag{12}$$

- *Outflow boundary condition:* These conditions were applied at the outlet of flow while the fluid passes the plate i.e. i=m. In this case, velocity remains constant i.e. change in velocity is zero. Outflow boundary condition is written as:

$$\frac{du}{dx} = 0,\ u_i = u_{i-1} \tag{13}$$

Or

$$u(i, j) = u(i\text{-}1, j), \quad v(i, j) = v(i\text{-}1, j) \tag{14}$$

- *Free slip boundary conditions:* Free slip boundary conditions were applied at top of boundary layer i.e. $j=n$. Free slip boundary conditions are:
  Hydrodynamic,

$$u(1,n) = u_o \ , v(1,n) = 0 \tag{15}$$

Thermal,

$$u(1,n) = u_o \ , v(1,n) = 0, \ T(1,n) = T_o \tag{16}$$

4. Numerical simulations

In present work, the numerical simulations were performed by developing a numerical code in FORTRAN language. For performing the numerical simulations, the governing equations were solved by using the Gauss-Seidel iterations for a fine grid spacing ($m \times n$). The grid size in the x-axis is about 1% of total length of flat plate and 0.1% of transverse dimension in the y-axis. Numeric iterations were run using a LINUX-based Ubuntu 2.6.321 operating system on Intel(R) core™ *i*5 machine having 2.67 GHz processing unit and 4 GB RAM. Average time taken by CPU to run a single iteration was 6-7 s. The properties of the air and water applied in numerical code are specified in Table 1. Surface temperature of plate (T_s) was taken as 300 K and temperature of fluid (T_o) was taken as 323 K. Grid taken for air/water have $1001 \times 101 (m \times n)$ nodes.

Table 1. Properties of air and water used in present study.

S. No.	Fluid	Kinematic viscosity (v)	Inflow velocity (u)
1	Air	17×10^{-6} m^2/s	3 m/s
2	Water	0.66×10^{-8} m^2/s	3 m/s

5. Results and Discussion

Numerical models were recently made in FORTRAN programming language were run to numerically predict the boundary layer thickness, velocity and temperature profiles. The detailed results are as follows:

5.1. Velocity profiles

In Figure 3(a), as the value of x increases as the velocity (u) decreases. It was seen that there is an initial sudden rise in velocity (u) at $x = 0.1$ m. This initial sudden rise reduces as value of x increases. At $x = 0.5$ m, this rise becomes minimum. In Figure 3(b), the effect of buoyancy force can be clearly observed. Velocity curve moves sharply above 3 m/s under the effect of buoyancy force. Moreover, the buoyancy force acts from downward position so at lower distance ($x = 0.1$), this effect was found maximum. As distance x increases, this effect rapidly decreases.

In case of water, characteristic of velocity curve remains same as air. However, the velocity curve of water travels a large vertical distance (y) as compared to air (In Figure 4). It was also observed that the initial sudden rise of velocity (up to 3 m/s) curve of water was higher than fluid air. Figure 5(b) shows the comparison of velocity profiles while buoyancy force is acting on the fluids. It was observed that characteristic curves moves more sharply and an initial sudden rise exists. Similar observation was made by Chang [15]. In case of air, characteristic curve has an initial sudden rise but it is not as sharper as in case of water.

IOP Conf. Series: Materials Science and Engineering **710** (2019) 012044 doi:10.1088/1757-899X/710/1/012044

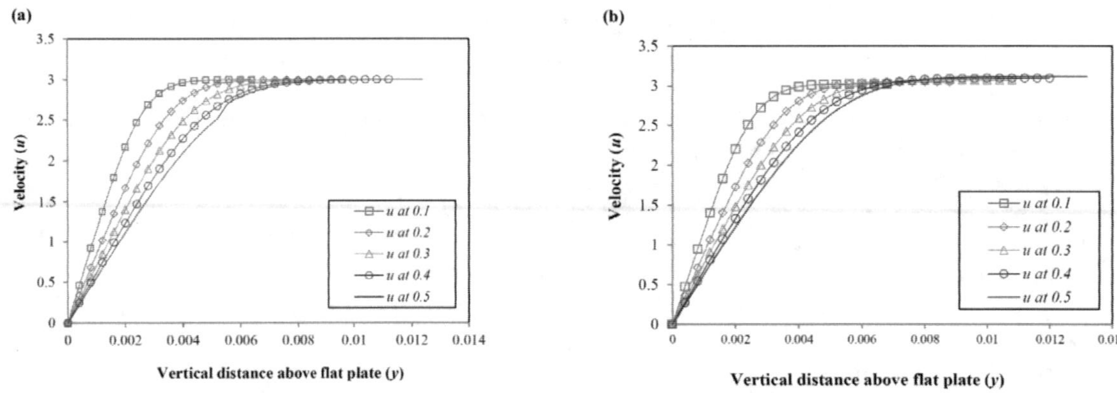

Figure 3. Velocity profiles for air at vertical distance above flat plate while **(a)** zero buoyant force, and **(b)** buoyant force is acting.

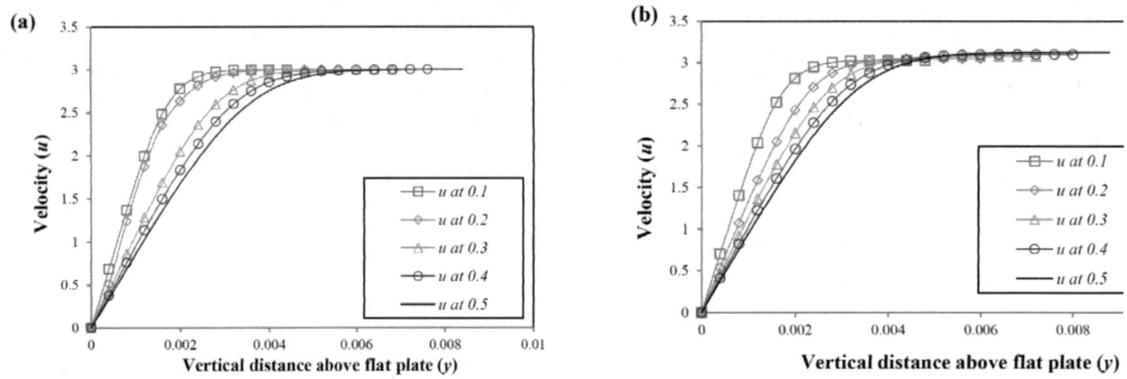

Figure 4. Velocity profiles for water at vertical distance above flat plate while **(a)** zero buoyant force, and **(b)** buoyant force is acting.

5.2. Hydrodynamic boundary layer (δ)

Validation of results is carried out by comparing the numerical model's results with the analytical results. The Blausius theory was applied to find the analytical results of boundary layer thickness. The Blausius equation for the laminar boundary layers over a flat plate is written as [16]:

$$\delta \approx 4.91\sqrt{\frac{\upsilon x}{u}} = 4.91\frac{\upsilon x}{\sqrt{\mathrm{Re}_x}} \tag{17}$$

Where, Re_x is the Reynolds Number at distance x on flat plate, δ is the overall thickness (or height) of the boundary layer (m), υ is the kinematic viscosity, $\upsilon=\mu/\rho$ (m^2/s), x is the distance downstream from the start of the boundary layer (m), ρ is the density of the fluid (kg/m^3), u_0 is the free stream velocity, and μ is the dynamic viscosity (Pa-s or kg/m-s).

The comparison between present numerical simulation results and Blausius model [16] is shown in Figure 6. By using the properties of air, the hydrodynamic boundary layer thickness from Blausius equation was found as 0.0035 m for air at x = 0.1 m. Generally, the hydrodynamic boundary layer thickness (δ) is defined normal to the wall, and the point where the free stream velocity (u) is essentially that of the free stream is customarily defined as the point where:

$$u(y) = 0.99u \tag{18}$$

Above equation indicates that the velocity of fluid equal to 99% of velocity at a particular distance (x).

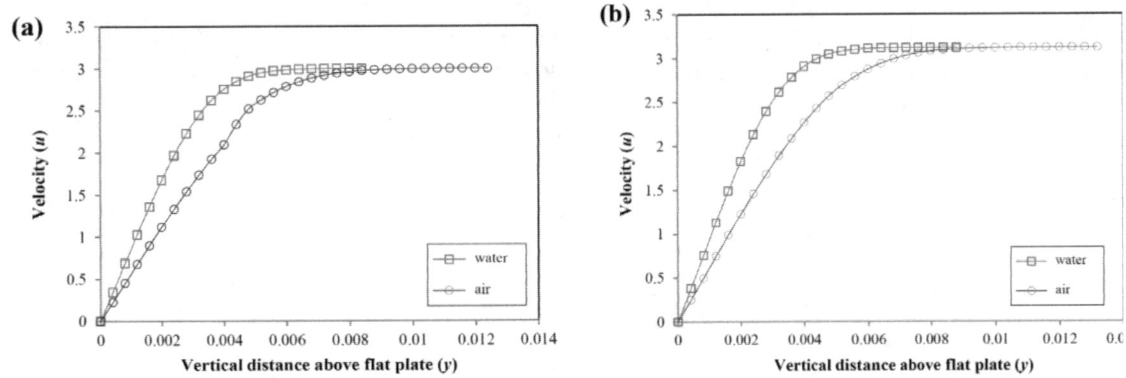

Figure 5. Comparison of velocity profiles between fluid air and water (at $x = 0.5$ m) while **(a)** zero buoyant force, and **(b)** buoyant force is acting.

In present study, the value of initial velocity is equal to 3 m/s so, the 99% of value u is calculated as 2.97 m/s. From Figure 6(a), at velocity 2.97 m/s value of y is approximately equal to 0.0032 m which is equal to the boundary layer thickness as calculated from Blausius equation [16] i.e. 0.0035 m. It also confirms the validity of model. By comparison of results, it could be said that the present numerical model holds a good agreement with the analytical model and lies within the standard deviation of ±6.5%.

Furthermore, the boundary layer thickness of water was found lesser as compared to water. The boundary layer thickness depends directly upon kinematic viscosity of fluid and distance of point from leading edge. The kinematic viscosity of water is lesser than air therefore, the boundary layer thickness of water was found lesser than that of air. Moreover, a stationary water particle near the stationary surface has lesser retarding effect as compared to the particles away from the surface [5]. Due to this, the effect of stationary surface in water stream forms up to a smaller distance than that of air stream.

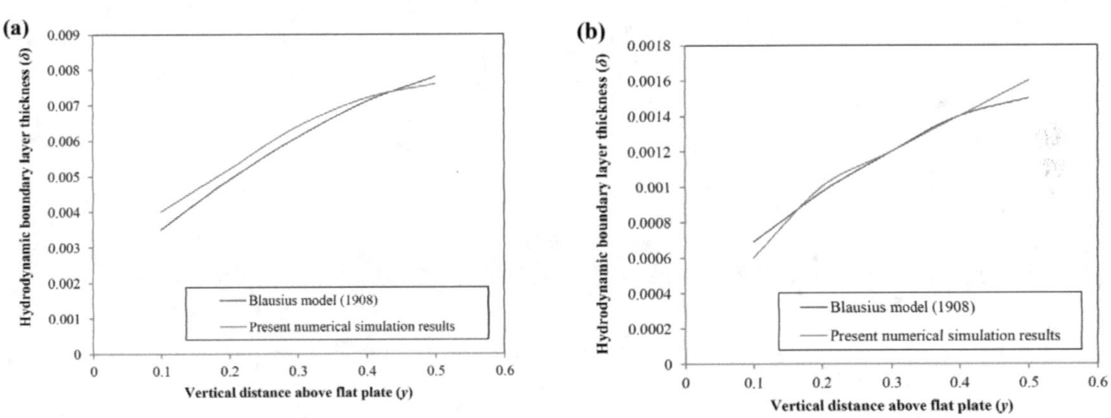

Figure 6. Comparison of numerical and analytical results of hydrodynamic boundary layer of (a) air and (b) water.

5.3. Reynolds's number (Re)

Reynolds's number is the defined by the ratio of inertia to viscous force which is given as [17]:

$$\mathrm{Re}_x = \frac{\rho u x}{\mu} \tag{19}$$

Symbol Re_x represents the value of Reynolds number at distance x on flat plate which was calculated as $\approx$ 17647 for air and $\approx$ 455000 for water at x = 0.1 m. From the value of Re, it was concluded that the boundary layer thickness remains laminar for the fluid air whereas turbulent for water. Similarly, the value of δ and Re_x at x = 0.2, 0.3, 0.4, 0.5 for air and water is shown in Figure 7.

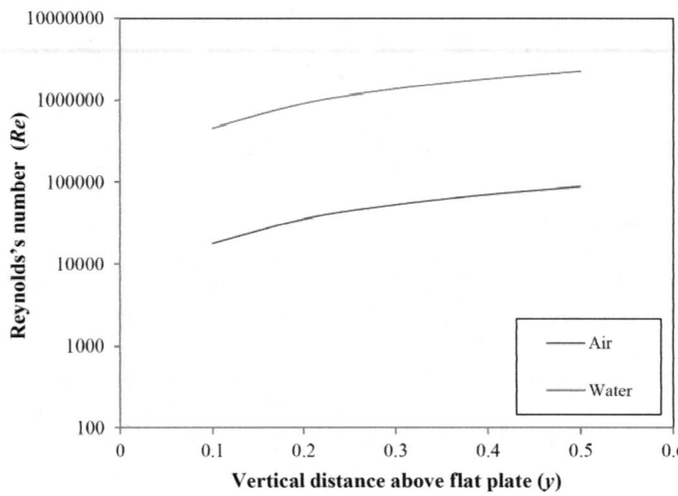

Figure 7. Reynolds's number of (a) air and (b) water at vertical distance above flat plate.

5.4. *Thermal boundary layer* (δ_T)

Thermal boundary layer can be calculated by using the temperature profiles. Prandtl number (Pr) is related to the velocity and temperature field, given by [18, 19]:

$$\frac{\delta_T}{\delta} = \frac{1}{\sqrt{\Pr}} \tag{20}$$

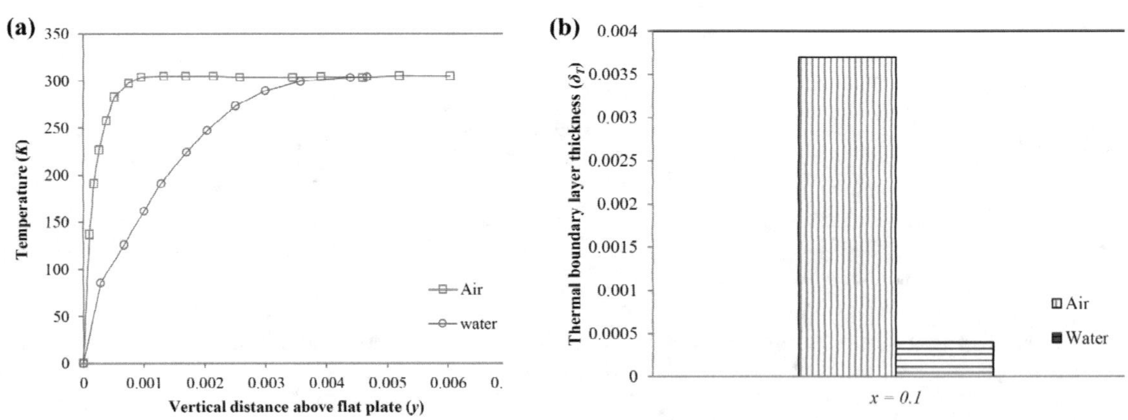

Figure 8. Numerically found thermal boundary layer thickness for air and water at x = 0.1.

From Figure 7, the temperature was evaluated as 322.46 K at 99% of velocity (u). Corresponding to this temperature, the boundary layer thickness (δ_T) was evaluated as 0.0037 m. By putting these values of δ and δ_T in Eq. 20, the value of Pr was calculated as 0.844 (approx.) which lies in the same range of Pr (0.7 to 0.8) for air. Similarly for water, value of δ and δ_T was evaluated as 0.00069 and 0.0004 respectively. Putting these values in Eq. 20, the value of Pr was found as 4 (approx.) for water. So, the present simulation results hold a good deal with the literature also.

IOP Conf. Series: Materials Science and Engineering **710** (2019) 012044 doi:10.1088/1757-899X/710/1/012044

5.5. Heat transfer coefficient

By using the results of boundary layer thickness and dimensionless numbers, various other parameters like shear stress, viscous force etc. can be evaluated. The slopes were plotted on the velocity and temperature profiles in order to evaluate the velocity and temperature gradients. From Figure 8, the value of gradient *du/dy* at a particular point was calculated by putting value of coefficient of friction using the shear stress formula, written as:

$$\tau = -\mu \frac{du}{dy} \tag{21}$$

In the Figure 8(a), change in velocity (*du*) and vertical distance (*dy*) was 2.75 m/s and 2 m respectively at a point A. Therefore, velocity gradient *du/dy* was evaluated as 1.375 s^{-1}. Here, the value of coefficient of viscosity (μ) is 690 kg/hr-m for air). The value of shear stress (τ) was calculated as 948 N/m^2 (approx.). Further, the shear force can be calculated by putting value of shear stress (τ) and area of plate on which shear stress occur (A = 0.25 m^2), by using the following empirical correlation:

$$F = \tau A \tag{22}$$

The value of shear force (F) was evaluated as 237 N.

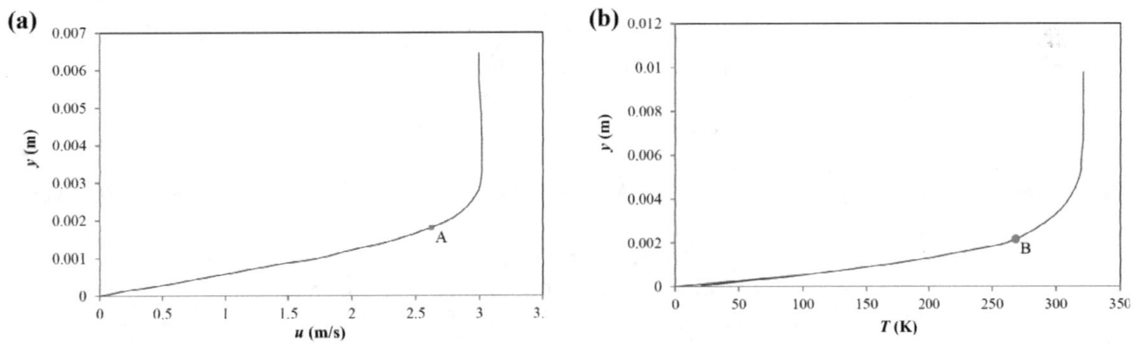

Figure 9. Variation of x-component of (a) velocity and (b) temperature, with the vertical distance above flat plate.

Similarly by using Figure 8(b), heat transfer (conductive and convective) was calculated by using temperature gradient *dT/dy* from graph of thermal boundary layer from graph [20]:

$$q = -k \frac{dT}{dy} \tag{23}$$

Where, k is the conductive heat transfer. At a point A, the change in temperature (*dT*) was 21 K (321-300), and change in vertical distance is (*dy*) was 0.0004 m. So therefore temperature gradient *dT/dy* = 52500 K/m. By putting value of heat transfer coefficient (k = 0.0268 W/m-K for air) and temperature gradient *dT/dy* in Eq. 23, heat transfer (q) was calculated as 1407 kJ. The convective heat transfer (h) was also calculated, by using following correlation [1]:

$$h = \frac{k}{T_o - T_s} \left(\frac{dT}{dy} \right)_A \tag{24}$$

IOP Conf. Series: Materials Science and Engineering **710** (2019) 012044 doi:10.1088/1757-899X/710/1/012044

In above equation, the value T_s = 300 K (temperature at surface of plate), T_o = 323 K (temperature of fluid), k = 0.0268 W/m-K (heat transfer coefficient for air). The value of convective heat transfer (h) was calculated as ≈ 60 W/m^{2}K.

6. Conclusions

In this paper, a numerical code was developed in FORTRAN programming language to analyse the forced convection across flat plate. Following conclusions are drawn on the basis of present work:

- Buoyancy force acts from downward position so it was observed maximum at lower distance of plate in x-axis. However, buoyancy effect reduces as the distance of across x-axis increases.
- Velocity profile curves moves more sharply and an initial sudden rise exists. In case of air, characteristic curve has an initial sudden rise but it is not as sharper as in case of water.
- The velocity curve of water travels a large vertical distance (in y-axis) as compared to air. There was an initial sudden rise of velocity (up to 3 m/s) curve of water was higher than fluid air.
- From the value of *Re*, it was concluded that the boundary layer thickness remains laminar for the fluid air whereas turbulent for water.
- Boundary layer thickness of water was found thinner as compared to fluid water.

7. References

[1] Singh J and Singh C 2017 *Proceed. IMeche. Part E J. Process Mech. Eng.* **231** 405.
[2] Singh J Singh C and Kumar S 2017 *Advance. Intelligent Sys. Comput.* **546** 331.
[3] Singh J Kumar S Singh JP Kumar P and Mohapatra SK 2019 *Particul. Sci. Technol.* **37** 275.
[4] Som SK 2008 *Introduction to heat transfer* (New Delhi: PHI learning Pvt. Ltd.).
[5] Sakiadis BC 1961 *American Inst. Chem. Eng.* **7** 221.
[6] Peters N 1976 *Int. J. Heat Mass Trans.* **19** 385.
[7] Janssen R and Armfield S 1996 *Int. J. Heat Fluid Flow* **17** 547.
[8] Mahgoub SE 2013. *Ain Shams Eng. J.* **4** 605.
[9] Kaw AK Kalu EK and Nguyen D 2011 *Numerical Methods with Applications* (Florida: University of South Florida).
[10] Liao SJ 1999 *Int. J. Non-Linear Mechan.* **34** 759.
[11] Crank J and Nicolson P 1947 *Math. Proceed. Cambridge Phil. Society* **43** 50.
[12] Singh RK Sahu KB and Mishra TD 2013. *Int. J. Eng. Res. App.* **3** 766.
[13] Totala NB Shimpi MV Shete NL and Bhopate VS 2013 *Int. J. Eng. Sci.* **3** 27.
[14] Subramanyam D Chandrasekhar M and Lokanadham R 2013 *Int. J. Mech. Eng. Technol.* **4** 54.
[15] Chang CL 2006 *J. Phys. D: Appl. Phys.* **39** 1132.
[16] Blasius H 1908 *J. Appl. Math. Phys.* **56** 1.
[17] Lindstedt M and Karvinen R 2012 *J. Phys.: Conf. Ser.* **395** 012059.
[18] Prandtl L 1904 *Int. Conf. of Mathematicians (Heidelberg: Germany)* vol 3 (*Verhandlinger) p* 484.
[19] Ishak A 2009 *J. Phys. A: Math. Theor.* **42** 195501.
[20] Mitra A 2017 *IOP Conf. Ser.: Mater. Sci. Eng.* **225** 012096.

IOP Conf. Series: Materials Science and Engineering **710** (2019) 012045 doi:10.1088/1757-899X/710/1/012045

Selection of the optimal unit of analysis in assessing the structure of terrestrial arthropods assemblages

M Kirichenko-Babko[1], Y Danko[2] and D Majerek[3]

[1] National Academy of Sciences of Ukraine, , Schmalhausen Institute of Zoology, Department Fauna and Systematics of Invertebrates, B. Khmelnitsky 15, 01030 Kyiv, Ukraine
[2] Sumy Makarenko State Pedagogical University, Department of General Biology and Ecology, Romenska 87, 40002 Sumy, Ukraine
[3] Lublin University of Technology, Faculty of Fundamentals of Technology, Nadbystrzycka 38, 20-618 Lublin, Poland.

d.majerek@pollub.pl, kirichenko@izan.kiev.ua, yaroslavdanko@gmail.com

Abstract. The article shows the use of statistical methods of multivariate analysis on the example of environmental data – the species composition of terrestrial arthropods. It is shown that the research results largely depend on the scale of observations made – whether it is a study conducted in a whole ecosystem or in a separate landscape element, type of habitat or, finally, an analysis of elementary samples. The results and their interpretations directly depend on the scale chosen by the researcher. We were mainly concerned with the reliability of using of multidimensional statistics in practical situations.

1. Introduction

Modern ideas about the quality of the environment rely on the study of its biological components. Attention is increasingly paid to indicators characterising the structure of biocenoses. Today, the most popular are index estimates based on species diversity [1; 2]. At the same time, problems with research coverage, choice of an adequate sample size, which can be considered representative with respect to one or another object, remain quite controversial [3]. Very often, different samples from the same ecosystem can produce very different results. This is quite fair, because ecosystems or their fragments are very complex, not to mention processes that keep them in dynamically stable condition [4]. The environmental and biodiversity assessment could be performed on territories of different size and materials with various depth of detailing [5]. Its interpretation depend on scale of compared objects and their deep study. Some researchers [6;7;8] emphasise that the scale of observations made — this study conducted in the landscape, the whole ecosystem, habitat or, finally, an elementary sample — significantly affect the results.

We analysed the possibility of interpretation of quantitative data (abundance) obtained at various levels of research. Thus, the subject of our work was to select the most appropriate scale of the territory covered by the research, to obtain the most representative results and their adequate interpretation.

2. Material and methods

2.1. Statistical analysis

Data were processed using R, version 3.6.1 [13]. Hierarchical clustering was performed with hclust (function ward.D2) from core R package stats. Two types of p-values (AU and BP) for the nodes of hierarchical clustering dendrograms were calculated with R package pvclust. According to the package documentation, "AU p-value, which is computed by multiscale bootstrap resampling, is a better approximation to unbiased p-value than BP value, computed by normal bootstrap resampling. Clusters with AU larger than 95% are highlighted by rectangles, which are strongly supported by data" [15]. K-mean clustering was performed with function kmens from package cluster [12] and plotted on the plain of principal components (PCA) with fviz_cluster from factoextra [11]. Plots were produced using packages ggplot2 [16], factoextra [11], ggrepel [14], directlabels [10]. Species data analysed with and without Hellinger transformation [18].

2.2. Region of study and collected data

The study was carried out in a steppe landscape at the regional landscape park (RLP) of Kleban-Byk, which is located about 10 km from Konstantinovka town (eastern part of Ukraine, Donetsk Oblast). The territory of the RLP Kleban-Byk is situated within the western outskirts of the Donetsk Ridge and is a wavy plateau, dissected by deep rivers and a ravine-beam system. This region has undergone many changes in the last 150 years. Petrophytic and crushed stipa steppes, real hygrophytic stipa steppe and stony outcrops occupy the largest area. Ravine and artificial forests occupy considerably smaller area.

The material was being collected from June to August 2005-2006 in seven locations of RLP (Tab. 1) [19]. Sampling has focused on a wide range of habitats, including steppes in gully and in plakor, shrubby vegetation on the steppe, forests, meadows, arable field and shore of ponds (Tab. 1).Carabid beetles were sampled using pitfall traps which were filled with preservative fluid – a propylene glycol [20]. In each sampling site 15 traps were used, arranged in a row with a spacing of 3 m between the traps. 70 species of ground beetles (Coleoptera, Carabidae) were identified at different habitats in the territory of the RLP Kleban-Byk.

Table 1. Characteristics of the studied sections of Kleban-Byk landscape park.

Code and Name of locations		Habitats & the area (ha) of locations
Loc-1	Druzhkovski skamjanili trees	Steppe, shore of pond, field; 110,5 ha
Loc-2	Gorilyj forest	Shrub steppe ("caragana"), wet meadow; 60 ha
Loc-3	Olenivskyj forest	Deciduous forest, meadow; 161,3 ha
Loc-4	Bila gora	Steppe, shore of pond; 134,7 ha
Loc-5	Ridkodub	Steppe with oak trees, meadow; 52,3 ha
Loc-6	Ravine Kravets'ka	Steppe, shore of pond; 28,6 ha
Loc-7	Kleban-Byk reservoir	Reeds on shore, forest belt, meadow; 650 ha

3. Results

Environmental studies involve the study of all components of ecosystems. Statistical methods provide an arsenal of techniques that allow to analyse data presented as "presence-absence" of species in samples or habitats or their changing abundances [21].

Due to the heterogeneity of the relief, the steppe landscape is represented by different types of habitats: the steppe proper localized, as a rule, on the open plateau; slopes of ravines overgrown with shrubs; ravine forests limited to low areas of relief; shore of the reservoir on the Kleban-Byk River and shores of small ponds. Materials were collected in 7 locations (Tabele 1) remote from each other but representing one type of ecosystem – the steppe and were analysed at two levels: (1) at the level of individual samples and (2) at the level of samples grouped by habitat. At the level of the working hypothesis, it could be assumed that different types of habitats should have a similar species composition of ground beetles. Therefore, according to this hypothesis, the presence of structures at the level of habitat types: open and shrubby steppe, ravine forest, shores of water bodies, etc. was expected. Obviously, expected was the presence of associations at the level of samples that were taken in various habitats. Using different levels of generalisation of data and methods of statistical analysis, an attempt is made to assess objectivity of the existence of the structure of the ground beetles' assemblage. And also, to evaluate which of the methods of analysis, and at what level will give a more objective idea of the structure of the assemblage.

The results of Ward's hierarchical clustering of the matrix with Kulczynski distances are presented in (Figure 1 A, B). The distance matrix was obtained on the basis of a Q-mode table in which the rows are individual samples, named for the corresponding habitats, and the columns are the abundance species present in these habitats. The results of cluster analysis at the level of local samples showed that our working hypothesis is not confirmed when using this method.

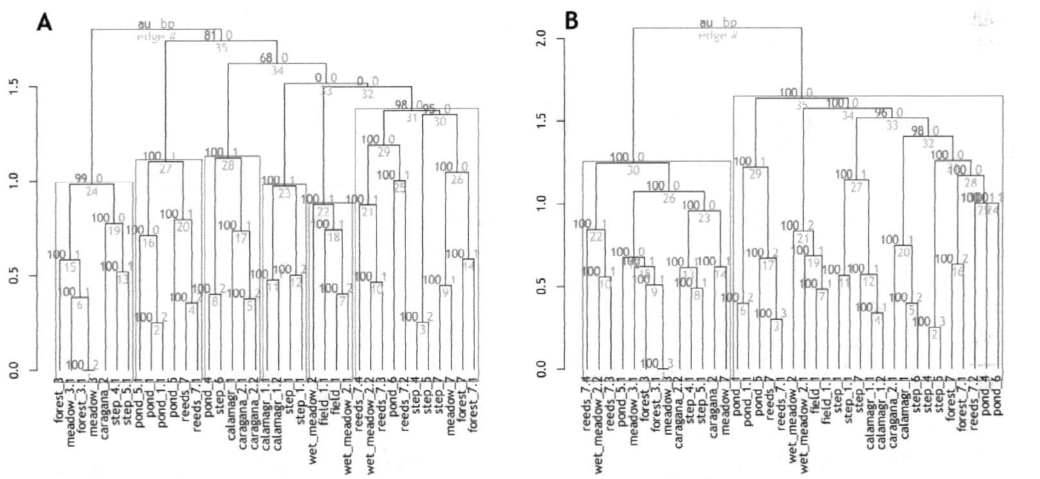

Figure 1. Ward's hierarchical clustering of a Kulczynski distances matrix among samples. A — untransformed, B — Hellinger-transformed data. Samples are named for their respective habitats. First number after the name of habitat corresponds to the number of locations. AU (Approximately Unbiased) and BP (Bootstrap Probability) p-values are shown next to the nodes.

Reliability of six clusters is confirmed by the AU (Approximately Unbiased) p-values obtained by bootstrap method (Figure 1 A). Although AU p-values are considered more reliable than BP (Bootstrap Probability) p-values, and it is on the basis of the first that a conclusion is made on reliability of the cluster, it should be noted that BP values are extremely low. The obtained cluster dendrogram showed that both these "reliable" clusters and higher and lower level clusters cannot be adequately interpreted biologically. Thus, hierarchical clustering on the basis of non-transformed data at the level of individual samples does not allow to draw conclusion about the structure of the ground beetles' assemblage. At this generalisation level, all studied plots and habitats are presented as a single territory on which the species composition is not structured.

Quite often, transformation of data, by one method or another, significantly improves the results of clustering. The dendrogram shows that transformation data by Hellinger has changed the composition of the clusters (Figure 1 B). Despite the fact that the application of the Hellinger's transformation before hierarchical clustering yielded two "reliable" according to p-values clusters, from a biological point of view, this separation of samples into two groups is just as uninterpretable as clustering obtained without transformation (Figure 1 A).

Further, a non-hierarchical K-mean clustering was applied to previously non-transformed sample data. This method requires pre-setting the desired number of clusters. To determine the optimal number of clusters, we used two methods: silhouette and Calinski-Harabasz [22]. Both of these methods consider two as the optimal number of clusters. Since two are, in fact, the minimum acceptable number of clusters, *a priori*, most likely, all clustering results will be banal. To visualise the clusters, they were depicted on the plane of the first two principal components of the PCA (Figure 2 A). Clustering by K-mean method without data transformation, as well as hierarchical clustering, showed that all samples make up a one unit. Only a sample taken on the shore of one of the reservoirs stood out in a separate cluster.

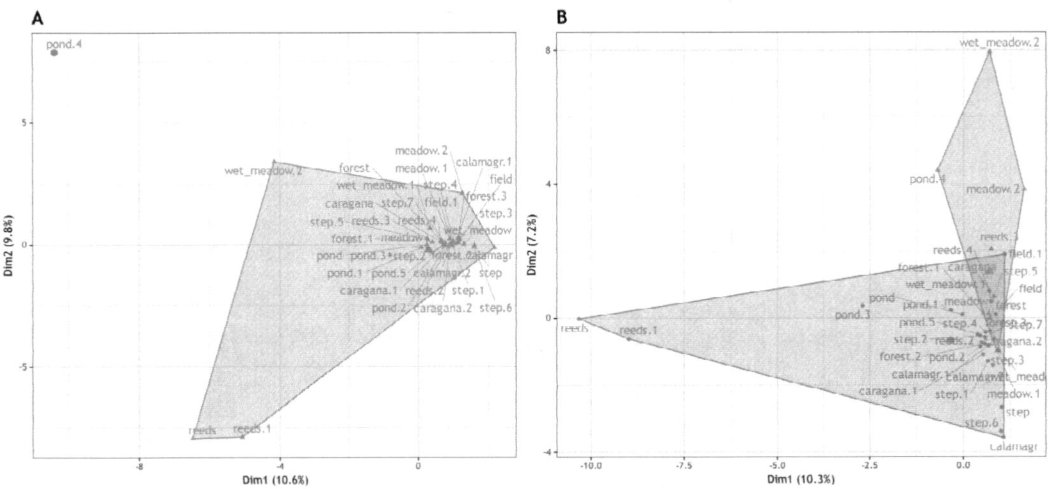

Figure 2. K-mean clustering of samples from various habitats on the plain of the first two principal components. A — untransformed, B — Hellinger-transformed data. The number after the habitat name indicates number of samples.

The methods for determining the optimal number of clusters for Hellinger-transformed data gave the following results. The silhouette method recommends 15 clusters; such a number of clusters seems senselessly large. The Calinski-Harabasz method recommends defining two clusters, alike in the case of untransformed data. Figure 2 B shows the results of clustering for a given number of clusters two. In this case, unlike non-transformed data, the second cluster is already represented not by one sample, but by twelve, which at first glance looks encouraging. Nevertheless, from a biological point of view, the interpretation of the allocation of groups is very difficult and cannot be associated with objective factors that determine the structure of the ground beetles' assemblage [23]. Transformed data demonstrate the presence of certain structures, the principle of separation of which, obviously, has a purely mathematical basis. Thus, we can conclude that the PCA also does not exhibit a biologically meaningful structure in our samples. In general, the first two principal components account for an extremely small fraction of the variability in data.

Examination of the cluster dendrogram obtained based on data grouped by habitat showed, that without transformation only a field and a wet meadow were found in the "reliable" cluster (Figure 3 A).

After Hellinger's transformation, two reliable clusters emerged (Figure 3 B). Such a division has an obvious biological meaning. Thus, the structure of the assemblages of ground beetles of the studied sites is represented by two components: one group of species is associated with xerophytic and petrophytic habitats of the steppe, and the other – with habitats along the shores of reservoir ("reeds") and ponds, meadows in ravines and fields. Fields as artificial habitats are usually grouped with adjacent habitats [24]. The studied fields were really close to the water bodies and adjoined to the meadows.

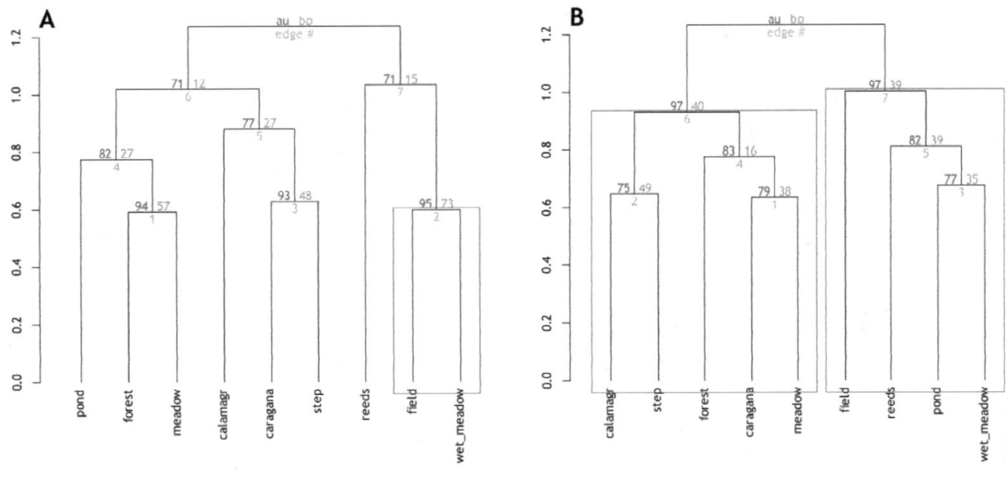

Figure 3. Ward's hierarchical clustering of a Kulczynski distances matrix among habitats. A — untransformed, B — Hellinger-transformed data. AU (Approximately Unbiased) and BP (Bootstrap Probability) p-values are shown next to the nodes.

The K-mean method applied to the habitat data without transformation did not show their structuredness in Figure 4 A. According to the silhouette method, the optimal number of clusters is two, the Calinski-Harabasz method – 7. Seven are obviously too many clusters, therefore, we chose two. On the plane of the first two principal components of the PCA (Figure 4 A), K-mean clustering without data transformation showed that all habitats, as well as samples form a whole. Only a sample taken on the shore of one of the reservoirs again stood out in a separate cluster.

Methods for determining the optimal number of clusters for Hellinger-transformed data for habitats gave the following results. The silhouette method recommends three clusters; the Calinski-Harabasz method recommends four clusters. More sensible is the option with three clusters (Figure 4 B). The shores open and covered with a reed stood out separately, which is quite logical. The grouping of fields and meadows is probably due to their artificial origin. Natural steppe areas are separated into a separate cluster, which makes biological sense, since these various habitats accumulate mainly xerophilic fauna. This article analyses faunistic data – samples collected in several fragmented areas of the steppe landscape. Since these samples were collected at the same time under almost identical conditions, the obtained results seem to be consistent with the neutral theory [25; 26]. Neutral theory considers rich and diverse biocenoses, such as tropical forests or coral reefs, where all events are stochastic. Environmentalists argue that factors limiting the number of species in communities, and even more so the question of whether they are truly stochastic or deterministic, have not yet been clarified [27]. It is possible that the answers to these questions can be obtained as a result of the study of different types of biocenoses.

Thus, we can conclude that the methods of K-mean [28] and PCA at the level of data habitat after Hellinger transformation exhibit a biologically meaningful, logical structuring. To establish and analyse the structure of assemblages, many researchers often use multivariate analysis such as principal component analysis (PCA) and cluster analysis [29; 30; 9].

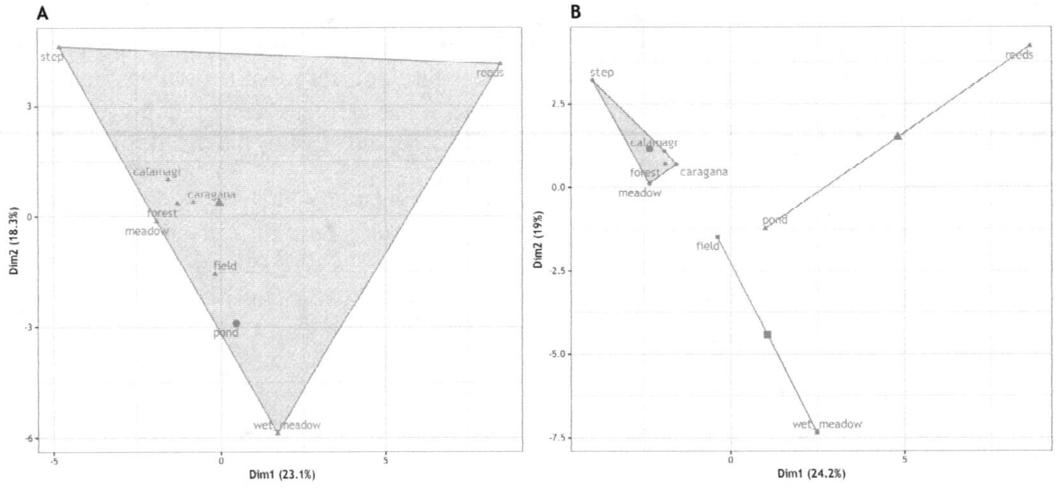

Figure 4. *K*-mean clustering of habitat types (without considering their location in seven territories of the Kleban-Byk RLP) on the plain of the first two principal components. A — untransformed, B — Hellinger-transformed data.

4. Conclusion

As an analysis of the data obtained during the study showed, the structure of the ground beetle's assemblages in the steppe landscape cannot be detected using one particular method. It is shown that the identification of structures of mobile biological objects assemblages involves a preliminary transformation of the data.

An important aspect for the systematisation of data are two points — the averaging of samples from similar habitats, and preliminary transformation of the data when analysing the structure at the habitat level. In our case, the application of Hellinger transformation was very successful.

5. References

[1] Clarke K R and Warwick R M 2001 A further biodiversity index applicable to species lists: variation in taxonomic distinctness *Marine Ecology Progress Series* ***216*** 265-78

[2] Shimadzu H and Darnell R. 2015 Attenuation of species abundance distributions by sampling *R. Soc. Open Sci.* **2** 140219 http://dx.doi.org/10.1098/rsos.140219

[3] Melo A S, Pereira R A S, Santos A J, Shepherd G J, Machado G, Medeiros H F and Sawaya R J 2003 Comparing species richness among assemblages using sample units: why not use extrapolation methods to standardize different sample sizes? *Oikos* ***101*** 398–410

[4] Banks-Leite C, Ewers R M and Metzger J P 2012 Unraveling the drivers of community dissimilarity and species extinction in fragmented landscapes *Ecology* ***93***(12) 2560–69

[5] Anderson M J, Ellingsen K E and McArdle B H 2006 Multivariate dispersion as a measure of beta diversity *Ecology Letters* **9** 683-93

[6] Ellwood M D F, Manica A and Foster W A 2009 Stochastic and deterministic processes jointly structure tropical arthropod communities *Ecol. Lett.* ***12*** 277– 84

[7] Ribera I, Foster G N and Vogler A P 2003 Does habitat use explain large scalespecies richness patterns of aquatic beetles in Europe? *Ecography* **26** 145 –52

[8] Pakulnicka J, Górski A, Bielecki A, Buczyński P, Tończyk G and Cichocka J M 2013 Relationships within aquatic beetle (Coleoptera) communities in the light of ecological theories *Fundam. Appl. Limnol* ***183/3*** 249–58

[9] Kirichenko-Babko M, Łagód G, Majerek D, Franus M and Babko R 2017 The effect of landscape on the diversity in urban green areas *Ecological Chemistry and Engineering* S ***24***(4) 613-625. DOI: 10.1515/eces-2017-0040
[10] Hocking T D 2018 directlabels: Direct Labels for Multicolor Plots. R package version 2018.05.22. https://CRAN.R-project.org/package=directlabels
[11] Kassambara A, Mundt F 2017 factoextra: Extract and Visualize the Results of Multivariate Data Analyses. R package version 1.0.5. https://CRAN.R-project.org/package=factoextra
[12] Maechler M, Rousseeuw P, Struyf A, Hubert M and Hornik K 2019. cluster: Cluster Analysis Basics and Extensions. R package version 2.0.8.
[13] R Core Team 2019 R: A language and environment for statistical computing. R Foundation for Statistical Computing, Vienna, Austria. https://www.R-project.org/
[14] Slowikowski K 2018 ggrepel: Automatically Position Non-Overlapping Text Labels with 'ggplot2'. R package version 0.8.0. https://CRAN.R-project.org/package=ggrepel
[15] Suzuki R, Shimodaira H 2015 pvclust: Hierarchical Clustering with P-Values via Multiscale Bootstrap Resampling. R package version 2.0-0. https://CRAN.R-project.org/package=pvclust
[16] Wickham H 2009 ggplot2: elegant graphics for data analysis. Springer New York.
[17] Huisman J, Olff H and Fresco L F M 1993 A hierarchical set of models for species response analysis *Journal of Vegetation Science* ***4*** 37–46, https://doi.org/10.2307/3235732
[18] Ruokolainen L and Blanchet G 2014 Introduction to ecological multivariate analysis Univ. Helsinki **79**
[19] Kirichenko M B and Babko R V 2007 Species diversity of ground beetles of the Kleban-Byk Regional Landscape Park. *Abstracts of 7th UET Congress* (September 14-18, 2007, Nizhyn) **52** [in Ukrainian]
[20] Jarošík V 1992 Pitfall trapping and species-abundance relationships: a value for carabid beetles (Coleoptera, Carabidae) Acta *Entomologica Bohemoslovaca* ***89*** 1–12.
[21] Green J L and Plotkin J B 2007 A statistical theory for sampling species abundance *Ecol. Lett.* ***10*** 1037–1045 (doi:10.1111/j.1461-0248.2007.01101.x)
[22] Kingrani S K, Levene M and Zhang D 2018 Estimating the number of clusters using diversity *Artificial Intelligence Research* *7*(1) 15–22.
[23] Trivellone V, Bougeard S, Giavi S, Krebs P, Balseiro D, Dray S and Moretti M 2017 Factors shaping community assemblages and species co- occurrence of different trophic levels. *Ecol Evol.*, *7* 4745–54.
[24] Wamser S, Dauber J, Birkhofer K and Wolters V 2011 Delayed colonisation of arable fields by spring breeding ground beetles (Coleoptera: Carabidae) in landscapes with a high availability of hibernation sites *Agriculture, Ecosystems and Environment* ***144*** 235–40
[25] Bell G 2001 Neutral macroecology *Science* ***293*** 2413–18.
[26] Chave J 2004 Neutral theory and community ecology *Ecol. Lett.* *7* 241–53
[27] Zhou S R and Zhang D Y 2006 Allee effects and the neutral theory of biodiversity *Funct. Ecol.* ***20*** 509–513
[28] Hill M O, Harrower C A and Preston C D 2013 Spherical k-means clustering is good for interpreting multivariate species occurrence data *Methods in Ecology and Evolution* ***4*** 542–51
[29] Jurkiewicz-Karnkowska E 2011 Effect of habitat conditions on the diversity and abundance of molluscs in floodplain water bodies of different permanence of flooding *Pol. J. Ecol.* **59** 165 –78
[30] Lenda M, Skórka P, Moron D, Rosin Z M and Tryjanowski P 2012 The importance of the gravel excavation industry for the conservation of grassland butterflies *Biol. Conserv.* **148** 180 –90.

FEM 3D analysis of rock cone failure range during pull-out of undercut anchors

J Jonak[1] and M Siegmund[2]

[1] Lublin University of Technology, Faculty of Mechanical Engineering, Nadbystrzycka 36, 20-618 Lublin, Poland
[2] KOMAG Institute of Mining Technology, Pszczyńska 37, 44-100 Gliwice, Poland

j.jonak@pollub.pl

Abstract. The paper presents the results from FEM (Finite Element Method) analysis of the process of pulling out of the undercut anchor. Anchors of this type are generally used to fasten steel structural elements in concrete buildings. However, the presented issue concerns a new area of application of these fasteners, i.e. in the aspect of the potential mining of rock in atypical situations, such as in mining rescue operations. Generally, in such situations there is no possibility of mechanical mining of rock with the use of mining machines and the use of explosives is even prohibited. There are only manual loosening methods whose effectiveness is unknown. Currently, the basic issue is to gain insight into the mechanism of loosening, the mechanics of the process in terms of the extent of loosening and shaping the force of pulling out anchors in a given rock. An effective tool in this type of analysis is FEM analysis, the results of which are presented in the paper.

1. Introduction

Numerous construction solutions for anchors are found among fastening methods (figure1, 4) [1,2]. Each is slightly different in terms of the mechanism of transferring the load from the anchor to the rock and the mechanism of the potential destruction of the structure of concrete or rock [1–19]. The anchors in figure 1a are used in the form of foundation bolts and are fixed in concrete when the concrete is poured. The remaining ones are fastened after drilling appropriate holes in the material in an adequate technological process.

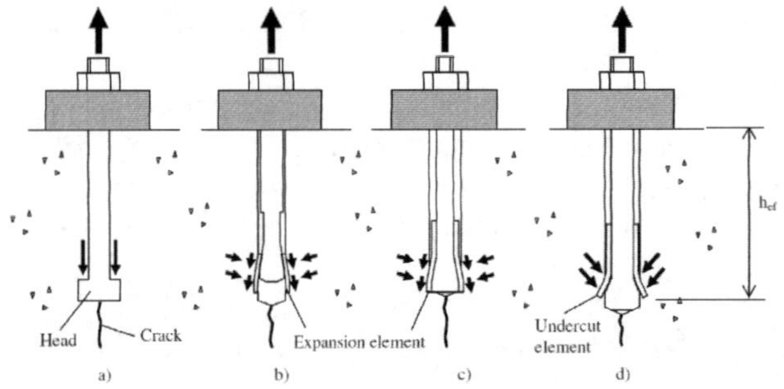

Figure 1. Common mechanical anchor types [1,2]: a) Cast-in headed stud; b) Torque-controlled expansion anchor (bolt type); c) Displacement-controlled expansion anchor; d) Undercut anchor.

In the engineering practice, there are two simplified mechanical models (figure 2) describing the potential course of destruction of the material of a structure in which the anchor is mounted [1,3–7].

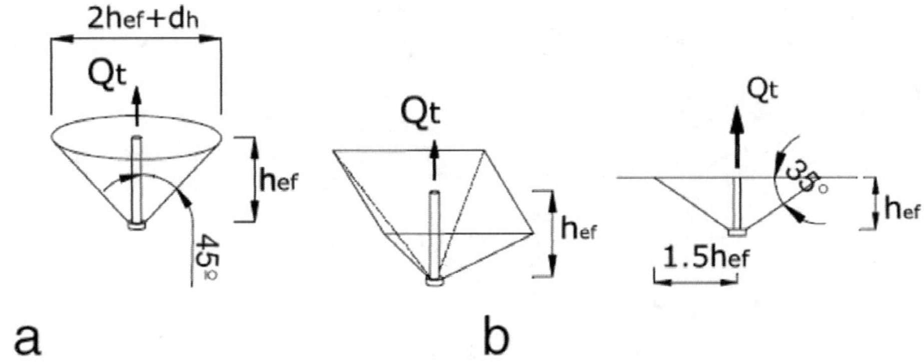

Figure 2. (a) Conical failure surface. (b) Four-sided pyramid failure surface [15,19,20].

In earlier models, conical damage was taken into account (concrete breakout in tension) of half-angle of the cone, equal to 45^0 [15,20]. Currently, the cone angle is slightly larger, since the mean value of the angle between the plane perpendicular to the axis of the cone (anchor) and the one creating a cone (figure 2, figure 3) is approx. 35^0 [1,3–7,21]. The practice and some analyses [22], however, indicate that the value of this angle changes within wide range and the loosening of the surface has a more complex shape, which translates into the extent of loosening measured on the free surface [2,9–19,22].

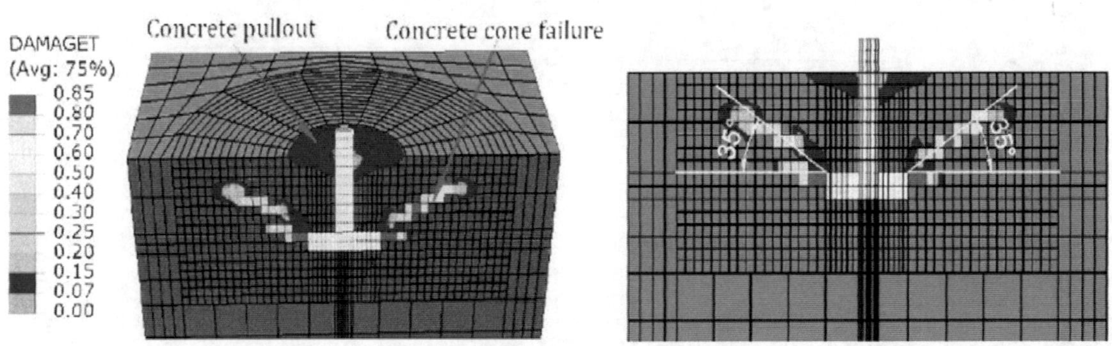

Figure 3. Concrete cone behaviour using the tensile damage parameter in ABAQUS [30].

The models above are used to analyse the extent of potential loosening, i.e. breakaway cone (damage), as well as estimating the maximum load-bearing capacity of the anchor for the assumed effective anchorage depth and strength parameters of the surface (e.g. concrete). This force is at the same time the minimum (under given conditions) force pulling out anchors from the surface.

Due to the number of construction solutions of anchors, the specificity of conducted numerical tests, numerical models implemented using, for example, FEM (e.g. [17], figure 3), also do not give unambiguous answers when this issue is concerned.

The aspect of precisely defining the extent of loosening is vital from the point of view of research because it determines the volume of loosening material being removed in one pull-out of the anchor. Therefore, there is a need for further research in the subject matter, especially with the use of FEM, as this method offers the most spectacular results in terms of fracture mechanics development in simulations.

IOP Conf. Series: Materials Science and Engineering **710** (2019) 012046 doi:10.1088/1757-899X/710/1/012046

2. Numerical tests using FEM

Actually, FEM systems are effectively used in many areas of research, e.g. in medical sciences (e.g. [23–25]), aviation (e.g. composite fracture [26–28] and other engineering activities.

The influence of undercut anchors type Hilti HAD-P [29] on rock has been tested (figure 4).

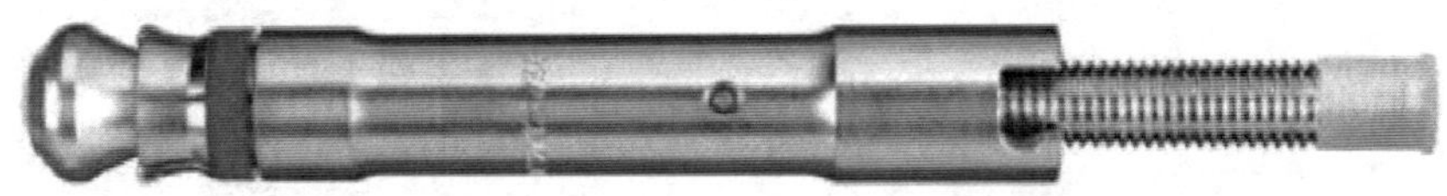

Figure 4. Pre-set undercut Hilti HDA-P anchor [14].

2.1. Numerical simulation assumptions

The numerical analysis was carried out using the FEM ABAQUS system (e.g. [30,31]). The geometric model of the analysed problem is shown in figure 5. The 3D model was obtained by rotating a flat model (figure 5a) around the longitudinal axis of the anchor. Figure 5b shows the quarter of the model obtained this way. This model was discretised with C3D8R elements (linear, continuum 3-D element, 8-node reduced integration). The impact of the anchor head (conical part) on rock was modelled in a simplified way by applying vectors of elemental forces in the nodes of the surface of hypothetical contact of the anchor with rock. Fixings: external vertical plane lower side and bottom - fixed, vertical middle plane - symmetry (figure 5c). The analysis was carried out for the geometrical parameters of the model: L = 700mm, h_{ef} = 100mm.

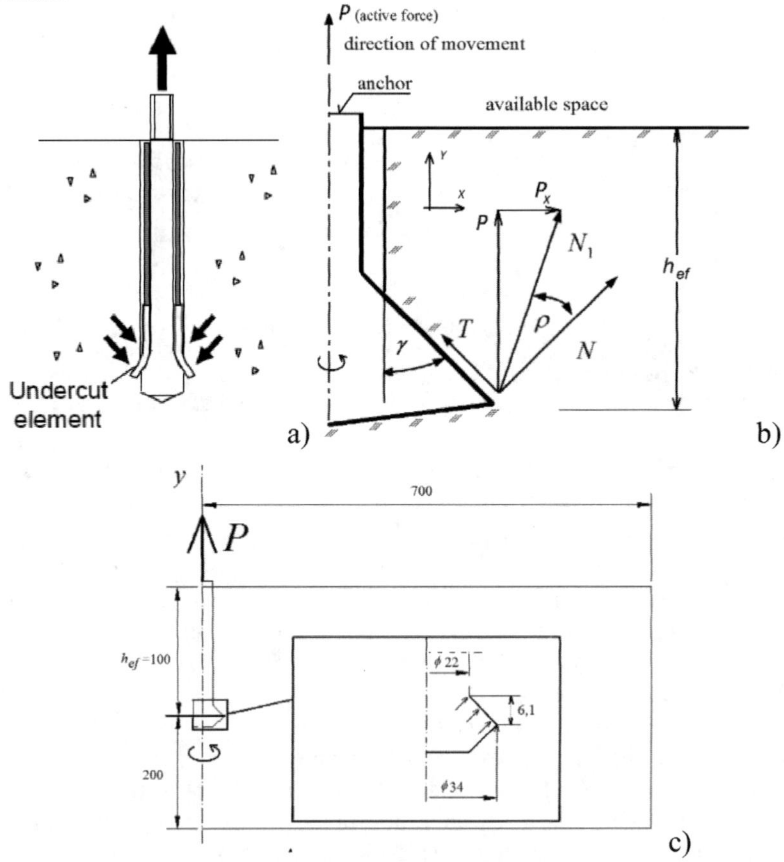

Figure 5. Scheme of the task.

The sandstone material was modelled as linear-elastic with Young modulus E = 14.276 GPa, Poisson's ratio ν = 0.247, tensile strength f_t = 7.74 MPa, and the fracture energy G_F = 0.335 N/mm.
The paper presents the results of simulation when the coefficient of friction of the anchor against rock μ =0 (elementary force vector in the node is perpendicular to the surface of the conical anchor, it is $\mu = tg\rho = 0 \Rightarrow T = 0, N_1 = N$, N=P/cos$\gamma$).
In the simulation, the active force (pulling the anchor) P, forcing the movement of the anchor along the OY axis (figure 5b), was replaced by the equivalent action of an equivalent load, distributed in the surface of the undercutting anchor's cone, as in figure 6. Otherwise, $P = \sum N_i cos\gamma$. The angle γ value was determined on the basis of the characteristic data of the anchor (figure 5c).

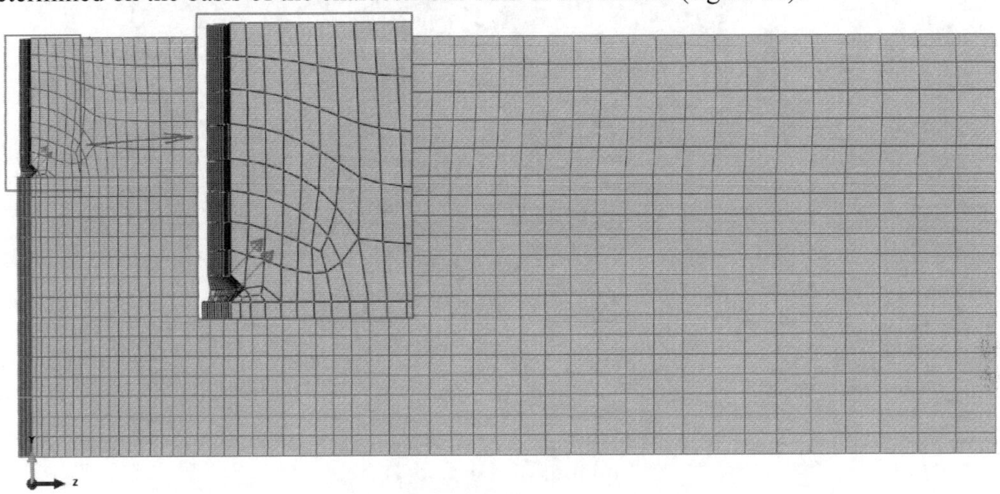

Figure 6. A fragment of the FEM mesh of an axially symmetrical 3D model of the tested undercut anchor.

Figure 7 presents a quarter of the model illustrating how to fix nodes located on its lateral external surface and in the base.

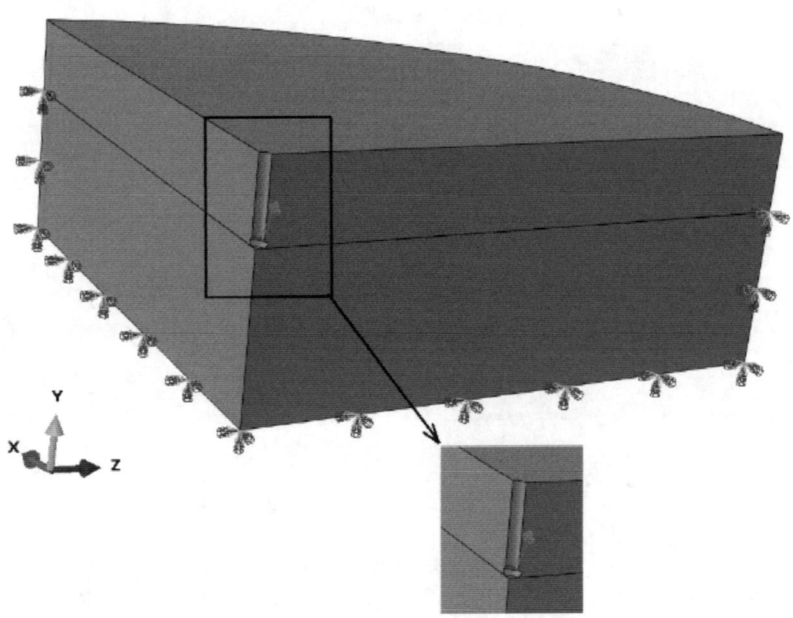

Figure 7. Boundary conditions of the model, restraint, load.

3. Computer simulations

Figure 8 illustrates the numerically obtained failure (cracking) surface and distribution of maximal principal stresses.

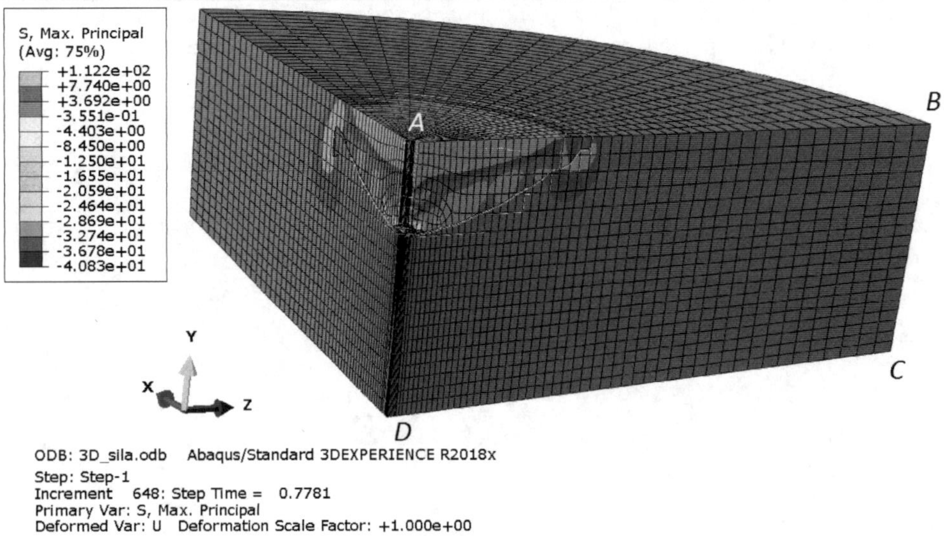

Figure 8. Maximal principal stress distribution and crack propagation path in the FEA model (a quarter of the model).

As it can be seen, the crack near the top edge began to distort and return. This is related to the limitations of ABAQUS procedures. The state of stress here was so complex that the programme probably could not decide how to lead the crack. For various programme settings and different mesh, it was not possible to cause the crack to go through to the end (see [16,30,31]).

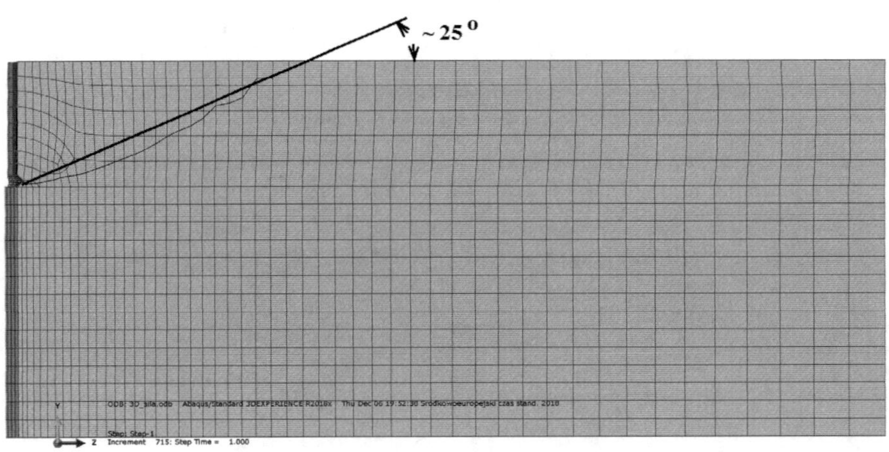

Figure 9. The average value of the angle of the concrete cone failure (section plane ABCD in figure 8).

Figure 9 illustrates the outline (in YZ-plane) axisymmetric failure surface and the average value of the angle of the concrete cone failure. The value is 25°, which is the lowest value among those available in literature sources (for example [16,19,32]).

IOP Conf. Series: Materials Science and Engineering **710** (2019) 012046 doi:10.1088/1757-899X/710/1/012046

4. Conclusions

The tests indicate that for the assumed mechanical parameters of rock and the depth of undercutting the rock with an anchor (h_{ef}, figure 1), the value of the angle of the cone failure is approximately 25°. In experimental studies carried out in the conditions of Polish stone mines, the obtained values of this angle were by almost 40% lower.

With a view to further verification and investigation, other simulation conditions are set to be investigated (material data, depth of undercutting h_{ef}, undercut anchor groups).

Acknowledgments

This research was sponsored by the Polish National Science Centre, project RODEST, No. 2015/19/B/ST10/02817.

5. References

[1] Eligehausen R and Sharma A seismic safety of anchorages in concrete construction – the latest perspective 11

[2] Mahrenholtz P and Eligehausen R 2015 Post-installed concrete anchors in nuclear power plants: Performance and qualification *Nuclear Engineering and Design* **287** 48–56

[3] Lennart Elfgren Development of the safety concept in design codes Anchor Bolts for Nuclear Power Plant Applications – Design and Assessment. Energiforsk Seminar, Stockholm, 3rd November, 2015.

[4] Eligehausen R, Fuchs W and Mayer B 2012 Tragverhalten von Dübelbefestigungen bei Zugbeanspruchung. Teil 1 Betonwerk + Fertigteil-Technik Heft 12/1981

[5] Eligehausen R, Hofacker I and Lettow S 2019 Fastening technique–current status and future trends 685

[6] Eligehausen R and Sippel T.M. Befestigungstecbnik.

[7] Eriksson D and Gasch T 2011 Load capacity of anchorage to concrete at nuclear facilities Numerical studies of headed studs and expansion anchors Master of Science Thesis (Stockholm, Sweden)

[8] Hirabayashi M., Suzuki T. and Kobayashi K. Basic Study on Pullout Resistance Mechanism of Taper-tipped Post-installed Anchors. JR EAST *Technical Review*-No 26

[9] Ljungberg J Pullout test of rock bolts at the Lima Hydropower station 58

[10] Momose M, Maruyama K and Shimizu K 1991 Load carrying mechanism of anchor bolt IABSE

[11] Nilsson M, Ohlsson U and Elfgren L 2011 Effects of surface reinforcement on bearing capacity of concrete with anchor bolts Nordic Concrete Research 161–74

[12] Marcon M, Ninčević K, Boumakis I, Czernuschka L-M and Wan-Wendner R 2018 Aggregate Effect on the Concrete Cone Capacity of an Undercut Anchor under Quasi-Static Tensile Load Materials 11 711

[13] Ninčević K, Boumakis I, Marcon M and Wan-Wendner R 2019 Aggregate effect on concrete cone capacity *Engineering Structures* **191** 358–69

[14] Ožbolt J, Kožar I, Eligehausen R and Periškić G 2004 Transient thermal 3D FE analysis of headed stud anchors exposed to fire

[15] Pallarés L and Hajjar J F 2010 Headed steel stud anchors in composite structures, Part II: Tension and interaction *Journal of Constructional Steel Research* **66** 213–28

[16] Periški G 3D Finite Element analysis of stud anchors with large head and embedment depth 7

[17] Tsavdaridis K D, Shaheen M A, Baniotopoulos C and Salem E 2016 Analytical approach of anchor rod stiffness and steel base plate calculation under tension *Structures* **5** 207–18

[18] Umehara H, Nakamura H, Shindo T, Sato Y and Furuichi K Establishment of the JSCE Recommendations for Design and Construction of Post-installed Anchors in Concrete 13

[19] Yang K-H and Ashour A F 2008 Mechanism analysis for concrete breakout capacity of single anchors in tension

IOP Conf. Series: Materials Science and Engineering **710** (2019) 012046 doi:10.1088/1757-899X/710/1/012046

[20] ACI Committee 349 2006 Code requirements for nuclear safety-related concrete structures: (ACI 349-06) and commentary, an ACI standard (Farmington Hills, Mich.: American Concrete Institute)

[21] Gallwoszus J 2014 Zur Ermüdung von Verbundkonstruktionen mit Verbunddübelleis (Aachen: Eigenverlag Lehrstuhl und Institut für Massivbau)

[22] Ahmed L T and Braimah A 2017 Behaviour of undercut anchors subjected to high strain rate loading *Procedia Engineering* **210** 326–33

[23] Karpiński R, Jaworski Ł and Zubrzycki J 2017 The design and structural analysis of the endoprosthesis of the shoulder joint *ITM* Web Conf. **15** 07015

[24] Karpiński R, Jaworski Ł, Jonak J and Krakowski P 2019 Stress distribution in the knee joint in relation to tibiofemoral angle using the finite element method ed M Kulisz, M Szala, M Badurowicz, W Cel, M Chmielewska, Z Czyż, K Falkowicz, J Kujawska and T Tulwin MATEC Web of Conferences 252 07007

[25] Karpiński R, Jaworski Ł, Jonak J and Krakowski P 2019 The influence of the nucleus pulposus on the stress distribution in the natural and prosthetic intervertebral disc ed M Kulisz, M Szala, M Badurowicz, W Cel, M Chmielewska, Z Czyż, K Falkowicz, J Kujawska and T Tulwin MATEC Web of Conferences 252 07006

[26] Falkowicz K and Debski H 2019 The work of a compressed, composite plate in asymmetrical arrangement of layers AIP Conference Proceedings **2078** p 020005

[27] Falkowicz K, Debski H and Teter A 2018 Design solutions for improving the lowest buckling loads of a thin laminate plate with notch Computer Methods In Mechanics (CMM2017): Proceedings of the 22nd International Conference on Computer Methods in Mechanics (Lublin, Poland) p 080004

[28] Falkowicz K, Dębski H, Wysmulski P and Różyło P 2019 The behaviour of compressed plate with a central cut-out, made of composite in an asymmetrical arrangement of layers *Composite Structures* **214** 406–13

[29] Anon Hilti. Technical Datasheet Update: Dec-17, HDA Undercut Anchor.

[30] Gontarz J, Podgórski J and Siegmund M 2018 Comparison of crack propagation analyses in a pull-out test Computer Methods In Mechanics (CMM2017): Proceedings of the 22nd International Conference on Computer Methods in Mechanics (Lublin, Poland) p 130011

[31] Gontarz J and Podgórski J 2019 Numerical analysis of crack propagation in a pull-out test ed M Kulisz, M Szala, M Badurowicz, W Cel, M Chmielewska, Z Czyż, K Falkowicz, J Kujawska and T Tulwin MATEC Web of Conferences 252 08001

[32] Jiabao Y 2012 Ultimate Strength Behaviour of Steel-Concrete-Steel Sandwich Composite Beams and Shells. Thesis.

The structure of higher aquatic vegetation in the genetic series of floodplain reservoirs

R Babko[1], T Diachenko[2], Y Danko[3], J Zaburko[4] and J Szulżyk-Cieplak[5]

[1] National Academy of Sciences of Ukraine, Schmalhausen Institute of Zoology, Department Fauna and Systematics of Invertebrates, B. Khmelnitsky 15, 01030 Kyiv, Ukraine
[2] National Academy of Sciences of Ukraine, Department of Ichthyology and Hydrobiology of River Systems, Institute of Hydrobiology, Heroiv Stalingrada 12, 04210 Kyiv, Ukraine
[3] Sumy Makarenko State Pedagogical University, Department of General Biology and Ecology, Romenska 87, 40002 Sumy, Ukraine
[4] Lublin University of Technology, Faculty of Environmental Engineering, Nadbystrzycka 40B, 20-618 Lublin, Poland
[5] Lublin University of Technology, Faculty of Fundamentals of Technology, Nadbystrzycka 38, 20-618 Lublin, Poland

rbabko@ukr.net, t_dyachenko@ukr.net, yaroslavdanko@gmail.com, jacek.zaburko@pollub.edu.pl, j.szulzyk-cieplak@pollub.pl

Abstract. The statistical analysis was carried out on the material collected from the Vorskla River (a first-order tributary of the Dnieper River) and the system of connected water bodies in the territory of the Getmanski National Nature Park (Sumy region, Ukraine). The methods of cluster analysis has enabled us to, first, isolate groups analogous to those obtained with the use of the method of dominants and, secondly, to compare the degree of their coincidence. From the comparison of the associations with the groups of the aquatic vegetation's species, a minimal overlapping could be seen in the composition of associations and clusters. In contrast, the groups identified by the hierarchical clustering method and by the k-means method showed significant similarities in the composition. For the objective assessment and classification of plant associations, it is most productive to use several methods with the coincidence of the results of which, it is possible to assert with more confidence that these combinations of species in nature are in fact not random.

1. Introduction

When studying vegetation, special attention is paid to the distribution of plant associations (i.e. the joint growth of groups of species) and individual plant populations in undisturbed natural sites [1]. Information on possible combinations of species in natural ecosystems is extremely important, since today, as a result of human activity, the territories of natural plant communities on all continents have been reduced to a critical minimum [2, 3, 4]. We are on the verge when some of the plants disappear, for the reasons including the lack of their natural habitats. As recent studies have shown, the rate of ongoing extinction is up to 500 times the background extinction rate for plants [5].

The problem of the conservation of aquatic plants is no less urgent, especially taking into account the extent of violation of various water bodies as a result of anthropogenic influence: changes in the riverbed morphology, pollution and pressure on hydroecosystems by seizing the production of industrial species of plants and animals [6]. Unlike terrestrial, aquatic and near-water vegetation is spatially limited because it requires a water body and has a narrow range of tolerance to the level of soil moisture. The diversity of the aquatic vegetation of natural reservoirs was largely determined by the variety of conditions that formed in the valleys of meandering rivers. It is the reclamation and management of river ecosystems that caused a reduction in the diversity of higher aquatic and near-water vegetation [7]. The issue of river renaturation is already being addressed in many European countries (e.g. England, Germany), and in the near future, is expected to dominate the scientific discourse, particularly given that the shortage of freshwater is already our present [8, 9]. Moreover, the surface water of fresh natural reservoirs is in most cases of very low quality, which is a consequence of both severe pollution of water bodies and violation of the river morphology and evolutionary established structure of their ecosystems [10].

Under these conditions, the role of aquatic vegetation is extremely high. Rivers are known to be most effectively cleaned in areas overgrown with aquatic vegetation. Plants, in addition to the ability to extract a number of biogens by accumulating them in their own biomass, are edificators of communities of numerous animal organisms. The composition of the population of thickets of aquatic vegetation is represented mainly by invertebrates and protozoa, which purify water with extreme efficiency [11]. It is on the thickets of aquatic vegetation that a trophic pyramid is based, and it is there that the passing water is most effectively conditioned.

All the elements presented in this introduction contribute to the fact that researching and studying natural associations of plants and the factors affecting their distribution becomes extremely relevant and of great practical importance.

In addition, the knowledge of the structural organisation of the plant component of aquatic ecosystems is extremely important for most European countries implementing river system restoration programmes.

2. Material and methods

2.1. *Statistical analysis*

Data were processed using R version 3.6.1 [12]. Hierarchical clustering was performed with hclust (function *ward.D2*) from core R package stats. Two types of p-values (AU and BP) for the nodes of hierarchical clustering dendrograms were calculated with R package pvclust. According to the package documentation, "AU p-value, which is computed by multiscale bootstrap resampling, is a better approximation to unbiased p-value than BP value computed by means of normal bootstrap resampling. Clusters with AU larger than 95% are highlighted by rectangles, which are strongly supported by data" [13]. *K*-mean clustering was performed with function *kmens* from package cluster [14] and plotted on the plain of principal components with *fviz_cluster* from factoextra [15], non-metric multidimensional scaling (NMDS) with *metaMDS* from vegan [16]. Plots were produced using packages ggplot2 [17], factoextra [15], ggrepel [18], directlabels [19]. The R software environment for statistical computing and graphics is widely applied in hydrobiology and environmental engineering [20, 21, 22]

2.2. *Region of study and collected data*

The statistical analysis was carried out on the material collected from the Vorskla River (a first-order tributary of the Dnieper River) and the system of its associated bodies of water in the territory of the Getmanski National Nature Park (Sumy region, Ukraine). The material was collected by methods generally accepted in biocenology. In each of the water bodies, the species composition of aquatic plants of all ecological groups was determined: semi-submerged, with floating leaves, and submerged higher water plants. A percentage cover of water surface by each species of the studied water body was estimated. The obtained percentages were evaluated on a 5-point scale. Associations of aquatic plants were classified on the basis of the ecological-phytocenotic approach [23]. Data on the species composition of the projective surface cover for 46 species of aquatic plants were obtained from 8 floodplain water bodies of various types and 1 river.

3. Results

One of the aspects of geobotanical research is the study of the spatial distribution of species, i.e. their choice of places for growth and, ultimately, the existence of any repeating cohabiting groups. Such groups are actively studied and classified and their principles of classification are different, for example, Shennikov's ecological-phytocenotic approach [23] or the popular Brown-Blanquet method [24, 25]. As for Braun-Blanquet, he restricted the use of relevés (a relevé is a list of species observed in a square plot together with estimates of their abundance/dominance or cover) to 'homogeneous' and 'typical' stands of communities. As a result, the relevé method is burdened by much bias, subjectivity, inconsistency, arbitrariness, circular argumentation and large sampling error [26, 27, 28]. Possibly the problems associated with sampling can be resolved by the use of various objective sampling designs and standardised square plot sizes [28, 29]. Moreover, the classification based on co-growth is not successful for all types of ecosystems. For terrestrial ecosystems, the most objective groupings are for the ecosystems dominated by woody vegetation (forests). The distribution of species of aquatic vegetation is determined by factors other than those that are important for terrestrial vegetation. The distribution of aquatic and near-water vegetation (hygrophilic) most depends on such factors as the speed of the current, the intensity of water exchange in weakly flowing water bodies, and the quality of the aquatic environment. A significant role is played by the content of nutrients and trace elements.

The classification of plant associations is according to the dominant species of each vegetation tier. Such associations do not account for the entire ecosystem but subjectively, based on the study of local sites. Thus, the allocation of associations without the use of statistical methods is largely subjective, which causes that its results debatable. This method involves the identification of combinations of species and is not aimed at typifying water bodies. The distinguished associations are objective, as their existence is results from specific conditions determined by the type of reservoir: whether it is a flowing body of water or not and as well as what trophic status the water body has. In other words, there can be different associations in lakes and rivers, and water bodies can be systematised on their basis. The adaptation of individual species to the water bodies of different types can potentially serve as the basis for their classification. Using factual material regarding the distribution and quantitative representation of species of hygrophilic plants in reservoirs of various types, we systematised reservoirs by the clustering method (Figure 1).

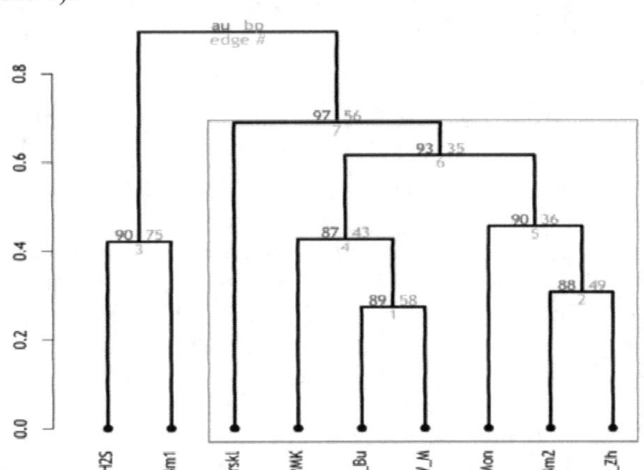

Figure 1. Ward's hierarchical clustering of objects based on Kulczynsky distances matrix of Hellinger-transformed data. AU (Approximately Unbiased) and BP (Bootstrap Probability) p-values are shown next to the nodes. For the codes of locations see Table 1.

According to the results obtained, most of the water bodies with high reliability combined with the river, which shows that the river and a number of its genetic derivatives form a kind of unity. At the same time, two lakes stood out in a separate cluster. Both lakes are specific because despite the shallow depth — from 1 to 3 m — the water column in them is stratified and the lower layers of the water and bottom sediments are anaerobic. Conventionally, we designate this category of reservoirs in the river valley as "hydrogen sulphide lakes". This type of reservoir does not grow idiosyncratic plant species and is distinguished by their small number.

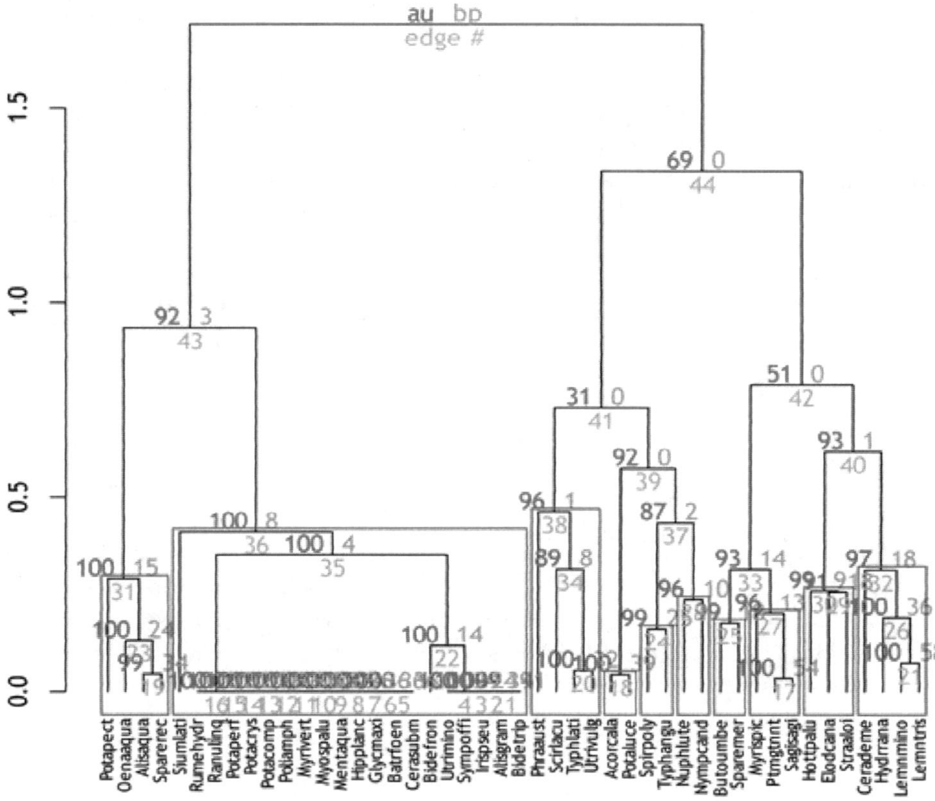

Figure 2. Ward's hierarchical clustering of species based on a Kulczynski distances matrix of Hellinger-transformed data. AU (Approximately Unbiased) and BP (Bootstrap Probability) p-values are shown next to the nodes. Abbreviations of plant species: Acorcala — *Acorus calamus*, Alisgram — *Alisma gramineum*, Alisaqua — *Alisma plantago aquatica*, Batrfoen — *Batrachium* foeniculaceum, Bidefron — *Bidens frondosa*, Bidetrip — *Bidens tripartita*, Butoumbe — *Butomus* umbellatus, Cerademe — *Ceratophyllum demersum*, Cerasubm — *Ceratophyllum submersum*, Elodcana — *Elodea canadensis*, Glycmaxi — *Glyceria maxima*, Hipplanc — *Hippuris lanceolata*, Hottpalu — *Hottonia palustris*, Hydrrana — *Hydrocharis morsus ranae*, Irispseu — *Iris pseudacorus*, Lemnmino — *Lemna minor*, Lemntris — *Lemna trisulca*, Mentaqua — *Mentha aquatica*, Myospalu — *Myosotis palustris*, Myrispic — *Myriophyllum spicatum*, Myrivert — *Myriophyllum verticillatum*, Nuphlute — *Nuphar lutea*, Nympcand — *Nymphaea candida*, Oenaaqua — *Oenanthe aquatica*, Phraaust — *Phragmites australis*, Poliamph — *Poligonum amphibium*, Potacomp — *Potamogeton compressus*, Potacrys — Potamogeton cryspus, Potaluce — *Potamogeton lucens*, Ptmgtnnt — *Potamogeton natans,* Potapect — *Potamogeton pectinatus*, Potaperf — *Potamogeton perfoliatus*, Ranulinq — *Ranunculus linqua*, Rumehydr — *Rumex hydrolapatum*, Sagisagi — *Sagittaria sagittifolia*, Scirlacu — *Scirpus lacustris*, Siumlati — *Sium latifolium*, Sparemer — *Sparganium emersum*, Sparerec — *Sparganium erectum*, Spirpoly — *Spirodela polyrrhyza*, Straaloi — *Stratiotes aloides*, Sympoffi — *Symphytum officinale*, Typhangu — *Typha angustifolia*, Typhlati — *Typha latifolia*, Utrimino — *Utricularia minor*, Utrivulg — *Utricularia vulgaris*

The hierarchical clustering of species based on the data on the distribution of all identified species in the studied reservoirs showed that their entire array is divided with a high level of reliability into 10 clusters, each representing 20 species (Figure 2). An analysis of the composition of this cluster showed that these are mainly species found in flowing waters.

The comparison of the associations established by the method of dominants with the groups of the aquatic vegetation's species, obtained from cluster analysis, showed a minimal overlapping in the composition of associations and clusters.

Furthermore, non-hierarchical k-means clustering was applied. This method requires presetting the desired number of clusters. To determine the optimal number of clusters, we used two methods: silhouette and Calinski-Harabasz. According to the silhouette method, the optimal number of clusters is 7, while according to Calinski-Harabasz — 2. A number of clusters 7 in our case appears, then, redundant. To visualise the clusters, they were depicted on the plane of the first two principal components of the PCA. The k-means clustering with 2 preset clusters showed that the floodplain PMK lake joined to the hydrogen sulphide lakes (Figure 3). Considering that the PMK Lake is stratified, most of the water column is oxygen-free, and the bottom sediments are represented by sapropel, whose presence in a reservoir in this group seems perfectly logical.

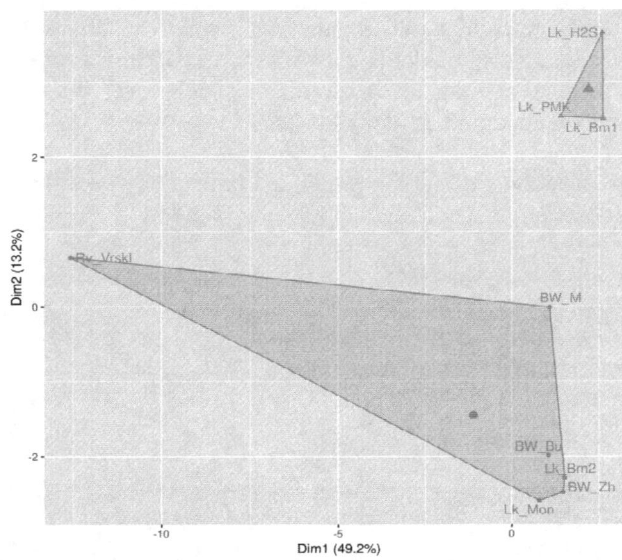

Figure 3. K-mean clustering of objects on the plain of the first two principal components. Data were Hellinger-transformed. For codes of objects see Table 1.

Both clustering methods show that in the genetic series of floodplain bodies of water, lakes in which secondary pollution causes eutrophication and triggers bogging processes in quantitative representation and number of species differ from reservoirs that maintain a connection with the river.

In the case of species, the silhouette method recommends for k-means clustering to choose seven clusters (Figure 4B), the Calinski-Harabasz method — two clusters (Figure 4A). The blue cluster (Figure 4A) united species that are tolerant to waterlogging. For our data, the most acceptable number of clusters is seven. The results of clustering make it possible to isolate groups similar to groups established using the method of dominants and compare the degree of their coincidence. In our case, the use of the k-means method allows us to make the partition into groups more objective. The results are shown in Figure 4B. Each of the obtained seven clusters combined the species closest in environmental requirements, with similar reactions to limiting factors including the degree of water exchange.

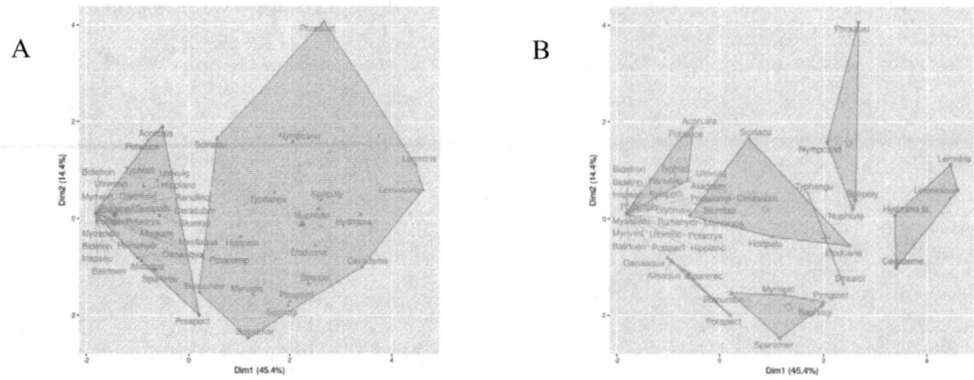

Figure 4. *K*-mean clustering of species on the plain of the first two principal components. Data were A – Calinski-Harabasz transformed and B – Hellinger-transformed. For codes of species see Figure 2.

4. Discussion

The grouping of aquatic vegetation based on the dominants method was tested by two statistical methods. Incorporation of statistical methods allows us to maximally avoid subjectivity in interpreting the results of field studies. According to the dominants technique, the following associations were identified in each of the examined bodies of water (Table 1).

Table 1. The associations of aquatics plants from Vorskla River and the floodplain lakes according to dominants classification.

Type of water body (Code)	Associations and their numbers	
Vorskla river (Rv_Vrskl)	1	*Sagittaria sagittifolia*
		Nuphar lutea
		Potamogeton perfoliatus
		Ceratophyllum demersum
Backwater Zhuravne (BW_Zh)	1	*Ceratophyllum demersum*
		Spirodela polyrrhyza
	2	*Nuphar lutea*
		Ceratophyllum demersum
		Lemna minor
	3	*Nuphar lutea*
		Sagittaria sagittifolia
		Nymphaea candida
		Lemna minor
	4	*Sagittaria sagittifolia*
Backwater Monastyrskie (BW_M)	1	*Ceratophyllum demersum*
		Spirodela polyrrhyza
	2	*Nuphar lutea*
		Ceratophyllum demersum
	3	*Nuphar lutea*
		Sagittaria sagittifolia
		Nymphaea candida
Backwater Bujmerovka (BW_Bu)	1	*Elodea canadensis*
		Lemna trisulca
		Lemna minor
	2	*Nuphar lutea*
		Ceratophyllum demersum
		Lemna trisulca
	3	*Sagittaria sagittifolia*
		Ceratophyllum demersum
		Elodea canadensis
	1	*Ceratophyllum demersum*
Lake Bujmerovka-1 (Lk_Bm1)		*Nuphar lutea*
		Stratiotes aloides
		Lemna minor
	2	*Ceratophyllum demersum*
		Spirodela polyrrhyza
		Lemna trisulca
	3	*Stratiotes aloides*
		Lemna minor
Lake Bujmerovka-2 (Lk-Bm2)	1	*Ceratophyllum demersum*
		Lemna minor
	2	*Lemna minor*
		Lemna trisulca
Lake Monastyrskie (Lk_Mon)	1	*Ceratophyllum demersum*
		Spirodela polyrrhyza
	2	*Potamogeton pectinatus*
		Ceratophyllum demersum
		Lemna minor
	3	*Stratiotes aloides*
		Lemna minor
	4	*Typha angustifolia*
		Ceratophyllum demersum
Lake PMK (Lk_PMK)	1	*Ceratophyllum demersum*
		Spirodela polyrrhyza
	2	*Nuphar lutea*
		Numphaea candida
	3	*Typha angustifola*
	4	*Phragmites australis*
	5	*Typha angustifola*
		Scirpus lacustris
Lake H2S (Lk_H2S)	1	*Lemna trisulca*
		Lemna minor
	2	*Lemna minor*
		Lemna trisulca

An analysis of the groups identified by the hierarchical clustering method and the k-means method showed significant similarities in the composition of these groups. Therefore, the composition of two groups coincided by 100% (Table 2). The first and second clusters obtained by hierarchical clustering coincided with the groups identified by the k-means method. Clusters 3 and 4 also formed a common group at a higher level of unification with group 3, isolated by the k-means method.

The coincidence of the groups identified by three different methods was observed in only one case. This association is represented by species from one ecological group — plants with floating leaves and represented by species *Nymphaea candida* and *Nuphar lutea*. It should be noted that the k-means method included two more species in the composition of this group: *Spirodea polyrrhyza* and *Phragmites australis*. A cluster obtained by hierarchical clustering fully coincided with the dominants association. Given that the dominants method allowed 23 associations to be distinguished, its coincidence with the clustering method amounted to approximately 5%.

Table 2. Plant species grouped of the basis of different statistical techniques.

Hierarchical clustering		*K*-mean		Braun-Blanket	
Cluster	Name species	Group	Name species	Association	Name species
1	*Lemna trisulca*	1	*Lemna trisulca*		
	Lemna minor		*Lemna minor*		
	Hydrocharis morsus ranae		*Hydrocharis morsus ranae*		
	Ceratophyllum demersum		*Ceratophyllum demersum*		
2	*Typha angustifolia*	2	*Typha angustifolia*		
	Spirodela polyrrhyza		*Spirodela polyrrhyza*		
3	*Sagittaria sagittifolia*	3	*Sagittaria sagittifolia*		
	Potamogeton natans		*Potamogeton natans*		
	Myriophyllum spicatum		*Myriophyllum spicatum*		
4	*Sparganium emersum*		*Sparganium emersum*		
	Butomus umbellatus		*Butomus umbellatus*		
5	*Nymphaea candida*	4	*Nymphaea candida*	1	*Nymphaea candida*
	Nuphar lutea		*Nuphar lutea*		*Nuphar lutea*
			Spirodel polyrrhyza		
			Phragmites australis		

5. Conclusion

Despite the widespread use of the dominants method for classification of aquatic plants communities, it should be noted that its results frequently do not coincide with the results of even simple statistical analysis. Apparently, the allocation of groups by the dominants method is associated with a high level of subjectivity.

IOP Conf. Series: Materials Science and Engineering **710** (2019) 012047 doi:10.1088/1757-899X/710/1/012047

It is likely that for an objective assessment and classification of plant associations, it is most productive to use several methods the combined results of which allow for a more confident assertion that these combinations of species in nature are, in fact, not random.

6. References

[1] Craft C, Krull K and Graham S 2007 Ecological indicators of nutrient enrichment, freshwater wetlands, Midwestern United States (US) *Ecol. Indic.* **7** (4) 733–50

[2] Camargo J A, Alonso A 2006 Ecological and toxicological effects of inorganic nitrogen pollution in aquatic ecosystems: a global assessment *Environ. Int.* **32** 831–49.

[3] Huijbregts M A J 2011 Implementing groundwater extraction in life cycle impact assessment: Characterization factors based on plant species richness for the Netherlands *Environ. Sci. Technol.* **45** (2) 629–35

[4] Lambert S J, Davy A J 2011 Water quality as a threat to aquatic plants: discriminating between the effects of nitrate, phosphate, boron and heavy metals on charophytes *New Phytol.* **189** 1051–59

[5] Humphreys A M, Govaerts R, Ficinski S Z, Lughadha N E and Vorontsova M S 2019 Global dataset shows geography and life form predict modern plant extinction and rediscovery *J Nat. Ecol. Evol.* **3** (7) 1043–47

[6] Kaenel B R, Buehrer H and Uehlinger U 2000 Effects of aquatic plant management on stream metabolism and oxygen balance in streams *Freshw. Biol.* **45** (1) 85–95

[7] Kim J H, Chung H Y, Kim S H, Kim J G 2016 The influence of water characteristics on the aquatic insect and plant assemblage in small irrigation ponds in Civilian Control Zone *Korea J. Wet. Res.* **18**(4) 331–41

[8] Milà I Canals L, Chenoweth J, Chapagain A, Orr S, Antón A and Clift R 2009 Assessing freshwater use impacts in LCA: Part I Inventory modelling and characterisation factors for the main impact pathways int *J. Life Cycle Assess* **14**(1) 28–42

[9] Pfister S, Koehler A and Hellweg S 2009 Assessing the environmental impacts of freshwater consumption in LCA *Environ. Sci. Technol.* **43**(11) 4098–104

[10] Report Millennium Ecosystem Assessment. In Ecosystems and Human Well-Being: Wetlands and Water Synthesis (Washington, DC: World Resources Institute) 2005.

[11] Babko R V and Kuzmina T N 2004 Ciliata (Protista, Ciliophora) of Epiphyton of Higher Aquatic Plants in a Small River *Hydrobiol. J. Begell House, Inc.* **40** (4) 22-38

[12] R Core Team 2019 R: A language and environment for statistical computing. R Foundation for Statistical Computing (Vienna: Austria) https://www.R-project.org/.

[13] Suzuki R and Shimodaira H 2015 pvclust: Hierarchical Clustering with P-Values via Multiscale Bootstrap Resampling. R package version 2.0-0. https://CRAN.R-project.org/ package= pvclust

[14] Maechler M, Rousseeuw P, Struyf A, Hubert M and Hornik K 2019 Cluster: Cluster Analysis Basics and Extensions. R package version 2.0.8.

[15] Kassambara A, Mundt F 2017 factoextra: Extract and Visualize the Results of Multivariate Data Analyses. R package version 1.0.5. https://CRAN.R-project.org/package=factoextra

[16] Oksanen J, Blanchet F G, Friendly M, Kindt R, Legendre P, McGlinn D, Minchin P R, O'Hara R B, Simpson G L, Solymos P, Henry M, Stevens H, Szoecs E and Wagner H 2019 vegan: Community Ecology Package. R package version 2.4-3. https://CRAN.R-project.org/package=vegan

[17] Wickham H 2009 ggplot2: elegant graphics for data analysis. Springer New York.

[18] Slowikowski K 2018 ggrepel: Automatically Position Non-Overlapping Text Labels with 'ggplot2'. R package version 0.8.0. https://CRAN.R-project.org/package=ggrepel

[19] Hocking T D 2018 directlabels: Direct Labels for Multicolor Plots. R package version 2018.05.22. https://CRAN.R-project.org/package=directlabels

IOP Conf. Series: Materials Science and Engineering **710** (2019) 012047 doi:10.1088/1757-899X/710/1/012047

[20] Pliashechnyk V, Danko Y, Łagód G, Drewnowski J, Kuzmina T and Babko R 2018 Ciliated protozoa in the impact zone of the Uzhgorod treatment plant *E3S Web of Conferences* **30** 02008. DOI: 10.1051/e3sconf/20183002008
[21] Babko R, Szulżyk-Cieplak J, Danko Y, Duda S, Kirichenko-Babko M and Łagód G 2019 Effect of stormwater system on the receiver *Journal of Ecological Engineering* **20**(6) 52–59 https://doi.org/10.12911/22998993/109433
[22] Łagód G, Duda S M, Majerek D, Szutt A and Dołhańczuk-Śródka A 2019 Application of electronic nose for evaluation of wastewater treatment process effects at full-scale WWTP *Processes* **7** 251 doi:10.3390/pr7050251
[23] Shennikov A P 1964 Vvedenie v geobotaniku. Leningrad LGU [in Russian]
[24] Westhoff V and Maarel E van der. 1973 The Braun-Blanquet approach: Handbookof vegetation science *Hague* **5**, 619−726
[25] Neuhauslova Z et al. 2001 Potential natural Vegetation of the Czech Republic. II Braun-Blanquetia, *Camerino* **30** 80
[26] Chytrý M 2001 Phytosociological data give biased estimates of species richness *J. Veg. Sci.* **12** 439−44
[27] Chytrý M and Otypková Z 2003 Plot sizes used for phytosociological sampling of European vegetation *J. Veg. Sci.* **14** 563−70
[28] Podani J 2006 Braun-Blanquet's legacy and data analysis in vegetation science *J. Veg. Sci.* **17** 113−17
[29] Bruelheide H and Chytrý M 2000 Towards unification of national vegetation classifications: A comparison of two methods for analysis of large data sets *J. Veg. Sci.* **11** 295−306.

AUTHOR INDEX